# The Definitive Guide to ARM® Cortex®-M3 and Cortex-M4 Processors

# The Definitive Guide to ARM® Cortex®-M3 and Cortex-M4 Processors

## Third Edition

**Joseph Yiu**

*ARM Ltd., Cambridge, UK*

AMSTERDAM • BOSTON • HEIDELBERG • LONDON
NEW YORK • OXFORD • PARIS • SAN DIEGO
SAN FRANCISCO • SINGAPORE • SYDNEY • TOKYO

Newnes is an imprint of Elsevier

**ELSEVIER**

**Newnes**

Newnes is an imprint of Elsevier
The Boulevard, Langford Lane, Kidlington, Oxford OX5 1GB, UK
225 Wyman Street, Waltham, MA 02451, USA

**Notice**

No responsibility is assumed by the publisher for any injury and/or damage to persons or
property as a matter of products liability, negligence or otherwise, or from any use or
operation of any methods, products, instructions or ideas contained in the material herein.
Because of rapid advances in the medical sciences, in particular, independent verification
of diagnoses and drug dosages should be made

**British Library Cataloguing in Publication Data**
A catalogue record for this book is available from the British Library

**Library of Congress Cataloging-in-Publication Data**
A catalog record for this book is available from the Library of Congress

ISBN−13: 978-0-12-408082-9

For information on all Newnes publications
visit our website at www.newnespress.com

Printed and bound by CPI Group (UK) Ltd, Croydon, CR0 4YY

Working together
to grow libraries in
developing countries

www.elsevier.com • www.bookaid.org

# Contents

Appendices A—I are available on the book's companion website

# Foreword

There is a revolution on the embedded market: Most new microcontrollers are nowadays based on the ARM architecture and specifically on the popular Cortex-M3 and Cortex-M4 processors. Recently we also saw the launch of several new ARM processors. At the low-end of the spectrum, the Cortex-M0+ processor has been introduced for applications that were previously dominated by 8-bit and 16-bit microcontrollers. The new 64-bit Cortex-A50 series processors address the high-end market such as servers. Along with the demand for standardized systems and energy efficient computing performance, the Internet-of-Things (IoT) is one driver for this revolution. In the year 2020, analysts are forecasting 50 billion devices that are connected to the IoT, and the ARM processors will span the whole application range from sensors to servers. Many devices will be based on Cortex-M3 and Cortex-M4 microcontrollers and may just use a small battery or even energy harvesting as power source.

Using ARM Cortex-M3 and Cortex-M4 processors based devices today is straightforward since a wide range of development tools, debug utilities, and many example projects are available. However, writing efficient applications could require in-depth knowledge about the hardware architecture and the software model. This book provides essential information for system architects and software engineers: It gives insight into popular software development tools along with extensive programming examples that are based on the Cortex Microcontroller Software Interface Standard (CMSIS). It also covers the Digital Signal Processing (DSP) features of the Cortex-M4 processor and the CMSIS-DSP library for interfacing with the analog world. And with many embedded applications becoming more complex and the wider availability of more capable microcontrollers, using of real-time operating systems is becoming common practice. All these topics are covered with easy-to-understand application examples.

I recommend this book to all type of users: From students that start with a small Cortex-M microcontroller project to system experts that need an in-depth understanding of processor features.

Reinhard Keil
Director of MCU Tools, ARM

# Preface

The last few years has seen the ARM® Cortex®-M3 processor continue to expand its market coverage and the adoption of the Cortex-M4 processor gaining momentum. At the same time the software development tools and various technologies surrounding the Cortex-M processors have also evolved. For example, the CMSIS-Core is now being used in almost all Cortex-M device driver libraries and the CMSIS project has expanded into areas such as the DSP library software.

In this edition, I have restructured my original book to enable beginners to quickly understand the M3 & M4 processor architecture, enabling them in the process to quickly develop software applications. I have also covered a number of advanced topics that numerous users have asked me to cover and which were missing from the previous editions — and were not covered in other books or in documentation created by ARM. In this edition I have also added a great deal of new information on the Cortex-M4 processor, for example, the detail uses of the floating point unit and the DSP instructions, and have extended the coverage of a number of topics. For example, this edition includes more microcontroller software development suites than previous editions, including a chapter on Real-Time Operating Systems (RTOS) based on the CMSIS-RTOS API, and additional information on a number of advanced topics.

Also included in this edition are two chapters on DSP written by Paul Beckmann, CEO of DSP Concepts, a company that has developed the CMSIS-DSP library for ARM. I am extremely pleased to have his contribution, since his in-depth knowledge of DSP applications and the CMSIS-DSP library make this book a worthwhile investment for any ARM-embedded software developer.

This book is for both embedded hardware system designers and software engineers. Because it has a wide range of chapters covering topics from "Getting Started" — to those detailing advanced information, it is suitable for a wide range of readers including programmers, embedded product designers, electronic enthusiasts, academic researchers, and even System-on-a-Chip (SoC) engineers. A chapter on software porting is also included to help readers who are porting software from other architectures or from ARM7TDMI™, a classic ARM processor, to Cortex-M microcontrollers.

Hopefully you will find this book useful and well worth reading.

Joseph Yiu

# Synopsis

This is the third edition of the Definitive Guide to the ARM® Cortex®-M3. The book name has been changed to reflect the addition of the details for the ARM Cortex-M4 processor. This third edition has been fully revised and updated, and now includes extensive information on the ARM Cortex-M4 processor, providing a complete up-to-date guide to both Cortex-M3 and Cortex-M4 processor but which also enables migration from various processor architectures to the exciting world of the Cortex-M3 and M4.

The book presents the background of the ARM architecture and outlines the features of the processors such as the instruction set and interrupt-handling and also demonstrates how to program and utilize various advanced features available such as the floating point unit.

Chapters on Getting Started with Keil™ MDK-ARM, IAR EWARM, gcc, and CooCox CoIDE tools are available to enable beginners to start developing program codes. The book then covers several important areas of software development such as input/output of information, using embedded OSs (CMSIS-RTOS), and mixed language projects with assembly and C.

Two chapters on DSP features and CMSIS-DSP libraries are contributed by Paul Beckmann, PhD, the founder and CEO of DSP Concepts. DSP Concepts is the company that developed the CMSIS-DSP library for ARM. These two chapters cover DSP fundamentals and how to write DSP software for the Cortex-M4 processor, including examples of using the CMSIS-DSP library, as well as useful information about the DSP capability of the Cortex-M4 processor.

Various debugging techniques are also covered in various chapters of the book, as well as topics on software porting from other architectures. This is the most comprehensive guide to the ARM Cortex-M3 and Cortex-M4 processors, written by an ARM engineer who helped to develop the core. It includes a full range of easy-to-understand examples, diagrams, quick reference appendices such as instruction sets, and CMSIS-Core APIs.

# About this Book

The source code of the example projects in this book can be download from the companion website from Elsevier: http://booksite.elsevier.com/9780124080829

# Contributor Bio-Paul Beckmann

Paul Beckmann is the founder of DSP Concepts, an engineering services company that specializes in DSP algorithm development and supporting tools. He has many years of experience developing and implementing numerically intensive algorithms for audio, communications, and video. Paul has taught industry courses on digital signal processing, and holds a variety of patents in processing techniques. Prior to founding DSP Concepts, Paul spent 9 years at Bose Corporation and was involved in R&D and product development activities.

# Acknowledgments

I would like to thank the following people for providing me with help, advice and feedback for the 3rd edition of this book:

First of all, a big thank you to Paul Beckmann, PhD, for contributing two chapters on the DSP subject. The DSP capability is an important part of the Cortex-M4 processor and the CMSIS-DSP library is a significant stepping stone for allowing microcontroller users to develop DSP applications. This book would not be complete without these two chapters.

Secondly, I would like to thanks my colleagues at ARM for their support. I have received much useful feedback from Joey Ye, Stephen Theobald, Graham Cunningham, Edmund Player, Drew Barbier, Chris Shore, Simon Craske, and Robert Boys. Also many thanks for the support from the ARM Embedded marketing team: Richard York, Andrew Frame, Neil Werdmuller, and Ian Johnson.

I would also like to thank Reinhard Keil, Robert Rostohar, and Martin Günther of Keil for answering my many questions on CMSIS, Anders Lundgren of IAR Systems for reviewing the materials related to EWARM, and Magnus Unemyr for reviewing materials related to Atollic TrueStudio®.

I also want to thank the following people for their help in assisting with the writing of the first and second editions of this book: Dominic Pajak, Alan Tringham, Nick Sampays, Dan Brook, David Brash, Haydn Povey, Gary Campbell, Kevin McDermott, Richard Earnshaw, Shyam Sadasivan, Simon Axford, Takashi Ugajin, Wayne Lyons, Samin Ishtiaq, Dev Banerjee, Simon Smith, Ian Bell, Jamie Brettle, Carlos O'Donell, Brian Barrera, and Daniel Jacobowitz.

And of course, I must express my gratitude to all the readers of my previous books that have provided me with their very useful feedback.

Also, many thanks to the staff at Elsevier for their professional work, which has enabled this book to be published

And finally, a special thank you to all of my friends for their support and understanding whist I was writing this book.

Regards,
Joseph Yiu

# Terms and Abbreviations

| Abbreviation | Meaning |
| --- | --- |
| ADK | AMBA Design Kit |
| AHB | Advanced High-Performance Bus |
| AHB-AP | AHB Access Port |
| AMBA | Advanced Microcontroller Bus Architecture |
| APB | Advanced Peripheral Bus |
| API | Application Programming Interface |
| ARM ARM | ARM Architecture Reference Manual |
| ASIC | Application Specific Integrated Circuit |
| ATB | Advanced Trace Bus |
| BE8 | Byte Invariant Big Endian Mode |
| CMSIS | Cortex Microcontroller Software Interface Standard |
| CPI | Cycles Per Instruction |
| CPU | Central Processing Unit |
| DAP | Debug Access Port |
| DSP | Digital Signal Processor/Digital Signal Processing |
| DWT | Data WatchPoint and Trace |
| EABI/ABI | Embedded Application Binary Interface |
| ETM | Embedded Trace Macrocell |
| FPB | Flash Patch and Breakpoint |
| FPGA | Field Programmable Gate Array |
| FPU | Floating Point Unit |
| FSR | Fault Status Register |
| ICE | In-Circuit Emulator |
| IDE | Integrated Development Environment |
| IRQ | Interrupt Request (normally refers to external interrupts) |
| ISA | Instruction Set Architecture |
| ISR | Interrupt Service Routine |
| ITM | Instrumentation Trace Macrocell |
| JTAG | Joint Test Action Group (a standard of test/debug interfaces) |
| JTAG-DP | JTAG Debug Port |
| LR | Link Register |
| LSB | Least Significant Bit |
| LSU | Load/Store Unit |
| MAC | Multiply Accumulate |
| MCU | Microcontroller Unit |
| MMU | Memory Management Unit |

*(Continued)*

| Abbreviation | Meaning |
| --- | --- |
| MPU | Memory Protection Unit |
| MSB | Most Significant Bit |
| MSP | Main Stack Pointer |
| NaN | Not-a-Number (floating point representation) |
| NMI | Non-maskable Interrupt |
| NVIC | Nested Vectored Interrupt Controller |
| OS | Operating System |
| PC | Program Counter |
| PMU | Power Management Unit |
| PSP | Process Stack Pointer |
| PPB | Private Peripheral Bus |
| PSR | Program Status Register |
| RTOS | Real-Time Operating System |
| SCB | System Control Block |
| SCS | System Control Space |
| SIMD | Single Instruction, Multiple Data |
| SP, MSP, PSP | Stack Pointer, Main Stack Pointer, Process Stack Pointer |
| SoC | System-on-a-Chip |
| SP | Stack Pointer |
| SRPG | State Retention Power Gating |
| SW | Serial-Wire |
| SW-DP | Serial-Wire Debug Port |
| SWJ-DP | Serial-Wire JTAG Debug Port |
| SWV | Serial-Wire Viewer (an operation mode of TPIU) |
| TCM | Tightly Coupled Memory (Cortex-M1 feature) |
| TPA | Trace Port Analyzer |
| TPIU | Trace Port Interface Unit |
| TRM | Technical Reference Manual |
| UAL | Unified Assembly Language |
| WIC | Wakeup Interrupt Controller |

# Conventions

Various typographical conventions have been used in this book, as follows:

- Normal assembly program codes:
  ```
  MOV  R0, R1  ; Move data from Register R1 to Register R0
  ```
- Assembly code in generalized syntax; items inside "< >" must be replaced by real register names:
  ```
  MRS <reg>, <special_reg>
  ```
- C program codes:
  ```
  for (i=0;i<3;i++) { func1(); }
  ```
- Values:
  1. 4'hC , 0x123 are both hexadecimal values
  2. *#3* indicates item number 3 (e.g., IRQ #3 means IRQ number 3)
  3. *#immed_12* refers to 12-bit immediate data
- Register bits:
  Typically used to illustrate a part of a value based on bit position. For example, bit[15:12] means bit number 15 down to 12.
- Register access types:
  1. R is Read only
  2. W is Write only
  3. R/W is Read or Write accessible
  4. R/Wc is Readable and clear by a Write access

# Introduction to ARM® Cortex®-M Processors

1

## CHAPTER OUTLINE

**The Definitive Guide to ARM® Cortex®-M3 and Cortex-M4 Processors. http://dx.doi.org/10.1016/B978-0-12-408082-9.00001-4**

1

## 1.1 What are the ARM® Cortex®-M processors?

### 1.1.1 The Cortex®-M3 and Cortex-M4 processors

The Cortex®-M3 and Cortex-M4 are processors designed by ARM®. The Cortex-M3 processor was the first of the Cortex generation of processors, released by ARM in 2005 (silicon products released in 2006). The Cortex-M4 processor was released in 2010 (released products also in 2010).

The Cortex-M3 and Cortex-M4 processors use a 32-bit architecture. Internal registers in the register bank, the data path, and the bus interfaces are all 32 bits wide. The Instruction Set Architecture (ISA) in the Cortex-M processors is called the Thumb® ISA and is based on Thumb-2 Technology which supports a mixture of 16-bit and 32-bit instructions.

The Cortex-M3 and Cortex-M4 processors have:

- Three-stage pipeline design
- Harvard bus architecture with unified memory space: instructions and data use the same address space
- 32-bit addressing, supporting 4GB of memory space
- On-chip bus interfaces based on ARM AMBA® (Advanced Microcontroller Bus Architecture) Technology, which allow pipelined bus operations for higher throughput
- An interrupt controller called NVIC (Nested Vectored Interrupt Controller) supporting up to 240 interrupt requests and from 8 to 256 interrupt priority levels (dependent on the actual device implementation)
- Support for various features for OS (Operating System) implementation such as a system tick timer, shadowed stack pointer
- Sleep mode support and various low power features
- Support for an optional MPU (Memory Protection Unit) to provide memory protection features like programmable memory, or access permission control
- Support for bit-data accesses in two specific memory regions using a feature called Bit Band
- The option of being used in single processor or multi-processor designs

The ISA used in Cortex-M3 and Cortex-M4 processors provides a wide range of instructions:

- General data processing, including hardware divide instructions
- Memory access instructions supporting 8-bit, 16-bit, 32-bit, and 64-bit data, as well as instructions for transferring multiple 32-bit data
- Instructions for bit field processing
- Multiply Accumulate (MAC) and saturate instructions
- Instructions for branches, conditional branches and function calls
- Instructions for system control, OS support, etc.

In addition, the Cortex-M4 processor also supports:

- Single Instruction Multiple Data (SIMD) operations
- Additional fast MAC and multiply instructions

- Saturating arithmetic instructions
- Optional floating point instructions (single precision)[1]

Both the Cortex-M3 and Cortex-M4 processors are widely used in modern microcontroller products, as well as other specialized silicon designs such as System on Chips (SoC) and Application Specific Standard Products (ASSP).

In general, the ARM Cortex-M processors are regarded as RISC (Reduced Instruction Set Computing) processors. Some might argue that certain characteristics of the Cortex-M3 and Cortex-M4 processors, such as the rich instruction set and mixed instruction sizes, are closer to CISC (Complex Instruction Set Computing) processors. But as processor technologies advance, the instruction sets of most RISC processors are also getting more complex, so much so that this traditional boundary between RISC and CISC processor definition can no longer be applied.

There are a lot of similarities between the Cortex-M3 and Cortex-M4 processors. Most of the instructions are available on both processors, and the processors have the same programmer's model for NVIC, MPU, etc. However, there are some differences in their internal designs, which allow the Cortex-M4 processor to deliver higher performance in DSP applications, and to support floating point operations. As a result, some of the instructions available on both processors can be executed in fewer clock cycles on the Cortex-M4.

## 1.1.2 The Cortex®-M processor family

The Cortex®-M3 and Cortex-M4 processors are two of the products in the ARM® Cortex-M processor family. The whole Cortex-M processor family is shown in Figure 1.1.

The Cortex-M3 and Cortex-M4 processors are based on ARMv7-M architecture. Both are high-performance processors that are designed for microcontrollers. Because the Cortex-M4 processor has SIMD, fast MAC, and saturate arithmetic instructions, it can also carry out some of the digital signal processing applications that traditionally have been carried out by a separate Digital Signal Processor (DSP).

The Cortex-M0, Cortex-M0+, and the Cortex-M1 processors are based on ARMv6-M, which has a smaller instruction set. Both Cortex-M0 and Cortex-M0+ are very small size in terms of gate count, with just about 12K gates[2] in minimum configuration, and are ideal for low-cost microcontroller products. The Cortex-M0+ processor has the most state-of-the-art low power optimizations, and has more available optional features.

The Cortex-M1 processor is designed specifically for FPGA applications. It has Tightly Coupled Memory (TCM) features that can be implemented using memories

---

[1]In many technical documents, and in command line option switches for a number of C compilers, the name Cortex-M4F is used for Cortex-M4 processor with the optional floating point unit.
[2]The silicon area is equivalent to approximately 12000 2-input NAND gates.

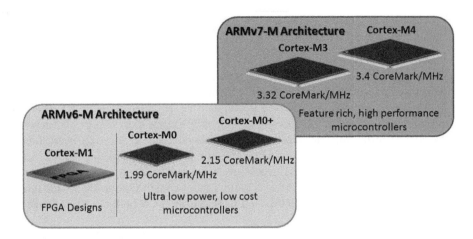

**FIGURE 1.1**

The Cortex-M processor family

inside the FPGA, and the design allows high clock frequency operations in advanced FPGA. For example, it can run at over 200 MHz in Altera Stratix III FPGA.

For general data processing and I/O control tasks, the Cortex-M0 and Cortex-M0+ processors have excellent energy efficiency due to the low gate count design. But for applications with complex data processing requirements, they may take more instructions and clock cycles. In this case, the Cortex-M3 or Cortex-M4 processor would be more suitable, because the additional instructions available in these processors allow the processing to be carried out with fewer instructions compared to ARMv6-M architecture. As a result, we need different processors for different applications.

It is worthwhile to note that the Cortex-M processors are not the only ARM processors to be used in generic microcontroller products. The venerable ARM7™ processor has been very successful in this market, with companies like NXP (formerly Philips Semiconductor), Texas Instruments, Atmel, OKI, and many other vendors delivering ARM-based microcontrollers using classic ARM processors like ARM7TDMI™. There are also wide ranges of microcontrollers designed with ARM9™ processors. The ARM7 processor is the most widely used 32-bit embedded processor in history, with over 2 billion processors produced each year in a huge variety of electronics products, from mobile phones to automotive systems.

## 1.1.3 Differences between a processor and a microcontroller

ARM® does not make microcontrollers. ARM designs processors and various components that silicon designers need and licenses these designs to various silicon design companies including microcontroller vendors. Typically we call these designs "Intellectualy Property" (IP) and the business model is called IP licensing.

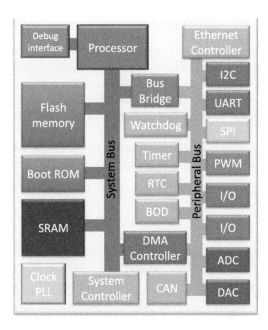

**FIGURE 1.2**

A microcontroller contains many different blocks

In a typical microcontroller design, the processor takes only a small part of the silicon area. The other areas are taken up by memories, clock generation (e.g., PLL) and distribution logic, system bus, and peripherals (hardware units like I/O interface units, communication interface, timers, ADC, DAC, etc.) as shown in Figure 1.2.

Although many microcontroller vendors use ARM Cortex-M processors as their choice of CPU, the memory system, memory map, peripherals, and operation characteristics (e.g., clock speed and voltage) can be completed differently from one product to another. This allows microcontroller manufacturers to add additional features in their products and differentiate their products from others on the market.

This book is focused on the Cortex-M3 and the Cortex-M4 processors. For details of the complete microcontroller system design, such as peripheral details, memory map, and I/O pin assignments, you still need to read the reference manuals provided by the microcontroller vendor.

### 1.1.4 ARM® and the microcontroller vendors

Currently there are more than 15 silicon vendors[3] using ARM® Cortex®-M3 or Cortex-M4 processors in microcontroller products. There are also some other

---

[3]Current Cortex-M3/M4 microcontroller vendors include: Analog Devices, Atmel, Cypress, EnergyMicro, Freescale, Fujitsu, Holtek, Infineon, Microsemi, Milandr, NXP, Samsung, Silicon Laboratories, ST Microelectronics, Texas Instrument, and Toshiba.

companies that use Cortex-M3 or Cortex-M4 for SoC designs, others companies that use only Cortex-M0 or Cortex-M0+ processors.

After a company licenses the Cortex-M processor design, ARM provides the design source code of the processor in a language called Verilog-HDL (Hardware Description Language). The design engineers in these companies then add their own design blocks like peripherals and memories, and use various EDA tools to convert the whole design from Verilog-HDL and various other forms into a transistor level chip layout.

ARM also provides other Intellectual Property (IP) products, and some can be used by these companies in their microcontroller products (see Figure 1.3). For example:

- Design of the cell libraries such as logic gates and memories (ARM Physical IP products)
- Peripherals and AMBA® infrastructure components (Cortex-M System Design Kit (CMSDK), ARM CoreLink™ IP products)
- Additional debug components for linking debug systems in multi-processor design (ARM CoreSight™ IP products)

For example, ARM provides a product called the Cortex-M System Design Kit (CMSDK), a design kit for Cortex-M processor with AMBA infrastructure components, baseline peripherals, example systems, and example software. This allows chip designers to start using the Cortex-M processors quickly and reduces the total chip development effort with reusable IP.

But of course, there is still a lot of work for the microcontroller chip designers to do. All of these microcontroller companies are working hard to develop better peripherals, lower power memories, and adding their own secret recipes to try to make their products better than others. In addition, they also need to develop example software and support materials to make it easier for the embedded product designers to use their chips.

On the software side, ARM has various software development platforms such as the Keil™ Microcontroller Development Kit (MDK-ARM) and ARM Development Studio 5 (DS-5™). These software development suites contain compilers, debuggers, and instruction set simulators. Designers can also use other third-party software development tools if they prefer. Since all of the Cortex-M microcontrollers have the same processor cores, the embedded product designers can use the same development suite for a massive range of microcontrollers from different vendors.

## 1.1.5 Selecting Cortex®-M3 and Cortex-M4 microcontrollers

On the market there are a wide range of Cortex®-M microcontroller products. These range from low-cost, off-the-shelf microcontroller products to high-performance multi-processor systems on a chip. There are many factors to be considered when selecting a microcontroller device for a product. For example:

- Peripherals and interface features
- Memory size requirements of the application

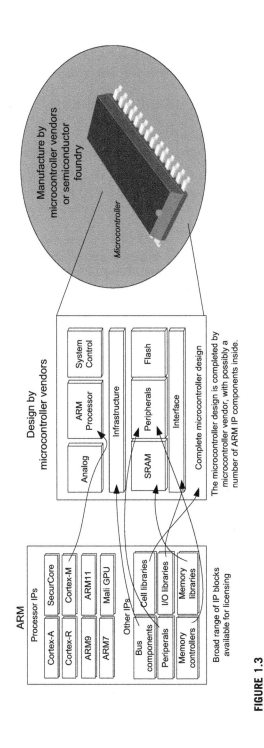

**FIGURE 1.3**

A microcontroller might contain multiple ARM IP products

- Low power requirements
- Performance and maximum frequency
- Chip package
- Operation conditions (voltage, temperature, electromagnetic interference)
- Cost and availability
- Software development tool support and development kits
- Future upgradability
- Firmware packages and firmware security
- Availability of application notes, design examples, and support

There are no golden rules on how to select the best microcontroller. All of the factors depend on your target applications as well as your project's situation. Some of the factors, like cost and product availability, might vary from time to time.

When developing projects based on off-the-shelf microcontrollers, usually the example projects and documentation from the microcontroller vendors are the best starting points. In addition, microcontroller vendors might also provide:

- Application notes
- Development kits
- Reference designs

You might also find additional examples from tools vendors and from various websites on the Internet.

When designing a Printed Circuit Board (PCB) for the Cortex-M microcontrollers or any other ARM-based microcontroller, it is best to bring out the debug interface with a standardize connector layout as documented in Appendix H. This makes debugging much easier.

## 1.2 Advantages of the Cortex®-M processors

The ARM® Cortex®-M processors have many technical and non-technical advantages compared to other architectures.

### 1.2.1 Low power

Compared to other 32-bit processor designs, Cortex®-M processors are relatively small. The Cortex-M processor designs are also optimized for low power consumption. Currently, many Cortex-M microcontrollers have power consumption of less than 200 μA/MHz, with some of them well under 100 μA/MHz. In addition, the Cortex-M processors also include support for sleep mode features and can be used with various advanced ultra-low power design technologies. All these allow the Cortex-M processors to be used in various ultra-low power microcontroller products.

### 1.2.2 **Performance**

The Cortex®-M3 and Cortex-M4 processors can deliver over 3 CoreMark/MHz and 1.25 DMIPS/MHz (based on the Dhrystone 2.1 benchmark). This allows Cortex-M3 and Cortex-M4 microcontrollers to handle many complex and demanding applications. Alternatively you can run the application with a much slower clock speed to reduce power consumption.

### 1.2.3 **Energy efficiency**

Combining low power and high-performance characteristics, the Cortex®-M3 and Cortex-M4 processors have excellent energy efficiency. This means that, you can still do a lot of processing, with a limited supply of energy. Or you can get tasks done quicker and allow the system to stay in sleep mode for longer durations of time, enabling longer battery life in portable products.

### 1.2.4 **Code density**

The Thumb® ISA provides excellent code density. This means that to achieve the same tasks, you need a smaller program size. As a result you can reduce cost and power consumption by using a microcontroller with smaller flash memory size, and chip manufacturers can produce microcontroller chips with smaller packages.

### 1.2.5 **Interrupts**

The Cortex®-M3 and Cortex-M4 processors have a configurable interrupt controller design, which can support up to 240 vectored interrupts and multiple levels of interrupt priorities (from 8 to 256 levels). Nesting of interrupts is automatically handled by hardware, and the interrupt latency is only 12 clock cycles for systems with zero wait state memory. The interrupt processing capability makes the Cortex-M processors suitable for many real-time control applications.[4]

### 1.2.6 **Ease of use, C friendly**

The Cortex®-M processors are very easy to use. In fact, they are easier than compared to many 8-bit processors because Cortex-M processors have a simple, linear memory map, and there are no special architectural restrictions, which you often find in 8-bit microcontrollers (e.g., memory banking, limited stack levels, non-re-entrant code, etc.). You can program almost everything in C including the interrupt handlers.

---

[4]There is always great debate as to whether we can have a "real-time" system using general processors. By definition, "real-time" means that the system can get a response within a guaranteed period. In any processor-based system, you may or may not be able to get this response due to choice of OS, interrupt latency, or memory latency, as well as if the CPU is running a higher priority interrupt.

### 1.2.7 Scalability

The Cortex®-M processor family allows easy scaling of designs from low-cost, simple microcontrollers costing less than a dollar to high-end microcontrollers running at 200 MHz or more. You can also find Cortex-M microcontrollers with multiprocessor designs. With all these, due to the consistency of the processor architecture, you only need one tool chain and you can reuse your software easily.

### 1.2.8 Debug features

The Cortex®-M processors include many debug features that allow you to analyze design problems easily. Besides standard design features, which you can find in most microcontrollers like halting and single stepping, you can also generate a trace to capture program flow, data changes, profiling information, and so on. In multiple processor designs, the debug system of each Cortex-M processor can be linked together to share debug connections.

### 1.2.9 OS support

The Cortex®-M processors are designed with OS applications in mind. A number of features are available to make OS implementation easier and make OS operations more efficient. Currently there are over 30 embedded OSs available for Cortex-M processors.

### 1.2.10 Versatile system features

The Cortex®-M3 and Cortex-M4 processors support a number of system features such as bit addressable memory range (bit band feature) and MPU (Memory Protection Unit).

### 1.2.11 Software portability and reusability

Since the architecture is very C friendly, you can program almost everything in standard ANSI C. One of ARM's initiatives called CMSIS (Cortex® Microcontroller Software Interface Standard) makes programming for Cortex-M processor based products even easier by providing standard header files and an API for standard Cortex-M processor functions. This allows better software reusability and also makes porting application code easier.

### 1.2.12 Choices (devices, tools, OS, etc.)

One of the best things about using Cortex®-M microcontrollers number amount of available choices. Besides the thousands of microcontroller devices available, you also have a wide range of coins on software development/debug tools, embedded OS, middleware, etc.

# 1.3 Applications of the ARM® Cortex®-M processors

With their wide range of powerful features, the ARM® Cortex®-M3 and Cortex-M4 processors are ideal for a wide variety of applications:

**Microcontrollers:** The Cortex-M processor family is ideally suited for microcontroller products.

This includes low-cost microcontrollers with small memory sizes and high-performance microcontrollers with high operation speeds. These microcontrollers can be used in consumer products, from toys to electrical appliances, or even specialized products for Information Technology (IT), industrial, or even medical systems.

**Automotive:** Another application for the Cortex-M3 and Cortex-M4 processors is in the automotive industry. As these processors offer great performance, very high energy efficiency, and low interrupt latency, they are ideal for many real-time control systems. In addition, the flexibility of the processor design (e.g., it supports up to 240 interrupt sources, optional MPU) makes it ideal for highly integrated ASSPs (Application Specific Standard Products) for the automotive industry. The MPU feature also provides robust memory protection, which is required in some of these applications.

**Data communications:** The processor's low power and high efficiency, coupled with instructions in Thumb®-2 for bit-field manipulation, make the Cortex-M3 and Cortex-M4 processors ideal for many communication applications, such as Bluetooth and ZigBee.

**Industrial control:** In industrial control applications, simplicity, fast response, and reliability are key factors. Again, the interrupt support features on Cortex-M3 and Cortex-M4 processors, including their deterministic behavior, automatic nested interrupt handling, MPU, and enhanced fault-handling, make them strong candidates in this area.

**Consumer products:** In many consumer products, a high-performance microprocessor (or several) is used. The Cortex-M3 and Cortex-M4 processors, being small, are highly efficient and low in power, and at the same time provide the performance required for handling complex GUIs on LCD panels and various communication protocols.

**Systems-on-Chips (SoC):** In some high-end application processor designs, Cortex-M processors are used in various subsystems such as audio processing engines, power management systems, FSM (Finite State Machine) replacement, I/O control task off loading, etc.

**Mixed signal designs:** In the IC design world, the digital and analog designs are converging. While microcontrollers contain more and more analogue components (e.g., ADC, DAC), some analog ICs such as sensors, PMIC (Power Management IC), and MEMS (Microelectromechanical Systems) now also include processors to provide additional intelligence. The low power capability and small gate count characteristics of the Cortex-M processors make it possible for them to be integrated on mixed signal IC designs.

There are already many Cortex-M3 and Cortex-M4 processor-based products on the market,[5] including low-end microcontrollers at less than 0.5 U.S. dollar, making the cost of ARM microcontrollers comparable to or lower than that of many 8-bit microcontrollers.

## 1.4 Resources for using ARM® processors and ARM microcontrollers

### 1.4.1 What can you find on the ARM® website

Although ARM® does not make or sell Cortex®-M3 or Cortex-M4 microcontrollers, there is quite a range of useful documentation on the ARM website. The documentation section of the ARM website (called Infocenter, http://infocenter.arm.com/) contains various specifications, application notes, knowledge articles, etc. Table 1.1 lists some of the reference documents containing details of the Cortex-M3 and Cortex-M4 processors.

Table 1.2 listed some of the Application Notes that can be useful for microcontroller software developers.

On the Infocenter you can also find manuals for ARM software products, such as the C compiler and linker, including Keil™ products.

For readers who are interested in the details of integrating Cortex-M processors into System-on-Chip designs or FPGA, the information listed in Table 1.3 might be useful.

### 1.4.2 Documentation from the microcontroller vendors

The documentation and resources from the microcontroller vendors are essential in embedded software development. Typically you can find:

- Reference manual for the microcontroller chip. This provides the programmer's model of the peripherals, memory maps and other information needed for software development.
- Data sheet of the microcontroller you use. This contains the information on package, pin layout, operation conditions (e.g., temperature), voltage and current characteristics, and other information you may need when designing the PCB.
- Application notes. These contain examples of using the peripherals or features on the microcontrollers, or information on handling specific task (e.g., flash programming).

You might also find additional resources on development kits, and additional firmware libraries.

---

[5]At the end of 3rd quarter of 2012, the accumulative shipment of Cortex-M3 and Cortex-M4 processors was 2.5 billion units.

**Table 1.1** Reference ARM Document on the Cortex-M3 and Cortex-M4 Processors

| Document | Reference |
| --- | --- |
| ARMv7-M Architecture Reference Manual<br>This is the specification of the architecture on which Cortex-M3 and Cortex-M processors are based. It contains detailed information about the instruction set, architecture defined behaviors, etc. This document can be accessed via the ARM website after a simple registration process. | 1 |
| Cortex-M3 Devices Generic User Guide<br>This is a user guide written for software developers using the Cortex-M3 processor. It provides information on the programmer's model, details on using core peripherals such as NVIC, and general information about the instruction set. | 2 |
| Cortex-M4 Devices Generic User Guide<br>This is a user guide written for software developers using the Cortex-M4 processor. It provides information on the programmer's model, details on using core peripherals such as NVIC, and general information about the instruction set. | 3 |
| Cortex-M3 Technical Reference Manual<br>This is a specification of the Cortex-M3 processor product. It contains implementation specific information such as instruction timing and some of the interface information (for silicon designers). | 4 |
| Cortex-M4 Technical Reference Manual<br>This is a specification of the Cortex-M3 processor product. It contains implementation specific information such as instruction timing and some of the interface information (for silicon designers). | 5 |
| Procedure Call Standard for the ARM Architecture<br>This document specifies how software code should work in procedure calls. This information is often needed for software projects with mixed assembly and C languages. | 13 |

**Table 1.2** ARM Application Notes That Can be Useful for Microcontroller Software Developers

| Document | Reference |
| --- | --- |
| AN179 – Cortex-M3 Embedded Software Development | 10 |
| AN210 – Running FreeRTOS on the Keil MCBSTM32 Board with RVMDK Evaluation Tools | 17 |
| AN234 – Migrating from PIC Microcontrollers to Cortex-M3 | 18 |
| AN237 – Migrating from 8051 to Cortex Microcontrollers | 19 |
| AN298 – Cortex-M4 Lazy Stacking and Context Switching | 11 |
| AN321 – ARM Cortex-M Programming Guide to Memory Barrier Instructions | 8 |

**Table 1.3** ARM Documents That Can be Useful for SoC/FPGA Designers

| Document | Reference |
|---|---|
| AMBA 3 AHB-Lite Protocol Specification<br>This is the specification for the AHB (Advanced High-performance Bus) Lite protocol, an on-chip bus protocol used on the bus interfaces of the Cortex-M processors. AMBA (Advanced Microcontroller Bus Architecture) is a collection of on-chip bus protocols developed by ARM and is used by many IC design companies. | 14 |
| AMBA 3 APB Protocol Specification<br>This is the specification for the APB (Advanced Peripheral Bus) Lite protocol, an on-chip bus protocol used for connecting peripherals to the internal bus system, and to connect debug components to the Cortex-M processors. APB is part of the AMBA specification. | 15 |
| CoreSight Technology System Design Guide<br>An introductory guide for silicon/FPGA designers who want to understand the basics of the CoreSight Debug Architecture. The debug system for the Cortex-M processors is based on the CoreSight Debug Architecture. | 17 |

**Table 1.4** Keil Application Notes That Can be Useful for Microcontroller Software Developers

| Document | Reference |
|---|---|
| Keil Application Note 202 – MDK-ARM Compiler Optimizations | 20 |
| Keil Application Note 209 – Using Cortex-M3 and Cortex-M4 Fault Exceptions | 21 |
| Keil Application Note 221 – Using CMSIS-DSP Algorithms with RTX | 22 |

### 1.4.3 Documentation from tool vendors

Very often the software development tool vendors also provide lots of useful information. In addition to tool chain manuals (e.g., compiler, linker) you can also find application notes. For example, on the Keil™ website (http://www.keil.com/appnotes/list/arm.h), you can find various for tutorials of using Keil MDK-ARM with Cortex-M development kits, as well as some application notes that cover some general programming information. Table 1.4 listed several application notes on the Keil website which are particularly useful for application development on the Cortex-M processors.

### 1.4.4 Other resources

On the ARM® website there are lots of other useful documents. For example, on the Infocenter you can find an ARM and Thumb®-2 Instruction Set quick reference card (reference 25). Although this quick reference card is not specific to the instruction set for the Cortex®-M processor, it can still be a handy reference for the majority of the instructions.

There are plenty of software vendors that provide software products like RTOS for Cortex-M processors. Often these companies also provide useful documentation on their websites that show how to use their products as well as general design guidelines.

On social media websites like YouTube, you can also find various tutorials on using Cortex-M based products, such as an introduction to microcontroller products and software tools.

There are a number of online discussion forums available that are focused on ARM Technologies. For example, the ARM website has a forum (http://forums.arm.com), and tool vendors and microcontroller vendors might also have their own online forums. In addition, some social media websites also have an ARM focused group; for example, LinkedIn has an ARM Based Group.[6]

There are already a number of books available on Cortex-M processors. Besides this book and "The Definitive Guide to the ARM Cortex-M0," Hitex also have a free online book on the STM32,[7] and you can also find quite a number of other books available in various online stores.

Don't forget that the distributor which provides you with the microcontroller chips can also be a useful source of information.

## 1.5 Background and history
### 1.5.1 A brief history of ARM®

Over the years, ARM® has designed many processors, and many features of the Cortex®-M3 and Cortex-M4 processors are based on the successful technologies which have evolved from some of the processors designed in the past. To help you understand the variations of ARM processors and architecture versions, let's look at a little bit of ARM history.

ARM was formed in 1990 as Advanced RISC Machines Ltd., a joint venture between Apple Computers, Acorn Computer Group, and VLSI Technology. In 1991, ARM introduced the ARM6 processor family (used in Apple Newton, see Figure 1.4), and VLSI became the initial licensee. Subsequently, additional companies, including Texas Instruments, NEC, Sharp, and ST Microelectronics, licensed the ARM processor designs, extending the applications of ARM processors into mobile phones, computer hard disks, personal digital assistants (PDAs), home entertainment systems, and many other consumer products.

Nowadays, ARM partners ship in excess of 5 billion chips with ARM processors each year (7.9 billion in 2011[8]). Unlike many semiconductor companies,

---

[6]http://www.linkedin.com/groups/ARM-Based-Group-85447
[7]http://www.hitex.co.uk/index.php?id=download-insiders-guides00
[8]Data from ARM Holdings — H1Q2 2012 presentation

**FIGURE 1.4**

The Apple Newton MessagePad H1000 PDA (based on ARM 610, released in 1993) placed next to an Apple iPhone 4, which is based on the Apple A4 processor that contains an ARM Cortex-A8 processor, released in 2010

ARM does not manufacture processors or sell the chips directly. Instead, ARM licenses the processor designs to business partners, including a majority of the world's leading semiconductor companies. Based on the ARM low-cost and power-efficient processor designs, these partners create their processors, micro-controllers, and system-on-chip solutions. This business model is commonly called IP licensing.

In addition to processor designs, ARM also licenses systems-level IP such as peripherals and memory controllers. To support the customers who use ARM products, ARM has developed a strong base of development tools, hardware, and software products to enable partners to develop their own products, and to enable software developers to develop software for ARM platforms.

## 1.5.2 ARM® processor evolution

Before the Cortex®-M3 processor was released, there were already quite a number of different ARM® processors available and some of them were already used in micro-controllers. One of the most successful processor products from ARM the ARM7TDMI™ processor, which is used in many 32-bit microcontrollers around the world. Unlike traditional 32-bit processors, the ARM7TDMI supports two instruction sets, one called the ARM instruction set with 32-bit instructions, and another 16-bit instruction set called Thumb®. By allowing both instruction sets to be used on the processor, the code density is greatly increased, hence reducing the memory footprint of the application code. At the same time, critical tasks can still execute with good speed. This enables ARM processors to be used in many portable devices, which require low power and small memory. As a result, ARM processors are the first choice for mobile devices like mobile phones.

Since then, ARM has continued to develop new processors to address the needs of different applications. For example, the ARM9™ processor family is used in a large number of high-performance microcontrollers and the ARM11™ processor family is used in a large number of mobile phones.

Following the introduction of the ARM11 family, it was decided that many of the new technologies, such as the optimized Thumb-2 instruction set, were just as applicable to the lower cost markets of microcontrollers and automotive components. It was also decided that although the architecture needed to be consistent from the lowest MCU to the highest performance application processor, there was a need to deliver processor architectures that best fit applications, enabling very deterministic and low gate count processors for cost-sensitive markets, and feature-rich and high-performance ones for high-end applications.

Over the past few years, ARM has extended its product portfolio by diversifying its CPU development, which resulted in the new processor family name "Cortex." In this Cortex processor range, the processors are divided into three profiles (Figure 1.5):

- The A profile is designed for high-performance open application platforms.
- The R profile is designed for high-end embedded systems in which real-time performance is needed.
- The M profile is designed for deeply embedded microcontroller-type systems.

Let's look at these profiles in a bit more detail.

**Cortex-A:** Application processors that are designed to handle complex applications such as high-end embedded operating systems (OSs) (e.g., iOS, Android, Linux, and Windows). These applications require the highest processing power, virtual memory system support with memory management units (MMUs), and, optionally, enhanced Java support and a secure program execution environment. Example products include high-end smartphones, tablets, televisions, and even computing servers.

**Cortex-R:** Real-time, high-performance processors targeted primarily at the higher end of the real-time market — these are applications, such as hard drive controllers, baseband controllers for mobile communications, and automotive systems, in which high processing power and high reliability are essential and for which low latency and determinism are important.

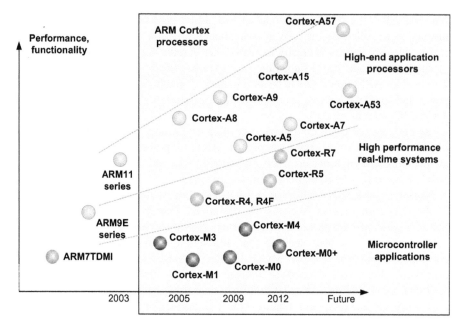

**FIGURE 1.5**

Diversity of processor products for three areas in the Cortex processor family

**Cortex-M:** Processors targeting smaller scale applications such as microcontrollers and mixed signal design, where criteria like low cost, low power, energy efficiency, and low interrupt latency are important. At the same time, the processor design has to be easy to use and able to provide deterministic behavior as required in many real-time control systems.

By creating this product range partitioning, the requirements of each marketing segment are addressed, allowing the ARM architecture to reach even more applications than before.

The Cortex processor families are the first products developed on ARM architecture v7, and the Cortex-M3 processor is based on one profile of ARMv7, called ARMv7-M, an architecture specification for microcontroller products.

## 1.5.3 Architecture versions and Thumb® ISA

ARM® develops new processors, new instructions, and architectural features are added from time to time (Figure 1.6). As a result, there are different versions of the architecture. For example, the successful ARM7TDMI™ is based on the architecture version ARMv4T (The "T" means Thumb instruction support). Note that architecture version numbers are independent of processor names.

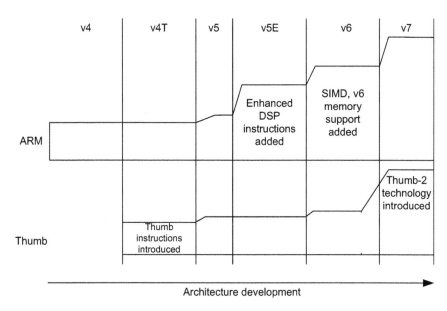

**FIGURE 1.6**

Instruction set enhancement

The ARMv5TE architecture was introduced with the ARM9E processor families, including the ARM926E-S and ARM946E-S processors. This architecture added "Enhanced" Digital Signal Processing (DSP) instructions for multimedia applications.

With the arrival of the ARM11 processor family, the architecture was extended to ARMv6. New features in this architecture included memory system features and Single Instruction Multiple Data (SIMD) instructions. Processors based on the ARMv6 architecture include the ARM1136J(F)-S, the ARM1156T2(F)-S, and the ARM1176JZ(F)-S.

In order to address different needs of a wide range of application areas, architecture version 7 is divided into three profiles (Figure 1.7):

- Cortex®-A Processors: ARMv7-A Architecture
- Cortex-R Processors: ARMv7-R Architecture
- Cortex-M Processors: ARMv7-M & ARMv6-M Architectures

Following the success of the Cortex-M3 processor, an additional architecture profile called ARMv6-M architecture was also created, to address the needs of ultra-low power designs. It uses the same programmer's model and exception handling methods as ARMv7-M (i.e., NVIC), but uses mostly just Thumb instructions from ARMv6 to reduce the complexity of the design. The Cortex-M0, Cortex-M0+, and Cortex-M1 processors are based on the ARMv6-M architecture.

The Cortex-M3 and Cortex-M4 processors are based on ARMv7-M, an architecture specification for microcontroller products. Please note that the enhanced DSP

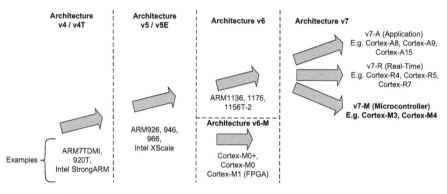

**FIGURE 1.7**

The evolution of ARM processor architecture

features in the Cortex-M4 processor are often referenced as ARMv7E-M, where the "E" reference to the "Enhanced" DSP instructions, as in ARMv5TE. The architecture details are documented in the ARMv7-M Architecture Reference Manual (reference 1). The architecture documentation contains the following key areas:

- Programmer's model
- Instruction set
- Memory model
- Debug architecture

Processor specific information, such as interface details and instruction timing, are documented in the product specific Technical Reference Manual (TRM) and other manuals from ARM.

In Figure 1.7 we can see that the Cortex-M0, Cortex-M0+, and Cortex-M1 processors are based on ARMv6-M. The ARMv6-M architecture is very similar to ARMv7-M in many ways, such as its interrupt handling, Thumb®-2 technology, and debug architecture. However, ARMv6-M has a smaller instruction set.

Following the success of the Cortex-M3 processor release, ARM decided to further expand its product range in microcontroller applications. The first step was to allow users to implement their ARM processor in an FPGA (Field Programmable Gate Array) easily, and the second step was to address the ultra-low power embedded processor. To do this, ARM took the Thumb instruction set from the existing ARMv6 architecture and developed a new architecture based on the exception and debug features in ARMv7-M architecture. As a result, ARMv6-M architecture was formed, and the processors based on this architecture are the Cortex-M0 processor (for microcontroller and ASICs) and the Cortex-M1 processor (for FPGA) (Figure 1.8).

The result of this development is a processor architecture that enables development of very small and energy efficient processors. At the same time, they are very easy to use, just like the Cortex-M3 and Cortex-M4.

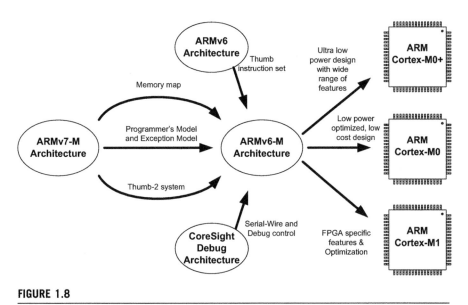

**FIGURE 1.8**

ARMv6-M architecture is based on many features from ARMv7-M

All the Cortex-M processors support Thumb-2 technology and support different subsets of the Thumb ISA (Instruction Set Architecture). Before Thumb-2 Technology was available, the Thumb ISA was a 16-bit only instruction set. Thumb-2 technology extended the Thumb Instruction Set Architecture (ISA) into a highly efficient and powerful instruction set that delivers significant benefits in terms of ease of use, code size, and performance (Figure 1.9).

With support for both 16-bit and 32-bit instructions in the Thumb-2 instruction set, there is no need to switch the processor between Thumb state (16-bit instructions) and ARM state (32-bit instructions). For example, in the ARM7 or ARM9™ family of processors, you might need to switch to ARM state if you want to carry out complex calculations or a large number of conditional operations and good performance is needed, whereas in the Cortex-M processors, you can mix 32-bit instructions with 16-bit instructions without switching state, getting high code density and high performance with no extra complexity.

Thumb-2 Technology is a very important feature of ARMv7. Compared with the instructions supported on ARM7 family processors (architecture ARMv4T), the Cortex-M3 and Cortex-M4 processor's instruction set has a large number of new features. For the first time, a hardware divide instruction is available on an ARM processor, and a number of multiply instructions are also available on the Cortex-M3 and Cortex-M4 processors to improve data-crunching performance. The Cortex-M3 and Cortex-M4 processors also support unaligned data accesses, a feature previously available only in high-end processors.

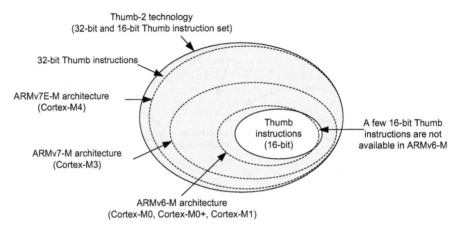

**FIGURE 1.9**

The relationship between the Thumb instruction set and the instruction set implemented in the Cortex-M processors

### 1.5.4 Processor naming

Traditionally, ARM® used a numbering scheme to name processors. In the early days (the 1990s), suffixes were also used to indicate features on the processors. For example, with the ARM7TDMI™ processor, the "T" indicates Thumb instruction support, "D" indicates JTAG debugging, "M" indicates fast multiplier, and "I" indicates an embedded ICE module. Subsequently, it was decided that these features should become standard features of future ARM processors; therefore, these suffixes are no longer added to the new processor family names. Instead, ARM created a new scheme for processor numbering to indicate variations in memory interface, cache, and tightly coupled memory (TCM).

For example, ARM processors with cache and MMUs are now given the suffix "26" or "36," whereas processors with MPUs are given the suffix "46" (e.g., ARM946E-S). In addition, other suffixes are added to indicate synthesizable[9] (S) and Jazelle (J) technology. Table 1.5 presents a summary of processor names.

With ARMv7, ARM has migrated away from these complex numbering schemes that needed to be decoded, moving to a consistent naming for families of processors, with Cortex as the overall brand. In addition to illustrating the compatibility across processors, this system removes confusion between architectural version and processor family number; for example, Cortex®-M always refers to processors for microcontroller applications, and this covers both ARMv7-M and ARMv6-M products.

---

[9]A synthesizable core design is available in the form of a hardware description language (HDL) such as Verilog or VHDL and can be converted into a design netlist using synthesis software.

**Table 1.5** Naming of Classic ARM Processors; "(F)" Means Optional Floating Point Unit

| Processor Name | Architecture Version | Memory Management Features | Other Features |
|---|---|---|---|
| ARM7TDMI | ARMv4T | | |
| ARM7TDMI-S | ARMv4T | | |
| ARM7EJ-S | ARMv5TEJ | | DSP, Jazelle |
| ARM920T | ARMv4T | MMU | |
| ARM922T | ARMv4T | MMU | |
| ARM926EJ-S | ARMv5TEJ | MMU | DSP, Jazelle |
| ARM946E-S | ARMv5TE | MPU | DSP |
| ARM966E-S | ARMv5TE | | DSP |
| ARM968E-S | ARMv5TE | | DMA, DSP |
| ARM966HS | ARMv5TE | MPU (optional) | DSP |
| ARM1020E | ARMv5TE | MMU | DSP |
| ARM1022E | ARMv5TE | MMU | DSP |
| ARM1026EJ-S | ARMv5TEJ | MMU or MPU | DSP, Jazelle |
| ARM1136J(F)-S | ARMv6 | MMU | DSP, Jazelle |
| ARM1176JZ(F)-S | ARMv6Z | MMU + TrustZone | DSP, Jazelle |
| ARM11 MPCore | ARMv6K | MMU + multiprocessor cache support | DSP, Jazelle |
| ARM1156T2(F)-S | ARMv6T2 | MPU | DSP |

### 1.5.5 About the ARM® ecosystem

Besides working with silicon vendors, ARM® is also working closely and actively with various parties that develop ARM solutions or use ARM products. For example, this includes vendors that provide software development suites, embedded OS and middleware, as well as design services providers, distributors, training providers, academic researchers, and so on (Figure 1.10). This close collaboration allows these parties to provide high-quality products or services, and allows more users to benefit from using the ARM architecture.

The ARM ecosystem also enables better knowledge sharing, which helps software developers to develop their applications faster and with better quality. For example, microcontroller users can get help and expert advice easily from various public forums on the Internet. Various microcontroller vendors, distributors, and other training service providers also organize regular ARM microcontroller training courses.

ARM also works closely with various open source projects to help the open source community to develop software for ARM platforms. For example, the Linaro organization (http://www.linaro.org) was set up by ARM as a not-for-profit engineering organization to enhance open source software such as GCC, Linux, and multimedia support.

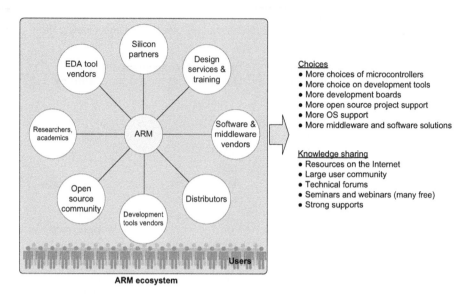

**ARM ecosystem**

**FIGURE 1.10**

The ARM ecosystem

Companies that develop ARM products or use ARM technologies can join the ecosystem by becoming a member of the ARM Connected Community. The ARM Connected Community is a global network of companies aligned to provide a complete solution, from design to manufacture and end use, for products based on the ARM architecture. ARM offers a variety of resources to Community members, including promotional programs and peer-networking opportunities that enable a variety of ARM Partners to come together to provide end-to-end customer solutions. Today, the ARM Connected Community has more than 1000[10] corporate members. Joining the ARM Connected Community is easy; details are on the ARM website http://cc.arm.com.

ARM also has a University Program that enables academic organizations like universities to access ARM technologies such as processor IP, reference materials, and so on. Details of the ARM University Program can be found on the ARM website (http://www.arm.com/support/university/).

---

[10]Information from Q4 2012.

# Introduction to Embedded Software Development

2

## CHAPTER OUTLINE

## 2.1 What are inside typical ARM® microcontrollers?

There are many different things inside a microcontroller. In many microcontrollers, the processor takes less than 10% of the silicon area, and the rest of the silicon die is occupied by other components such as:

- Program memory (e.g., flash memory)
- SRAM
- Peripherals

- Internal bus infrastructure
- Clock generator (including Phase Locked Loop), reset generator, and distribution network for these signals
- Voltage regulator and power control circuits
- Other analog components (e.g., ADC, DAC, voltage reference circuits)
- I/O pads
- Support circuits for manufacturing tests, etc.

While some of these components are directly visible to programmers, some others could be invisible to software developers (e.g., support circuit for manufacturing tests). Don't worry; to use a Cortex®-M microcontroller, we only need to have basic understanding of the processors (e.g., how to use the interrupt features), as well as the detailed programmer's model of the peripherals. Since the peripherals from different microcontroller vendors are different, you need to download and read the user manuals (or similar documents) from microcontroller vendors. This book is focused on the processors, although a number of examples on using peripherals are also covered.

Peripherals and control registers for system management are accessible from the memory map. To make it easier for software developers, most microcontroller vendors provide C header files and driver libraries for their microcontrollers. In most cases, these files are developed with the Cortex Microcontroller Software Interface Standard (CMSIS), which means it used a set of standardized headers for accessing processor features. We will cover more on this later in this chapter.

In most cases, the processor does all the work of controlling the peripherals and handles the system management. This book will cover a few examples of using a number of popular Cortex-M3/M4-based microcontrollers. In some microcontrollers there are also some smart peripherals that can do small amounts of processing without processor intervention. This depends on the vendor-specific peripherals on the microcontrollers and is beyond the scope of this book, but you can find the details in user manuals on the microcontroller vendor's website.

## 2.2 What you need to start

### 2.2.1 Development suites

With more than 10 different vendors selling C compiler suites for Cortex®-M microcontrollers, deciding which one to use can be a difficult choice. The development suites range from open-source free tools, to budget low-cost tools, to high-end commercial packages. The current available choices included various products from the following vendors:

- Keil™ Microcontroller Development Kit (MDK-ARM)
- ARM® DS-5™ (Development Studio 5)
- IAR Systems (Embedded Workbench for ARM Cortex-M)
- Red Suite from Code Red Technologies (acquired by NXP in 2013)
- Mentor Graphics Sourcery CodeBench (formerly CodeSourcery Sourcery g++)
- mbed.org

- Altium Tasking VX-toolset for ARM Cortex-M
- Rowley Associates (CrossWorks)
- Coocox
- Texas Instruments Code Composer Studio (CCS)
- Raisonance RIDE
- Atollic TrueStudio
- GNU Compiler Collection (GCC)
- ImageCraft ICCV8
- Cosmic Software C Cross Compiler for Cortex-M
- mikroElektronika mikroC
- Arduino

Some development boards also include a basic or evaluation edition of the development suites. In addition, there are development suites for other languages. For example:

- Oracle Java ME Embedded
- IS2T MicroEJ Java virtual machine
- mikroElektronika mikroBasic, mikroPascal

The illustrations in this book are mostly based on the Keil Microcontroller Development Kit (MDK-ARM) because of its popularity, but most of the example code can be used with the other development suites.

## 2.2.2 Development boards

There are already a large number of development kits for the Cortex®-M3/M4 microcontrollers from various microcontroller vendors and their distributors. Many of them are offered at an excellent price. For example, you can get a Cortex-M3 evaluation board for less than $12.

You can also get development kits from software tool vendors; for example, companies like Keil™ (an example is show in Figure 2.1), IAR Systems, and Code Red Technologies all have a number of development boards available.

A number of low-cost development boards are designed to work with particular development suites. For example, the "mbed.org" development boards, a low-cost solution for rapid software prototyping, are designed to work with the mbed development platform.

To start learning about ARM® Cortex-M microcontrollers, it is not always necessary to get an actual development board. Several development suites include an instruction set simulator, and the Keil MDK-ARM even supports device-level simulation for some of the popular Cortex-M microcontrollers. So you can learn Cortex-M programming just by using simulation.

## 2.2.3 Debug adaptor

In order to download your program code to the microcontroller, and to carry out debug operations like halting and single stepping, you might need a debug adaptor

**FIGURE 2.1**

A Cortex-M3 development board from Keil (MCBSTM32)

to convert a USB connection from your PC to a debug communication protocol used by the microcontrollers. Most C compiler vendors have their own debug adaptor products. For example, Keil™ has the ULINK product family (Figure 2.2), and IAR provides the I-Jet product. Most development suites also support third-party debug adaptors. Note that different vendors might have different terminologies for these debug adaptors, for example, debug probe, USB-JTAG adaptor, JTAG/SW Emulator, JTAG In-Circuit Emulator (ICE), etc.

Some of the development kits already have a USB debug adaptor built-in on the board. This includes some of the low-cost evaluation boards from Texas Instruments,

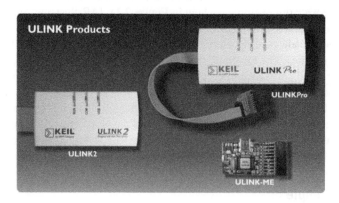

**FIGURE 2.2**

The Keil ULINK debug adaptor family

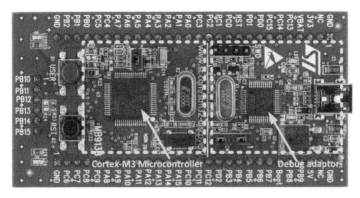

**FIGURE 2.3**

An example of development board with USB debug adaptor — STM32 Value Line Discovery

ST Microelectronics (e.g., STM32 Value Line Discovery; Figure 2.3), NXP, Energy-Micro, etc. Many of these onboard USB adaptors are also supported by mainstream commercial development suites. So you can start developing software for the Cortex®-M microcontrollers with a tiny budget.

In a number of evaluation/development boards, the built-in USB debug adaptor can also be used to connect to other development boards. You can also find "open-source" versions of such debug adaptors. The CMSIS-DAP from ARM and CoLink from Coocox are two examples.

While these low-cost debug adaptors work for most debug operations, there might be some features that are not well supported. There are a number of commercial USB debug adaptor products that offer a large number of useful features.

### 2.2.4 Software device driver

The term device driver here is quite different from its meaning in a PC environment. In order to help microcontroller software developers, microcontroller vendors usually provide header files and C codes that include:

- Definitions of peripheral registers
- Access functions for configuring and accessing the peripherals

By adding these files to your software projects, you can access various peripheral functions via function calls and access peripheral registers easily. If you want to, you can also create modified versions of the access functions based on the methods shown in the driver code and optimize them for your application.

### 2.2.5 Examples

Don't forget to download some example code from the microcontroller vendor's website. Most of the microcontroller vendors put their device-driver codes and examples on their websites as free downloads. This can save you a lot of time in developing new applications.

### 2.2.6 Documentation and other resources

Aside from user manuals of the microcontrollers, often you can also find application notes, FAQs, and online discussion forums on microcontroller vendor websites. The user manuals are essential, as they provides the details of the peripherals' programmer models.

On the ARM® website, the documentation is placed in a section called Info Center (http://infocenter.arm.com). From there you can find the Cortex®-M3/M4 Devices Generic User Guides (references 2 and 3), which covers the programming model of the processors, as well as various application notes.

Finally, you can also find a number of useful application notes and online discussion forums from tool vendor websites.

### 2.2.7 Other equipment

Depending on the applications you are developing and the development board you are using, you might need additional hardware that interfaces to the development boards, such as external LCD display modules or communication interface adaptors. Also you might need some hardware development tools like a laboratory power supply, logic analyzer/oscilloscope, signal generator, etc.

## 2.3 Software development flow

The software development flow depends on the compiler suite you use. Assuming that you are using a compiler suite with Integrated Development Environment (IDE), the software development flow (as shown in Figure 2.4) usually involves:

Create project — you need to create a project that will specify the location of source files, compile target, memory configurations, compilation options, and so on. Many IDEs have a project creation wizard for this step.

Add files to project — you need to add the source code files required by the project. You might also need to specify the path of any included header files in the project options. Obviously you might also need to create new program source code files and write the program. Note that you should be able to reuse a number of files from the device-driver library to reduce the effort in writing new files. This includes startup code, header files, and some of the peripheral control functions.

Setup project options — In most cases, the project file created allows a number of project options such as compiler optimization options, memory map, and output file types. Based on the development board and debug adaptor you have, you might also need to setup options for debug and code download.

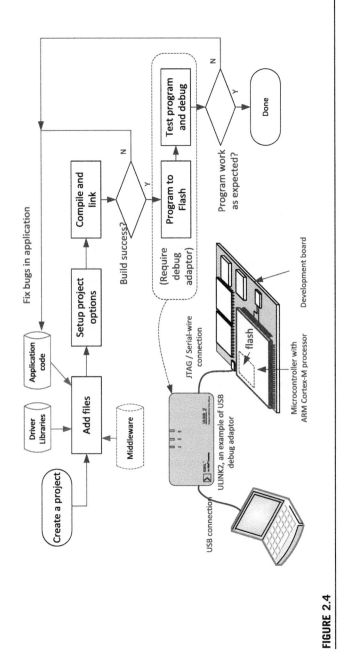

**FIGURE 2.4**

A simplified software development flow

Compile and link — In most cases, a project contains a number of files that are compiled separately. After the compilation process, each source file will have a corresponding object file. In order to generate the final combined executable image, a separate linking process is required. After the link stage, the IDE can also generate the program image in other file formats for the purpose of programming the image to the device.

Flash programming — Almost all of the Cortex®-M microcontrollers use flash memories for program storage. After a program image is created, we need to download the program to the flash memory of the microcontroller. To do this, you need a debug adaptor if the microcontroller board you use does not have one built in. The actual flash programming procedures can be quite complex, but these are usually fully handled by the IDE and you can carry out the whole programming process with a single mouse click. Note that if you want to, you can also download applications to SRAM and execute them from there.

Execute program and debug — After the compiled program is downloaded to the microcontroller, you can then run the program and see if it works. You can use the debug environment in the IDE to stop the processor (commonly referred as halt) and check the status of the system to ensure it is working properly. If it doesn't work correctly, you can use various debug features like single stepping to examine the program operations in detail. All these operations will require a debug adaptor (or the one built in to the development kit if available) to link up the IDE and the microcontroller being tested. If a software bug is found, then you can edit your program code, recompile the project, download the code to the microcontroller, and test it again.

If you are using open source toolchain, you might not have an IDE and might need to handle the compile and link process using scripts or makefile. Depending on the microcontroller product you are using, there can be third-party tools that can be used to download the compiled program image to the flash memory in the microcontroller.

During execution of the compiled program, you can check the program execution status and results by outputting information via various I/O mechanisms such as a UART interface or an LCD module. A number of examples in this book will show how some of these methods can be implemented. See Chapter 18 for some of the examples.

## 2.4 Compiling your applications

The procedure for compiling an embedded program depends on the development tools you use. Later in this book, a number of chapters cover the use of a couple of development tools to compile simple applications (Chapters 15 to 17). Here we will first have a look at some basic concepts of the compilation process.

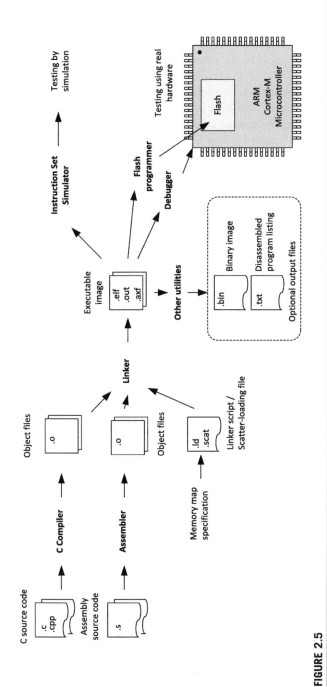

**FIGURE 2.5**

Common software compilation flow

First, we assume that you are developing your project using C programming language. This is the most commonly used programming language for microcontroller software development. Your project might also contain some assembly language files; for example, startup code that is supplied by microcontroller vendors. In most cases, the compilation process will be similar to the one shown in Figure 2.5.

Most development suites contain the tools listed in Table 2.1.

Different development tools have different ways to specify the layout of the program and data memory in the microcontroller system. In ARM® toolchains, you can use a file type called scatter-loading file, or in the case of Keil™ MDK-ARM, the scatter-loading file can be generated automatically by the μVision development environment. For some other ARM toolchains, you can also use command line options to specify the locations of ROM and RAM.

In a GNU-based toolchain, the memory specification is handled by linker scripts. These scripts are typically included in the installation of commercial gcc toolchains. However, some gcc users might have to create these files themselves. A later chapter of this book contains examples for compiling programs using gcc, which covers more information on linker scripts.

When using the GNU gcc toolchain, it is common to compile the whole application in one go instead of separating the compilation and linking stages (Figure 2.6).

The gcc compilation automatically invokes the linker and assembler if needed. This arrangement ensures that the details of the required parameters and libraries are passed on to the linker correctly. Using the linker as a separate step can be error prone and therefore is not recommended by most gcc tool vendors.

**Table 2.1** Various Tools You Can Find in a Development Suite

| Tools | Descriptions |
| --- | --- |
| **C compiler** | To compile C program files into object files |
| **Assembler** | To assemble assembly code files into object files |
| **Linker** | A tool to join multiple object files together and define memory configuration |
| **Flash programmer** | A tool to program the compiled program image to the flash memory of the microcontroller |
| **Debugger** | A tool to control the operation of the microcontroller and to access internal operation information so that status of the system can be examined and the program operations can be checked |
| **Simulator** | A tool to allow the program execution to be simulated without real hardware |
| **Other utilities** | Various tools, for example, file converters to convert the compiled files into various formats |

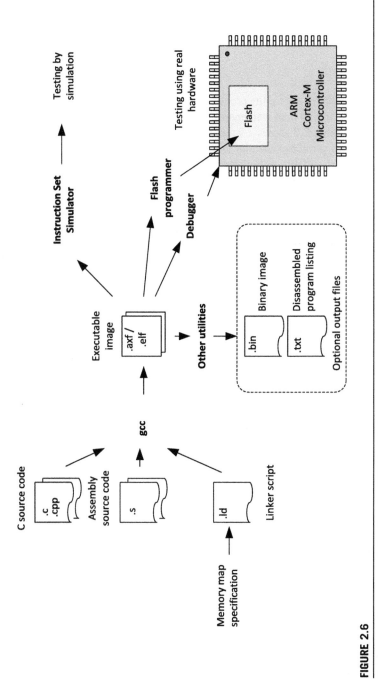

**FIGURE 2.6**

Common software compilation flow for GNU toolchain

## 2.5 Software flow

There are many ways to construct program flow for an application. Here we will cover some of the basic concepts.

### 2.5.1 Polling

For very simple applications, the processor can wait until there is data ready for processing, process it, and then wait again. This is very easy to setup and works fine for simple tasks. Figure 2.7 shows a simple polling program flow chart.

In most cases, a microcontroller will have to serve multiple interfaces and therefore be required to support multiple processes. The polling program flow method can be expanded to support multiple processes easily (Figure 2.8). This arrangement is sometimes called a "super-loop."

The polling method works well for simple applications, but it has several disadvantages. For example, when the application gets more complex, the polling loop design might get very difficult to maintain. Also, it is difficult to define priorities between different services using polling — you might end up with poor responsiveness, where a peripheral requesting service might need to wait a long time while the processor is handling less important tasks.

### 2.5.2 Interrupt driven

Another main disadvantage of the polling method is that it is not energy efficient. Lots of energy is wasted during the polling when service is not required. To solve this

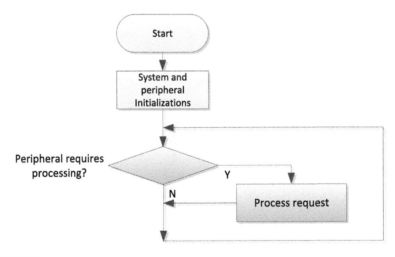

**FIGURE 2.7**

Polling method for simple application processing

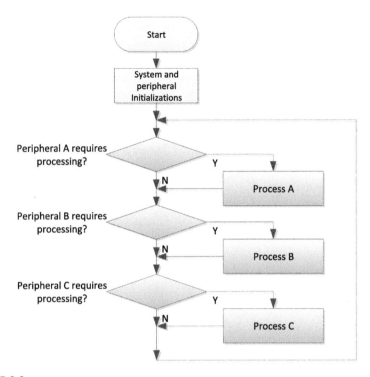

**FIGURE 2.8**

Polling method for application with multiple devices that need processing

problem, almost all microcontrollers have some sort of sleep mode support to reduce power, in which the peripheral can wake up the processor when it requires a service (Figure 2.9). This is commonly known as an interrupt-driven application.

In an interrupt-driven application, interrupts from different peripherals can be assigned with different interrupt priority levels. For example, important/critical peripherals can be assigned with a higher priority level so that if the interrupt arrives when the processor is servicing a lower priority interrupt, the execution of the lower priority interrupt service is suspended, allowing the higher priority interrupt service to start immediately. This arrangement allows much better responsiveness.

In some cases, the processing of data from peripheral services can be partitioned into two parts: the first part needs to be done quickly, and the second part can be carried out a little bit later. In such situations we can use a mixture of interrupt-driven and polling methods to construct the program. When a peripheral requires service, it triggers an interrupt request as in an interrupt-driven application. Once the first part of the interrupt service is carried out, it updates some software variables so that the second part of the service can be executed in the polling-based application code (Figure 2.10).

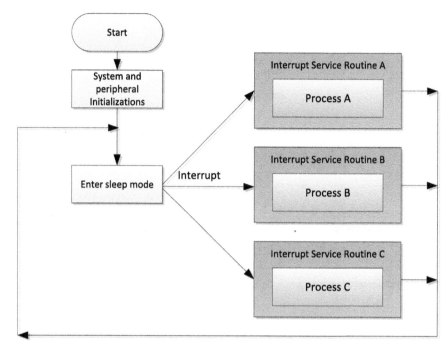

**FIGURE 2.9**

Simple interrupt-driven application

Using this arrangement, we can reduce the duration of high-priority interrupt handlers so that lower priority interrupt services can get served quicker. At the same time, the processor can still enter sleep mode to save power when no servicing is needed.

## 2.5.3 Multi-tasking systems

When the applications get more complex, a polling and interrupt-driven program structure might not be able to handle the processing requirements. For example, some tasks that can take a long time to execute might need to be processed concurrently. This can be done by dividing the processor's time into a number of time slots and allocating the time slots to these tasks. While it is technically possible to create such an arrangement by manually partitioning the tasks and building a simple scheduler to handle this, it is often impractical to do this in real projects as it is time consuming and can make the program much harder to maintain and debug.

In these applications, a Real-Time Operating System (RTOS) can be used to handle the task scheduling (Figure 2.11). An RTOS allows multiple processes to be executed concurrently, by dividing the processor's time into time slots and allocating the time slots to the processes that require services. A timer is need to handle the timekeeping for the RTOS, and at the end of each time slot, the timer generates a timer interrupt, which triggers the task scheduler and decides if context switching should be carried

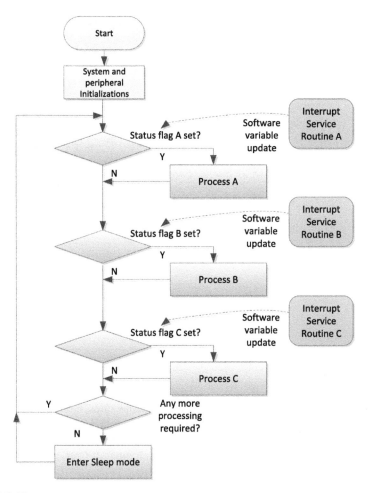

**FIGURE 2.10**

Application with both polling method and interrupt-driven arrangement

out. If yes, the current executing process is suspended and the processor executes another process.

Besides task scheduling, RTOSs also have many other features such as semaphores, message passing, etc. There are many RTOSs developed for the Cortex®-M processors, and many of them are completely free of charge.

## 2.6 Data types in C programming

The C programming language supports a number of "standard" data types. However, the way a data item is represented in hardware depends on the processor architecture

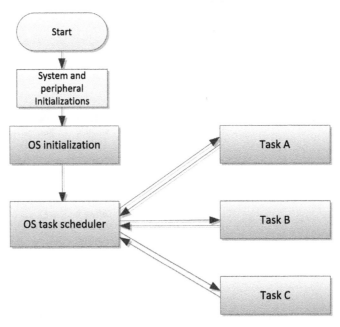

**FIGURE 2.11**

Using an RTOS to handle multiple tasks

as well as the C compiler. In different processor architectures, the size of certain data types can be different. For example, the integer is often 16-bit in 8-bit or 16-bit microcontrollers, and is always 32-bit in the ARM® architecture. Table 2.2 shows the common data types in ARM architecture, including all Cortex®-M processors. These data types are supported by all C compilers.

Because of differences of size in certain data types, it might be necessary to modify the source code when porting an application from an 8-bit or 16-bit microcontroller to an ARM Cortex-M microcontroller. More details on porting software from 8-bit and 16-bit architectures are covered in Chapter 24.

In ARM programming, we also refer to the size of a data item as BYTE, HALF WORD, WORD, and DOUBLE WORD, as shown in Table 2.3.

These terms are very common in ARM documentation, including the instruction set descriptions as well as hardware descriptions.

## 2.7 Inputs, outputs, and peripherals accesses

Almost all microcontrollers have various Input/Output (I/O) interfaces and peripherals such as timers, Real-time Clock (RTC), and so on. For microcontroller products based on the ARM® Cortex®-M3, and M4 processors, as well as common

**Table 2.2** Size and Range of Data Types in ARM Architecture Including Cortex-M Processors

| C and C99 (stdint.h) Data Type | Number of Bits | Range (Signed) | Range (Unsigned) |
|---|---|---|---|
| char, int8_t, uint8_t | 8 | −128 to 127 | 0 to 255 |
| short int16_t, uint16_t | 16 | −32768 to 32767 | 0 to 65535 |
| int, int32_t, uint32_t | 32 | −2147483648 to 2147483647 | 0 to 4294967295 |
| Long | 32 | −2147483648 to 2147483647 | 0 to 4294967295 |
| long long, int64_t, uint64_t | 64 | $-(2^{63})$ to $(2^{63} - 1)$ | 0 to $(2^{64} - 1)$ |
| Float | 32 | $-3.4028234 \times 10^{38}$ to $3.4028234 \times 10^{38}$ | |
| Double | 64 | $-1.7976931348623157 \times 10^{308}$ to $1.7976931348623157 \times 10^{308}$ | |
| long double | 64 | $-1.7976931348623157 \times 10^{308}$ to $1.7976931348623157 \times 10^{308}$ | |
| Pointers | 32 | 0x0 to 0xFFFFFFFF | |
| Enum | 8 / 16/ 32 | Smallest possible data type, except when overridden by compiler option | |
| bool (C++ only), _Bool (C only) | 8 | True or false | |
| wchar_t | 16 | 0 to 65535 | |

**Table 2.3** Data Size Definition in ARM Processor

| Terms | Size |
|---|---|
| **Byte** | 8-bit |
| **Half word** | 16-bit |
| **Word** | 32-bit |
| **Double word** | 64-bit |

interface peripherals such as GPIO, SPI, UART, I2C, you can also find many advanced interface peripherals like USB, CAN, Ethernet, and analogue interfaces like ADCs (Analog to Digital Converters) and DACs (Digital to Analog Converters). Most of these interface peripherals are vendor specific, so you need to read the user manuals provided by the microcontroller vendors to learn how to use them. In most cases you can also find programming examples on microcontroller vendor websites.

On these microcontrollers, the peripherals are memory-mapped, which means the registers are accessible from the system memory map. In order to access these peripherals registers in C programs, we can use pointers. We will see some examples of how this can be done in the following sections.

Typically, a peripheral requires an initialization process before it can be used. This might include some of the following steps:

- Programming the clock control circuitry to enable the clock signal connection to the peripheral, and clock signal connection to corresponding I/O pins if needed. Many modern microcontrollers allow fine tuning of clock signal distribution, such as enabling/disabling the clock connection to each individual peripheral for better energy saving. Typically the clocks to peripherals are turned off by default and you need to enable the clock before programming the peripheral. In some cases you might also need to enable the clock to the peripheral bus system.
- In some cases you might need to configure the operation mode of the I/O pins. Most microcontrollers have multiplexed I/O pins that can be used for multiple purposes. In order to use a peripheral, it might be necessary to configure its I/O pins to match the usage (e.g., input/output direction, function, etc.). In addition, you might also need to program additional configuration registers to define the expected electrical characteristics such as output type (voltage, pull up/down, open drain, etc.).
- Peripheral configuration. Most peripherals contain a number of programmable registers that need configuration before using the peripheral. In some cases, you can find the programming sequence a bit more complex than that of a 8-bit microcontroller, because the peripherals on 32-bit microcontrollers are often much more sophisticated than peripherals on 8-bit/16-bit systems. On the other hand, often the microcontroller vendors will have provided device-driver library code and you can use these driver functions to reduce the programming work required.
- Interrupt configuration. If a peripheral is to be used with interrupt operations, you will need to program the interrupt controller on the Cortex-M3/M4 processor (NVIC) to enable the interrupt and to configure the interrupt priority level.

All these initialization steps are carried out by programming peripheral registers in various peripheral blocks. As mentioned, peripheral registers are memory-mapped and therefore can be accessed using pointers. For example, you can define a General Purpose Input Output (GPIO) register set as a number of pointers as:

```
/* STM32F 100RBT6B — GPIO A Port Configuration Register Low */
#define GPIOA_CRL (*((volatile unsigned long *) (0x40010800)))
/* STM32F 100RBT6B — GPIO A Port Configuration Register High */
#define GPIOA_CRH (*((volatile unsigned long *) (0x40010804)))
/* STM32F 100RBT6B — GPIO A Port Input Data Register */
#define GPIOA_IDR (*((volatile unsigned long *) (0x40010808)))
/* STM32F 100RBT6B — GPIO A Port Output Data Register */
#define GPIOA_ODR (*((volatile unsigned long *) (0x4001080C)))
/* STM32F 100RBT6B — GPIO A Port Bit Set/Reset Register */
#define GPIOA_BSRR(*((volatile unsigned long *) (0x40010810)))
/* STM32F 100RBT6B — GPIO A Port Bit Reset Register */
#define GPIOA_BRR (*((volatile unsigned long *) (0x40010814)))
/* STM32F 100RBT6B — GPIO A Port Configuration Lock Register */
#define GPIOA_LCKR (*((volatile unsigned long *) (0x40010818)))
```

Then we can use the definitions directly. For example:

```
void GPIOA_reset(void) /* Reset GPIO A */
{
  // Set all pins as analog input mode
  GPIOA_CRL = 0; // Bit 0 to 7, all set as analog input
  GPIOA_CRH = 0; // Bit 8 to 15, all set as analog input
  GPIOA_ODR = 0; // Default output value is 0
  return;
}
```

This method is fine for a small number of peripheral registers. However, as the number of peripheral registers increases, this coding style can be problematic because:

- For each register address definition, the program needs to store the 32-bit address constant, resulting in increased code size.
- When there are multiple instantiations of the same peripheral, for example, the STM32 microcontroller has five GPIO peripherals, and the same definition has to be repeated for each of the instantiations. This is not scalable and makes it hard for software maintenance.
- It is not easy to create a function that can be shared between multiple instantiations of the same peripheral. For example, with the above example definition we might have to create the same GPIO reset function for each of the GPIO ports, resulting in increased code size.

In order to solve these problems, the common practice is to define the peripheral registers as data structures. For example, in the device-driver software package from the microcontroller vendors, we can find:

```
typedef struct
{
  __IO uint32_t CRL;
  __IO uint32_t CRH;
  __IO uint32_t IDR;
  __IO uint32_t ODR;
  __IO uint32_t BSRR;
  __IO uint32_t BRR;
  __IO uint32_t LCKR;
} GPIO_TypeDef;
```

Then each peripheral base address (GPIO A to GPIO G) is defined as pointers to the data structure:

```
#define PERIPH_BASE  ((uint32_t)0x40000000)
    /*!< Peripheral base address in the bit-band region */
...
#define APB2PERIPH_BASE (PERIPH_BASE + 0x10000)
...
```

```
#define GPIOA_BASE  (APB2PERIPH_BASE + 0x0800)
#define GPIOB_BASE  (APB2PERIPH_BASE + 0x0C00)
#define GPIOC_BASE  (APB2PERIPH_BASE + 0x1000)
#define GPIOD_BASE  (APB2PERIPH_BASE + 0x1400)
#define GPIOE_BASE  (APB2PERIPH_BASE + 0x1800)
...
#define GPIOA  ((GPIO_TypeDef *) GPIOA_BASE)
#define GPIOB  ((GPIO_TypeDef *) GPIOB_BASE)
#define GPIOC  ((GPIO_TypeDef *) GPIOC_BASE)
#define GPIOD  ((GPIO_TypeDef *) GPIOD_BASE)
#define GPIOE  ((GPIO_TypeDef *) GPIOE_BASE)
...
```

In these code snippets, there are a number of new things we have not covered:

The "__IO" is defined in a standardized header file in CMSIS. It implies a volatile data item (e.g., a peripheral register), which can be read or written to by software. Aside from "__IO," a peripheral register can also be defined as "__I" (read only) and "__O" (write only).

```
#ifdef __cplusplus
  #define __I volatile        /*!< defines 'read only' permissions */
#else
  #define __I volatile const /*!< defines 'read only' permissions */
#endif
#define __O volatile  /*!< defines 'write only' permissions */
#define __IO volatile /*!< defines 'read / write' permissions */
```

The "uint32_t" (unsigned 32-bit integer) is a data type supported in C99. This ensures the data size is 32-bit, independent of the processor architecture, which can help the software to be more portable. To use this data type, the project needs to include the standard data type header (Note: if you are using a CMSIS-compliant device header file this is already done for you in the device header file):

```
#include <stdint.h>/* Include standard types */
  /* C99 standard data types:
     uint8_t  : unsigned 8-bit, int8_t  : signed 8-bit,
     uint16_t : unsigned 16-bit, int16_t : signed 16-bit,
     uint32_t : unsigned 32-bit, int32_t : signed 32-bit,
     uint64_t : unsigned 64-bit, int64_t : signed 64-bit
  */
```

When peripherals are declared using such a method, we can create functions that can be used for each instance of the peripheral easily. For example, the code to reset the GPIO port can be written as:

```
void GPIO_reset(GPIO_TypeDef* GPIOx)
{
  // Set all pins as analog input mode
  GPIOx->CRL = 0; // Bit 0 to 7, all set as analog input
```

```
GPIOx->CRH = 0; // Bit 8 to 15, all set as analog input
GPIOx->ODR = 0; // Default output value is 0
return;
}
```

To use this function, we just need to pass the peripheral base pointer to the function:

```
GPIO_reset(GPIOA); /* Reset GPIO A */
GPIO_reset(GPIOB); /* Reset GPIO B */
...
```

This method for declaring peripheral registers is used by almost all of the Cortex-M microcontroller device-driver packages.

## 2.8 Microcontroller interfaces

The applications running in the microcontroller connect with external world using various peripheral interfaces. While usage of peripheral interfaces is not the main focus of this book, a few basic examples will be covered. In most cases, you can use device-driver library software packages from the microcontroller vendors to simplify the software development, and you can find examples and application notes on the Internet for such information.

Unlike programming for PCs, most embedded applications do not have a rich GUI. Some development boards might have an LCD screen, but many others just have a couple of LEDs and buttons. While the application itself might not require a user interface, often a simple text-based communication method is very useful for software development. For example, it can be handy to able to use printf to display a value captured by the Analog-to-Digital Converter (ADC) during program execution.

A number of methods can be used to handle such message display:

- Using a character LCD display module connected to the I/O pins of the microcontroller
- Using a simple UART to communicate with a terminal program running on a PC
- Set up a USB interface on the microcontroller as a virtual COM port to communicate with a terminal program running on a PC
- Use the Instrumentation Trace Macrocell (ITM), a standard debug feature on the Cortex®-M3/M4, to communicate with the debugger software

In some cases, a character LCD might be part of the embedded product, so using this hardware to display information can be convenient. However, the size of the screen limits the amount of information that can be displayed at a time.

A UART is easy to use, and allows more information to be passed to the developer quickly. The Cortex-M3/M4 processor does not have a UART as standard, but most microcontroller vendors have included a UART peripheral in their microcontroller designs. However, most modern computers do not have a UART interface (COM port) anymore, so you might need to use a USB-to-UART adaptor cable to

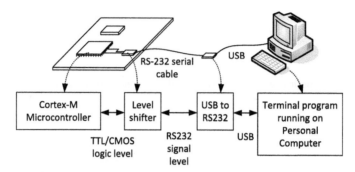

**FIGURE 2.12**

Using a UART to communicate with a PC via USB

handle this communication. In addition, you need to have a TTL-to-RS232 adaptor in your development setup to convert the signal's voltage (see Figure 2.12).

In some development boards (e.g., Texas Instruments Stellaris LaunchPad), the onboard debug adaptor has the feature of converting UART communications to USB.

If the microcontroller you use has a USB interface, you can use this to communicate with a PC using USB. For example, you can use a Virtual COM port solution for text-based communication with a terminal program running on a computer. It requires more effort in setting up the software but allows the microcontroller hardware to interface with the PC directly, avoiding the cost of the RS232 adaptors.

If you are using commercial debug adaptors like the Keil ULINK2, Segger J-LINK, or similar, you can use a feature called Instrumentation Trace Macrocell (ITM) to transfer messages to the debug host (the PC running the debugger) and display the messages in the development environment. This does not require any extra hardware and does not require much software overhead. It allows the peripheral interfaces to be free for other purposes. Examples of using the ITM are covered in Chapter 18.

The technique to redirect text messages from a "printf" (in C language) to specific hardware (e.g., UART, character LCD, etc.) is commonly referred as "retargeting." Retargeting can also be used to handle user inputs and system functions. The C code for retargeting is toolchain specific. Examples of retargeting for a couple of development tools will be covered in Chapter 18.

## 2.9 The Cortex® microcontroller software interface standard (CMSIS)

### 2.9.1 Introduction to CMSIS

Earlier in this chapter we mentioned CMSIS. CMSIS was developed by ARM® to allow microcontroller and software vendors to use a consistent software infrastructure to develop software solutions for Cortex®-M microcontrollers. As such, you can see that many software products for Cortex-M microcontrollers are CMSIS-compliant.

Currently the Cortex-M microcontroller market comprises:

- More than 15 microcontroller vendors shipping Cortex-M microcontroller products (see section 1.1.4 for the list of Cortex-M3 and Cortex-M4 microcontroller vendors), with some other silicon vendors providing Cortex-M based FPGA and ASICs
- More than 10 toolchain vendors
- More than 30 embedded operating systems
- Additional Cortex-M middleware software providers for codecs, communication protocol stacks, etc.

With such a large ecosystem, some form of standardization of the way the software infrastructure works becomes necessary to ensure software compatibility with various development tools and between different software solutions.

At the same time, embedded systems are also becoming more and more complex, and the amount of effort in developing and testing the software has increased substantially. In order to reduce development time as well as reducing the risk of having defects in products, software reuse is becoming more and more common. In addition, the complexity of the embedded systems has also increased the use of third-party software solutions. For example, an embedded software project might involve software components from many different sources:

- Software developed by in house developers
- Software reused from other projects
- Device-driver libraries from microcontroller vendors
- Embedded OSs
- Other third-party software products such as communication protocol stacks

In such scenarios, the interoperability of various software components becomes critical. For all these reasons, ARM worked with various microcontroller vendors, tools vendors, and software solution providers to develop CMSIS, a software framework covering most Cortex-M processors and Cortex-M microcontroller products.

The aims of CMSIS include:

- Enhanced software reusability — makes it easier to reuse software code in different Cortex-M projects, reducing time to market and verification efforts.
- Enhanced software compatibility — by having a consistent software infrastructure (e.g., API for processor core access functions, system initialization method, common style for defining peripherals), software from various sources can work together, reducing the risk in integration.
- Easy to learn — the CMSIS allows easy access to processor core features from the C language. In addition, once you learn to use one Cortex-M microcontroller product, starting to use another Cortex-M product is much easier because of the consistency in software setup.
- Toolchain independent — CMSIS-compliant device drivers can be used with various compilation tools, providing much greater freedom.

- Openness — the source code for CMSIS core files can be downloaded and accessed by everyone, and everyone can develop software products with CMSIS.

CMSIS is an evolving project. It started out as a way to establish consistency in device-driver libraries for the Cortex-M microcontrollers, and this has become CMSIS-Core. Since then additional CMSIS projects have started:

- CMSIS-Core (Cortex-M processor support) — a set of APIs for application or middleware developers to access the features on the Cortex-M processor regardless of the microcontroller devices or toolchain used. Currently the CMSIS processor support includes the Cortex-M0, Cortex-M0+, Cortex-M3, and Cortex-M4 processors and SecurCore products like SC000 and SC300. Users of the Cortex-M1 can use the Cortex-M0 version because they share the same architecture.
- CMSIS-DSP library — in 2010 the CMSIS DSP library was released, supporting many common DSP operations such as FFT and filters. The CMSIS-DSP is intended to allow software developers to create DSP applications on Cortex-M microcontrollers easily.
- CMSIS-SVD — the CMSIS System View Description is an XML-based file format to describe peripheral set in microcontroller products. Debug tool vendors can then use the CMSIS SVD files prepared by the microcontroller vendors to construct peripheral viewers quickly.
- CMSIS-RTOS — the CMSIS-RTOS is an API specification for embedded OS running on Cortex-M microcontrollers. This allows middleware and application code to be developed for multiple embedded OS platforms, and allows better reusability and portability.
- CMSIS-DAP — the CMSIS-DAP (Debug Access Port) is a reference design for a debug interface adaptor, which supports USB to JTAG/Serial protocol conversions. This allows low-cost debug adaptors to be developed which work for multiple development toolchains.

In this chapter we will first look at the processor support in CMSIS (CMSIS-Core). The CMSIS DSP library will be covered in Chapter 22. The CMSIS-SVD and CMSIS-DAP topics are beyond the scope of this book.

## 2.9.2 Areas of standardization in CMSIS-Core

From a software development point of view, the CMSIS-Core standardizes a number of areas:

Standardized definitions for the processor's peripherals — These include the registers in the Nested Vector Interrupt Controller (NVIC), a system tick timer in the processor (SysTick), an optional Memory Protection Unit (MPU), various programmable registers in the System Control Block (SCB), and some software programmable registers related to debug features. Note: Some of the registers in the Cortex®-M4 are not available in Cortex-M3, and similarly, some registers in Cortex-M3 and Cortex-M4 are not available in the Cortex-M0.

Standardized access functions to access processor's features — These include various functions for interrupt control using NVIC, and functions for accessing special registers in the processors. It is still possible to access the registers directly if needed, but for general programming using the access functions (or sometimes referred as Application Programming Interface, API, in some literature) can help software portability. More details of these functions are covered in Appendix E.

Standardized functions for accessing special instructions easily — The Cortex-M processors support a number of instructions for special purposes (e.g., Wait-For-Interrupt, WFI, for entering sleep mode). These instructions cannot be generated using generic IEC/ISO C[1] language. Instead, CMSIS implements a set of functions to allow these instructions to be accessed within C program code. Without these functions, the users would have to rely on toolchain specific solutions such as intrinsic functions or inline assembly to inject special instructions into the application, which make the software less reusable and might require certain in-depth knowledge of the toolchain in order to handle them correctly. CMSIS provides a standardized API for these features so that they can be easily used by application developers.

Standardized function names for system exception handlers — A number of system exception types are presented in the architecture for the Cortex-M processors. By giving the corresponding system exception handlers standardized names, it makes it much easier to develop software solutions that can be applied to multiple Cortex-M products. This is especially important for embedded OS developers, as the embedded OS requires the use of several types of system exception.

Standardized functions for system initialization — Most modern feature-rich microcontroller products require some configuration of clock circuitry and power management registers before the application starts. In CMSIS-compliant device-driver libraries, these configuration steps are placed in a function called "SystemInit()." Obviously, the actual implementation of this function is device specific and might need adaption for various project requirements. However, having a standardized function name, a standardized way that this function is used and a standardized location where this function can be found makes it much easier for a designer to pick up and start using a new Cortex-M microcontroller device.

Standardized software variables for clock speed information — This might not be obvious, but often our application code does need to know what clock frequency the system is running at. For example, such information might be needed for setting up the baud rate divider in a UART, or to initialize the SysTick timer for an embedded OS. A software variable called "SystemCoreClock" (for CMSIS 1.3 or newer versions, or "SystemFreq" in older versions of CMSIS) is defined in the CMSIS-Core.

In addition, the CMSIS-Core also provides:

A common platform for device-driver libraries — Each device-driver library has the same look and feel, making it easier for beginners to learn how to use the devices.

---

[1]C/C++ features are specified in a standard document "ISO/IEC 14882" prepared by the International Organisation for Standards (ISO) and the International Electrotechnical Commission (IEC).

This also makes it easier for software developers to develop software for multiple Cortex-M microcontroller products.

### 2.9.3 Organization of CMSIS-Core

The CMSIS files are integrated into device-driver library packages from microcontroller vendors. Some of the files in the device-driver library are prepared by ARM® and are common to various microcontroller vendors. Other files are vendor/device specific. In a general sense, we can define the CMSIS into multiple layers:

- Core Peripheral Access Layer — Name definitions, address definitions, and helper functions to access core registers and core peripherals. This is processor specific and is provided by ARM.
- Device Peripheral Access Layer — Name definitions, address definitions of peripheral registers, as well as system implementations including interrupt assignments, exception vector definitions, etc. This is device specific (note: multiple devices from the same vendor might use the same file set).
- Access Functions for Peripherals — The driver code for peripheral accesses. This is vendor specific and is optional. You can choose to develop your application using the peripheral driver code provided by the microcontroller vendor, or you can program the peripherals directly if you prefer.

There is also a proposed additional layer for peripheral accesses:

Middleware Access Layer — This layer does not exist in current version of CMSIS. The idea is to develop a set of APIs for interfacing common peripherals such as UART, SPI, and Ethernet. If this layer exists, developers of middleware can develop their applications based on this layer to allow software to be ported between devices easily.

The roles of the various layers are summarized in Figure 2.13.

Note that in some cases, the device-driver libraries might contain additional vendor-specific functions for the NVIC implemented by the microcontroller vendor. The aim of CMSIS is to provide a common starting point, and the microcontroller vendors can add additional functions if they prefer. But software using these functions will need porting if the software design is to be reused on another microcontroller product.

### 2.9.4 How do I use CMSIS-Core?

The CMSIS files are included in the device-driver packages provided by the microcontroller vendors. So when you are using CMSIS-compliant device-driver libraries provided by the microcontroller vendors, you are already using CMSIS.

Typically, you need to do the following.

- Add source files to project. This includes:
  - Device-specific, toolchain-specific startup code, in the form of assembly or C
  - Device-specific device initialization code (e.g., *system_<device>.c*)
  - Additional vendor-specific source files for peripheral access functions. This is optional.

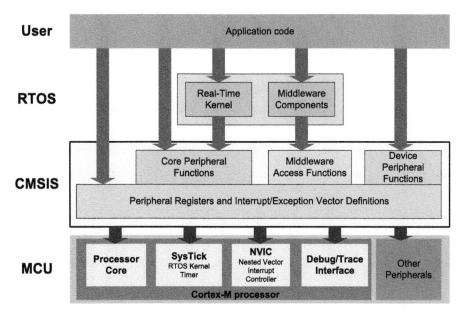

**FIGURE 2.13**

CMSIS-Core structure

- For CMSIS 2.00 or older versions of CMSIS-Core libraries, you might also need to add a processor-specific C program file (e.g., core_cm3.c) to the project for some of the core register access functions. This is not required from CMSIS-Core 2.10.
- Add header files into search path of the project. This includes:
  - A device-specific header file for peripheral registers definitions and interrupt assignment definitions. (e.g., <device>.h)
  - A device-specific header file for functions in device initialization code (e.g., system_<device>.h)
  - A number of processor-specific header files (e.g., core_cm3.h, core_cm4.h; they are generic for all microcontroller vendors)
  - Optionally additional vendor-specific header files for peripheral access functions
  - In some cases the development suites might also have some of the generic CMSIS support files pre-installed.

Figure 2.14 shows a typical project setup using a CMSIS device-driver package. Inside the device-driver package obtained from the microcontroller vendor, you will find the various files you need, including the CMSIS generic files. The names of some of these files depend on the actual microcontroller device name chosen by the microcontroller vendor (indicated as <device> in the diagram).

When the device-specific header file is included in the application code, it automatically includes additional header files, therefore you need to set up the project search path for the header files in order to compile the project correctly.

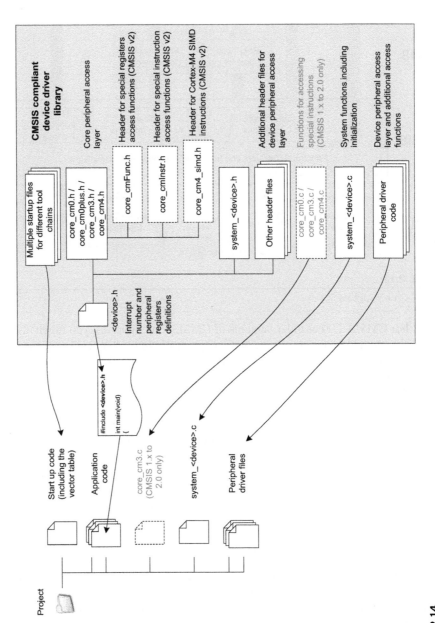

**FIGURE 2.14**

Using CMSIS-Core in a project

In some cases, the Integrated Development Environment (IDE) automatically sets up the startup code for you when you create a new project. Otherwise you just need to add the startup code from the device-driver library to the project manually. Startup code is required for the starting sequence of the processor, and it also includes the exception vector table definition that is required for interrupt handling.

### 2.9.5 Benefits of CMSIS-Core

So what does CMSIS mean to users?

The main advantage is much better software portability and reusability:

*   A project for a Cortex®-M microcontroller device can be migrated to another device from the same vendor with a different Cortex-M processor very easily. Often microcontroller vendors provide devices with Cortex-M0/M0+/M3/M4 with the same peripheral and same pin out, and the change required is just replacing a couple of CMSIS files in the project.
*   CMSIS-Core made it easier for a Cortex-M microcontroller project to be migrated to another device from a different vendor. Obviously, peripheral setup and access code will need to be modified, but processor core access functions are based on the same CMSIS source code and do not require changes.
    CMSIS allows software to be much more future proof because embedded software developed today can be reused on other Cortex-M products in the future.

The CMSIS-Core also allows faster time to market because:

*   It is easier to reuse software code from previous projects.
*   Since all CMSIS-compliant device drivers have a similar structure, learning to use a new Cortex-M microcontroller is much easier.
*   The CMSIS code has been tested by many silicon vendors and software developers around the world. It is compliant with Motor Industry Software Reliability Association (MISRA). Therefore it reduces the validation effort required, as there is no need to develop and test your own processor feature access functions.
*   Starting from CMSIS 2.0, a DSP library is included that provides tested, optimized DSP functions. The DSP library code is available as a free download and can be used by software developers free of charge.

There are also a number of other advantages:

*   CMSIS is supported by multiple compiler toolchain vendors.
*   CMSIS has a small memory footprint (less than 1KB for all core access functions and a few bytes of RAM for several variables).
*   CMSIS files contain Doxygen tags (http://www.doxygen.org) to enable easy automatic generation of documentation.

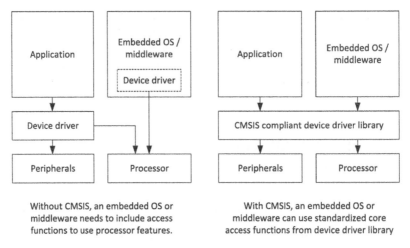

Without CMSIS, an embedded OS or middleware needs to include access functions to use processor features.

With CMSIS, an embedded OS or middleware can use standardized core access functions from device driver library

**FIGURE 2.15**

CMSIS-Core avoids the need for middleware or OS to carry their own driver code

For developers of embedded OS and middleware, the advantage of CMSIS is significant:

- By using processor core access functions from CMSIS, embedded OS, and middleware can work with device-driver libraries from various microcontroller vendors, including future products that are yet to be released.
- Since CMSIS is designed to work with various toolchains, many software products can be designed to be toolchain independent.
- Without CMSIS, middleware might need to include a small set of driver functions for accessing processor peripherals such as the interrupt controller. Such an arrangement increases the program size, and might cause compatibility issues with other software products (Figure 2.15).

## 2.9.6 Various versions of CMSIS

The CMSIS project is evolving. Over the last few years several versions of CMSIS have been released, bringing wider processor support and improvements. Apart from coding improvement, there have also been a number of other changes:

| Version | Main Changes |
|---------|--------------|
| 1.0 | Nov 2008<br>Initial release. Support Cortex®-M3 processor only. |
| 1.10 | Feb 2009<br>Support for Cortex-M0 added. |
| 1.20 | May 2009<br>Add support for TASKING compiler.<br>Add more functions to manage priority settings in NVIC. |

| Version | Main Changes |
|---------|--------------|
| 1.30 | Oct 2009<br>The system initialization function *SystemInit()* is called in startup code instead of beginning of *main()*.<br>SystemFrequency variable renamed to SystemCoreClock to reflect the processor clock definition. Additional functions "void SystemCoreClockUpdate(void)" added.<br>Add support for data receive for debug communication. (Previous versions use ITM for data output in debug communication.)<br>Add bit definition for processor's peripheral registers.<br>Directory structure changed. |
| 2.0 | Nov 2010<br>Support for Cortex-M4 added.<br>Included a CMSIS DSP library (CMSIS-DSP) for Cortex-M4 and Cortex-M3.<br>New header files *core_cm4_simd.h*, *core_cmFunc.h* and *core_cmInst.h* introduced, with a number of core access functions are moved to these files and become inlined.<br>Add CMSIS System View Description |
| 2.10 | July 2011<br>CMSIS-DSP library for Cortex-M0 added.<br>Added big endian support for DSP library.<br>Directory structure simplified.<br>Processor specific C program files (e.g., *core_cm3.c*, *core_cm4.c*) are no longer required and are removed.<br>Reworded CMSIS-DSP library example.<br>Documentation update. |
| 3.0 | October 2011<br>Added support for GNU Tools for ARM Embedded Processors.<br>Added function __ROR.<br>Added Register Mapping for TPIU, DWT.<br>Added support for SC000 and SC300 processors.<br>Corrected ITM_SendChar function.<br>Corrected the functions __STREXB, __STREXH, __STREXW for the GNU GCC compiler section.<br>Documentation restructured. |
| 3.01 | March 2012<br>Added support for Cortex-M0+ processor.<br>Integration of CMSIS DSP Library version 1.1.0. |

In normal cases, embedded applications can work with different versions of the CMSIS source files without problems. Most microcontroller vendors keep their device-driver library up to date with the most recent versions of CMSIS, but there is always the chance that the device-driver library package from microcontroller vendors could be a couple of releases behind the latest CMSIS version. This is not usually a problem, as the functionalities of the driver functions remain unchanged.

In a few cases, application code might need to be updated to allow it to be used with a newer version of the CMSIS driver package (e.g., when the "SystemFrequency" variable is used, which is replaced by "SystemCoreClock" from CMSIS 1.3).

You can download the latest version of the CMSIS source package from http://www.arm.com/cmsis.

# Technical Overview

## CHAPTER OUTLINE

## 3.1 General information about the Cortex®-M3 and Cortex-M4 processors

### 3.1.1 Processor type

All the ARM® Cortex®-M processors are 32-bit RISC (Reduced Instruction Set Computing) processors. They have:

- 32-bit registers
- 32-bit internal data path
- 32-bit bus interface

In addition to 32-bit data, the Cortex-M processors (as well as any other ARM processors) can also handle 8-bit, and 16-bit data efficiently. The Cortex-M3 and

The Definitive Guide to ARM® Cortex®-M3 and Cortex-M4 Processors. http://dx.doi.org/10.1016/B978-0-12-408082-9.00003-8

M4 processors also support a number of operations involving 64-bit data (e.g., multiply, accumulate).

The Cortex-M3 and Cortex-M4 processors both have a three-stage pipeline design (instruction fetch, decode, and execution), and both have a Harvard bus architecture, which allows simultaneous instruction fetches and data accesses.

The memory system of the ARM Cortex-M processors uses 32-bit addressing, which allows a maximum 4GB address space. The memory map is unified, which means that although there can be multiple bus interfaces, there is only one 4GB memory space. The memory space is used by the program code, data, peripherals, and some of the debug support components inside the processors.

Just like any other ARM processors, the Cortex-M processors are based on a load-store architecture. This means data needs to be loaded from the memory, processed, and then written back to memory using a number of separate instructions. For example, to increment a data value stored in SRAM, the processor needs to use one instruction to read the data from SRAM and put it in a register inside the processor, a second instruction to increment the value of the register, and then a third instruction to write the value back to memory. The details of the registers inside the processors are commonly known as a programmer's model.

### 3.1.2 Processor architecture

As explained in Chapter 1, the processor is only a part of a microcontroller chip. The memory system, peripherals, and various interface features are developed by the microcontroller vendors. As a result, you can find Cortex®-M processors being used in a wide range of devices, from low-cost microcontroller products to high-end multi-processor products. But these devices share the same architecture. In ARM® processors, the term architecture can refer to two areas:

- Architecture: Instruction Set Architecture (ISA), programmer's model (what the software sees), and debug methodology (what the debugger sees).
- Micro-architecture: Implementation-specific details such as interface signals, instruction execution timing, pipeline stages. Micro-architecture is processor design-specific.

Various versions of the ARM Architecture exist for the different ARM processors released over the years. For example, the Cortex-M3 and Cortex-M4 processors are both implementations of ARMv7-M Architecture. An Instruction Set Architecture can be implemented with various implementations of micro-architecture; for example, different numbers of pipeline stages, different types of bus interface protocol, etc.

The details of the ARMv7-M architecture are documented in the ARMv7-M Architecture Reference Manual (also known as ARMv7-M ARM). This document covers:

- Instruction set details
- Programmer's model

- Exception model
- Memory model
- Debug architecture

This document can be obtained from ARM after a simple registration process. However, for general programming, it is not necessary to have the full architecture reference manual. ARM provides alternate documents for software developers called Cortex-M3/M4/M0 Devices Generic User Guides. This can be found on the ARM website:

http://infocenter.arm.com
→ Cortex-M series processors
→ Cortex-M0/M0+/M3/M4
→ Revision number
→ Cortex-M4/M3/M0/M0+ Devices Generic User Guide

Some of the microarchitecture information such as instruction execution timing information can be found in the Technical Reference Manuals (TRM) of the Cortex-M processors, which can be found on the ARM website. Other microarchitecture information like the processor interface details are documented in other Cortex-M product documentation, which is normally accessible only by silicon chip designers.

Theoretically, a software developer does not necessarily need to know anything about the micro-architecture to develop software for the Cortex-M products. But in some cases, knowing some of the micro-architecture details could help. This is particularly true for optimizing software or even C compilers for best performance.

### 3.1.3 Instruction set

The instruction set used by the Cortex®-M processors is called Thumb (this covers both the 16-bit Thumb instructions and the newer 32-bit Thumb® instructions). The Cortex-M3 and Cortex-M4 processors incorporate Thumb®-2 Technology,[1] which allow mixture of 16-bit and 32-bit instructions for high code density and high efficiency.

In classic ARM® processors, for example, the ARM7TDMI™, the processor has two operation states: a 32-bit ARM state and a 16-bit Thumb state. In the ARM state, the instructions are 32-bit and the core can execute all supported instructions with very high performance. In Thumb state, the instructions are 16-bit, which provides excellent code density, but Thumb instructions do not have all the functionality of ARM instructions and more instructions may be needed to complete certain types of operation.

To get the best of both worlds, many applications for classic ARM processors have mixed ARM and Thumb code. However, the mixed-code arrangement does not always work ideally. There is overhead (in terms of both execution time and

---

[1]From trademark point of view, "Thumb-2" is a technology to support mixture of 16-bit and 32-bit Thumb instructions. Officially the whole instruction set is called "Thumb."

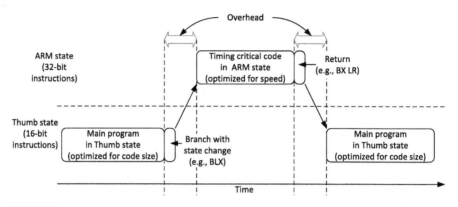

**FIGURE 3.1**

Switching between ARM code and Thumb code in class ARM processors such as the ARM7TDMI

instruction count; see Figure 3.1) to switch between the states, and the separation of two states can increase the complexity of the software compilation process and make it harder for inexperienced developers to optimize the software.

With the introduction of Thumb-2 technology, the Thumb instruction set has been extended to support both 16-bit and 32-bit instruction encoding. It is now possible to handle all processing requirements without switching between the two different operation states. In fact, the Cortex-M processors do not support 32-bit ARM instructions at all (Figure 3.2). Even interrupt processing is handled entirely in Thumb state, whereas in classic ARM processors interrupt handlers are entered in ARM state. With Thumb-2 technology, the Cortex-M processor has a number of advantages over classic ARM processors, such as:

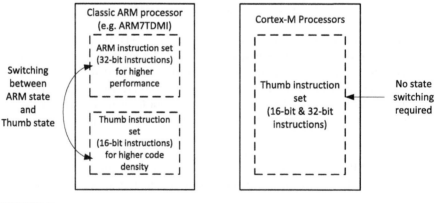

**FIGURE 3.2**

Instruction set comparison between Cortex-M processors and ARM7TDMI

**Table 3.1** Range of Instructions in Different Cortex-M Processors

| Instruction Groups | Cortex-M0, M1 | Cortex-M3 | Cortex-M4 | Cortex-M4 with FPU |
|---|---|---|---|---|
| 16-bit ARMv6-M instructions | ● | ● | ● | ● |
| 32-bit Branch with Link instruction | ● | ● | ● | ● |
| 32-bit system instructions | ● | ● | ● | ● |
| 16-bit ARMv7-M instructions | | ● | ● | ● |
| 32-bit ARMv7-M instructions | | ● | ● | ● |
| DSP extensions | | | ● | ● |
| Floating point instructions | | | | ● |

- No state switching overhead, saving both execution time and instruction space.
- No need to specify ARM state or Thumb state in source files, making software development easier.
- It is easier to get the best code density, efficiency, and performance at the same time.
- With Thumb-2 technology, the Thumb instruction set has been extended by a wide margin when compared to a classic processor like the ARM7TDMI. Note that although all of the Cortex-M processors support Thumb-2 technology, they implement various subsets of the Thumb ISA (Table 3.1).

Some instructions defined in the Thumb instruction set are not available in the current Cortex-M processors. For example, the co-processor instructions are not supported (though separate memory-mapped data processing engines could be added). Also, a few other Thumb instructions from classic ARM processors are not supported, such as Branch with Link and Exchange (BLX) with immediate (used to switch processor state from Thumb to ARM), a couple of change process state (CPS) instructions, and the SETEND (Set Endian) instruction, which were introduced in architecture v6. For a complete list of supported instructions, refer to Appendix A.

### 3.1.4 Block diagram

From a high-level point of view, Cortex®-M3 and Cortex-M4 are very similar to each other. Although there are significant differences in the internal data path designs, some parts of the processors such as instruction fetch buffer, parts of the instruction decode and execution stages, and the NVIC are similar to each other. In addition, the components outside the "core" level are almost identical.

The Cortex-M3 and the Cortex-M4 processors contain the core of the processor, the Nested Vectored Interrupt Controller (NVIC), the SysTick timer, and optionally the floating point unit (for Cortex-M4). Apart from these, the processors also contain

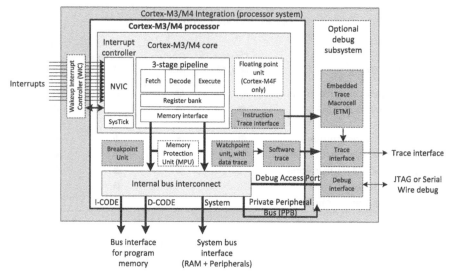

**FIGURE 3.3**

Block diagram of the Cortex-M3 and Cortex-M4 processor

some internal bus systems, an optional Memory Protection Unit (MPU), and a set of components to support software debug operations. The internal bus interconnect is needed to route transfers from the processor and the debugger to various parts of the design.

The Cortex-M3 and Cortex-M4 processors are highly configurable. For example, the debug features are optional, allowing system-on-chip designers to remove debug components if debug support is not required in the product. This allows the silicon area of the design to be reduced significantly. In some cases, silicon designers can also choose to reduce the number of hardware instruction breakpoint and data watchpoint comparators to reduce the gate count. Many system features like the number of interrupt inputs, number of interrupt priority levels supported, and the MPU are also configurable.

The integration level in Figure 3.3 is a reference design ARM® provides to silicon designers. This integration level can be modified by the silicon vendors to customize debug support such as the debug interface and to support device-specific low-power features (e.g., adding a customized Wake-up Interrupt Controller).

The top level of the Cortex-M3 and Cortex-M4 processors has a number of bus interfaces, as shown in Table 3.2.

---

[2]In many ARM document, and in command line option switches for some C compilers, a Cortex®-M4 processor with the floating point unit is referred to as Cortex-M4F.

| Table 3.2 Various Bus Interfaces on the Cortex-M3 and Cortex-M4 Processors | |
| --- | --- |
| **Bus Interface** | **Descriptions** |
| **I-CODE** | Primarily for program memory: Instruction fetch and vector fetch for address 0x0 to 0x1FFFFFFF. Based on AMBA 3.0 AHB Lite bus protocol. |
| **D-CODE** | Primarily for program memory: Data and debugger accesses for address 0x0 to 0x1FFFFFFF. Based on AMBA 3.0 AHB Lite bus protocol. |
| **System** | Primarily for RAM and peripherals: Any accesses from address 0x20000000 to 0xFFFFFFFF (apart from PPB regions). Based on AMBA 3.0 AHB Lite bus protocol. |
| **PPB** | External Private Peripheral Bus (PPB): For private debug components on system level from address 0xE0040000 to 0xE00FFFFF. Based on AMBA 3.0 APB protocol. |
| **DAP** | Debug Access Port (DAP) interface: For debugger accesses generated from the debug interface module to any memory locations including system memory and debug components. Based on the ARM CoreSight™ debug architecture. |

### 3.1.5 Memory system

The Cortex®-M3 and M4 processors themselves do not include memories (i.e., they do not have program memory, SRAM, or cache). Instead, they come with a generic on-chip bus interface, so microcontroller vendors can add their own memory system to their design. Typically, the microcontroller vendor will need to add the following items to the memory system:

- Program memory, typically flash
- Data memory, typically SRAM
- Peripherals

In this way, different microcontroller products can have different memory configurations, different memory sizes and types, and different peripherals.

The bus interfaces on the Cortex-M processors are 32-bit, and based on the Advanced Microcontroller Bus Architecture (AMBA®) standard. AMBA contains a collection of several bus protocol specifications. The AMBA specifications can be downloaded from the ARM® website, and any silicon designers can freely use these protocol standards. Due to the low hardware cost, efficiency, and openness of these standards, they are very popular among silicon designers.

The main bus interface protocol used by the Cortex-M3 and M4 processors is the AHB Lite (Advanced High-performance Bus), which is used in program memory and system bus interfaces. The AHB Lite protocol is a pipelined bus protocol, allowing high operation frequency and low hardware area cost. Another bus protocol used is the Advanced Peripheral Bus (APB) interface, commonly used in the peripheral systems of ARM®-based microcontrollers. In addition, the APB protocol is used inside the Cortex-M3 and Cortex-M4 processor for debug support.

Unlike off-chip bus protocols, the AHB Lite and APB protocols are fairly simple, as the hardware configuration inside a chip is fixed, so there is no need to have a complex initialization protocol to handle various possible configurations (e.g., no need for "plug and play" support as in computer technology).

The use of an open and generic bus architecture allows various silicon designers to develop peripherals, memory controllers, and on-chip memory modules for ARM processors. These designs are commonly referred to as IP, and microcontroller vendors can use their own peripheral designs as well as IP licensed from other companies in their microcontroller products. With a standardized bus protocol, these IPs can easily be integrated together in a large-scale design. Today, the AMBA protocols are a de facto standard for on-chip bus systems. You can find these designs in many system-on-chip devices, including those that use processors from other processor design companies.

For a software developer writing software for the Cortex-M microcontrollers, there is no need to understand the bus protocol details. However, their nature can affect the programmer's view in certain ways such as in data alignment and cycling timing.

### 3.1.6 Interrupt and exception support

The Cortex®-M3 and Cortex-M4 processors include an interrupt controller called the Nested Vectored Interrupt Controller (NVIC). It is programmable and its registers are memory mapped. The address location of the NVIC is fixed and the programmer's model of the NVIC is consistent across all Cortex-M processors.

Beside interrupts from peripherals and other external inputs, the NVIC also supports a number of system exceptions, including a Non-Maskable Interrupt (NMI) and other exception sources within the processor.

The Cortex-M3 and Cortex-M4 processors are configurable. Microcontroller vendors can determine how many interrupt signals the NVIC should provide, and how many programmable interrupt priority levels are supported in the NVIC design. Although some of the details of NVIC in different Cortex-M3/M4 microcontrollers can be different, the handling of interrupt/exception and the programmer's model of NVIC are the same and are defined in the architecture reference manual.

## 3.2 Features of the Cortex®-M3 and Cortex-M4 processors

Today, most major microcontroller vendors ship microcontrollers based on the ARM® Cortex®-M3/M4 processors. What are the advantages of the Cortex®-M processors which have made them so popular? The strength of the Cortex-M3/M4 processors and their benefits are summarized in this section.

### 3.2.1 **Performance**

The Cortex®-M processors deliver high performance in microcontroller products.

- The three-stage pipeline allows most instructions, including multiply, to execute in a single cycle, and at the same time allows high clock frequencies for microcontroller devices — typically over 100 MHz, and up to approx 200 MHz[3] in modern semiconductor manufacturing processes. Even when running at the same clock frequency as most other processor products, the Cortex-M3 and Cortex-M4 processors have a better Clock Per Instruction (CPI) ratio. This allows more work to be done per MHz or allows the designs to run at lower clock frequency for reduced power consumption.
- Multiple bus interfaces allow simultaneous instruction and data accesses to be performed.
- The pipelined bus interface allows a higher clock frequency in the memory system.
- The highly efficient instruction set allows complex operations to be carried out in a low numbers of instructions.
- Each instruction fetch is 32-bit, and most instructions are 16-bit. Therefore up to two instructions can be fetched at a time, allowing extra bandwidth on the memory interface for better performance and better energy efficiency.

With the current compiler technologies, the performance of the Cortex-M3 and Cortex-M4 processors are given in Table 3.3.

This high performance makes it possible to develop products that previously couldn't be done with legacy 8-bit/16-bit low-cost microcontroller products. For example, it is now possible to add low-cost graphical interfaces to embedded devices without switching to a high-end microprocessor.

### 3.2.2 **Code density**

The Thumb instruction set used on the ARM® Cortex®-M processors provides excellent code density compared to other processor architectures. Many software developers migrating from 8-bit microcontrollers will see a significant reduction in the required program size, while performance will also be improved significantly. The

**Table 3.3** Performance of the Cortex-M Processors in Commonly Used Benchmark[4]

| Processor | Dhrystone 2.1/MHz | CoreMark/MHz |
|---|---|---|
| **Cortex-M3** | 1.25 DMIPS/MHz | 3.32 |
| **Cortex-M4** | 1.25 DMIPS/MHz | 3.38 |
| **Cortex-M4 with FPU** | 1.25 DMIPS/MHz | 3.38 |

[3]When this book was written the maximum clock speed available was 204 MHz.
[4]Certified CoreMark result in December 2012.

code density of the Cortex-M processors is also better than many commonly used 16-bit and 32-bit architectures. There are also additional advantages:

- Thumb-2 technology allows 16-bit instructions and 32-bit instructions to work together without any state switching overhead. Most simple operations can be carried out with a 16-bit instruction.
- Various memory addressing modes for efficient data accesses
- Multiple memory accesses can be carried out in a single instruction
- Support for hardware divide instructions and Multiply-and-Accumulate (MAC) instructions exist in both Cortex-M3 and Cortex-M4
- Instructions for bit field processing in Cortex-M3/M4
- Single Instruction, multiple data (SIMD) instruction support exists in Cortex-M4
- Optional single precision floating point instructions are available in Cortex-M4

Besides lower system cost, high code density also reduces power consumption, because you can use a device with less flash memory. You can also copy some parts of the program code (e.g., interrupt handlers) into SRAM for high speed execution without worrying that this will take up too much SRAM space.

### 3.2.3 Low power

The Cortex®-M processors are designed for low power implementations. Many Cortex-M3 and Cortex-M4 microcontroller products can run at under 200 μA/ MHz (approximately 0.36 mW/MHz for a supply voltage of 1.8 volt) and some of them can even run at under 100 μA/MHz. Low power characteristics of Cortex-M processors includes:

- The Cortex-M3 is designed to target low-cost microcontrollers in which a small silicon area (low gate count) is essential. The Cortex-M4 is slightly larger due to the additional SIMD instructions and the optional floating point unit. The three-stage pipeline design provides a good balance between performance and silicon die size.
- The high code density of the Cortex-M processor allows software developers to use a microcontroller device with smaller program memory to implement their products to reduce power consumption.
- The Cortex-M processors provide a number of low power features. These include multiple sleep modes defined in the architecture, and integrated architectural clock gating support, which allows clock circuits for parts of the processor to be deactivated when the section is not in use.
- The fully static, synchronous, and synthesizable design enables the processors to be manufactured using any low power or standard semiconductor process technology. Starting in revision 2 of the Cortex-M3 design, and on all current revisions of Cortex-M4, the processors also have additional optional hardware support called the Wakeup Interrupt Controller (WIC) to enable advanced low power technologies such as State Retention Power Gating (SRPG). This is covered in Chapter 9.

With all of these low power features, the Cortex-M processors are very popular with embedded product designers, who are constantly looking for new ways to improve battery life in their portable products.

In addition to longer battery life, lower power in the microcontroller can also help reduce Electro-Magnetic Interference (EMI), and potentially simplify the power supply (or reduce the battery size) and hence reduce system cost.

### 3.2.4 Memory system

The Cortex®-M3/M4 processors support a wide range of memory features:

- Total of 4GB of addressable memory space with linear 32-bit addressing, with no need to use memory paging.
- Architectural memory map definition consistency across all Cortex-M processors. The predefined memory map allows processor designs to be optimized for Harvard bus architecture, and allows easy access to memory-mapped peripherals (such as the NVIC) inside the processors.
- Pipelined AHB Lite bus interface that allows high speed, low latency transfers. The AHB Lite interface supports efficient transfers of 32-bit, 16-bit, and 8-bit data. The bus protocol also allows insertion of wait states, supports bus error conditions, and allows multiple bus masters to share a bus.
- Optional bit band feature: two bit addressable regions in SRAM and peripheral regions. Bit value modifications via bit band alias addresses are converted into atomic Read-Modify-Write operations to bit band regions. (See section 6.7 for details.)
- Exclusive accesses for multi-processor system designs. This is important for semaphore operation in multi-processor systems.
- Support of little endian or big endian memory systems. The Cortex-M3/M4 processors can operate in both little endian or big endian mode. However, almost all microcontrollers will be designed for either little endian or big endian, but not both. The majority of the Cortex-M microcontroller products use little endian.
- Optional Memory Protection Unit (MPU). (See the next section.)

### 3.2.5 Memory protection unit

The MPU is an optional feature available on the Cortex®-M3 and Cortex-M4 processors. Microcontroller vendors can decide whether to include the MPU or not. The MPU is a programmable device that monitors the bus transactions and needs to be configured by software, typically an embedded OS. If an MPU is included, applications can divide the memory space into a number of regions and define the access permissions for each of them. When an access rule is violated, a fault exception is generated and the fault exception handler will be able to analyze the problem and, if possible, correct it.

The MPU can be used in various ways. In common scenarios, an OS can set up the MPU to protect data used by the OS kernel and other privileged tasks, preventing untrusted user programs from corrupting them. Optionally, the OS can also isolate memory regions between different user tasks. These measures allow better detection of system failures and allow systems to be more robust in handling error conditions.

The MPU can also be used to make memory regions read-only, to prevent accidental erasure of data in SRAM or overwriting of instruction code.

By default the MPU is disabled and applications that do not require a memory protection feature do not have to initialize it.

### 3.2.6 Interrupt handling

The Cortex®-M3 and Cortex-M4 processors come with a sophisticated interrupt controller called the Nested Vectored Interrupt Controller (NVIC). The NVIC provides a number of features:

- Supports up to 240 interrupt inputs, a Non-Maskable Interrupt (NMI) input, and a number of system exceptions. Each interrupt (apart from the NMI) can be individually enabled or disabled.
- Programmable priority levels for interrupts and a number of system exceptions. In Cortex-M3 and Cortex-M4, the priority levels can be changed dynamically at run time (note: dynamic changing of priority level is not supported in the Cortex-M0/M0+).
- Automatic handling of interrupt/exception prioritization and nested interrupt/ exception handling.
- Vectored interrupt/exception. This means the processor automatically fetches interrupt/exception vectors without the need for software to determine which interrupt/exception needs to be served.
- Vector table can be relocated to various areas in the memory.
- Low interrupt latency. With zero wait state memory system, the interrupt latency is only 12 cycles.
- Interrupts and a number of exceptions can be triggered by software.
- Various optimizations to reduce interrupt processing overhead when switching between different exception contexts.
- Interrupt/exception masking facilities allow all interrupts and exceptions (apart from the NMI) to be masked, or to mask interrupt/exceptions below a certain priority level.

In order to support these features, the NVIC has a number of programmable registers. These registers are memory mapped, and CMSIS-Core provides the required register definitions and access functions (API) for most common interrupt control tasks. These access functions are very easy to use and most can be used on other Cortex-M processors such as the Cortex-M0.

The vector table, which holds the starting addresses of interrupts and system exceptions, is a part of the system memory. By default the vector table is located at the beginning of the memory space (address 0x0), but the vector table offset can be changed at runtime if needed. In most applications, the vector table can be set up during compile-time as a part of the application program image and remain unchanged at runtime.

The number of interrupts supported by each Cortex-M3 or Cortex-M4 device is determined by the microcontroller vendors when the chips are designed.

### 3.2.7 OS support and system level features

The Cortex®-M3 and Cortex-M4 processors are designed to support embedded OSs efficiently. They have a built-in system tick timer called SysTick, which can be set up to generate regular timer interrupts for OS timekeeping. Since the SysTick timer is available in all Cortex-M3 and Cortex-M4 devices, source code for the embedded OS can easily be used on all of these devices without modification for device-specific timers.

The Cortex-M3 and Cortex-M4 processors also have banked stacked pointers: for OS kernel and interrupts, the Main Stack Pointer (MSP) is used; for application tasks, the Process Stack Pointer (PSP) is used. In this way, the stack used by the OS kernel can be separated from that use by application tasks, enabling better reliability as well as allowing optimum stack space usage. For simple applications without an OS, the MSP can be used all the time.

To improve system reliability further, the Cortex-M3 and Cortex-M4 processors support the separation of privileged and non-privileged operation modes. By default, the processors start in privileged mode. When an OS is used and user tasks are executed, the execution of user tasks can be carried out in non-privileged operation mode so that certain restrictions can be enforced, such as blocking access to some NVIC registers. The separation of privileged and non-privileged operation modes can also be used with the MPU to prevent non-privileged tasks from accessing certain memory regions. In this way a user task cannot corrupt data used by the OS kernel or other tasks, thus enhancing the system's stability.

Most simple applications do not require the use of non-privileged mode at all. But when building an embedded system that requires high reliability, the separation of privileged and non-privileged tasks may allow the system to continue operation even if a non-privileged task has failed.

The Cortex-M processors also have a number of fault handlers. When a fault is detected (e.g., accessing of invalid memory address), a fault exception will be triggered and this can be used as a measure to prevent further system failures, and to diagnose the problem.

### 3.2.8 Cortex®-M4 specific features

The Cortex®-M4 processor is very similar to Cortex-M3 in many aspects. However, it has a number of features that the Cortex-M3 does not. This includes the DSP extensions and the optional single precision floating point unit.

The DSP extensions of the Cortex-M4 cover:

- 8-bit and 16-bit Single Instruction Multiple Data (SIMD) instructions. These instructions allow multiple data operations to be carried out in parallel. The most common application of SIMD is audio processing, where the calculations for the left and right channel can be carried out at the same time. It can also be used in image processing, where R-G-B or C-M-Y-K elements of image pixels can be represented as an 8-bit SIMD data set and processed in parallel.

- A number of saturated arithmetic instructions including SIMD versions are also supported. This prevents massive distortion of calculation results when overflow/underflow occurs.
- Single-cycle 16-bit, dual 16-bit, and 32-bit Multiply and Accumulate (MAC). While the Cortex-M3 also supports a couple of MAC instructions, the MAC instructions in Cortex-M4 provide more options, including multiplication for various combinations of upper and lower 16-bits in the registers and a SIMD version of 16-bit MAC. In addition, the MAC operation can be carried out in a single cycle in the Cortex-M4 processor, while in the Cortex-M3 it takes multiple cycles.

The optional floating point unit (FPU) in the Cortex-M4 covers:

- A single precision floating point unit compliant to IEEE 754 standard. In order to support floating point operations, the Cortex-M4 processor supports a number of floating point instructions. There are also a number of instructions to convert between single precision and half precision floating point data.
- The floating point unit supports fused MAC operations; this allows better precision in the MAC result.
- The floating point unit can be disabled when not in use, allowing for a reduction in power consumption.

In order to support the additional instructions and the high performance DSP requirements, the internal data path of the Cortex-M4 is different from that of the Cortex-M3 processor. As a result of these differences, some of the instructions take fewer clock cycles in the Cortex-M4.

In order to allow users access to the full potential of the Cortex-M4 DSP capability, ARM® provides a DSP library though the CMSIS-DSP project. This DSP library is free and can be used on Cortex-M4, Cortex-M3 processors, and even the Cortex-M0+ and Cortex-M0 processors. More details of the DSP library are covered in Chapter 22.

### 3.2.9 **Ease of use**

Compared to other 32-bit processor architectures, the Cortex®-M processors are very easy to use. The programmer's model and the instruction set is very C-friendly. Therefore you can develop your applications entirely in C code without using any assembly and yet very easily get high performance. With the help of CMSIS compliant device driver libraries, developing your application is even easier. For example, system initialization code is provided by microcontroller vendors and usually interrupt controller functions are embedded in the CMSIS-Core files as part of the device driver libraries.

Most of the features of the Cortex-M3 and Cortex-M4 processors are controlled by memory-mapped registers. Therefore you can access almost all of the features via

C pointers. Because there is no need to use compiler specific data types or directives to access these features, the program code is extremely portable.

In the Cortex-M processors, interrupt handlers can be written as normal C functions. Since interrupt prioritization and nesting of interrupts is handled by the NVIC and the exception entry is vectored, there is no need to use the software to check which interrupt needs to be served, or to handle nested interrupts explicitly. All you need to do is to assign a priority level to each interrupt and system exception.

### 3.2.10 Debug support

The Cortex®-M3 and Cortex-M4 processors come with comprehensive debug features to make software development much easier. Besides standard debug features like halting and single stepping, you can also use various trace features to find out details of the program execution without using expensive equipment.

To start with, the Cortex-M3 and Cortex-M4 processors support up to eight hardware comparators for breakpoints (six for instruction addresses, two for literal data addresses) in the Flash Patch and BreakPoint Unit (FPB). When triggered, the processor can be halted or the transfers can be remapped to a SRAM location. The remapping feature allows a read-only program memory location to be modified; for example, to patch the program in a masked ROM with a small programmable memory. This enables bugs to be rectified or enhancements made even when the main code is in masked ROM. (See section 23.10 for details.)

The Cortex-M3 and Cortex-M4 processors also have up to four hardware data watchpoint comparators in the Data Watchpoint and Trace (DWT) unit. These can be used to generate watchpoint events to halt the processor when selected data is accessed, or to generate trace information that can be collected by the trace interface without stopping the processor. The data value and additional information can then be presented by the debugger in an Integrated Development Environment (IDE) to visualize the change of data values over time. The DWT can also be used to generate exception event traces and basic profiling information, which is again output through the trace interface.

The Cortex-M3 and Cortex-M4 processors also have an optional Embedded Trace Macrocell (ETM) module that can be used to generate instruction traces. This allows full visibility of the program flow during execution, which is very useful for debugging complex software issues and also can be used for detailed profiling and code coverage analysis.

Debugging the Cortex-M3 and Cortex-M4 processors can be handled by a JTAG connection, or a two-wire interface called a Serial-Wire Debug (SWD) interface. Both JTAG and SWD protocols are widely supported by many development tool vendors. Trace information can be collected using a single wire Serial-Wire Viewer (SWV) interface, or a trace port interface (typically 5-pin) if high-trace bandwidth is required (e.g., when instruction trace is used). The debug and trace interfaces can be combined into a single connector (see Appendix H).

### 3.2.11 Scalability

The Cortex®-M processors are not just for low-cost microcontroller products. Nowadays you can find various multi-processor products containing Cortex-M3 or Cortex-M4 processors. These include:

- Microcontrollers with multiple Cortex-M processors, such as LPC4300 from NXP.
- High-end Digital Signal Processing devices with one or more Cortex-M processors as the main processor and an additional DSP for the data processing engine. For example, the Concerto product series from Texas Instruments, which combines a Cortex-M3 processor with a DSP core.
- Complex System-on-Chips with one or more Cortex-M processors as companion processors. For example, the Texas Instrument OMAP5 combines a Cortex-A15 and two Cortex-M4 processors into a single device.
- Complex System-on-Chips with one or more Cortex-M processors for power management and system control.
- Complex System-on-Chips with one or more Cortex-M processors for Finite State Machine (FSM) replacement.

Using various AMBA® bus infrastructure solutions such as the Cortex-M System Design Kit from ARM®, the bus system of the Cortex-M processors can be expanded to support multi-processor systems. In addition, the Cortex-M3 and Cortex-M4 support the following features to support multi-processor system design:

- Exclusive access instructions — The Cortex-M3 and Cortex-M4 processors support a number of exclusive access instructions. These are special memory access instructions that work in pairs for load and store operations of variables for semaphore or mutually exclusive operations. With additional hardware support in the bus infrastructure, the processor can determine if it has successfully carried out exclusive access to a shared data memory area (i.e., no another processor has accessed the same area during the operation).
- Scalable debug support — The debug systems of the Cortex-M processors are based on the CoreSight™ Architecture. This can be expanded to support multiple processors, sharing just one debug connection and one trace interface.
- Event communication interface — The Cortex-M3 and Cortex-M4 processors support a simple event communication interface. This allows multi-processor systems to reduce power by having some of the processors enter sleep mode and wake up when certain events have occurred, such as completion of semaphore operations in one of the processors.

Another aspect of scalability is the range of microcontroller products you can find with the Cortex-M processor. Since all the Cortex-M processors are very similar to each other in terms of the programmer's model, interrupt handling, and software development including debug, you can easily switch between different processors for your embedded systems to satisfy different performance requirements, system level requirements, and price levels.

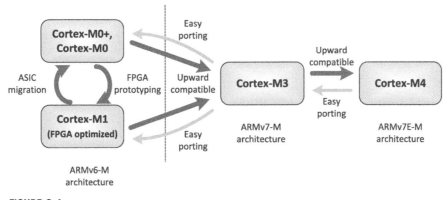

**FIGURE 3.4**

Cortex-M compatibility

### 3.2.12 **Compatibility**

One advantage of using the ARM® Cortex®-M3 and Cortex-M4 processors is that they have great compatibility with a wide family of other ARM devices (Figure 3.4). For instance, there are thousands of different Cortex-M3 and Cortex-M4 devices from different microcontroller vendors for you to choose from. If you want to reduce cost, you can easily transfer your program code to a microcontroller based on the Cortex-M0 processor. If your application needs a bit more processing power, often you can find a faster Cortex-M3/M4 microcontroller, or even migrate your design to a Cortex-R or Cortex-A processor based product.

Besides easy migration between ARM Cortex processor families, you can also reuse software developed for ARM7™ and ARM9™ microcontrollers. For C source code, often you only need to recompile the code to target Cortex-M3 or Cortex-M4. Some assembly code files can be reused with minor modifications. This includes some codec applications developed for the ARM9E processor family, which often contain optimized assembly code for Digital Signal Processing (DSP).

There is still some software porting work to be done when migrating from ARM7 or ARM9 to Cortex-M microcontrollers: due to differences in the processor architecture, such as processor modes and interrupt/exception models, interrupt handlers need to be changed and some assembly code will need to be changed or removed. Normally the migration to Cortex-M processors will simplify the application code as the initialization is simpler and nested exceptions/interrupts are automatically handled by hardware.

Besides migration of hardware, ARM has a well-established software architecture that allows different software development tools to work together. In addition, the CMSIS ecosystem also enables applications developed for the Cortex-M microcontrollers to be compiled using various different tool chains with no or little modification, which further protects your software IP investment.

# Architecture

4

## CHAPTER OUTLINE

# 4.1 Introduction to the architecture

The Cortex®-M3 and Cortex-M4 Processors are based on the ARMv7-M architecture. The original ARMv7-M architecture was defined when the Cortex-M3 processor was developed, and when the Cortex-M4 was released, the architecture was extended to included additional instructions and architectural features. The extended architecture is sometimes called ARMv7E-M architecture. Both ARMv7-M and ARMv7E-M features are documented in the same architecture specification document: the ARMv7-M Architecture Reference Manual (reference 1).

The ARMv7-M Architecture Reference Manual is a massive document of over 1000 pages. It provides very detailed architectural requirements of the processor's behavior, from instruction set, memory system to debug support. While it is useful for experts like processor designers, designers of C compilers, and development tools, this document is not easy to read, especially for readers new to the ARM® architecture.

To use a Cortex-M microcontroller in typical applications, there is no need to have detailed knowledge of this architecture. You only need to have a basic understanding of the programmer's model, how exceptions (such as interrupts) are handled, the memory map, how to use the peripherals, and how to use the software driver library files from the microcontroller vendors.

In the next few chapters of this book, we will look at the architecture from software developer's point of view. First, we will look at the programmer's model of the processor, which covers operation modes, register banks, and special registers.

# 4.2 Programmer's model
## 4.2.1 Operation modes and states

The Cortex®-M3 and Cortex-M4 processors have two operation states and two modes. In addition, the processors can have privileged and unprivileged access levels. These are shown in Figure 4.1. The privileged access level can access all resources in the processor, while unprivileged access level means some memory regions are inaccessible, and a few operations cannot be used. In some documents, the unprivileged access level might also be referred as "User" state, a term inherited from ARM7TDMI™.

### Operation states
- Debug state: When the processor is halted (e.g., by the debugger, or after hitting a breakpoint), it enters debug state and stops executing instructions.

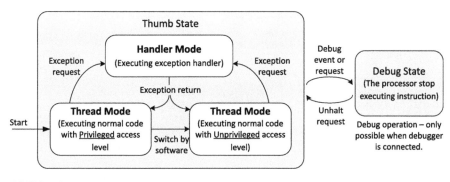

**FIGURE 4.1**

Operation states and modes

- Thumb state: If the processor is running program code (Thumb instructions), it is in the Thumb state. Unlike classic ARM® processors like ARM7TDMI, there is no ARM state because the Cortex-M processors do not support the ARM instruction set.

### Operation modes

- Handler mode: When executing an exception handler such as an Interrupt Service Routine (ISR). When in handler mode, the processor always has privileged access level.
- Thread mode: When executing normal application code, the processor can be either in privileged access level or unprivileged access level. This is controlled by a special register called "CONTROL." We will cover this more in section 4.2.3.

Software can switch the processor in privileged Thread mode to unprivileged Thread mode. However, it cannot switch itself back from unprivileged to privileged. If this is needed, the processor has to use the exception mechanism to handle the switch.

The separation of privileged and unprivileged access levels allows system designers to develop robust embedded systems by providing a mechanism to safeguard memory accesses to critical regions and by providing a basic security model. For example, a system can contain an embedded OS kernel that executes in privileged access level, and application tasks which execute in unprivileged access level. In this way, we can set up memory access permissions using the Memory Protection Unit (MPU) to prevent an application task from corrupting memory and peripherals used by the OS kernel and other tasks. If an application task crashes, the remaining application tasks and the OS kernel can still continue to run.

Besides the differences in memory access permission and access to several special instructions, the programmer's model of the privileged access level and unprivileged access level are almost the same. Note that almost all of the NVIC registers are privileged access only.

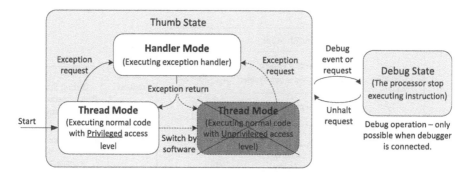

**FIGURE 4.2**

In simple applications, the unprivileged Thread mode can be unused

Similarly, Thread mode and Handler mode have very similar programmer's models. However, Thread mode can switch to using a separate shadowed Stack Pointer (SP). Again, this allows the stack memory for application tasks to be separated from the stack used by the OS kernel, thus allowing better system reliability.

By default, the Cortex-M processors start in privileged Thread mode and in Thumb state. In many simple applications, there is no need to use the unprivileged Thread model and the shadowed SP at all (see Figure 4.2). Unprivileged Thread model is not available in the Cortex-M0 processor, but is optional in the Cortex-M0+ processor.

The debug state is used for debugging operations only. This state is entered by a halt request from the debugger, or by debug events generated from debug components in the processor. This state allows the debugger to access or change the processor register values. The system memory, including peripherals inside and outside the processor, can be accessed by the debugger in either Thumb state or debug state.

## 4.2.2 Registers

Similarly to almost all other processors, the Cortex®-M3 and Cortex-M4 processors have a number of registers inside the processor core to perform data processing and control. Most of these registers are grouped in a unit called the register bank. Each data processing instruction specifies the operation required, the source register(s), and the destination register(s) if applicable. In the ARM® architecture, if data in memory is to be processed, it has to be loaded from the memory to registers in the register bank, processed inside the processor, and then written back to the memory, if needed. This is commonly called a "load-store architecture." By having a sufficient number of registers in the register bank, this arrangement is easy to use, and allows efficient program code to be generated using C compilers. For instance, a number of data variables can be stored in the register bank for a short period of time while other data processing takes place, without the need to be updated to the system memory and read back every time they are used.

**Register bank**

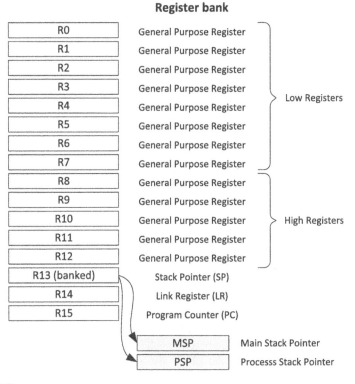

**FIGURE 4.3**

Registers in the register bank

The register bank in the Cortex-M3 and Cortex-M4 processors has 16 registers. Thirteen of them are general purpose 32-bit registers, and the other three have special uses, as can be seen in Figure 4.3.

### R0 – R12

Registers R0 to R12 are general purpose registers. The first eight (R0 – R7) are also called low registers. Due to the limited available space in the instruction set, many 16-bit instructions can only access the low registers. The high registers (R8 – R12) can be used with 32-bit instructions, and a few with 16-bit instructions, like MOV (move). The initial values of R0 to R12 are undefined.

### R13, stack pointer (SP)

R13 is the Stack Pointer. It is used for accessing the stack memory via PUSH and POP operations. Physically there are two different Stack Pointers: the Main Stack Pointer (MSP, or SP_main in some ARM documentation) is the default Stack Pointer. It is selected after reset, or when the processor is in Handler Mode. The other Stack Pointer is called the Process Stack Pointer (PSP, or SP_process in some ARM documentation). The PSP can only be used in Thread Mode. The selection of Stack Pointer is determined by a special register called CONTROL, which will be explained in section 4.2.3. In normal programs, only one of these Stack Pointers will be visible.

Both MSP and PSP are 32-bit, but the lowest two bits of the Stack Pointers (either MSP or PSP) are always zero, and writes to these two bits are ignored. In ARM Cortex-M processors, PUSH and POP are always 32-bit, and the addresses of the transfers in stack operations must be aligned to 32-bit word boundaries.

For most cases, it is not necessary to use the PSP if the application doesn't require an embedded OS. Many simple applications can rely on the MSP completely. The PSP is normally used when an embedded OS is involved, where the stack for the OS kernel and application tasks are separated. The initial value of PSP is undefined, and the initial value of MSP is taken from the first word of the memory during the reset sequence.

### R14, link register (LR)

R14 is also called the Link Register (LR). This is used for holding the return address when calling a function or subroutine. At the end of the function or subroutine, the program control can return to the calling program and resume by loading the value of LR into the Program Counter (PC). When a function or subroutine call is made, the value of LR is updated automatically. If a function needs to call another function or subroutine, it needs to save the value of LR in the stack first. Otherwise, the current value in LR will be lost when the function call is made.

During exception handling, the LR is also updated automatically to a special EXC_RETURN (Exception Return) value, which is then used for triggering the exception return at the end of the exception handler. This will be covered in more depth in Chapter 8.

Although the return address values in the Cortex-M processors are always even (bit 0 is zero because the instructions must be aligned to half-word addresses), bit 0 of LR is readable and writeable. Some of the branch/call operations require that bit zero of LR (or any register being used) be set to 1 to indicate Thumb state.

### R15, program counter (PC)

R15 is the Program Counter (PC). It is readable and writeable: a read returns the current instruction address plus 4 (this is due to the pipeline nature of the design, and compatibility requirement with the ARM7TDMI™ processor). Writing to PC (e.g., using data transfer/processing instructions) causes a branch operation.

Since the instructions must be aligned to half-word or word addresses, the Least Significant Bit (LSB) of the PC is zero. However, when using some of the branch/ memory read instructions to update the PC, you need to set the LSB of the new PC value to 1 to indicate the Thumb state. Otherwise, a fault exception can be triggered, as it indicates an attempt to switch to use ARM instructions (i.e., 32-bit ARM in-structions as in ARM7TDMI), which is not supported. In high-level programming languages (including C, C++), the setting of LSB in branch targets is handled by the compiler automatically.

**Table 4.1** Allowed Register Names as Assembly Code

| Register | Possible Register Names | Notes |
|---|---|---|
| R0-R12 | R0, R1 ... R12, r0, r1 ... r12 | |
| R13 | R13, r13, SP, sp | Register name MSP and PSP are used in special register access instructions (MRS, MSR) |
| R14 | R14, r14, LR, lr | |
| R15 | R15, r15, PC, pc | |

In most cases, branches and calls are handled by instructions dedicated to such operations. It is less common to use data processing instructions to update the PC. However, the value of PC is useful for accessing literal data stored in program memory. So you can frequently find memory read operations with PC as base address register with address offsets generated by immediate values in the instructions.

### Register names in programming

With most assembly tools, you can use a variety of names for accessing the registers in the register bank. In some assembly tools, such as the ARM assembly (supported in DS-5™ Professional, Keil™ MDK-ARM), you can use either uppercase, or lowercase, or mixed cases (Table 4.1).

## 4.2.3 Special registers

Besides the registers in the register bank, there are a number of special registers (Figure 4.4). These registers contain the processor status and define the operation states and interrupt/exception masking. In the development of simple applications with high level programming languages such as C, there are not many scenarios that require access to these registers. However, they are needed for development of an embedded OS, or when advanced interrupt masking features are needed.

Special registers are not memory mapped, and can be accessed using special register access instructions such as MSR and MRS.

```
MRS <reg>, <special_reg>; Read special register into register
MSR <special_reg>, <reg>; write to special register
```

CMSIS-Core also provides a number of C functions that can be used to access special registers. Do not confuse these special registers with "special function registers (SFR)" in other microcontroller architectures, which are commonly referred to as registers for I/O control.

### Program status registers

The Program Status Register is composed of three status registers:

- Application PSR (APSR)
- Execution PSR (EPSR)
- Interrupt PSR (IPSR)

**Special Registers**

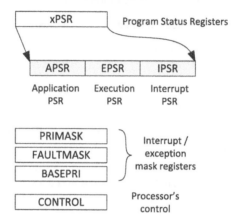

**FIGURE 4.4**

Special Registers

| | 31 | 30 | 29 | 28 | 27 | 26:25 | 24 | 23:20 | 19:16 | 15:10 | 9 | 8 | 7 | 6 | 5 | 4:0 |
|---|---|---|---|---|---|---|---|---|---|---|---|---|---|---|---|---|
| APSR | N | Z | C | V | Q | | | | GE* | | | | | | | |
| IPSR | | | | | | | | | | | | | Exception Number | | | |
| EPSR | | | | | | ICI/IT | T | | | ICI/IT | | | | | | |

*GE is available in ARMv7E-M processors such as the Cortex-M4. It is not available in the Cortex-M3 processor.

**FIGURE 4.5**

APSR, IPSR, and EPSR

| | 31 | 30 | 29 | 28 | 27 | 26:25 | 24 | 23:20 | 19:16 | 15:10 | 9 | 8 | 7 | 6 | 5 | 4:0 |
|---|---|---|---|---|---|---|---|---|---|---|---|---|---|---|---|---|
| xPSR | N | Z | C | V | Q | ICI/IT | T | | GE* | ICI/IT | | | Exception Number | | | |

*GE is available in ARMv7E-M processors such as the Cortex-M4. It is not available in the Cortex-M3 processor.

**FIGURE 4.6**

Combined xPSR

These three registers (Figure 4.5) can be accessed as one combined register, referred to as xPSR in some documentation. In ARM® assembler, when accessing xPSR (Figure 4.6), the symbol PSR is used. For example:

```
MRS   r0, PSR   ; Read the combined program status word
MSR   PSR, r0   ; Write combined program state word
```

**Table 4.2** Bit Fields in Program Status Registers

| Bit | Description |
|---|---|
| N | Negative flag |
| Z | Zero flag |
| C | Carry (or NOT borrow) flag |
| V | Overflow flag |
| Q | Sticky saturation flag (not available in ARMv6-M) |
| GE[3:0] | Greater-Than or Equal flags for each byte lane (ARMv7E-M only; not available in ARMv6-M or Cortex®-M3). |
| ICI/IT | Interrupt-Continuable Instruction (ICI) bits, IF-THEN instruction status bit for conditional execution (not available in ARMv6-M). |
| T | Thumb state, always 1; trying to clear this bit will cause a fault exception. |
| Exception Number | Indicates which exception the processor is handling. |

You can also access an individual PSR (Figure 4.5). For example:

```
MRS    r0, APSR    ; Read Flag state into R0
MRS    r0, IPSR    ; Read Exception/Interrupt state
MSR    APSR, r0    ; Write Flag state
```

Please note:

- The EPSR cannot be accessed by software code directly using MRS (read as zero) or MSR
- The IPSR is read only and can be read from combined PSR (xPSR).

Figure 4.5 shows the definition of the various PSRs in ARMv7-M, and Table 4.2 lists the definition of the bit fields in the PSRs.

Please note that some of the bit fields in the APSR and EPSR are not available in ARMv6-M architecture (e.g., the Cortex®-M0 processor). Also, it is quite different from classic ARM processors such as the ARM7TDMI™. If you compare this with the Current Program Status Register (CPSR) in ARM7™, you might find that some of the bit fields used in ARM7 are gone. The Mode (M) bit field is gone because the Cortex-M3 does not have the operation mode as defined in ARM-7. Thumb-bit (T) is moved to bit 24. Interrupt status (I and F) bits are replaced by the new interrupt mask registers (PRIMASKs), which are separated from PSR. For comparison, the CPSR in traditional ARM processors is shown in Figure 4.7.

Detailed behavior of the APSR is covered in a later part of this chapter (section 4.3).

### PRIMASK, FAULTMASK, and BASEPRI registers

The PRIMASK, FAULTMASK, and BASEPRI registers are all used for exception or interrupt masking. Each exception (including interrupts) has a priority level where a

| | 31 | 30 | 29 | 28 | 27 | 26:25 | 24 | 23:20 | 19:16 | 15:10 | 9 | 8 | 7 | 6 | 5 | 4:0 |
|---|---|---|---|---|---|---|---|---|---|---|---|---|---|---|---|---|
| ARM general (Cortex-A/R) | N | Z | C | V | Q | IT | J | Reserved | GE[3:0] | IT | E | A | I | F | T | M[4:0] |
| ARM7TDMI (ARMv4) | N | Z | C | V | Reserved | | | | | | | | I | F | T | M[4:0] |
| ARMv7-M (Cortex-M3) | N | Z | C | V | Q | ICI/IT | T | | | ICI/IT | | | Exception Number | | | |
| ARMv7E-M (Cortex-M4) | N | Z | C | V | Q | ICI/IT | T | | GE[3:0] | ICI/IT | | | Exception Number | | | |
| ARMv6-M (Cortex-M0) | N | Z | C | V | | | T | | | | | | | | | Exception Number |

**FIGURE 4.7**

Comparing PSR of various ARM architectures

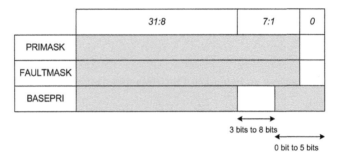

**FIGURE 4.8**

PRIMASK, FAULTMASK, and BASEPRI registers

smaller number is a higher priority and a larger number is a lower priority. These special registers are used to mask exceptions based on priority levels. They can only be accessed in the privileged access level (in unprivileged state writes to these registers are ignored and reads return zero). By default, they are all zero, which means the masking (disabling of exception/interrupt) is not active. Figure 4.8 shows the programmer's model of these registers.

The PRIMASK register is a 1-bit wide interrupt mask register. When set, it blocks all exceptions (including interrupts) apart from the Non-Maskable Interrupt (NMI) and the HardFault exception. Effectively it raises the current exception priority level to 0, which is the highest level for a programmable exception/interrupt.

The most common usage for PRIMASK is to disable all interrupts for a time critical process. After the time critical process is completed, the PRIMASK needs to be cleared to re-enable interrupts. Details for using PRIMASK are given in section 7.10.1.

The FAULTMASK register is very similar to PRIMASK, but it also blocks the HardFault exception, which effectively raises the current exception priority level to [minus]1. FAULTMASK can be used by fault handling code to suppress the triggering of further faults (only several types of them) during fault handling. For example, FAULTMASK can be used to bypass MPU or suppress bus fault (these are configurable). This potentially makes it easier for fault handling code to carry out remedial actions. Unlike PRIMASK, FAULTMASK is cleared automatically at exception return. Details for using FAULTMASK are given in section 7.10.2.

In order to allow more flexible interrupt masking, the ARMv7-M architecture also provides the BASEPRI, which masks exceptions or interrupts based on priority level. The width of the BASEPRI register depends on how many priority levels are implemented in the design, which is determined by the microcontroller vendors. Most Cortex-M3 or Cortex-M4 microcontrollers have eight programmable exception priority levels (3-bit width) or 16 levels, and in these cases the width of BASEPRI will be 3 bits or 4 bits, respectively. When BASEPRI is set to 0, it is disabled. When it is set to a non-zero value, it blocks exceptions (including interrupts) that have the same or lower priority level, while still allowing exceptions with a higher priority level to be accepted by the processor. Details of using BASEPRI are covered in section 7.10.3.

CMSIS-Core provides a number of functions for accessing the PRIMASK, FAULTMASK, and BASEPRI registers in the C programming environment (note: these registers can only be accessed in the privileged access level).

```
x = __get_BASEPRI();    // Read BASEPRI register
x = __get_PRIMARK();    // Read PRIMASK register
x = __get_FAULTMASK();  // Read FAULTMASK register
__set_BASEPRI(x);       // Set new value for BASEPRI
__set_PRIMASK(x);       // Set new value for PRIMASK
__set_FAULTMASK(x);     // Set new value for FAULTMASK
__disable_irq();        // Set PRIMASK, disable IRQ
__enable_irq();         // Clear PRIMASK, enable IRQ
```

Alternatively, you can also access these exception masking registers with assembly code:

```
MRS   r0, BASEPRI    ; Read BASEPRI register into R0
MRS   r0, PRIMASK    ; Read PRIMASK register into R0
MRS   r0, FAULTMASK ; Read FAULTMASK register into R0
MSR   BASEPRI, r0   ; Write R0 into BASEPRI register
MSR   PRIMASK, r0   ; Write R0 into PRIMASK register
MSR   FAULTMASK, r0 ; Write R0 into FAULTMASK register
```

In addition, the Change Processor State (CPS) instructions allow the value of the PRIMASK and FAULTMASK to be set or cleared with a simple instruction.

```
CPSIE i    ; Enable interrupt (clear PRIMASK)
CPSID i    ; Disable interrupt (set PRIMASK)
CPSIE f    ; Enable interrupt (clear FAULTMASK)
CPSID f    ; Disable interrupt (set FAULTMASK)
```

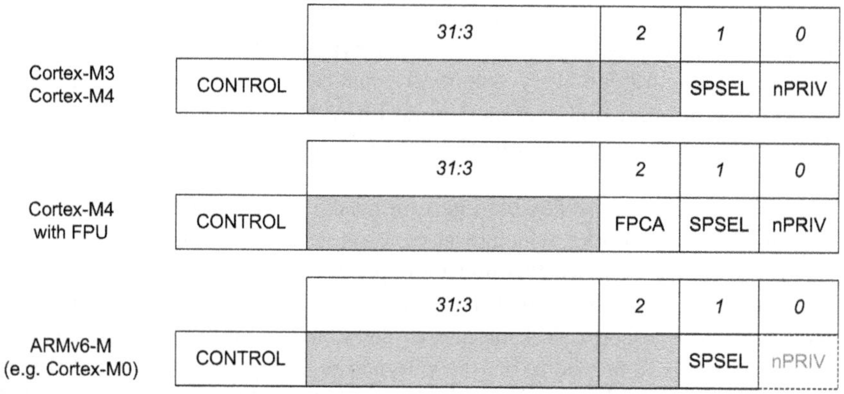

**FIGURE 4.9**

CONTROL register in Cortex-M3, Cortex-M4, Cortex-M4 with FPU. The bit nPRIV is not available in the Cortex-M0 and is optional in the Cortex-M0+ processor

Note: The FAULTMASK and BASEPRI registers are not available in ARMv6-M (e.g., Cortex-M0).

### CONTROL register

The CONTROL register (Figure 4.9) defines:

- The selection of stack pointer (Main Stack Point/Process Stack Pointer)
- Access level in Thread mode (Privileged/Unprivileged)

In addition, for Cortex-M4 processor with a floating point unit, one bit of the CONTROL register indicates if the current context (currently executed code) uses the floating point unit or not.

Note: The CONTROL register for ARMv6-M (e.g., Cortex-M0) is also shown for comparison. In ARMv6-M, support of nPRIV and unprivileged access level is implementation dependent, and is not available in the first generation of the Cortex-M0 products and Cortex-M1 products. It is optional in the Cortex-M0+ processor.

The CONTROL register can only be modified in the privileged access level and can be read in both privileged and unprivileged access levels. The definition of each bit field in the CONTROL register is shown in Table 4.3.

After reset, the CONTROL register is 0. This means the Thread mode uses the Main Stack Pointer as Stack Pointer and Thread mode has privileged accesses. Programs in privileged Thread mode can switch the Stack Pointer selection or switch to unprivileged access level by writing to CONTROL (Figure 4.10). However, once nPRIV (CONTROL bit 0) is set, the program running in Thread can no longer access the CONTROL register.

**Table 4.3** Bit Fields in CONTROL Register

| Bit | Function |
|-----|----------|
| nPRIV (bit 0) | Defines the privileged level in Thread mode:<br>When this bit is 0 (default), it is privileged level when in Thread mode.<br>When this bit is 1, it is unprivileged when in Thread mode.<br>In Handler mode, the processor is always in privileged access level. |
| SPSEL (bit 1) | Defines the Stack Pointer selection:<br>When this bit is 0 (default), Thread mode uses Main Stack Pointer (MSP).<br>When this bit is 1, Thread mode uses Process Stack Pointer (PSP).<br>In Handler mode, this bit is always 0 and write to this bit is ignored. |
| FPCA (bit 2) | Floating Point Context Active – This bit is only available in Cortex-M4 with floating point unit implemented. The exception handling mechanism uses this bit to determine if registers in the floating point unit need to be saved when an exception has occurred.<br>When this bit is 0 (default), the floating point unit has not been used in the current context and therefore there is no need to save floating point registers.<br>When this bit is 1, the current context has used floating point instructions and therefore need to save floating point registers.<br>The FPCA bit is set automatically when a floating point instruction is executed. This bit is clear by hardware on exception entry.<br>There are several options for handling saving of floating point registers. This will be covered in Chapter 13. |

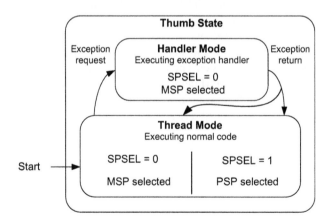

**FIGURE 4.10**

Stack Pointer selection

A program in unprivileged access level cannot switch itself back to privileged access level. This is essential in order to provide a basic security usage model. For example, an embedded system might contain untrusted applications running in unprivileged access level and the access permission of these applications must

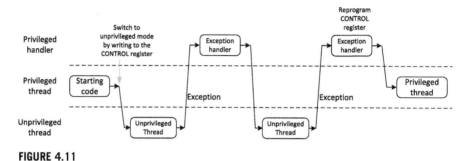

**FIGURE 4.11**

Switching between privileged thread mode and unprivileged thread mode

be restricted to prevent security breaches or to prevent an unreliable application from crashing the whole system.

If it is necessary to switch the processor back to using privileged access level in Thread mode, then the exception mechanism is needed. During exception handling, the exception handler can clear the nPRIV bit (Figure 4.11). When returning to Thread mode, the processor will be in privileged access level.

When an embedded OS is used, the CONTROL register could be reprogrammed at each context switch to allow some application tasks to run with privileged access level and the others to run with unprivileged access level.

The settings of nPRIV and SPSEL are orthogonal. Four different combinations of nPRIV and SPSEL are possible, although only three of them are commonly used in real world applications, as shown in Table 4.4.

In most simple applications without an embedded OS, there is no need to change the value of the CONTROL register. The whole application can run in privileged access level and use only the MSP (Figure 4.12).

To access the CONTROL register in C, the following functions are available in CMSIS-compliant device-driver libraries:

```
x = __get_CONTROL(); // Read the current value of CONTROL
__set_CONTROL(x);    // Set the CONTROL value to x
```

There are two points that you need to be aware of when changing the value of the CONTROL register:

- For the Cortex-M4 processor with floating point unit (FPU), or any variant of ARMv7-M processors with (FPU), the FPCA bit can be set automatically due to the presence of floating point instructions. If the program contains floating point operations and the FPCA bit is cleared accidentally, and subsequently an interrupt occurs, the data in registers in the floating point unit will not be saved by the exception entry sequence and could be overwritten by the interrupt handler. In this case, the program will not be able to continue correct processing when resuming the interrupted task.

**Table 4.4** Different Combinations of nPRIV and SPSEL

| nPRIV | SPSEL | Usage Scenario |
|---|---|---|
| 0 | 0 | Simple applications – the whole application is running in privileged access level. Only one stack is used by the main program and interrupt handlers. Only the Main Stack Pointer (MSP) is used. |
| 0 | 1 | Applications with an embedded OS, with current executing task running in privileged Thread mode. The Process Stack Pointer (PSP) is selected in current task, and the MSP is used by OS Kernel and exception handlers. |
| 1 | 1 | Applications with an embedded OS, with current executing task running in unprivileged Thread mode. The Process Stack Pointer (PSP) is selected in current task, and the MSP is used by OS Kernel and exception handlers. |
| 1 | 0 | Thread mode tasks running with unprivileged access level and use MSP. This can be observed in Handler mode but is less likely to be used for user tasks because in most embedded OS, the stack for application tasks is separated from the stack used by OS kernel and exception handlers. |

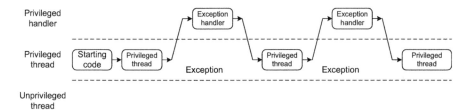

**FIGURE 4.12**

Simple applications do not require unprivileged Thread mode

- After modifying the CONTROL register, architecturally an Instruction Synchronization Barrier (ISB) instruction (or __ISB() function in CMSIS compliant driver) should be used to ensure the effect of the change applies to subsequent code. Due to the simple nature of the Cortex-M3, Cortex-M4, Cortex-M0+, Cortex-M0, and Cortex-M1 pipeline, omission of the ISB instruction does not cause any problem.

To access the Control register in assembly, the MRS and MSR instructions are used:

```
MRS   r0, CONTROL ; Read CONTROL register into R0
MSR   CONTROL, r0 ; Write R0 into CONTROL register
```

You can detect if the current execution level is privileged by checking the value of IPSR and CONTROL:

```
int in_privileged(void)
{
if (__get_IPSR() != 0) return 1; // True
else
 if ((__get_CONTROL() & 0x1)==0) return 1; // True
 else return 0; // False
}
```

### 4.2.4 Floating point registers

The Cortex-M4 processor has an optional floating point unit. This provides additional registers for floating point data processing, as well as a Floating Point Status and Control Register (FPSCR) (Figure 4.13).

#### S0 to S31/D0 to D15

Each of the 32-bit registers S0 to S31 ("S" for single precision) can be accessed using floating point instructions, or accessed as a pair, in the symbol of D0 to D15 ("D" for double-word/double-precision). For example, S1 and S0 are paired together to become D0, and S3 and S2 are paired together to become D1. Although

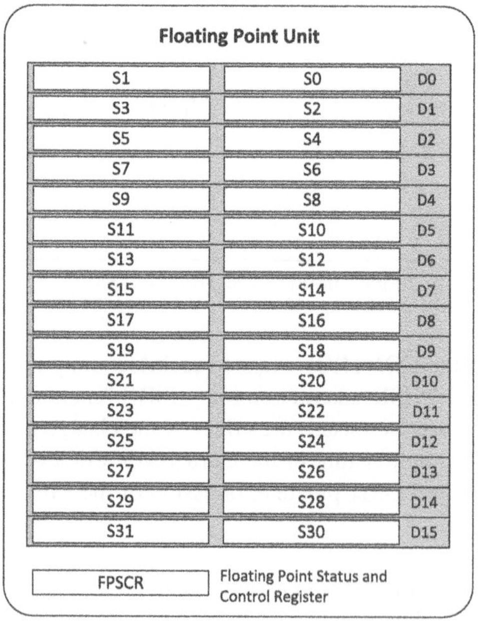

**FIGURE 4.13**

Registers in the floating point unit

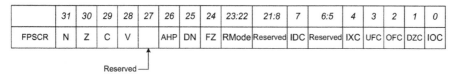

| | 31 | 30 | 29 | 28 | 27 | 26 | 25 | 24 | 23:22 | 21:8 | 7 | 6:5 | 4 | 3 | 2 | 1 | 0 |
|---|---|---|---|---|---|---|---|---|---|---|---|---|---|---|---|---|---|
| FPSCR | N | Z | C | V | | AHP | DN | FZ | RMode | Reserved | IDC | Reserved | IXC | UFC | OFC | DZC | IOC |

Reserved ⏎

**FIGURE 4.14**

Bit field in FPSCR

the floating point unit in the Cortex-M4 does not support double precision floating point calculations, you can still use floating point instructions for transferring double precision data.

### Floating point status and control register (FPSCR)

The FPSCR contains various bit fields (Figure 4.14) for a couple of reasons:

- To define some of the floating point operation behaviors
- To provide status information about the floating point operation results

By default, the behavior is configured to be compliant with IEEE 754 single precision operation. In normal applications there is no need to modify the settings of the floating point operation control. Table 4.5 lists the descriptions for the bit fields in FPSCR.

Note: The exception bits in FPSCR can be used by software to detect abnormalities in floating point operations. Bit fields in FPSCR are covered in Chapter 13.

### Memory-mapped floating point unit control registers

In addition to the floating point register bank and FPSCR, the floating point unit also introduces several additional memory-mapped registers into the system. For example, the Coprocessor Access Control Register (CPACR) is used to enable or disable the floating point unit. By default the floating point unit is disabled to reduce power consumption. Before using any floating point instructions, the floating point unit must be enabled by programming the CPACR register (Figure 4.15).

In the C programming environment with a CMSIS-compliant device-driver:

```
SCB->CPACR |= 0xF << 20; // Enable full access to the FPU
```

In assembly language programming environment, you can use the following code:

```
LDR R0,=0xE000ED88 ; R0 set to address of CPACR
LDR R1,=0x00F00000 ; R1 = 0xF << 20
LDR R2 [R0]        ; Read current value of CPACR
ORRS R2, R2, R1    ; Set bit
STR R2,[R0]        ; Write back modified value to CPACR
```

The other memory-mapped floating point unit registers will be covered in Chapter 13, which also covers details of the floating point unit.

**Table 4.5** Bit Fields in FPSCR

| Bit | Description |
|-----|-------------|
| N | Negative flag (update by floating point comparison operations) |
| Z | Zero flag (update by floating point comparison operations) |
| C | Carry/borrow flag (update by floating point comparison operations) |
| V | Overflow flag (update by floating point comparison operations) |
| AHP | Alternate half-precision control bit:<br>0 – IEEE half-precision format (default)<br>1 – Alternative half-precision format |
| DN | Default NaN (Not a Number) mode control bit:<br>0 – NaN operands propagate through to the output of a floating point operation (default)<br>1 – Any operation involving one or more NaN(s) returns the default NaN |
| FZ | Flush-to-zero model control bit:<br>0 – Flush-to-zero mode disabled (default). (IEEE 754 standard compliant)<br>1 – Flush-to-zero mode enabled |
| RMode | Rounding Mode Control field. The specified rounding mode is used by almost all floating-point instructions:<br>00 – Round to Nearest (RN) mode (default)<br>01 – Round towards Plus Infinity (RP) mode<br>10 – Round towards Minus Infinity (RM) mode<br>11 – Round towards Zero (RZ) mode |
| IDC | Input Denormal cumulative exception bit. Set to 1 when floating point exception occurred, clear by writing 0 to this bit. (Result not within normalized value range; see section 13.1.2.) |
| IXC | Inexact cumulative exception bit. Set to 1 when floating point exception occurred, clear by writing 0 to this bit. |
| UFC | Underflow cumulative exception bit. Set to 1 when floating point exception occurred, clear by writing 0 to this bit. |
| OFC | Overflow cumulative exception bit. Set to 1 when floating point exception occurred, clear by writing 0 to this bit. |
| DZC | Division by Zero cumulative exception bit. Set to 1 when floating point exception occurred, clear by writing 0 to this bit. |
| IOC | Invalid Operation cumulative exception bit. Set to 1 when floating point exception occurred, clear by writing 0 to this bit. |

## 4.3 Behavior of the application program status register (APSR)

The APSR contains several groups of status flags:

- Status flags for integer operations (N-Z-C-V bits)
- Status flags for saturation arithmetic (Q bit)
- Status flags for SIMD operations (GE bits)

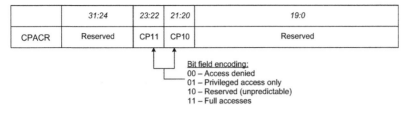

| CPACR | 31:24 | 23:22 | 21:20 | 19:0 |
|---|---|---|---|---|
| | Reserved | CP11 | CP10 | Reserved |

Bit field encoding:
00 – Access denied
01 – Privileged access only
10 – Reserved (unpredictable)
11 – Full accesses

**FIGURE 4.15**

Bit field in CPACR

### 4.3.1 **Integer status flags**

The integer status flags are very similar to ALU status flags in many other processor architectures. These flags are affected by general data processing instructions, and are essential for controlling conditional branches and conditional executions. In addition, one of the APSR flags, the C (Carry) bit, can also be used in add and subtract operations.

There are four integer flags in the Cortex®-M processors, shown in Table 4.6. A few examples of the ALU flag results are shown in Table 4.7.

In the ARMv7-M and ARMv7E-M architecture, most of the 16-bit instructions affect these four ALU flags. In most of the 32-bit instructions one of the bits in the instruction encoding defines if the APSR flags should be updated or not. Note that some of these instructions do not update the V flag or the C flag. For example, the MULS (multiply) instruction only changes the N flag and the Z flag.

**Table 4.6** ALU Flags on the Cortex-M Processors

| Flag | Descriptions |
|---|---|
| N (bit 31) | Set to bit[31] of the result of the executed instruction. When it is "1," the result has a negative value (when interpreted as a signed integer). When it is "0," the result has a positive value or equal zero. |
| Z (bit 30) | Set to "1" if the result of the executed instruction is zero. It can also be set to "1" after a compare instruction is executed if the two values are the same. |
| C (bit 29) | Carry flag of the result. For unsigned addition, this bit is set to "1" if an unsigned overflow occurred. For unsigned subtract operations, this bit is the inverse of the borrow output status. This bit is also updated by shift and rotate operations. |
| V (bit 28) | Overflow of the result. For signed addition or subtraction, this bit is set to "1" if a signed overflow occurred. |

**Table 4.7** ALU Flags Example

| Operation | Results, Flags |
|---|---|
| 0x70000000 + 0x70000000 | Result = 0xE0000000, N= 1, Z=0, C = 0, V = 1 |
| 0x90000000 + 0x90000000 | Result = 0x30000000, N= 0, Z=0, C = 1, V = 1 |
| 0x80000000 + 0x80000000 | Result = 0x00000000, N= 0, Z=1, C = 1, V = 1 |
| 0x00001234 − 0x00001000 | Result = 0x00000234, N= 0, Z=0, C = 1, V = 0 |
| 0x00000004 − 0x00000005 | Result = 0xFFFFFFFF, N= 1, Z=0, C = 0, V = 0 |
| 0xFFFFFFFF − 0xFFFFFFFC | Result = 0x00000003, N= 0, Z=0, C = 1, V = 0 |
| 0x80000005 − 0x80000004 | Result = 0x00000001, N= 0, Z=0, C = 1, V = 0 |
| 0x70000000 − 0xF0000000 | Result = 0x80000000, N= 1, Z=0, C = 0, V = 1 |
| 0xA0000000 − 0xA0000000 | Result = 0x00000000, N= 0, Z=1, C = 1, V = 0 |

In addition to conditional branch or conditional execution code, the Carry bit of APSR can also be used to extend add and subtract operations to over 32 bits. For example, when adding two 64-bit integers together, we can use the carry bit from the lower 32-bit add operation as an extra input for the upper 32-bit add operation:

```
// Calculating Z = X + Y, where X, Y and Z are all 64-bit
Z[31:0] = X[31:0] + Y[31:0]; // Calculate lower word addition,
                            // carry flag get updated
Z[63:32] = X[63:32] + Y[63:32] + Carry; // Calculate upper
                                        // word addition
```

The N-Z-C-V flags are available in all ARM® processors including the Cortex-M0 processor.

### 4.3.2 Q status flag

The Q is used to indicate an occurrence of saturation during saturation arithmetic operations or saturation adjustment operations. It is available in ARMv7-M (e.g., Cortex®-M3 and Cortex-M4 processors), but not ARMv6-M (e.g., Cortex-M0 processor). After this bit is set, it remains set until a software write to the APSR clears the Q bit. Saturation arithmetic/adjustment operations do not clear this bit. As a result, you can use this bit to determine if saturation occurred at the end of a sequence of Saturation arithmetic/adjustment operations, without the need to check the saturation status during each step.

Saturation arithmetic is useful for digital signal processing. In some cases, the destination register used to hold a calculation result might not have sufficient bit width and as a result, overflow or underflow occurs. If normal data arithmetic instructions are used, the MSB of the result would be lost and can cause a serious distortion in the output. Instead of just cutting off the MSB, saturation arithmetic

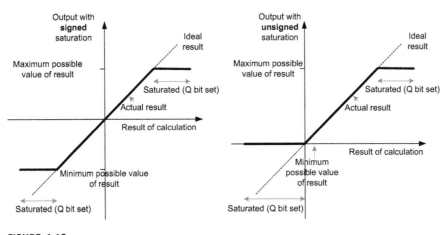

**FIGURE 4.16**

Signed saturation and unsigned saturation

forces the result to the maximum value (in case of overflow) or minimum value (in case of underflow) to reduce the impact of signal distortion (figure 4.16).

The actual maximum and minimum values that trigger the saturation depend on the instructions being used. In most cases, the instructions for saturation arithmetic are mnemonic starting with "Q," for example "QADD16." If saturation occurred, the Q bit is set; otherwise, the value of the Q bit is unchanged.

The Cortex-M3 processor provides a couple of saturation adjustment instructions, and the Cortex-M4 provides a full set of saturation arithmetic instructions, as well as those saturation adjustment instructions available in the Cotex-M3 processor.

### 4.3.3 **GE bits**

The "Greater-Equal" (GE) is a 4-bit wide field in the APSR in the Cortex®-M4, and is not available in the Cortex-M3 processor. It is updated by a number of SIMD instructions where, in most cases, each bit represents positive or overflow of SIMD operations for each byte (Table 4.8). For SIMD instructions with 16-bit data, bit 0 and bit 1 are controlled by the result or lower half-word, and bit 2 and bit 3 are controlled by the result of upper half-word.

The GE flags are used by the SEL instruction(Figure 4.17), which multiplexes the byte values from two source registers based on each GE bit. When combining SIMD instructions with the SEL instruction, simple conditional data selection can be created in SIMD arrangement for better performance.

You can also read back the GE bits by reading APSR into a general purpose register for additional processing. More details of the SIMD and SEL instructions are given in Chapter 5.

**Table 4.8** GE Flags Results

| SIMD Operation | Results |
|---|---|
| SADD16, SSUB16, USUB16, SASX, SSAX | If lower half-word result >= 0 then GE[1:0] = 2'b11 else GE[1:0] = 2'b00<br>If upper half-word result >= 0 then GE[3:2] = 2'b11 else GE[3:2] = 2'b00 |
| UADD16 | If lower half-word result >= 0x10000 then GE[1:0] = 2'b11 else GE[1:0] = 2'b00<br>If upper half-word result >= 0x10000 then GE[3:2] = 2'b11 else GE[3:2] = 2'b00 |
| SADD8, SSUB8, USUB8 | If byte 0 result >= 0 then GE[0] = 1'b1 else GE[0] = 1'b0<br>If byte 1 result >= 0 then GE[1] = 1'b1 else GE[1] = 1'b0<br>If byte 2 result >= 0 then GE[2] = 1'b1 else GE[2] = 1'b0<br>If byte 3 result >= 0 then GE[3] = 1'b1 else GE[3] = 1'b0 |
| UADD8 | If byte 0 result >= 0x100 then GE[0] = 1'b1 else GE[0] = 1'b0<br>If byte 1 result >= 0x100 then GE[1] = 1'b1 else GE[1] = 1'b0<br>If byte 2 result >= 0x100 then GE[2] = 1'b1 else GE[2] = 1'b0<br>If byte 3 result >=0x100 then GE[3] = 1'b1 else GE[3] = 1'b0 |
| UASX | If lower half-word result >= 0 then GE[1:0] = 2'b11 else GE[1:0] = 2'b00<br>If upper half-word result >= 0x10000 then GE[3:2] = 2'b11 else GE[3:2] = 2'b00 |
| USAX | If lower half-word result >= 0x10000 then GE[1:0] = 2'b11 else GE[1:0] = 2'b00<br>If upper half-word result >= 0x0 then GE[3:2] = 2'b11 else GE[3:2] = 2'b00 |

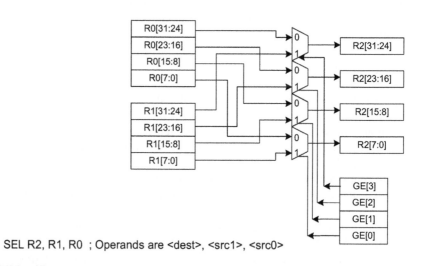

SEL R2, R1, R0  ; Operands are <dest>, <src1>, <src0>

**FIGURE 4.17**

SEL operation

## 4.4 **Memory system**
### 4.4.1 **Memory system features**

The Cortex®-M3 and Cortex-M4 processors have the following memory system features:

- 4GB linear address space — With 32-bit addressing, the ARM® processors can access up to 4GB of memory space. While many embedded systems do not need more than 1MB of memory, the 32-bit addressing capability ensures future upgrade and expansion possibilities. The Cortex-M3 and Cortex-M4 processors provide 32-bit buses using a generic bus protocol called AHB LITE. The bus allows connections to 32/16/8-bit memory devices with suitable memory interface controllers.
- Architecturally defined memory map — The 4GB memory space is divided into a number of regions for various predefined memory and peripheral uses. This allows the processor design to be optimized for performance. For example, the Cortex-M3 and Cortex-M4 processors have multiple bus interfaces to allow simultaneous access from the CODE region for program code and data operations to SRAM or peripheral regions.
- Support for little endian and big endian memory systems — The Cortex-M4 and Cortex-M4 processors can work with either little endian or big endian memory systems. In practice, a microcontroller product is normally designed with just one endian configuration.
- Bit band accesses (optional) — When the bit-band feature is included (determined by microcontroller/System-on-Chip vendors), two 1MB regions in the memory map are bit addressable via two bit-band regions. This allows atomic access to individual bits in SRAM or peripheral address space.
- Write buffer — When a write transfer to a bufferable memory region will take multiple cycles, the transfer can be buffered by the internal write buffer in the Cortex-M3 or Cortex-M4 processor so that the processor can continue to execute the next instruction, if possible. This allows higher program execution speed.
- Memory Protection Unit (Optional) — The MPU is a programmable unit which defines access permissions for various memory regions. The MPU in the Cortex-M3 and Cortex-M4 processor supports eight programmable regions, and can be used with an embedded OS to provide a robust system.
- Unaligned transfer support — All processors supporting ARMv7-M architecture (including Cortex-M3 and Cortex-M4 processors) support unaligned data transfers.

The bus interfaces on the Cortex-M processors are generic bus interfaces, and can be connected to different types and sizes of memory via different memory controllers. The memory systems in microcontrollers often contain two or more types of memories: flash memory for program code, static RAM (SRAM) for data, and in some cases Electrical Erasable Read Only Memory (EEPROM). In most cases, these memories are on-chip and the actual memory interface details are transparent to software developers. Hence, software developers only need to know the address and size of the program memory and SRAM.

### 4.4.2 Memory map

The 4GB address space of the Cortex®-M processors is partitioned into a number of memory regions (Figure 4.18). The partitioning is based on typical usages so that different areas are designed to be used primarily for:

- Program code accesses (e.g., CODE region)
- Data accesses (e.g., SRAM region)
- Peripherals (e.g., Peripheral region)
- Processor's internal control and debug components (e.g., Private Peripheral Bus)

The architecture also allows high flexibility to allow memory regions to be used for other purposes. For example, programs can be executed from the CODE as well as the SRAM region, and a microcontroller can also integrate SRAM blocks in CODE region.

In practice, many microcontroller devices only use a small portion of each region for program flash, SRAM, and peripherals. Some of the regions can be unused. Different microcontrollers have different memory sizes and peripheral address locations. This information is usually outlined in user manuals or datasheets from microcontroller vendors.

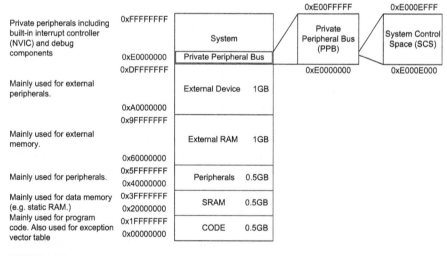

**FIGURE 4.18**

Memory map

The memory map arrangement is consistent between all of the Cortex-M processors. For example, the PPB address space hosts the registers for the Nested Vectored Interrupt Controller (NVIC), processor's configuration registers, as well as registers for debug components. This is the same across all Cortex-M devices. This makes it easier to port software from one Cortex-M device to another, and allows better software reusability. It also makes it easier for tool vendors, as the debug control for the Cortex-M3 and Cortex-M4 devices work in the same way.

### 4.4.3 Stack memory

As in almost all processor architectures, the Cortex®-M processors need stack memory to operate and have stack pointers (R13). Stack is a kind of memory usage mechanism that allows a portion of memory to be used as Last-In-First-Out data storage buffer. ARM® processors use the main system memory for stack memory operations, and have the PUSH instruction to store data in stack and the POP instruction to retrieve data from stack. The current selected stack pointer is automatically adjusted for each PUSH and POP operation.

Stack can be used for:

- Temporary storage of original data when a function being executed needs to use registers (in the register bank) for data processing. The values can be restored at the end of the function so the program that called the function will not lose its data.
- Passing of information to functions or subroutines.
- For storing local variables.
- To hold processor status and register values in the case of exceptions such as an interrupt.

The Cortex-M processors use a stack memory model called "full-descending stack." When the processor is started, the SP is set to the end of the memory space reserved for stack memory. For each PUSH operation, the processor first decrements the SP, then stores the value in the memory location pointed by SP. During operations, the SP points to the memory location where the last data was pushed to the stack (Figure 4.19).

In a POP operation, the value of the memory location pointed by SP is read, and then the value of SP is incremented automatically.

The most common uses for PUSH and POP instructions are to save contents of register banks when a function/subroutine call is made. At the beginning of the function call, the contents of some of the registers can be saved to the stack using the PUSH instruction, and then restored to their original values at the end of the function using the POP instruction. For example, in Figure 4.20 a simple function/subroutine named function1 is called from the main program. Since function1 needs to use and modify R4, R5, and R6 for data processing, and these registers hold values that the main program need later, they are saved to the stack using PUSH, and restored using POP at the end of function1. In this way, the

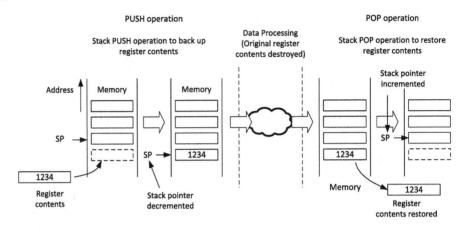

**FIGURE 4.19**

Stack PUSH and POP

**Main Program**

```
        ...
    ; R4 = X, R5 = Y, R6 = Z         Subroutine
    BL    function1
                                     function1
                                             PUSH    {R4} ; store R4 to stack & adjust SP
                                             PUSH    {R5} ; store R5 to stack & adjust SP
                                             PUSH    {R6} ; store R6 to stack & adjust SP
                                             ... ; Executing task (R4, R5 and R6
                                              ; could be changed)
                                             POP     {R4} ; restore R4 and SP re-adjusted
                                             POP     {R5} ; restore R5 and SP re-adjusted
                                             POP     {R6} ; restore R6 and SP re-adjusted
                                             BX      LR   ; Return
    ; Back to main program
    ; R4 = X, R5 = Y, R6 = Z
    ... ; next instructions
```

**FIGURE 4.20**

Simple PUSH and POP usage in functions — one register in each stack operation

program code that called the function will not lose any data and can continue to execute. Note that for each PUSH (store to memory) operation, there must be a corresponding POP (read from memory), and the address of the POP should match that of the PUSH operation.

Each PUSH and POP instruction can transfer multiple data to/from the stack memory. This is shown in Figure 4.21. Since the registers in the register bank are 32 bits, each memory transfer generated by stack PUSH and stack POP transfers at least 1 word (4 bytes) of data, and the addresses are always aligned to 4-byte boundaries. The lowest two bits of the SP are always zero.

Main Program

```
        ...
    ; R4 = X, R5 = Y, R6 = Z        Subroutine
    BL    function1
                                    function1
                                        PUSH    {R4-R6} ; Store R4, R5, R6 to stack
                                        ... ; Executing task (R4, R5 and R6
                                          ; could be changed)
                                        POP     {R4-R6} ; restore R4, R4, R6
                                        BX      LR   ; Return

    ; Back to main program
    ; R4 = X, R5 = Y, R6 = Z
    ... ; next instructions
```

**FIGURE 4.21**

Simple PUSH and POP usage in functions — Multiple register stack operations

Main Program

```
        ...
    ; R4 = X, R5 = Y, R6 = Z        Subroutine
    BL     function1
                                    function1
                                        PUSH    {R4-R6, LR} ; Save registers
                                                  ; including link register
                                        ... ; Executing task (R4, R5 and R6
                                          ; could be changed)
                                        POP     {R4-R6, PC} ; Restore registers and
                                                  ; return
    ; Back to main program
    ; R4 = X, R5 = Y, R6 = Z
    ... ; next instructions
```

**FIGURE 4.22**

Combining stack POP and return

You can also combine the return with a POP operation. This is done by first pushing the value of LR (R14) to the stack memory, and popping it back to PC (R15) at the end of the subroutine/function, as shown in Figure 4.22.

Physically there are two stack pointers in the Cortex-M processors. They are the:

- Main Stack Pointer (MSP) — This is the default stack pointer used after reset, and is used for all exception handlers.
- Process Stack Pointer (PSP) — This is an alternate stack point that can only be used in Thread mode. It is usually used for application tasks in embedded systems running an embedded OS.

As mentioned previously (Table 4.3 and Figure 4.10), the selection between MSP and PSP can be controlled by the value of SPSEL in bit 1 of the CONTROL register. If this bit is 0, Thread mode uses MSP for the stack operation. Otherwise, Thread

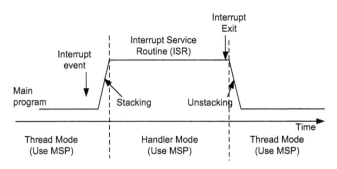

**FIGURE 4.23**

SPSEL = 0. Both Thread Level and Handler use the Main Stack Pointer

mode uses the PSP. In addition, during exception return from Handler mode to Thread mode, the selection can be controlled by the value of EXC_RETURN (exception return) value. In that case the value of SPSEL will be updated by the processor hardware accordingly.

In simple applications without an OS, both Thread mode and Handler mode can use MSP only. This is shown in Figure 4.23: After an interrupt event is triggered, the processor first pushes a number of registers into the stack before entering the Interrupt Service Routine (ISR). This register state saving operation is called "Stacking," and at the end of the ISR, these registers are restored to the register bank and this operation is called "Unstacking."

When embedded systems use an embedded OS, they often use separate memory areas for application stack and the kernel stack. As a result, the PSP is used and switching of SP selection takes place in exception entry and exception exit. This is shown in Figure 4.24. Note that the automatic "Stacking" and "Unstacking" stages use PSP. The separating stack arrangement can prevent a stack corruption or error in

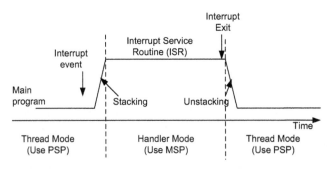

**FIGURE 4.24**

SPSEL = 1. Thread Level uses the Process Stack and Handler uses the Main Stack an application task from damaging the stack use by the OS. It also simplifies the OS design and hence allows faster context switching.

Although only one of the SPs is visible at a time (when using SP or R13 to access it), it is possible to read/write directly to the MSP and PSP, without any confusion over which SP/R13 you are referring to. Provided that you are in privileged level, you can access MSP and PSP using the following CMSIS functions:

```
x = __get_MSP(); // Read the value of MSP
__set_MSP(x); // Set the value of MSP
x = __get_PSP(); // Read the value of PSP
__set_PSP(x); // Set the value of PSP
```

In general it is not recommended to change the value of the current selected SP in a C function, as part of the stack memory could be used for storing local variables or other data. To access MSP and PSP in assembly code, you can use the MSR and MRS instructions:

```
MRS R0, MSP ; Read Main Stack Pointer to R0
MSR MSP, R0 ; Write R0 to Main Stack Pointer
MRS R0, PSP ; Read Process Stack Pointer to R0
MSR PSP, R0 ; Write R0 to Process Stack Pointer
```

Most application code does not need to access MSP and PSP explicitly. Access to MSP and PSP is often required for embedded OSs. For example, by reading the PSP value using an MRS instruction, the OS can read data pushed to the stack from API calls in application tasks (such as register contents before execution of an SVC instruction). Also, the value of PSP is updated by context switching code in the OS during context switching.

After power up, the processor hardware automatically initializes the MSP by reading the vector table. More information on vector tables will be covered in section 4.5.3. The PSP is not initialized automatically and must be initialized by the software before being used.

### 4.4.4 Memory protection unit (MPU)

The MPU is optional in the Cortex-M3 and Cortex®-M4 processors. Therefore not all Cortex-M3 or Cortex-M4 microcontrollers have the MPU feature. In the majority of applications, the MPU is not used and can be ignored. In embedded systems that require high reliability, the MPU can be used to protect memory regions by means of defining access permissions in privileged and unprivileged access states.

The MPU is programmable, and the MPU design in the Cortex-M3 and Cortex-M4 processors supports eight programmable regions. The MPU can be used in different ways. In some cases the MPU is controlled by an embedded OS, and memory permissions are configured for each task. In other cases the MPU is configured just to protect a certain memory region; for example, to make a memory range read only.

More information about the MPU is covered in Chapter 11.

## 4.5 Exceptions and interrupts
### 4.5.1 What are exceptions?

Exceptions are events that cause changes to program flow. When one happens, the processor suspends the current executing task and executes a part of the program called the exception handler. After the execution of the exception handler is completed, the processor then resumes normal program execution. In the ARM® architecture, interrupts are one type of exception. Interrupts are usually generated from peripheral or external inputs, and in some cases they can be triggered by software. The exception handlers for interrupts are also referred to as Interrupt Service Routines (ISR).

In Cortex®-M processors, there are a number of exception sources:

Exceptions are processed by the NVIC. The NVIC can handle a number of Interrupt Requests (IRQs) and a Non-Maskable Interrupt (NMI) request. Usually IRQs are generated by on-chip peripherals or from external interrupt inputs though I/O ports. The NMI could be used by a watchdog timer or brownout detector (a voltage monitoring unit that warns the processor when the supply voltage drops below a certain level). Inside the processor there is also a timer called SysTick, which can generate a periodic timer interrupt request, which can be used by embedded OSs for timekeeping, or for simple timing control in applications that don't require an OS.

The processor itself is also a source of exception events. These could be fault events that indicate system error conditions, or exceptions generated by software to support embedded OS operations. The exception types are listed in Table 4.9.

Each exception source has an exception number. Exception numbers 1 to 15 are classified as system exceptions, and exceptions 16 and above are for interrupts. The design of the NVIC in the Cortex-M3 and Cortex-M4 processors can support up to 240 interrupt inputs. However, in practice the number of interrupt inputs implemented in the design is far less, typically in the range of 16 to 100. In this way the silicon size of the design can be reduced, which also reduces power consumption.

The exception number is reflected in various registers, including the IPSR, and it is used to determine the exception vector addresses. Exception vectors are stored in a vector table, and the processor reads this table to determine the starting address of an

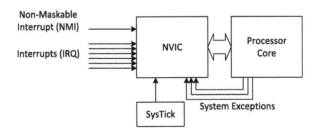

**FIGURE 4.25**

Various exception sources

**Table 4.9** Exception Types

| Exception Number | CMSIS Interrupt Number | Exception Type | Priority | Function |
|---|---|---|---|---|
| 1 | — | Reset | −3 (Highest) | Reset |
| 2 | −14 | NMI | −2 | Non-Maskable interrupt |
| 3 | −13 | HardFault | −1 | All classes of fault, when the corresponding fault handler cannot be activated because it is currently disabled or masked by exception masking |
| 4 | −12 | MemManage | Settable | Memory Management fault; caused by MPU violation or invalid accesses (such as an instruction fetch from a non-executable region) |
| 5 | −11 | BusFault | Settable | Error response received from the bus system; caused by an instruction prefetch abort or data access error |
| 6 | −10 | Usage fault | Settable | Usage fault; typical causes are invalid instructions or invalid state transition attempts (such as trying to switch to ARM state in the Cortex-M3) |
| 7–10 | — | — | — | Reserved |
| 11 | −5 | SVC | Settable | Supervisor Call via SVC instruction |
| 12 | −4 | Debug monitor | Settable | Debug monitor – for software based debug (often not used) |
| 13 | — | — | — | Reserved |
| 14 | −2 | PendSV | Settable | Pendable request for System Service |
| 15 | −1 | SYSTICK | Settable | System Tick Timer |
| 16–255 | 0–239 | IRQ | Settable | IRQ input #0–239 |

exception handler during the exception entrance sequence. Note that the exception number definitions are different from interrupt numbers in the CMSIS device-driver library. In the CMSIS device-driver library, interrupt numbers start from 0, and system exception numbers have negative values.

As opposed to classic ARM processors such as the ARM7TDMI™, there is no FIQ (Fast Interrupt) in the Cortex-M processor. However, the interrupt latency of the Cortex-M3 and Corex-M4 is very low, only 12 clock cycles, so this does not cause problems.

Reset is a special kind of exception. When the processor exits from a reset, it executes the reset handler in Thread mode (rather than Handler mode as in other exceptions). Also the exception number in IPSR is read as zero.

### 4.5.2 Nested vectored interrupt controller (NVIC)

The NVIC is a part of the Cortex®-M processor. It is programmable and its registers are located in the System Control Space (SCS) of the memory map (see Figure 4.18). The NVIC handles the exceptions and interrupt configurations, prioritization, and interrupt masking. The NVIC has the following features:

- Flexible exception and interrupt management
- Nested exception/interrupt support
- Vectored exception/interrupt entry
- Interrupt masking

#### Flexible exception and interrupt management

Each interrupt (apart from the NMI) can be enabled or disabled and can have its pending status set or cleared by software. The NVIC can handle various types of interrupt sources:

- Pulsed interrupt request — the interrupt request is at least one clock cycle long. When the NVIC receives a pulse at its interrupt input, the pending status is set and held until the interrupt gets serviced.
- Level triggered interrupt request — the interrupt source holds the request high until the interrupt is serviced.

The signal level at the NVIC input is active high. However, the actual external interrupt input on the microcontroller could be designed differently and is converted to an active high signal level by on-chip logic.

#### Nested exception/interrupt support

Each exception has a priority level. Some exceptions, such as interrupts, have programmable priority levels and some others (e.g., NMI) have a fixed priority level. When an exception occurs, the NVIC will compare the priority level of this exception to the current level. If the new exception has a higher priority, the current running task will be suspended. Some of the registers will be stored on the stack memory, and the processor will start executing the exception handler of the new exception. This process is called "preemption." When the higher priority exception handler is complete, it is terminated with an exception return operation and the processor automatically restores the registers from stack and resumes the task that was running previously. This mechanism allows nesting of exception services without any software overhead.

### Vectored exception/interrupt entry

When an exception occurs, the processor will need to locate the starting point of the corresponding exception handler. Traditionally, in ARM® processors such as the ARM7TDMI™, software handles this step. The Cortex-M processors automatically locate the starting point of the exception handler from a vector table in the memory. As a result, the delays from the start of the exception to the execution of the exception handlers are reduced.

### Interrupt masking

The NVIC in the Cortex-M3 and Cortex-M4 processors provide several interrupt masking registers such as the PRIMASK special register. Using the PRIMASK register you can disable all exceptions, excluding HardFault and NMI. This masking is useful for operations that should not be interrupted, like time critical control tasks or real-time multimedia codecs. Alternatively you can also use the BASEPRI register to select mask exceptions or interrupts which are below a certain priority level.

The CMSIS-Core provides a set of functions to make it easy to access various interrupt control functions. The flexibility and capability of the NVIC also make the Cortex-M processors very easy to use, and provide better a system response by reducing the software overhead in interrupt processing, which also leads to smaller code size.

## 4.5.3 Vector table

When an exception event takes place and is accepted by the processor core, the corresponding exception handler is executed. To determine the starting address of the exception handler, a vector table mechanism is used. The vector table is an array of word data inside the system memory, each representing the starting address of one exception type (Figure 4.26). The vector table is relocatable and the relocation is controlled by a programmable register in the NVIC called the Vector Table Offset Register (VTOR). After reset, the VTOR is reset to 0; therefore, the vector table is located at address 0x0 after reset.

For example, if the reset is exception type 1, the address of the reset vector is 1 times 4 (each word is 4 bytes), which equals 0x00000004, and the NMI vector (type 2) is located at 2 x 4 = 0x00000008. The address 0x00000000 is used to store the starting value of the MSP.

The LSB of each exception vector indicates whether the exception is to be executed in the Thumb state. Since the Cortex®-M processors can support only Thumb instructions, the LSB of all the exception vectors should be set to 1.

| Exception Type | CMSIS Interrupt Number | Address Offset | Vectors | |
|---|---|---|---|---|
| 18 - 255 | 2 - 239 | 0x48 – 0x3FF | IRQ #2 - #239 | 1 |
| 17 | 1 | 0x44 | IRQ #1 | 1 |
| 16 | 0 | 0x40 | IRQ #0 | 1 |
| 15 | -1 | 0x3C | SysTick | 1 |
| 14 | -2 | 0x38 | PendSV | 1 |
| NA | NA | 0x34 | Reserved | |
| 12 | -4 | 0x30 | Debug Monitor | 1 |
| 11 | -5 | 0x2C | SVC | 1 |
| NA | NA | 0x28 | Reserved | |
| NA | NA | 0x24 | Reserved | |
| NA | NA | 0x20 | Reserved | |
| NA | NA | 0x1C | Reserved | |
| 6 | -10 | 0x18 | Usage fault | 1 |
| 4 | -11 | 0x14 | Bus Fault | 1 |
| 4 | -12 | 0x10 | MemManage Fault | 1 |
| 3 | -13 | 0x0C | HardFault | 1 |
| 2 | -14 | 0x08 | NMI | 1 |
| 1 | NA | 0x04 | Reset | 1 |
| NA | NA | 0x00 | Initial value of MPS | |

**FIGURE 4.26**

Exception types (LSB of exception vectors should be set to 1 to indicate Thumb state)

### 4.5.4 Fault handling

Several types of exceptions in the Cortex-M3 and Cortex®-M4 processors are fault handling exceptions. Fault exceptions are triggered when the processor detects an error such as the execution of an undefined instruction, or when the bus system returns an error response to a memory access. The fault exception mechanism allows errors to be detected quickly, and potentially allows the software to carry out remedial actions (Figure 4.27).

By default the Bus Fault, Usage Fault, and Memory Management Fault are disabled and all fault events trigger the HardFault exception. However, the configurations are programmable and you can enable the three programmable fault exceptions individually to handle different types of faults. The HardFault exception is always enabled.

Fault exceptions can also be useful for debugging software issues. For example, the fault handler can automatically collect information and report to the user or other systems that an error has occurred and provide debug information. A number of fault

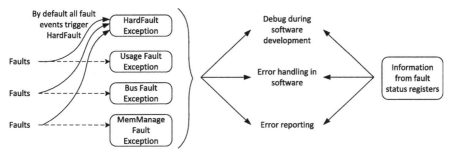

**FIGURE 4.27**

Fault exceptions usages

status registers are available in the Cortex-M3 and Cortex-M4 processors, which provide hints about the error sources. Software developers can also examine these fault status registers using the debugger during software development.

## 4.6 System control block (SCB)

One part of the processor that is merged into the NVIC unit is the SCB. The SCB contains various registers for:

- Controlling processor configurations (e.g., low power modes)
- Providing fault status information (fault status registers)
- Vector table relocation (VTOR)

The SCB is memory-mapped. Similar to the NVIC registers, the SCB registers are accessible from the System Control Space (SCS). More information about SCB registers is covered in Chapters 7 and 9.

## 4.7 Debug

As software gets more complex, debug features are becoming more and more important in modern processor architectures. Although their designs are compact, the Cortex®-M3 and Cortex-M4 processors include comprehensive debugging features such as program execution controls, including halting and stepping, instruction breakpoints, data watchpoints, registers and memory accesses, profiling, and traces.

There are two types of interfaces provided in the Cortex-M processors: debug and trace.

The debug interface allows a debug adaptor to connect to a Cortex-M microcontroller to control the debug features and access the memory space on the chip. The

Cortex-M processor supports the traditional JTAG protocol, which uses either 4 or 5 pins, or a newer 2-pin protocol called Serial Wire Debug (SWD). The SWD protocol was developed by ARM®, and can handle the same debug features as in JTAG in just two pins, without any loss of debug performance. Many commercially available debug adaptors, such as the ULINK 2 or ULINK Pro products from Keil™, support both protocols. The two protocols can use the same connector, with JTAG TCK shared with the Serial Wire clock, and JTAG TMS shared with the Serial Wire Data, which is bidirectional (Figure 4.28). Both protocols are widely supported by different debug adaptors from different companies.

The trace interface is used to collect information from the processor during runtime such as data, event, profiling information, or even complete details of program execution. Two types of trace interface are supported: a single pin protocol called Serial Wire Viewer (SWV) and a multi-pin protocol called Trace Port (Figure 4.29).

SWV is a low-cost solution that has a lower trace data bandwidth limit. However, the bandwidth is still large enough to handle capturing of selective data trace, event trace, and basic profiling. The output signal, which is called Serial Wire Output (SWO), can be shared with the JTAG TDO pin so that you only need one standard JTAG/SWD connector for both debug and trace. (Obviously, the trace data can only be captured when the two-pin SWD protocol is used for debugging.)

The Trace Port mode requires one clock pin and several data pins. The number of data pins used is configurable, and in most cases the Cortex-M3 or Cortex-M4 microcontrollers support a maximum of four data pins (a total of five pins including the clock). The Trace Port mode supports a much higher trace data bandwidth than SWV. You can also use Trace Port mode with fewer pins if needed; for example, when some of the Trace Data pins are multiplexed with I/O functions and you need to use some of these I/O pins for your application.

The high trace data bandwidth of the Trace Port model allows real-time recording of program execution information, in addition to the other trace information you can collect using SWV. The real-time program trace requires a companion component called Embedded Trace Macrocell (ETM) in the chip. This is an optional component for the Cortex-M3 and Cortex-M4 processors. Some of the Cortex-M3 and Cortex-M4 microcontrollers do not have ETM and therefore do not provide program/instruction trace.

To capture trace data, you can use a low-cost debug adaptor such as Keil ULINK-2 or Segger J-Link, which can capture data through the SWV interface. Or you can use advanced products such as Keil ULINK Pro or Segger J-Trace to capture trace data in trace port mode.

There are a number of other debug components inside the Cortex-M3 and Cortex-M4 processors. For example, the Instrumentation Trace Macrocell (ITM) allows program code running on the microcontroller to generate data to be output through the trace interface. The data can then be displayed on a debugger window. More information about various debug features are covered in Chapter 14. Appendix H also provides information about standard debug connectors used by various debug adaptors.

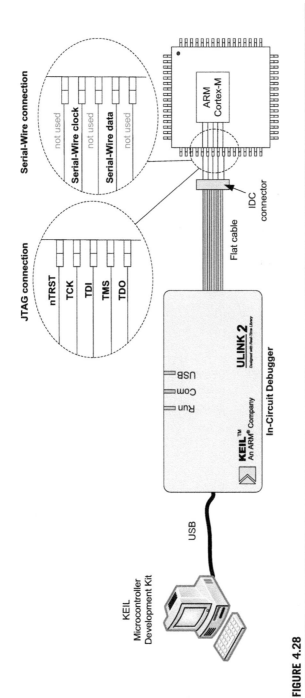

**FIGURE 4.28**

Debug connection

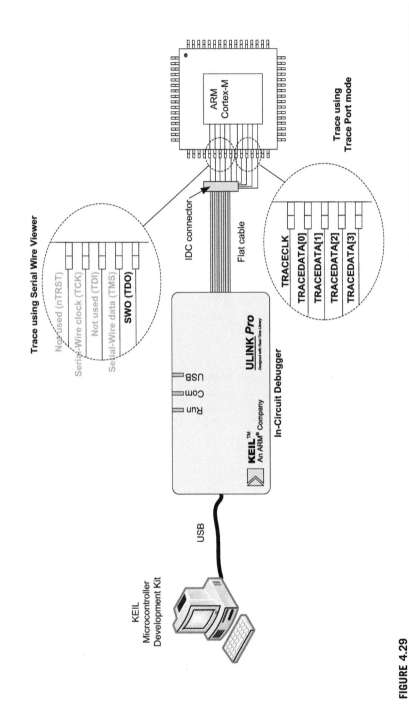

**FIGURE 4.29**

Trace connection (SWO or Trace Port mode)

## 4.8 Reset and reset sequence

In typical Cortex®-M microcontrollers, there can be three types of reset:

- Power on reset — reset everything in the microcontroller. This includes the processor and its debug support component and peripherals.
- System reset — reset just the processor and peripherals, but not the debug support component of the processor.
- Processor reset — reset the processor only.

During system debug or processor reset operations, the debug components in the Cortex-M3 or Cortex-M4 processors are not reset so that the connection between the debug host (e.g., debugger software running on a computer) and the microcontroller can be maintained. The debug host can generate a system reset or processor reset via a register in the System Control Block (SCB). This is covered in section 7.9.4.

The duration of Power on reset and System reset depends on the microcontroller design. In some cases the reset lasts a number of milli seconds as the reset controller needs to wait for a clock source such as a crystal oscillator to stabilize.

After reset and before the processor starts executing the program, the Cortex-M processors read the first two words from the memory (Figure 4.30). The beginning of the memory space contains the vector table, and the first two words in the vector table are the initial value for the Main Stack Pointer (MSP), and the reset vector, which is the starting address of the reset handler (as described in section 4.5.3 and Figure 4.26). After these two words are read by the processor, the processor then sets up the MSP and the Program Counter (PC) with these values.

The setup of the MSP is necessary because some exceptions such as the NMI or HardFault handler could potentially occur shortly after the reset, and the stack memory and hence the MSP will then be needed to push some of the processor status to the stack before exception handling.

Note that for most C development environments, the C startup code will also update the value of the MSP before entering the main program main(). This two-step stack initialization allows a microcontroller device with external memory to use the external memory for the stack. For example, it can boot up with the stack placed in a

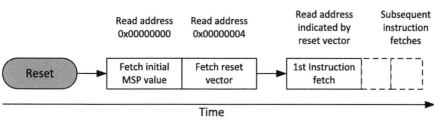

**FIGURE 4.30**

Reset sequence

small internal on-chip SRAM, and initialize an external memory controller while in the reset handler, and then execute the C startup code, which then sets up the stack memory to the external memory.

The Stack Pointer initialization behavior is different from classic ARM® processors such as the ARM7TDMI™, where upon reset the processor executes instructions from address zero, and the stack pointers must be initialized by software. In classic ARM processors, the vector table holds instruction code rather than address values.

Because the stack operations in the Cortex-M3 or Cortex-M4 processors are based on full descending stack (SP decrement before store), the initial SP value should be set to the first memory after the top of the stack region. For example, if you have a stack memory range from 0x20007C00 to 0x20007FFF (1Kbytes), the initial stack value should be set to 0x20008000, as shown in Figure 4.31.

Notice that in the Cortex-M processors, vector addresses in the vector table should have their LSB set to 1 to indicate that they are Thumb code. For that reason, the example in Figure 4.31 has 0x101 in the reset vector, whereas the boot code starts at address 0x100. After the reset vector is fetched, the Cortex-M processor can then

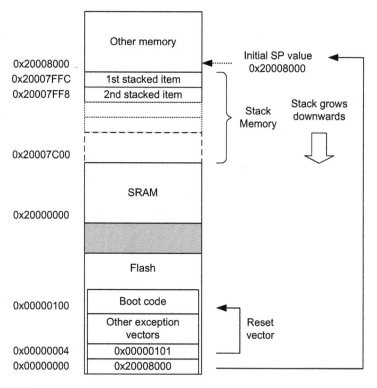

**FIGURE 4.31**

Initial Stack Pointer value and Initial Program Counter value example start to execute the program from the reset vector address and begin normal operations.

Various software development tools might have different ways to specify the starting stack pointer value and reset vector. If you need more information on this topic, it's best to look at project examples provided with the development tools. Some information is provided in section 15.9 (for Keil™ MDK-ARM) and section 16.9 (for IAR toolchain) of this book.

# Instruction Set

# 5

The Definitive Guide to ARM® Cortex®-M3 and Cortex-M4 Processors. http://dx.doi.org/10.1016/B978-0-12-408082-9.00005-1

# 5.1 Background to the instruction set in ARM® Cortex®-M processors

The design of the instruction set is one of the most important parts of a processor's architecture. In ARM's terminology, it is commonly referred as the Instruction Set Architecture (ISA). All the ARM® Cortex®-M processors are based on Thumb®-2 technology, which allows a mixture of 16-bit and 32-bit instructions to be used within one operating state. This is different from classic ARM processors such as the ARM7TDMI™. To help understand the differences between the different instruction sets available in the ARM processors, we include a quick review of the history of the ARM ISA.

Early ARM processors (prior to the ARM7TDMI processor) supported a 32-bit instruction set called the ARM instruction set. It evolved for a few years, progressing from ARM architecture version 1 to version 4. It is a powerful instruction set, which supports conditional execution of most instructions and provides good performance. However, it often requires more program memory when compared to 8-bit and 16-bit architecture. As demand for 32-bit processors started to increase in mobile phone applications, where power and cost are often both critical, a solution was needed to reduce the program size.

In 1995, ARM introduced the ARM7TDMI processor, which supports a new operation state that runs a new 16-bit instruction set (Figure 5.1). This 16-bit instruction set is called "Thumb" (it is a play on words to indicate that it has smaller size than the ARM instruction set). The ARM7TDMI can operate in the ARM state, the default state, and also in the Thumb state. During operation, the processor switches between ARM state and Thumb state under software control. Parts of the application program are compiled with ARM instructions for higher performance, and the remaining parts are compiled as Thumb instructions for better code density. By providing this two-state mechanism, the applications can be squeezed into a smaller program size, while

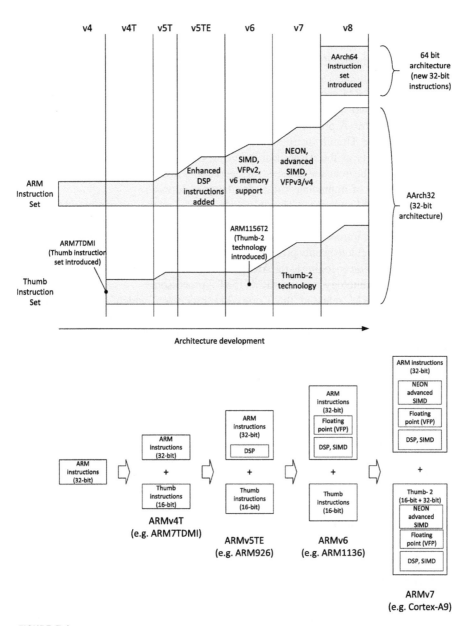

**FIGURE 5.1**

Evolution of the ARM Instruction Set Architecture

maintaining high performance when needed. In some cases, the Thumb code provides a code size reduction of 30% compared to the equivalent ARM code.

The Thumb instruction set provides a subset of the ARM instruction set. In the ARM7TDMI processor design, a mapping function is used to translate Thumb instructions into ARM instructions for decoding so that only one instruction decoder is needed. The two states of operation are still supported in newer ARM processors, such as the Cortex-A processor family and the Cortex-R processor family.

Although the Thumb instruction set can provide most of the same commonly used functionality as the ARM instructions, it does have some limitations, such as restrictions on the register choices for operations, available addressing modes, or a reduced range of immediate values for data or addresses.

In 2003, ARM announced Thumb-2 technology, a method to combine 16-bit and 32-bit instruction sets in one operation state. In Thumb-2, a new superset of the Thumb instructions were introduced, with many as 32-bit size, hence they can handle most of the operations previously only possible in the ARM instruction set. However, they have different instruction encoding to the ARM instruction set. The first processor supporting the Thumb-2 technology was the ARM1156T-2 processor.

In 2006, ARM released the Cortex-M3 processor, which utilizes Thumb-2 technology and supports just the Thumb operation state. Unlike earlier ARM processors, it does not support the ARM instruction set. Since then, more Cortex-M processors have been introduced, implementing different ranges of the Thumb instruction set for different markets. Since the Cortex-M processors do not support ARM instructions, they are not backward compatible with classic ARM processors such as the ARM7TDMI. In other words, you cannot run a binary image for ARM7TDMI processors on a Cortex-M3 processor. Nevertheless, the Thumb instruction set in the Cortex-M3 processor (ARMv7-M) is a superset of the Thumb instructions in ARM7TDMI (ARMv4T), and many ARM instructions can be ported to equivalent 32-bit Thumb instructions, making application porting fairly easy.

The evolution of the ARM ISA is a continuing process. In 2011, ARM announced the ARMv8 architecture, which has a new instruction set for 64-bit operations. Currently the support for the ARMv8 architecture is limited to Cortex-A processors only, and does not cover Cortex-M processors.

## 5.2 Comparison of the instruction set in ARM® Cortex®-M processors

One of the differences between the Cortex®-M processors is the instruction set features. In order to reduce the circuit size to a minimum, the Cortex-M0, Cortex-M0+ and the Cortex-M1 processors only support most of the 16-bit Thumb instructions and a few 32-bit Thumb instructions. The Cortex-M3 processor supports more 32-bit instructions, and a few more 16-bit instructions. The Cortex-M4 processor supports the remaining DSP enhancing instructions such as SIMD (Single Instruction Multiple Data), MAC (Multiply Accumulate), and the optional floating point instructions. The instruction set support in the current Cortex-M processors is illustrated in Figure 5.2.

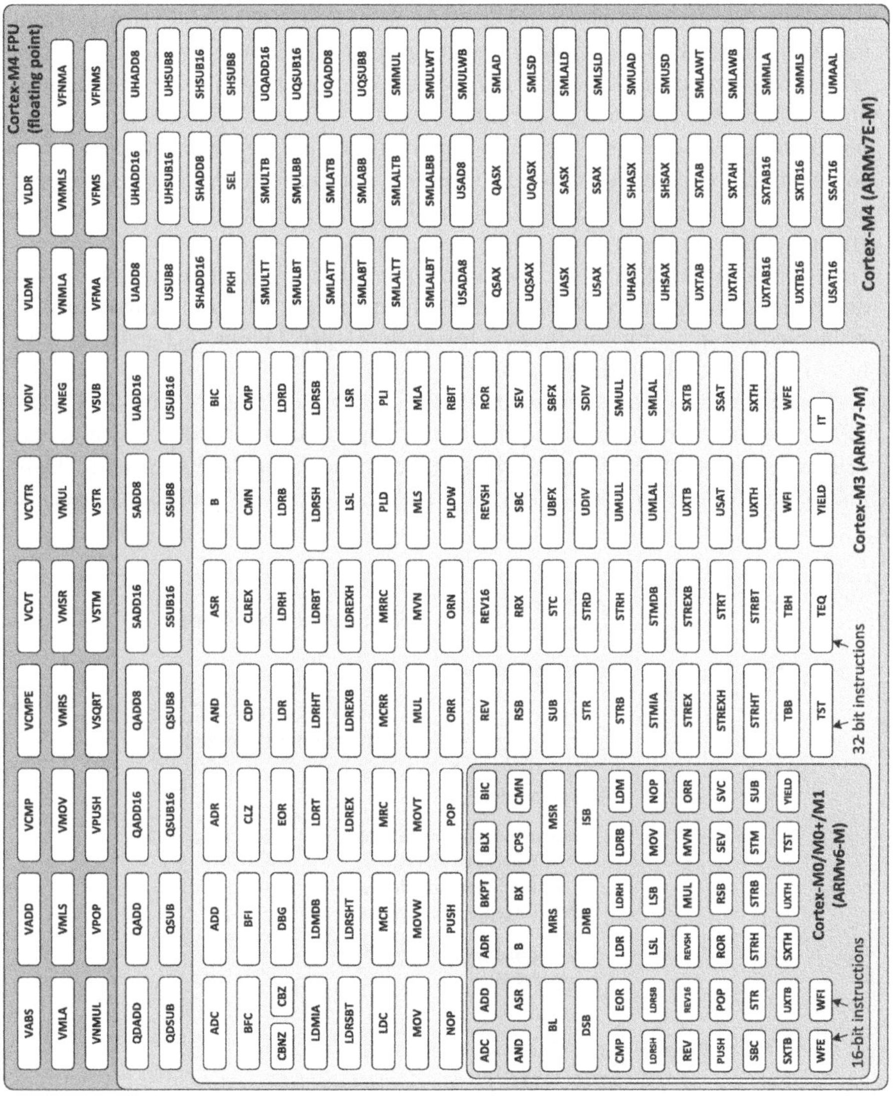

**FIGURE 5.2**

Instruction set of the Cortex-M processors

As you can see in Figure 5.2, the instruction set design of the Cortex-M processors is upward compatible from Cortex-M0, to Cortex-M3, and then to the Cortex-M4. Therefore code compiled for the Cortex-M0/M0+/M1 processor can run on the Cortex-M3 or Cortex-M4 processors, and code compiled for Cortex-M3 can also run on the Cortex-M4 processor.

Another observation which can be made about Figure 5.2 is that most of the instructions in ARMv6-M are 16-bit, and some are available in both 16-bit and 32-bit format. When an operation can be carried out in 16-bit, the compiler will normally choose the 16-bit version to give a smaller code size. The 32-bit version might support a greater choice of registers (e.g., high registers), larger immediate data, longer address range, or a larger choice of addressing modes. However, for the same operation, the 16-bit version and the 32-bit version of an instruction will take the same amount of time to execute.

As you can see, there are lots of instructions in the Thumb instruction set, and different Cortex-M processors support different ranges of these instructions. So what does this mean for embedded software developers? Figure 5.3 gives a simplified view of what it means to users.

For general data processing and I/O control tasks, the Cortex-M0 and the Cortex-M0+ processors are entirely adequate. For example, the Cortex-M0+ processor can deliver 2.15 CoreMark/MHz, which is approximately double that of other 16-bit microcontrollers at the same operating frequency. If your application needs to

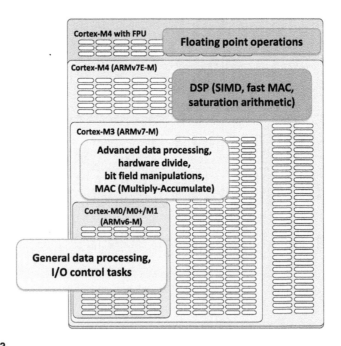

**FIGURE 5.3**

Simplified view of the instruction sets supported by Cortex-M processors process more complex data, perform faster divide operations, or requires the data processing to be done faster, then you might need to upgrade to the Cortex-M3 or Cortex-M4 processor. If you need to have the best performance in DSP applications or floating point operations, then the Cortex-M4 is a better choice.

Although there are quite a lot of instructions in the Cortex-M processors, there is no need to learn them all in detail, as C compilers are good enough to generate efficient code. Also, the free CMSIS-DSP library and various middleware (e.g., software libraries) help software developers to implement high-performance DSP applications without the need to dig into the details of each instruction.

In the rest of this chapter we will briefly go through the instruction set, which can be useful for helping you to understand a program when debugging your projects. Appendix A also provides a summary of each of the instructions.

## 5.3 **Understanding the assembly language syntax**

In most situations, application code will be written in C or other high-level languages and therefore it is not necessary for most software developers to know the details of the instruction set. However, it is still useful to have a general overview of what instructions are available, and of assembly language syntax; for example, knowledge in this area can be very useful for debugging. Most of the assembly examples in this book are written in ARM® assembler (armasm), which is used in the Keil™ Microcontroller Development Kit for ARM (MDK-ARM). Assembly tools from different vendors (e.g., the GNU toolchain) have different syntaxes. In most cases, the mnemonics of the assembly instructions are the same, but assembly directives, definitions, labeling, and comment syntax can be different.

With ARM assembly (applies to ARM RealView® Compilation Toolchain, DS-5™, and Keil Microcontroller Development Kit), the following instruction formatting is used:

```
label
  mnemonic operand1, operand2, ... ; Comments
```

The "label" is used as a reference to an address location. It is optional; some instructions might have a label in front of them so that the address of the instruction can be obtained by using the label. Labels can also be used to reference data addresses. For example, you can put a label for a lookup table inside the program. After the "label" you can find the "mnemonic," which is the name of the instruction, followed by a number of operands:

- For data processing instructions written for the ARM assembler, the first operand is the destination of the operation.
- For a memory read instruction (except multiple load instructions), the first operand is the register which data is loaded into.
- For a memory write instruction (except multiple store instructions), the first operand is the register that holds the data to be written to memory.
  Instructions that handle multiple loads and stores have a different syntax.

The number of operands for each instruction depends on the instruction type. Some instructions do not need any operand and some might need just one.

Note that some mnemonics can be used with different types of operands, and this can result in different instruction encodings. For example, the MOV (move) instruction can be used to transfer data between two registers, or it can be used to put an immediate constant value into a register.

The number of operands in an instruction depends on what type of instruction it is, and the syntax for the operands can also be different in each case. For example, immediate data are usually prefixed with "#":

```
MOVS R0, #0x12 ; Set R0 = 0x12 (hexadecimal)
MOVS R1, #'A'  ; Set R1 = ASCII character A
```

The text after each semicolon ";" is a comment. Comments do not affect program operation, but should make programs easier for humans to understand.

In the GNU toolchain, the common assembly syntax is:

```
label:
mnemonic operand1, operand2,... /* Comments */
```

The opcode and operands are the same as the ARM assembler syntax, but the syntax for labels and comments are different. For the same instructions as above, the GNU version is:

```
MOVS R0, #0x12 /* Set R0 = 0x12 (hexadecimal) */
MOVS R1, #'A'  /* Set R1 = ASCII character A */
```

An alternate way to insert comments in gcc is to make use of the inline comment character "@." For example:

```
MOVS R0, #0x12 @ Set R0 = 0x12 (hexadecimal)
MOVS R1, #'A'  @ Set R1 = ASCII character A
```

One of the commonly required features in assembly code is the ability to define constants. By using constant definitions, the program code can be made more readable and this can make code maintenance much easier. In ARM assembly, an example of defining a constant is:

```
NVIC_IRQ_SETEN   EQU 0xE 000E100
NVIC_IRQ0_ENABLE EQU 0x1

  ...

LDR R0,=NVIC_IRQ_SETEN      ; Put 0xE000E100 into R0
     ; LDR here is a pseudo instruction that will be converted
     ; to a PC relative literal data load by the assembler
MOVS R1, #NVIC_IRQ0_ENABLE ; Put immediate data (0x1) into
                           ;          register R1
 STR R1, [R0] ; Store 0x1 to 0xE000E100, this enable external
              ; interrupt IRQ#0
```

In the code above, the address value of an NVIC register is loaded into register R0 using the pseudo instruction LDR. The assembler will place the constant value into a location in the program code, and insert a memory read instruction to read the value into R0. The use of a pseudo instruction is needed because the value is too large to be encoded in a single move immediate instruction. When using LDR pseudo instructions to load a value into a register, the value requires an "=" prefix. In the normal case of loading an immediate data into a register (e.g., with MOV), the value should be prefixed by "#."

Similarly, the same code can be written in GNU toolchain assembler syntax:

```
.equ NVIC_IRQ_SETEN,  0xE000E100
.equ NVIC_IRQ0_ENABLE, 0x1
...
LDR R0,=NVIC_IRQ_SETEN /* Put 0xE000E100 into R0
     LDR here is a pseudo instruction that will be
     converted to a PC relative load by the assembler */
MOVS R1, #NVIC_IRQ0_ENABLE /* Put immediate data (0x1) into
                               register R1 */
STR R1, [R0] /* Store 0x1 to 0xE000E100, this enable
                external interrupt IRQ#0 */
```

Another typical feature of most assembly tools is allowing data to be inserted inside the program. For example, we can define data in a certain location in the program memory and access it with memory read instructions. In ARM assembler, an example is:

```
LDR R3,=MY_NUMBER ; Get the memory location of MY_NUMBER
LDR R4, [R3]      ; Read the value 0x12345678 into R4
...
LDR R0,=HELLO_TEXT ; Get the starting address of HELLO_TEXT
BL  PrintText      ; Call a function called PrintText to
                   ; display string
...
ALIGN 4
MY_NUMBER  DCD 0x12345678
HELLO_TEXT DCB "Hello\n", 0 ; Null terminated string
```

In the above example, "DCD" is used to insert a word-sized data item, and "DCB" is used to insert byte-size data into the program. When inserting word-size data in program, we should use the "ALIGN" directive before the data. The number after the ALIGN directive determines the alignment size. In this case, the value 4 forces the following data to be aligned to a word boundary. By ensuring the data placed at MY_NUMBER is word aligned, the program will be able to access the data with just a single bus transfer, and the code can be more portable (unaligned accesses are not supported in the Cortex®-M0/M0+/M1 processors).

Again, this example can be rewritten in GNU toolchain assembler syntax:

```
LDR R3,=MY_NUMBER /* Get the memory location of MY_NUMBER */
LDR R4, [R3]       /* Read the value 0x12345678 into R4 */
...
LDR R0,=HELLO_TEXT /* Get the starting address of
                      HELLO_TEXT */
BL  PrintText      /* Call a function called PrintText to
                      display string */
...
.align 4
MY_NUMBER:
.word 0x12345678
HELLO_TEXT:
.asciz "Hello\n"  /* Null terminated string */
```

A number of different directives are available in both ARM assembler and GNU assembler for inserting data into a program. Table 5.1 gives a few commonly used examples.

**Table 5.1** Commonly Used Directives for Inserting Data Into a Program

| Type of Data to Insert | ARM Assembler (e.g., Keil MDK-ARM) | GNU Assembler |
|---|---|---|
| Byte | DCB<br>E.g., DCB 0x12 | .byte<br>E.g., .byte 0x012 |
| Half-word | DCW<br>E.g., DCW 0x1234 | .hword / .2byte<br>E.g., .hword 0x01234 |
| Word | DCD<br>E.g., DCD 0x01234567 | .word / .4byte<br>E.g., .word 0x01234567 |
| Double-word | DCQ<br>E.g., DCQ<br>0x12345678FF0055AA | .quad/.octa<br>E.g., .quad<br>0x12345678FF0055AA |
| Floating point (single precision) | DCFS<br>E.g., DCFS 1E3 | .float<br>E.g., .float 1E3 |
| Floating point (double precision) | DCFD<br>E.g., DCFD 3.14159 | .double<br>E.g., .double 3f14159 |
| String | DCB<br>E.g., DCB "Hello\n" 0, | .ascii / .asciz (with NULL termination)<br>E.g., .ascii "Hello\n"<br>.byte 0 /* add NULL character */<br>E.g., .asciz "Hello\n" |
| Instruction | DCI<br>E.g., DCI 0xBE00 ;<br>Breakpoint (BKPT 0) | .word / .hword<br>E.g., .hword 0xBE00<br>/* Breakpoint (BKPT 0) */ |

In most cases, you can also add a label before the directive so that the addresses of the data can be determined using the label.

There are a number of other useful directives that are often used in assembly language programming. For example, some of the ARM assembler directives given in Table 5.2 are commonly used and some are used in the examples in this book.

Additional information about directives in ARM assembler can be found in the "ARM Compiler Toolchain Assembler Reference," (reference 6, section 6.3, Data, Data Definition Directives[1]).

| **Table 5.2** Commonly Used Directives | |
|---|---|
| **Directive (GNU assembler equivalent)** | **ARM Assembler** |
| THUMB (.thumb) | Specify assembly code as Thumb instruction in Unified Assembly Language (UAL) format. |
| CODE16 (.code 16) | Specify assembly code as Thumb instruction in legacy pre-UAL syntax. |
| AREA <section_name>{,<attr>} {,attr}... (.section <section_name>) | Instructs the assembler to assemble a new code or data section. Sections are independent, named, indivisible chunks of code or data that are manipulated by the linker. |
| SPACE <num of bytes> (.zero <num of bytes>) | Reserves a block of memory and fills it with zeros. |
| FILL <num of bytes>{, <value> {, <value_sizes>}} (.fill <num of bytes>{, <value> {, <value_sizes>}}) | Reserves a block of memory and fills it with the specified value. The size of the value can be byte, half-word, or word, specified by value_sizes (1/2/4). |
| ALIGN {<expr>{,<offset>{,<pad> {,<padsize>}}}} (.align <alignment>{,<fill>{,<max}}}) | Aligns the current location to a specified boundary by padding with zeros or NOP instructions. E.g., ALIGN 8 ; make sure the next instruction or      ; data is aligned to 8 byte boundary |
| EXPORT <symbol> (.global <symbol>) | Declare a symbol that can be used by the linker to resolve symbol references in separate object or library files. |
| IMPORT <symbol> | Declare a symbol reference in separate object or library files that is to be resolved by linker. |
| LTORG (.pool) | Instructs the assembler to assemble the current literal pool immediately. Literal pool contains data such as constant values for LDR pseudo instruction. |

---

[1]http://infocenter.arm.com/help/topic/com.arm.doc.dui0489c/Cacgadfj.html

## 5.4 Use of a suffix in instructions

In assembler for ARM® processors, some instructions can be followed by suffixes. For Cortex®-M processors, the available suffixes are shown in Table 5.3.

For the Cortex-M3/M4 processors, a data processing instruction can optionally update the APSR (flags). If using the Unified Assembly Language (UAL) syntax, we can specify if the APSR update should be carried out or not. For example, when moving a data from one register to another, it is possible to use

```
MOVS R0, R1   ; Move R1 into R0 and update APSR
```
Or
```
MOV  R0, R1 ; Move R1 into R0, and not update APSR
```

The second type of suffix is for conditional execution of instructions. The Cortex-M3 and Cortex-M4 processors support conditional branches, as well as conditional execution of instructions by putting the conditional instructions in an IF-THEN (IT) instruction block. By updating the APSR using data operations, or instructions like test (TST) or compare (CMP), the program flow can be controlled based on conditions of operation results.

**Table 5.3** Suffixes for Cortex-M Assembly Language

| Suffixes | Descriptions |
|---|---|
| S | Update APSR (Application Program Status Register, such as Carry, Overflow, Zero and Negative flags); for example:<br>ADDS R0, R1 ; this ADD operation will update APSR |
| EQ, NE, CS, CC, MI, PL, VS, VC, HI, LS, GE, LT, GT, LE | Conditional execution. EQ = Equal, NE = Not Equal, LT = Less Than, GT = Greater Than, etc. On the Cortex-M processors these conditions can be applied to conditional branches; for example:<br>BEQ label ; Branch to label if previous operation result in<br>; equal status<br>or conditionally executed instructions (see IF-THEN instruction in section 5.6.9); for example:<br>ADDEQ R0, R1, R2 ; Carry out the add operation if the previous<br>; operation results in equal status |
| .N, .W | Specify the use of 16-bit (narrow) instruction or 32-bit (wide) instruction. |
| .32, .F32 | Specify the operation is for 32-bit single-precision data. In most toolchains, the .32 suffix is optional. |
| .64, F64 | Specify the operation is for 64-bit double-precision data. In most toolchains, the .64 suffix is optional. |

## 5.5 **Unified assembly language (UAL)**

Several years ago, before Thumb®-2 technology was developed, the features available in the Thumb instruction set were limited, and the Thumb instruction syntax was more relaxed. For example, in ARM7TDMI™, almost all data processing instructions in Thumb mode will update the APSR anyway, so the "S" suffix is not strictly required for the Thumb instruction, and omitting it would still result in an instruction that updates the APSR.

When Thumb-2 technology arrived, almost all Thumb instructions were available in a version that updates APSR and a version that does not. As a result, traditional Thumb syntax can be problematic in Thumb-2 software development.

In order to allow better portability between architectures, and to use a single Assembly language syntax in ARM® processors with various architectures, recent ARM development tools have been updated to support the Unified Assembler Language (UAL). For users who have been using ARM7TDMI in the past, the most noticeable differences are:

- Some data operation instructions use three operands even when the destination register is the same as one of the source registers. In the past (pre-UAL), the syntax might only use two operands for these instructions.
- The "S" suffix becomes more explicit. In the past, when an assembly program file was assembled into Thumb code, most data operations are encoded as instructions that update the APSR. As a result, the "S" suffix was not essential. With the UAL syntax, instructions that update the APSR should have the "S" suffix to clearly indicate the expected operation. This prevents program code failing when being ported from one architecture to another.

For example, a pre-UAL ADD instruction for 16-bit Thumb code is

```
ADD R0, R1 ; R0 = R0 + R1, update APSR
```

In UAL syntax, this should be written as follows, being more specific about register usage and APSR update operations:

```
ADDS R0, R0, R1 ; R0 = R0 + R1, update APSR
```

However, in most cases (depending on the toolchain being used), you can still write the instruction with a pre-UAL style (only two operands), but the use of "S" suffix will be more explicit:

```
ADDS R0, R1 ; R0 = R0 + R1, update APSR
```

The pre-UAL syntax is currently still accepted by most development tools, including the Keil™ Microcontroller Development Kit for ARM (MDK-ARM) and the ARM Compiler toolchain. However, using UAL is recommended in new projects. For assembly development with Keil MDK, you can specify the use of UAL syntax with the "THUMB" directive, and pre-UAL syntax with the "CODE16" directive.

The choice of assembler syntax depends on which tool you use. Please refer to the documentation of your development suite to determine which syntax is suitable.

One thing you need to be careful about when reusing code with traditional Thumb is that some instructions change the flags in APSR, even if the S suffix is not used. However, if you copy and paste the same instruction to a project using UAL syntax, the instruction becomes one that does not change the flags in APSR. For example:

```
CODE16
...
AND R0, R1 ; R0=R0 AND R1, update APSR (Traditional Thumb syntax)
```

If this line of code is used in a project using UAL, the result will become R0=R0 AND R1 with no APSR update.

With the new instructions in Thumb-2 technology, some of the operations can be handled by either a Thumb instruction or a Thumb-2 instruction. For example, R0 = R0 + 1 can be implemented as a 16-bit Thumb instruction or a 32-bit Thumb-2 instruction. With UAL, you can specify which instruction you want by adding suffixes:

```
ADDS R0, #1   ; Use 16-bit Thumb instruction by default
              ; for smaller size
ADDS.N R0, #1 ; Use 16-bit Thumb instruction (N=Narrow)
ADDS.W R0, #1 ; Use 32-bit Thumb-2 instruction (W=wide)
```

The .W (wide) suffix specifies a 32-bit instruction. If no suffix is given, the assembler tool can choose either instruction but usually defaults to the smaller option to get the best code density. Depending on tool support, you may also use the .N (narrow) suffix to specify a 16-bit Thumb instruction.

Again, this syntax is for ARM assembler tools. Other assemblers might have slightly different syntax. If no suffix is given, the assembler might choose the instruction for you which gives the minimum code size.

In most cases, applications will be coded in C, and the C compilers will use 16-bit instructions if possible due to their smaller code size. However, when the immediate data exceeds a certain range, or when the operation can be better handled with a 32-bit Thumb-2 instruction, the 32-bit instruction will be used. When the compilation is optimized for speed, the C compiler might also use 32-bit instructions to adjust the branch target addresses to 32-bit aligned for better performance.

32-bit Thumb-2 instructions can be half-word aligned. For example, you can have a 32-bit instruction located in a half-word location (unaligned) (Figure 5.4):

```
0x1000 : LDR r0,[r1] ;a 16-bit instructions (occupy 0x1000-0x1001)
0x1002 : RBIT.W r0   ;a 32-bit Thumb-2 instruction (occupy
                     ; 0x1002-0x1005)
```

Most 16-bit instructions can only access registers R0 to R7; 32-bit Thumb-2 instructions do not have this limitation. However, use of PC (R15) might not be allowed in some of the instructions. Refer to the ARM v7-M Architecture Reference Manual (reference 1) or Cortex®-M3/M4 Devices Generic User Guides (section 3.3.2, references 2 and 3) if you need to find out more detail in this area.

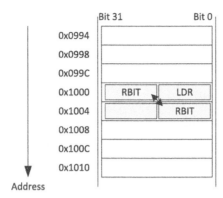

**FIGURE 5.4**

An unaligned 32-bit instruction

## 5.6 Instruction set

The instructions in the Cortex®-M3 and Cortex-M4 processors can be divided into various groups based on functionality:

- Moving data within the processor
- Memory accesses
- Arithmetic operations
- Logic operations
- Shift and Rotate operations
- Conversion (extend and reverse ordering) operations
- Bit field processing instructions
- Program flow control (branch, conditional branch, conditional execution, and function calls)
- Multiply accumulate (MAC) instructions
- Divide instructions
- Memory barrier instructions
- Exception-related instructions
- Sleep mode-related instructions
- Other functions

In addition, the Cortex-M4 processor supports the Enhanced DSP instructions:

- SIMD operations and packing instructions
- Adding fast multiply and MAC instructions
- Saturation algorithms
- Floating point instructions (if the floating point unit is present)

Details of each instruction are covered in the Cortex-M3/Cortex-M4 Devices Generic User Guides (reference 2 & 3, available on the ARM® website). In the rest of this section we will look into some of the basic concepts of assembly language programming.

To make it easier for beginners, in this part we will skip the conditional suffix for now. Most of the instructions can be executed conditionally when used together with the IF-THEN (IT) instruction, which will require the suffix to indicate the condition.

### 5.6.1 Moving data within the processor

The most basic operation in a microprocessor is to move data around inside the processor. For example, you might want to:

- Move data from one register to another
- Move data between a register and a special register (e.g., CONTROL, PRIMASK, FAULTMASK, BASEPRI)
- Move an immediate constant into a register

For the Cortex®-M4 processor with the floating point unit, you can also:

- Move data between a register in the core register bank and a register in the floating point unit register bank
- Move data between registers in the floating point register bank
- Move data between a floating point system register (such as the FPSCR — Floating point Status and Control Register) and a core register
- Move immediate data into a floating point register

Table 5.4 shows some examples of these operations.

The instructions in Table 5.5 are available for Cortex-M4 with floating point unit only.

**Table 5.4** Instructions for Transferring Data within the Processor

| Instruction | Dest | Source | Operations |
|---|---|---|---|
| MOV | R4, | R0 | ; Copy value from R0 to R4 |
| MOVS | R4, | R0 | ; Copy value from R0 to R4 with APSR (flags) update |
| MRS | R7, | PRIMASK | ; Copy value of PRIMASK (special register) to R7 |
| MSR | CONTROL, | R2 | ; Copy value of R2 into CONTROL (special register) |
| MOV | R3, | #0x34 | ; Set R3 value to 0x34 |
| MOVS | R3, | #0x34 | ; Set R3 value to 0x34 with APSR update |
| MOVW | R6, | #0x1234 | ; Set R6 to a 16-bit constant 0x1234 |
| MOVT | R6, | #0x8765 | ; Set the upper 16-bit of R6 to 0x8765 |
| MVN | R3, | R7 | ; Move NOT value of R7 into R3 |

**Table 5.5** Instructions for Transferring Data between the Floating Point Unit and Core Registers

| Instruction | Dest | Source | Operations |
|---|---|---|---|
| VMOV | R0, | S0 | ; Copy floating point register S0 to general purpose register R0 |
| VMOV | S0, | R0 | ; Copy general purpose register R0 to floating point register S0 |
| VMOV | S0, | S1 | ; Copy floating point register S1 to S0 (single precision) |
| VMRS.F32 | R0, | FPSCR | ; Copy value in FPSCR, a floating point unit system register to R0 |
| VMRS | APSR_nzcv, | FPSCR | ; Copy flags from FPSCR to the flags in APSR |
| VMSR | FPSCR, | R3 | ; Copy R3 to FPSCR, a floating point unit system register |
| VMOV.F32 | S0, | #1.0 | ; Move single-precision value into floating point register S0 |

The MOVS instruction is similar to the MOV instruction, apart from the fact that it updates the flags in the APSR, hence the "S" suffix is used. For setting a register in the general purpose register bank to an 8-bit immediate value, the MOVS instruction is sufficient and can be carried out with a 16-bit Thumb instruction if the destination register is a low register (R0 to R7). For moving an immediate value into a high register, or if the APSR must not be updated, the 32-bit version of the MOV/MOVS instructions would be used.

To set a register to a larger immediate value (between 9-bit and 16-bit), the MOVW instruction can be used. Depending on the assembler tool you are using, it might automatically convert a MOV or MOVS instruction into MOVW if the immediate data is between 9-bit and 16-bit.

If you need to set a register to a 32-bit immediate data value, there are several ways of doing this.

The most common method is to use a pseudo instruction called "LDR"; for example:

```
LDR R0, =0x12345678 ; Set R0 to 0x12345678
```

This is not a real instruction. The assembler converts this instruction into a memory transfer instruction and a literal data item stored in the program image:

```
LDR R0, [PC, #offset]
....
DCD 0x12345678
```

The LDR instruction reads the memory at [PC+offset] and stores the value into R0. Note that due to the pipeline nature of the processor, the value of PC is not exactly the address of the LDR instruction. However, the assembler will calculate the offset for you so you don't have to worry about it.

---

**LITERAL POOL**

---

Usually the assembler groups various literal data (e.g., DCD 0x12345678 in the above example) together into data blocks called literal pools. Since the value of the offset in the LDR instruction is limited, a program will often need a number of literal pools so that the LDR instruction can access the literal data. Therefore we need to insert assembler directives like LTORG (or .pool) to tell the assembler where it can insert literal pools. Otherwise the assembler will try to put all the literal data after the end of the program code, which might be too far away for the LDR instruction to access it.

---

If the operation needs to set the register to an address in the program code within a certain address range, you can use the ADR pseudo instruction, which will be converted into a single instruction, or ADRL pseudo instruction, which can provide a wider address range but needs two instructions to implement. For example:

```
ADR R0, DataTable
...
ALIGN
DataTable
DCD 0, 245, 132, ...
```

The ADR instruction will be converted into an "add" or "subtract" operation based on the program counter value.

Another way to generate a 32-bit immediate data value is to use a combination of MOVW and MOVT instructions. For example:

```
MOVW R0, #0x789A ; Set R0 to 0x0000789A
MOVT R0, #0x3456 ; Set upper 16-bit of R0 to 0x3456,
                 ; now R0 = 0x3456789A
```

When comparing this method to using the LDR pseudo instruction, the LDR method gives better readability, and the assembler might be able to reduce code size by reusing the same literal data, if the same constant value is used in several places of the assembly code. However, depending on the memory system design, in some cases the MOVW + MOWT method can result in faster code if a system-level cache is used and if the LDR resulted in a data cache miss.

## 5.6.2 Memory access instructions

There are large numbers of memory access instructions in the Cortex®-M3 and Cortex-M4 processors. This is due to the combination of support of various addressing modes, as well as data size and data transfer direction. For normal data transfers, the instructions available are given in Table 5.6

**Table 5.6** Memory Access Instructions for Various Data Sizes

| Data Type | Load (Read from Memory) | Store (Write to Memory) |
|---|---|---|
| 8-bit unsigned | LDRB | STRB |
| 8-bit signed | LDRSB | STRB |
| 16-bit unsigned | LDRH | STRH |
| 16-bit signed | LDRSH | STRH |
| 32-bit | LDR | STR |
| Multiple 32-bit | LDM | STM |
| Double-word (64-bit) | LDRD | STRD |
| Stack operations (32-bit) | POP | PUSH |

Note: The LDRSB and the LDRSH automatically perform a sign extend operation on the loaded data to convert it to a signed 32-bit value. For example, if 0x83 is read in a LDRB instruction, the value is converted into 0xFFFFFF83 before being placed in the destination register.

If the floating point unit is present, the instructions in Table 5.7 are also available to transfer data between the register bank in the floating point unit and memory.

There are also a number of addressing modes available. In some of these modes, you can also optionally update the register holding the address (write back).

### *Immediate offset (pre-index)*

The memory address of the data transfer is the sum of a register value and an immediate constant value (offset). Sometimes this is referred to as "pre-index" addressing. For example:

```
LDRB R0, [R1, #0x3] ; Read a byte value from address R1+0x3, and
store the read data in R0.
```

The offset value can be positive or negative. Table 5.8 shows a list of commonly used load and store instructions.

**Table 5.7** Memory Access Instructions for the Floating Point Unit

| Data Type | Read from Memory (Load) | Write to Memory (Store) |
|---|---|---|
| Single-precision data (32-bit) | VLDR.32 | VSTR.32 |
| Double-precision data (64-bit) | VLDR.64 | VSTR.64 |
| Multiple data | VLDM | VSTM |
| Stack operations | VPOP | VPUSH |

**Table 5.8** Memory Access Instructions with Immediate Offset

| Example of Pre-index Accesses<br>Note: the #offset field is optional | Description |
| --- | --- |
| `LDRB  Rd, [Rn, #offset]` | Read byte from memory location Rn + offset |
| `LDRSB Rd, [Rn, #offset]` | Read and signed extend byte from memory location Rn + offset |
| `LDRH  Rd, [Rn, #offset]` | Read half-word from memory location Rn + offset |
| `LDRSH Rd, [Rn, #offset]` | Read and signed extended half-word from memory location Rn + offset |
| `LDR   Rd, [Rn, #offset]` | Read word from memory location Rn + offset |
| `LDRD  Rd1,Rd2, [Rn, #offset]` | Read double-word from memory location Rn + offset |
| `STRB  Rd, [Rn, #offset]` | Store byte to memory location Rn + offset |
| `STRH  Rd, [Rn, #offset]` | Store half-word to memory location Rn + offset |
| `STR   Rd, [Rn, #offset]` | Store word to memory location Rn + offset |
| `STRD  Rd1,Rd2, [Rn, #offset]` | Store double-word to memory location Rn + offset |

This addressing mode supports write back of the register holding the address. For example:

```
LDR R0, [R1, #0x8]! ; After the access to memory[R1+0x8], R1 is updated to R1+0x8
```

The exclamation mark (!) in the instruction specifies whether the register holding the address should be updated (write back) when the instruction is completed. The address used for the data transfer uses the sum of R1+0x8 calculated regardless of whether the exclamation mark (!) is stated. The write back operation can be used with a number of load and store instructions as shown in Table 5.9.

Please note that some of these instructions cannot be used with R15(PC) or R14(SP). In addition, the 16-bit versions of these instructions only support low registers (R0-R7) and do not provide write back.

If the floating point unit is present, the instructions in Table 5.10 are also available to perform LDM and STM operations to the registers in the floating point unit.

Note that many floating point instructions use the .32 and .64 suffixes to specify the floating data type. In most toolchains, the .32 and .64 suffixes are optional.

### PC-related addressing (Literal)

A memory access can generate the address value from the current PC value and an offset value (Table 5.11). This is commonly needed for loading immediate values into a register, also known as literal pool accesses, as mentioned earlier in this chapter (LDR pseudo instruction).

If the floating point unit is present, the instructions in Table 5.12 are also available.

**Table 5.9** Memory Access Instructions with Immediate Offset and Write Back

| Example of Pre-index with Write Back<br>Note: the #offset field is optional | Description |
| --- | --- |
| LDRB Rd, [Rn, #offset]! | Read byte with write back |
| LDRSB Rd, [Rn, #offset]! | Read and signed extend byte with write back |
| LDRH Rd, [Rn, #offset]! | Read half-word with write back |
| LDRSH Rd, [Rn, #offset]! | Read and signed extended half-word with write back |
| LDR Rd, [Rn, #offset]! | Read word with write back |
| LDRD Rd1,Rd2, [Rn, #offset]! | Read double-word with write back |
| STRB Rd, [Rn, #offset]! | Store byte to memory with write back |
| STRH Rd, [Rn, #offset]! | Store half-word to memory with write back |
| STR Rd, [Rn, #offset]! | Store word to memory with write back |
| STRD Rd1,Rd2, [Rn, #offset]! | Store double-word to memory with write back |

**Table 5.10** Memory Access Instructions for Floating Point Unit

| Examples<br>Note: the #offset field is optional | Description |
| --- | --- |
| VLDR.32 Sd, [Rn, #offset] | Read single-precision data from memory to single-precision register $Sd$ |
| VLDR.64 Dd, [Rn, #offset] | Read double-precision data from memory to double-precision register $Dd$ |
| VSTR.32 Sd, [Rn, #offset] | Write single-precision data from single-precision register $Sd$ to memory |
| VSTR.64 Dd, [Rn, #offset] | Write double-precision data from double precision register $Dd$ to memory |

**Table 5.11** Memory Access Instructions with PC Related Addressing

| Example of Literal Read | Description |
| --- | --- |
| LDRB Rt,[PC, #offset] | Load unsigned byte into Rt using PC offset |
| LDRSB Rt,[PC, #offset] | Load and signed extend a byte data into Rt using PC offset |
| LDRH Rt,[PC, #offset] | Load unsigned half-word into Rt using PC offset |
| LDRSH Rt,[PC, #offset] | Load and signed extend a half-word data into Rt using PC offset |
| LDR Rt, [PC, #offset] | Load a word data into Rt using PC offset |
| LDRD Rt,Rt2,[PC, #offset] | Load a double-word into Rt and Rt2 using PC offset |

**Table 5.12** Floating Point Unit Memory Access Instructions with PC-related Addressing

| Example of Literal Read | Description |
|---|---|
| VLDR.32 Sd,[PC, #offset] | Load single-precision data into single-precision register Sd using PC offset |
| VLDR.64 Dd,[PC, #offset] | Load double-precision data into double-precision register Dd using PC offset |

### Register offset (pre-index)

Another useful address mode is the register offset. This is often used in the processing of data arrays where the address is a combination of a base address and an offset calculated from an index value. To make this address calculation even more efficient, the index value can be shifted by a distance of 0 to 3 bits before being added to the base register. For example:

```
LDR R3, [R0, R2, LSL #2] ; Read memory[R0+(R2 << 2)] into R3
```

The shift operation is optional. You can have a simple operation like

```
STR R5, [R0,R7] ; Write R5 into memory[R0+R7]
```

Similarly to immediate offset, there are various forms for different data size, as shown in Table 5.13.

### Post-index

Memory access instructions with post-index addressing mode also have an immediate offset value. However, the offset is not used during the memory access, but is

**Table 5.13** Memory Access Instructions with Register Offset

| Example of Register Offset Accesses | Description |
|---|---|
| LDRB Rd, [Rn, Rm{, LSL #n}] | Read byte from memory location Rn + (Rm << n) |
| LDRSB Rd, [Rn, Rm{, LSL #n}] | Read and signed extend byte from memory location Rn + (Rm << n) |
| LDRH Rd, [Rn, Rm{, LSL #n}] | Read half-word from memory location Rn + (Rm << n) |
| LDRSH Rd, [Rn, Rm{, LSL #n}] | Read and signed extended half-word from memory location Rn + (Rm << n) |
| LDR Rd, [Rn, Rm{, LSL #n}] | Read word from memory location Rn + (Rm << n) |
| STRB Rd, [Rn, Rm{, LSL #n}] | Store byte to memory location Rn + (Rm << n) |
| STRH Rd, [Rn, Rm{, LSL #n}] | Store half-word to memory location Rn + (Rm << n) |
| STR Rd, [Rn, Rm{, LSL #n}] | Store word to memory location Rn + (Rm << n) |

used to update the address register after the data transfer is completed. For example:

```
LDR R0, [R1], #offset ; Read memory[R1], then R1 updated to R1+offset
```

When the post-index memory addressing mode is used, there is no need to use the exclamation mark (!) sign because the base address register is always updated if the data transfer is completed successfully. Table 5.14 lists various form of post indexing memory access instructions.

The post-index address mode can be very useful for processing data in an array. As soon as an element in the array is accessed, the address register can be adjusted to the next element automatically to save code size and execution time.

Please note that post-index instructions cannot be used with R15(PC) o R14(SP). The post-index memory access instructions are 32-bit. The offset value can be positive or negative.

### Multiple load and multiple store

One of the key advantages of the ARM architecture is that it allows you to read or write multiple data that are contiguous in memory. The LDM (Load Multiple

**Table 5.14** Memory Access Instructions with Post-Indexing

| Example of Post Index Accesses | Description |
|---|---|
| LDRB Rd,[Rn], #offset | Read byte from memory[Rn] to Rd, then update Rn to Rn+offset |
| LDRSB Rd,[Rn], #offset | Read and signed extended byte from memory[Rn] to Rd, then update Rn to Rn+offset |
| LDRH Rd,[Rn], #offset | Read half-word from memory[Rn] to Rd, then update Rn to Rn+offset |
| LDRSH Rd,[Rn], #offset | Read and signed extended half-word from memory[Rn] to Rd, then update Rn to Rn+offset |
| LDR Rd,[Rn], #offset | Read word from memory[Rn] to Rd, then update Rn to Rn+offset |
| LDRD Rd1,Rd2,[Rn], #offset | Read double-word from memory[Rn] to Rd1, Rd2, then update Rn to Rn+offset |
| STRB Rd,[Rn], #offset | Store byte to memory[Rn] then update Rn to Rn+offset |
| STRH Rd,[Rn], #offset | Store half-word to memory[Rn] then update Rn to Rn+offset |
| STR Rd,[Rn], #offset | Store word to memory[Rn] then update Rn to Rn+offset |
| STRD Rd1,Rd2,[Rn], #offset | Store double-word to memory[Rn] then update Rn to Rn+offset |

registers) and STM (Store Multiple registers) instructions only support 32-bit data. They support two types of pre-indexing:

- IA: Increment address After each read/write
- DB: Decrement address Before each read/write

The LDM and STM instructions can be used without base address write back (Table 5.15).

The <reg list> in Table 5.15 is the register list. It contains at least one register, and:

- Start with "{" and end with "}"
- Use "-" (hypen) to indicate range. For example, R0-R4 means R0, R1, R2, R3 and R4.
- Use "," (comma) to separate each register

For example, the following instructions read address 0x20000000 to 0x2000000F (four words) into R0 to R3:

```
LDR    R4,=0x20000000 ; Set R4 to 0x20000000 (address)
LDMIA R4, {R0-R3}    ; Read 4 words and store them to R0 - R3
```

The register list can be non-contiguous such as {R1, R3, R5-R7, R9, R11-12}, which contains R1, R3, R5, R6, R7, R8, R11, R12.

Similar to other load/store instructions, you can use write back with STM and LDM. For example:

```
LDR    R8,=0x8000   ; Set R8 to 0x8000 (address)
STMIA R8!, {R0-R3} ; R8 change to 0x8010 after the store
```

Instructions with multiple Load/Store memory access instructions with write back are listed in Table 5.16. The 16-bit versions of the LDM and STM instructions are limited to low registers only and always have write back enabled, except when the base register is one of the destination registers to be updated by the memory read.

If the floating point unit is present, the instructions in Table 5.17 are also available to perform load multiple and store multiple operations to the registers in the floating point unit.

**Table 5.15** Multiple Load/Store Memory Access Instructions

| Examples of Multiple Load/Store | Description |
|---|---|
| LDMIA Rn,<reg list> | Read multiple words from memory location specified by *Rn*. Address Increment After (IA) each read. |
| LDMDB Rn,<reg list> | Read multiple words from memory location specified by *Rn*. Address Decrement Before (DB) each read. |
| STMIA Rn,<reg list> | Write multiple words to memory location specified by *Rn*. Address increment after each write. |
| STMDB Rn,<reg list> | Write multiple words to memory location specified by *Rn*. Address Decrement Before each write. |

**Table 5.16** Multiple Load/Store Memory Access Instructions with Write Back

| Example of Multiple Load / Store with Write Back | Description |
|---|---|
| LDMIA Rn!,<reg list> | Read multiple words from memory location specified by *Rd*. Address Increment After (IA) each read. Rn writes back after the transfer is done. |
| LDMDB Rn!,<reg list> | Read multiple words from memory location specified by *Rd*. Address Decrement Before (DB) each read. Rn writes back after the transfer is done. |
| STMIA Rn!,<reg list> | Write multiple words to memory location specified by *Rd*. Address increment after each write. Rn writes back after the transfer is done. |
| STMDB Rn!,<reg list> | Write multiple words to memory location specified by *Rd*. Address Decrement Before each write Rn writes back after the transfer is done. |

**Table 5.17** Multiple Load/Store Memory Access Instructions for Floating Point Unit with Write Back

| Example of Stack Operations | Description |
|---|---|
| VLDMIA.32 Rn, <s_reg list> | Read multiple single-precision data. Address Increment After (IA) each read. |
| VLDMDB.32 Rn, <s_reg list> | Read multiple single-precision data. Address Decrement Before (DB) each read. |
| VLDMIA.64 Rn, <d_reg list> | Read multiple double-precision data. Address Increment After (IA) each read. |
| VLDMDB.64 Rn, <d_reg list> | Read multiple double-precision data. Address Decrement Before (DB) each read. |
| VSTMIA.32 Rn, <s_reg list> | Write multiple single-precision data. Address increment after each write. |
| VSTMDB.32 Rn, <s_reg list> | Write multiple single-precision data. Address decrement before each write. |
| VSTMIA.64 Rn, <d_reg list> | Write multiple double-precision data. Address increment after each write. |
| VSTMDB.64 Rn, <d_reg list> | Write multiple double-precision data. Address decrement before each write. |
| VLDMIA.32 Rn!, <s_reg list> | Read multiple single-precision data. Address Increment After (IA) each read. *Rn* writes back after the transfer is done. |
| VLDMDB.32 Rn!, <s_reg list> | Read multiple single-precision data. Address Decrement Before (DB) each read. *Rn* writes back after the transfer is done. |
| VLDMIA.64 Rn!, <d_reg list> | Read multiple double-precision data. Address Increment After (IA) each read. *Rn* writes back after the transfer is done. |

*(Continued)*

**Table 5.17** Multiple Load/Store Memory Access Instructions for Floating Point Unit with Write Back—*Cont'd*

| Example of Stack Operations | Description |
|---|---|
| VLDMDB.64 Rn!, <d_reg list> | Read multiple double-precision data. Address Decrement Before (DB) each read. *Rn* writes back after the transfer is done. |
| VSTMIA.32 Rn!, <s_reg list> | Write multiple single-precision data. Address increment after each write. *Rn* writes back after the transfer is done. |
| VSTMDB.32 Rn!, <s_reg list> | Write multiple single-precision data. Address decrement before each write. *Rn* writes back after the transfer is done. |
| VSTMIA.64 Rn!, <d_reg list> | Write multiple double-precision data. Address increment after each write. *Rn* writes back after the transfer is done. |
| VSTMDB.64 Rn!, <d_reg list> | Write multiple double-precision data. Address decrement before each write. *Rn* writes back after the transfer is done. |

**Table 5.18** Stack Push and Stack POP Instructions for Core Registers

| Example of Stack Operations | Description |
|---|---|
| PUSH <reg list> | Store register(s) in stack. |
| POP <reg list> | Restore register(s) from stack. |

### Stack push and pop

Stack push and pop are another form of the store multiple and load multiple. They use the currently selected stack pointer for address generation. The currently selected stack pointer can either be the Main Stack Pointer (MSP), or the Process Stack Pointer (PSP), depending on the current mode of the processor and the value in the CONTROL special register (see Chapter 4). Instructions for stack push and stack pop are shown in Table 5.18.

The register list syntax is the same as LDM and STM. For example:

```
PUSH {R0, R4-R7, R9} ; PUSH R0, R4, R5, R6, R7, R9 into stack
POP  {R2, R3}        ; POP R2 and R3 from stack
```

Usually a PUSH instruction will have a corresponding POP with the same register list, but this is not always necessary. For example, a common exception is when POP is used as a function return:

```
PUSH {R4—R6, LR} ; Save R4 to R6 and LR (Link Register) at the
                 ; beginning of a subroutine. LR contains the
                 ; return address
...              ; processing in the subroutine
```

**Table 5.19** Stack Push and Stack POP Instructions for Floating Point Unit Registers

| Example of Stack Operations | Description |
|---|---|
| VPUSH.32 <s_reg list> | Store single-precision register(s) in stack. (i.e., s0-s31) |
| VPUSH.64 <d_reg list> | Store double-precision register(s) in stack. (i.e., d0-d15) |
| VPOP.32 <s_reg list> | Restore single-precision register(s) from stack. |
| VPOP.64 <d_reg list> | Restore double-precision register(s) from stack. |

```
POP {R4-R6, PC} ; POP R4 to R6, and return address from stack.
                ; the return address is stores into PC directly,
                ; this triggers a branch (subroutine return)
```

Instead of popping the return address into LR, and then writing it to the program counter (PC), we can write the return address directly to PC to save instruction count and cycle count.

The 16-bit versions of PUSH and POP are limited to low registers (R0 to R7), LR (for PUSH), and PC (for POP). Therefore if a high register is modified in a function and the contents of the register need to be saved, you need to use a pair of 32-bit PUSH and POP instructions.

If the floating point unit is present, the instructions in Table 5.19 are also available to perform stack operations to the registers in the floating point unit.

Unlike PUSH and POP, VPUSH and VPOP instructions require that:

- The registers in the register list are consecutive
- The maximum number of registers stacked/unstacked for each VPUSH or VPOP is 16

If it is necessary to save more than 16 single-precision floating point registers, you can use double-precision instruction, or use two pairs of VPUSH and VPOP.

### SP-relative addressing

Besides being used for the temporary storage of registers in functions or subroutines, the stack memory is very often also used for local variables, and accessing these variables requires SP-relative addressing. There is no special 32-bit version of SP-relative addressing as this is already covered by the load and store instructions with immediate offset. However, most 16-bit Thumb instructions can only use low registers. As a result, there is a pair of dedicated 16-bit version of LDR and STR instructions with SP-relative addressing.

An example of using SP-relative addressing mode (Figure 5.5) can be: at the beginning of a function the SP value can be decremented to reserve space for local variables and then the local variables can be accessed using SP-related addressing. At the end of the function, the SP is incremented to return to the original value, which frees the allocated stack space before returning to the calling code.

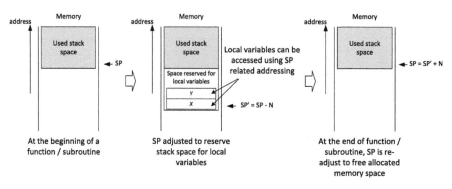

**FIGURE 5.5**

Local variable space allocation and accesses in stack

### *Load and store with unprivileged access level*

There is a set of load and store instructions to allow program code executing in privileged access level to access memory with unprivileged access rights, as shown in Table 5.20.

These instructions might be needed in some OS environments where an unprivileged application can access an API function (running within the privileged access level) with a data pointer as an input parameter, and this API operates on memory data specified by the pointer. If the data access is carried out using normal load and store instructions, the unprivileged application task will then have the ability to modify data that is used by other tasks or OS kernel using this API. By coding the API using these special Load and Store instructions with unprivileged access level, the API can only access the data which the application task can access.

**Table 5.20** Memory Access Instructions with Unprivileged Access Level

| Example of LDR/STR with Unprivileged Access Level | Description<br>Note: the #offset field is optional |
| --- | --- |
| `LDRBT  Rd, [Rn, #offset]` | Read byte from memory location Rn + offset |
| `LDRSBT Rd, [Rn, #offset]` | Read and signed extend byte from memory location Rn + offset |
| `LDRHT  Rd, [Rn, #offset]` | Read half word from memory location Rn + offset |
| `LDRSHT Rd, [Rn, #offset]` | Read and signed extended half word from memory location Rn + offset |
| `LDRT   Rd, [Rn, #offset]` | Read word from memory location Rn + offset |
| `STRBT  Rd, [Rn, #offset]` | Store byte to memory location Rn + offset |
| `STRHT  Rd, [Rn, #offset]` | Store half-word to memory location Rn + offset |
| `STRT   Rd, [Rn, #offset]` | Store word to memory location Rn + offset |

### Exclusive accesses

The exclusive access instructions are a special group of memory access instructions for implementing semaphores or MUTEX (Mutual Exclusive) operations. They are normally used within embedded OS where a resource (often hardware, but can also be software) has to be shared between multiple application tasks, or even multiple processors.

Exclusive access instructions include exclusive loads and exclusive stores. Special hardware inside the processor and optionally in the bus interconnect are needed to monitor exclusive accesses. Inside the processor, a single bit register is present to record an on-going exclusive access sequence: we call it the local exclusive access monitor. On the system bus level, an exclusive access monitor might also be present to check if a memory location (or memory device) used by an exclusive access sequence has been accessed by another processor or bus master. The processor has extra signals in the bus interface to indicate that a transfer is an exclusive access and to receive a response from the system bus level exclusive access monitor.

In a semaphore or a MUTEX operation, a data variable in RAM is used to represent a token. It can be used to indicate, for example, that a hardware resource has been allocated to an application task. For example, assume that if the variable is 0, it indicates the resource is available, and 1 indicates that it is already allocated to a task. The exclusive access sequence for requesting the resource might be:

1. The variable is accessed with an exclusive load (read). The local exclusive access monitor inside the processor is updated to indicate an active exclusive access transfer and, if a bus level exclusive access monitor is present, it will also be updated.
2. The variable is checked by the application code to determine whether the hardware resource has already been allocated. If the value is 1 (already allocated), then it can retry later or give up. If the value is 0 (resource free), then it can try to allocate the resource in the next step.
3. The task uses an exclusive store to write a value of 1 to the variable. If the local exclusive access monitor is set and there is no error reported by the bus level exclusive access monitor, the variable will be updated and the exclusive store will get a success return status. If something happened between the exclusive load and exclusive store that could affect the exclusiveness of the access to the variable, the exclusive store will get a failed return status and the variable will not be updated (either cancelled by the processor itself or the store is blocked by the bus level exclusive access monitor).
4. From the return status, the application task knows that if it has allocated the hardware resource successfully. If not, it can retry later or give up.

The exclusive store fails if:

- The bus level exclusive access monitor returns an exclusive fail response (e.g., the memory location or memory range has been accessed by another processor)

**Table 5.21** Exclusive Access Instructions

| Example of Exclusive Access | Description |
|---|---|
| LDREXB Rt, [Rn] | Exclusive read byte from memory location Rn |
| LDREXH Rt, [Rn] | Exclusive read half-word from memory location Rn |
| LDREX  Rt, [Rn, #offset] | Exclusive read word from memory location Rn + offset |
| STREXB Rd, Rt, [Rn] | Exclusive store byte in Rt to memory location Rn. Return status in Rd. |
| STREXH Rd, Rt, [Rn] | Exclusive store half word in Rt to memory location Rn. Return status in Rd. |
| STREX  Rd, Rt, [Rn, #offset] | Exclusive store word in Rt from to location Rn + offset. Return status in Rd. |
| CLREX | Force the local exclusive access monitor to clear so that next exclusive store must fail. This is not a memory access instruction, but is listed here due to its usage. |

- The local exclusive access monitor is not set. This can be caused by:
  - **a)** Incorrect exclusive access sequence
  - **b)** An interrupt entry/exit between the exclusive load and exclusive store (the memory location or memory range could have been accessed by an interrupt handler or another application task).
  - **c)** Execution of a special instruction CLREX that clears the local exclusive access monitor.

The instructions for exclusive accesses are given in Table 5.21.

## 5.6.3 Arithmetic operations

The Cortex®-M3 and Cortex-M4 processors provide many different instructions for arithmetic operations. A few basic ones are introduced here. Many data processing instructions can have multiple instruction formats. For example, an ADD instruction can operate between two registers or between one register and an immediate data value:

```
ADD   R0, R0,   R1  ; R0 = R0 + R1
ADDS  R0, R0, #0x12 ; R0 = R0 + 0x12 with APSR (flags) update
ADC   R0, R1,   R2  ; R0 = R1 + R2 + carry
```

These are all ADD instructions, but they have different syntaxes and binary coding.

With the traditional Thumb instruction syntax (pre-UAL), when 16-bit Thumb code is used, an ADD instruction can change the flags in the PSR. However, the 32-bit Thumb-2 instruction can either change the flags or leave them unchanged.

To distinguish between the two different operations, in Unified Assembly Language (UAL) syntax, the S suffix should be used if the following operation depends on the flags:

```
ADD  R0, R1, R2 ; Flag unchanged
ADDS R0, R1, R2 ; Flag change
```

Aside from ADD instructions, the arithmetic functions that the Cortex-M3 supports include SUB (subtract), MUL (multiply), and UDIV/SDIV (unsigned and signed divide). Table 5.22 shows some of the most commonly used arithmetic instructions.

These instructions can be used with or without the "S" suffix to specify whether the APSR should be updated.

By default, if a divide by zero takes place, the result of the UDIV and SDIV instructions will be zero. You can set up the DIVBYZERO bit in the NVIC Configuration Control Register so that when a divide by zero occurs, a fault exception (usage fault) takes place.

Both the Cortex-M3 and Cortex-M4 processors support 32-bit multiply instructions and multiply accumulate (MAC) instructions that give 32-bit and 64-bit results.

**Table 5.22** Instructions for Arithmetic Data Operations

| Commonly Used Arithmetic Instructions (optional suffixes not shown) | Operation |
|---|---|
| `ADD Rd, Rn, Rm       ; Rd = Rn + Rm` | ADD operation |
| `ADD Rd, Rn, #immed ; Rd = Rn + #immed` | |
| `ADC Rd, Rn, Rm       ; Rd = Rn + Rm + carry` | ADD with carry |
| `ADC Rd, #immed       ; Rd = Rd + #immed + carry` | |
| `ADDW Rd, Rn,#immed ; Rd = Rn + #immed` | ADD register with 12-bit immediate value |
| `SUB Rd, Rn, Rm       ; Rd = Rn - Rm` | SUBTRACT |
| `SUB Rd, #immed       ; Rd = Rd - #immed` | |
| `SUB Rd, Rn,#immed   ; Rd = Rn - #immed` | |
| `SBC Rd, Rn, #immed   ; Rd = Rn - #immed - borrow` | SUBTRACT with borrow (not carry) |
| `SBC Rd, Rn, Rm       ; Rd = Rn - Rm - borrow` | |
| `SUBW Rd, Rn,#immed ; Rd = Rn - #immed` | SUBTRACT register with 12-bit immediate value |
| `RSB Rd, Rn, #immed   ; Rd = #immed - Rn` | Reverse subtract |
| `RSB Rd, Rn, Rm       ; Rd = Rm - Rn` | |
| `MUL Rd, Rn, Rm       ; Rd = Rn * Rm` | Multiply (32-bit result) |
| `UDIV Rd, Rn, Rm     ; Rd = Rn / Rm` | Unsigned and signed divide |
| `SDIV Rd, Rn, Rm     ; Rd = Rn / Rm` | |

**Table 5.23** Instructions for Multiply and MAC (Multiply Accumulate)

| Instruction (no "S" suffix because APSR is not updated) | Operation |
|---|---|
| MLA Rd, Rn, Rm, Ra ; Rd = Ra + Rn * Rm | 32-bit MAC instruction, 32-bit result |
| MLS Rd, Rn, Rm, Ra ; Rd = Ra - Rn * Rm | 32-bit multiply with subtract instruction, 32-bit result |
| SMULL RdLo, RdHi, Rn, Rm ;{RdHi,RdLo} = Rn * Rm | 32-bit multiply & MAC instructions for signed values, 64-bit result |
| SMLAL RdLo, RdHi, Rn, Rm ;{RdHi,RdLo} += Rn * Rm | |
| UMULL RdLo, RdHi, Rn, Rm ;{RdHi,RdLo} = Rn * Rm | 32-bit multiply & MAC instructions for unsigned values, 64-bit result |
| UMLAL RdLo, RdHi, Rn, Rm ;{RdHi,RdLo} += Rn * Rm | |

These instructions support signed or unsigned values (Table 5.23). The APSR flags are not affected by these instructions.

Additional MAC instructions are supported by the Cortex-M4 processor. This will be introduced later in section 5.7.3 Multiply and MAC instructions.

### 5.6.4 Logic operations

The Cortex®-M3 and Cortex-M4 processors support various instructions for logic operations such as AND, OR, exclusive OR and so on. Like the arithmetic instructions, the 16-bit versions of these instructions update the flags in APSR. If the "S" suffix is not specified, the assembler will convert them into 32-bit instructions.

The logic operation instructions are given in Table 5.24.

To use the 16-bit versions of these instructions, the operation must be between two registers with the destination being one of the source registers. Also, the registers used must be low registers (R0-R7), and the S suffix should be used (APSR update). The ORN instruction is not available in 16-bit form.

### 5.6.5 Shift and rotate instructions

The Cortex®-M3 and Cortex-M4 processors support various shift and rotate instructions, as shown in Table 5.25, and illustrated in Figure 5.6.

If the S suffix is used, these rotate and shift instructions also update the Carry flag in the APSR. If the shift or rotate operation shifts the register position by multiple bits, the value of the carry flag C will be the last bit that shifts out of the register.

You might wonder why there are rotate right instructions but no instructions for rotate left. Actually, a rotate left operation can be replaced by a rotate right operation

**Table 5.24** Instructions for Logical Operations

| Instruction (optional S suffix not shown) | | Operation |
|---|---|---|
| AND Rd, Rn | ; Rd = Rd & Rn | Bitwise AND |
| AND Rd, Rn,#immed | ; Rd = Rn & #immed | |
| AND Rd, Rn, Rm | ; Rd = Rn & Rm | |
| ORR Rd, Rn | ; Rd = Rd \| Rn | Bitwise OR |
| ORR Rd, Rn,#immed | ; Rd = Rn \| #immed | |
| ORR Rd, Rn, Rm | ; Rd = Rn \| Rm | |
| BIC Rd, Rn | ; Rd = Rd & (~Rn) | Bit clear |
| BIC Rd, Rn,#immed | ; Rd = Rn &(~ #immed) | |
| BIC Rd, Rn, Rm | ; Rd = Rn &(~Rm) | |
| ORN Rd, Rn,#immed | ; Rd = Rn \| (~#immed) | Bitwise OR NOT |
| ORN Rd, Rn, Rm | ; Rd = Rn \| (~Rm) | |
| EOR Rd, Rn | ; Rd = Rd ^ Rn | Bitwise Exclusive OR |
| EOR Rd, Rn,#immed | ; Rd = Rn \| #immed | |
| EOR Rd, Rn, Rm | ; Rd = Rn \| Rm | |

**Table 5.25** Instructions for Shift and Rotate Operations

| Instruction (optional "S" suffix not shown) | | Operation |
|---|---|---|
| ASR | Rd, Rn,#immed ; Rd = Rn >> immed | Arithmetic shift right |
| ASR | Rd, Rn       ; Rd = Rd >> Rn | |
| ASR | Rd, Rn, Rm   ; Rd = Rn >> Rm | |
| LSL | Rd, Rn,#immed ; Rd = Rn << immed | Logical shift left |
| LSL | Rd, Rn       ; Rd = Rd << Rn | |
| LSL | Rd, Rn, Rm   ; Rd = Rn << Rm | |
| LSR | Rd, Rn,#immed ; Rd = Rn >> immed | Logical shift right |
| LSR | Rd, Rn       ; Rd = Rd >> Rn | |
| LSR | Rd, Rn, Rm   ; Rd = Rn >> Rm | |
| ROR | Rd, Rn       ; Rd rot by Rn | Rotate right |
| ROR | Rd, Rn, Rm   ; Rd = Rn rot by Rm | |
| RRX | Rd, Rn       ; {C, Rd} = {Rn, C} | Rotate right extended |

with a different rotate amount. For example, a rotate left by 4 bits can be written as a rotate right by 28 bits. This gives you the same result in the destination register (note that the C flag will be different from rotate left) and takes same amount of time to execute.

To use the 16-bit version of these instructions, the registers used must be low registers (R0-R7), and the S suffix should be used (APSR update). The RRX instruction is not available in 16-bit form.

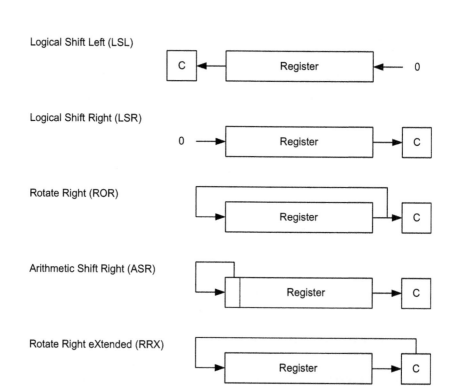

Logical Shift Left (LSL)

Logical Shift Right (LSR)

Rotate Right (ROR)

Arithmetic Shift Right (ASR)

Rotate Right eXtended (RRX)

**FIGURE 5.6**

Shift and Rotate operations

## 5.6.6 Data conversion operations (extend and reverse ordering)

In the Cortex®-M3 and Cortex-M4 processors, a number of instructions are available for handling signed and unsigned extensions of data; for example, to convert an 8-bit value to 32-bit, or from 16-bit to 32-bit. The signed and unsigned instructions are available in both 16-bit and 32-bit forms (Table 5.26). The 16-bit form of the instructions can only access low registers (R0 to R7).

| Table 5.26 Signed and Unsigned Extension | |
|---|---|
| **Instruction** | **Operation** |
| SXTB Rd, Rm    ; Rd = signed_extend(Rn[7:0]) | Signed extend byte data into word |
| SXTH Rd, Rm    ; Rd = signed_extend(Rn[15:0]) | Signed extend half-word data into word |
| UXTB Rd, Rm    ; Rd = unsigned_extend(Rn[7:0]) | Unsigned extend byte data into word |
| UXTH Rd, Rm    ; Rd = unsigned_extend(Rn[15:0]) | Unsigned extend half-word data into word |

**Table 5.27** Signed and Unsigned Extension with Optional Rotate

| Instruction | Operation |
|---|---|
| SXTB Rd, Rm {, ROR #n} ; n = 8 / 16/ 24 | Signed extend byte data into word |
| SXTH Rd, Rm {, ROR #n} ; n = 8 / 16/ 24 | Signed extend half-word data into word |
| UXTB Rd, Rm {, ROR #n} ; n = 8 / 16/ 24 | Unsigned extend byte data into word |
| UXTH Rd, Rm {, ROR #n} ; n = 8 / 16/ 24 | Unsigned extend half-word data into word |

The 32-bit form of these instructions can access high registers, and optionally rotate the input data before the signed extension operations, as shown in Table 5.27.

For SXTB/SXTH, the data are sign extended using bit[7]/bit[15] of Rn. With UXTB and UXTH, the value is zero extended to 32-bit.

For example, if R0 is 0x55AA8765:

```
SXTB R1, R0 ; R1 = 0x00000065
SXTH R1, R0 ; R1 = 0xFFFF8765
UXTB R1, R0 ; R1 = 0x00000065
UXTH R1, R0 ; R1 = 0x00008765
```

These instructions are useful for converting between different data types. Sometimes the signed extend or unsigned extend operation can be taking place on the fly when loading data from memory (e.g., LDRB for unsigned data and LDRSB for signed data).

Another group of data conversion operations is used for reversing data bytes in a register, listed in Table 5.28 and illustrated in Figure 5.7. These instructions are usually used for converting data between little endian and big endian.

The 16-bit form of these instructions can only access low registers (R0 to R7).

REV reverses the byte order in a data word, and REVH reverses the byte order inside a half-word. For example, if R0 is 0x12345678, in executing the following:

```
REV  R1, R0
REVH R2, R0
```

R1 will become 0x78563412, and R2 will be 0x34127856.

**Table 5.28** Instructions for Reversing Data

| Instruction | Operation |
|---|---|
| REV   Rd, Rn ; Rd = rev(Rn) | **Reverse bytes in word** |
| REV16 Rd, Rn ; Rd = rev16(Rn) | **Reverse bytes in each half-word** |
| REVSH Rd, Rn ; Rd = revsh(Rn) | **Reverse bytes in bottom half-word and sign extend the result** |

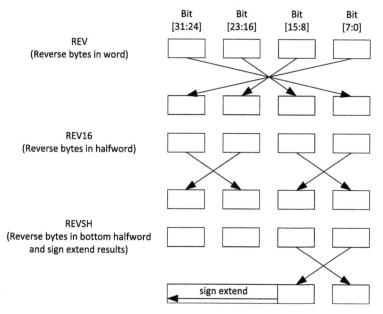

**FIGURE 5.7**

Reverse operations

REVSH is similar to REVH except that it only processes the lower half-word and then sign extends the result. For example, if R0 is 0x33448899, running:

```
REVSH R1, R0
```

R1 will become 0xFFFF9988.

## 5.6.7 Bit-field processing instructions

To make the Cortex®-M3 and Cortex-M4 processor an excellent architecture for control applications, these processors support a number of bit-field processing operations, as listed in Table 5.29.

**Table 5.29** Instructions for Bit-Field Processing

| Instruction | Operation |
| --- | --- |
| BFC Rd, #<lsb>, #<width> | Clear bit field within a register |
| BFI Rd, Rn, #<lsb>, #<width> | Insert bit field to a register |
| CLZ Rd, Rm | Count leading zero |
| RBIT Rd, Rn | Reverse bit order in register |
| SBFX Rd, Rn, #<lsb>, #<width> | Copy bit field from source and sign extend it |
| UBFX Rd, Rn, #<lsb>, #<width> | Copy bit field from source register |

BFC (Bit Field Clear) clears 1 to 31 adjacent bits in any position of a register. The syntax of the instruction is:

```
BFC <Rd>, <#lsb>, <#width>
```

For example:

```
LDR R0,=0x1234FFFF
BFC R0, #4, #8
```

This will give R0 = 0x1234F00F.

BFI (Bit Field Insert) copies 1 to 31 bits (#width) from one register to any location (#lsb) in another register. The syntax is:

```
BFI <Rd>, <Rn>, <#lsb>, <#width>
```

For example:

```
LDR R0,=0x12345678
LDR R1,=0x3355AACC
BFI R1, R0, #8, #16 ; Insert R0[15:0] to R1[23:8]
```

This will give R1 = 0x335678CC.

The CLZ instruction counts the number of leading zeros. If no bits are set the result is 32, and if all bits are set the result is 0. It is commonly used to determine the number of bit shifts required to normalize a value so that the leading one is shifted to bit 31. It is often used in floating point calculations.

The RBIT instruction reverses the bit order in a data word. The syntax is:

```
RBIT <Rd>, <Rn>
```

This instruction is very useful for processing serial bit streams in data communications. For example, if R0 is 0xB4E10C23 (binary value 1011_0100_1110_0001_0000_1100_0010_0011), executing:

```
RBIT R0, R1
```

R0 will become 0xC430872D (binary value 1100_0100_0011_0000_1000_0111_0010_1101).

UBFX and SBFX are the Unsigned and Signed Bit Field Extract instructions. The syntax of the instructions are:

```
UBFX <Rd>, <Rn>, <#lsb>, <#width>
SBFX <Rd>, <Rn>, <#lsb>, <#width>
```

UBFX extracts a bit field from a register starting from any location (specified by the <#lsb> operand) with any width (specified by the <#width> operand), zero extends it, and puts it in the destination register. For example:

```
LDR  R0,=0x5678ABCD
UBFX R1, R0, #4, #8
```

This will give R1 = 0x000000BC (zero extend of 0xBC).

Similarly, SBFX extracts a bit field but it sign extends it before putting it in a destination register. For example:

```
LDR  R0,=0x5678ABCD
SBFX R1, R0, #4, #8
```

This will give R1 = 0xFFFFFFBC (signed extend of 0xBC).

### 5.6.8 Compare and test

The compare and test instructions are used to update the flags in the APSR, which may then be used by a conditional branch or conditional execution (this will be covered in the next section). Table 5.30 listed these instructions.

Note that these instructions do not have the "S" suffix because the APSR is always updated.

### 5.6.9 Program flow control

There are several types of instructions for program flow control:

- Branch
- Function call

**Table 5.30** Instructions for Compare and Test

| Instruction | Operation |
|---|---|
| CMP <Rn>, <Rm> | Compare: Calculate Rn-Rm. APSR is updated but the result is not stored. |
| CMP <Rn>, #<immed> | Compare: Calculate Rn – immediate data. |
| CMN <Rn>, <Rm> | Compare negative: Calculate Rn + Rm. APSR is updated but the result is not stored. |
| CMN <Rn>, #<immed> | Compare negative: Calculate Rn + immediate data. APSR is updated but the result is not stored. |
| TST <Rn>, <Rm> | Test (bitwise AND): Calculate AND result between Rn and Rm. N bit and Z bit in APSR are updated but the AND result is not stored. C bit can be updated if barrel shifter is used. |
| TST <Rn>, #<immed> | Test (bitwise AND): Calculate AND result between Rn and immediate data. N bit and Z bit in APSR are updated but the AND result is not stored. |
| TEQ <Rn>, <Rm> | Test (bitwise XOR): Calculate XOR result between Rn and Rm. N bit and Z bit in APSR are updated but the AND result is not stored. C bit can be updated if barrel shifter is used. |
| TEQ <Rn>, #<immed> | Test (bitwise XOR): Calculate XOR result between Rn and immediate data. N bit and Z bit in APSR are updated but the AND result is not stored. |

- Conditional branch
- Combined compare and conditional branch
- Conditional execution (IF-THEN instruction)
- Table branch

### Branches

A number of instructions can cause branch operations:

- Branch instructions (e.g., B, BX)
- A data processing instruction that updates R15 (PC) (e.g., MOV, ADD)
- A memory read instruction that writes to PC (e.g., LDR, LDM, POP)

In general, although it is possible to use any of the above operations to create branches, it is more common to use B (Branch), BX (Branch with Exchange), and POP instructions (commonly used for function return). Sometimes the other methods are used in table branches for older ARM® processors, which are not required in Cortex®-M3/M4 as these processors have special instructions for table branches.

In this section we will focus on just the branch instructions. The most basic branch instructions are given in Table 5.31.

### Function calls

To call a function, the Branch and Link (BL) instruction or Branch and Link with eXchange (BLX) instructions can be used (Table 5.32). They execute the branch and at the same time save the return address to the Link Register (LR), so that the processor can branch back to the original program after the function call is completed.

**Table 5.31** Unconditional Branch Instructions

| Instruction | Operation |
| --- | --- |
| B <label><br>B.W <label> | Branch to label. If a branch range of over +/-2KB is needed, you might need to specify B.W to use 32-bit version of branch instruction for wider range. |
| BX <Rm> | Branch and exchange. Branch to an address value stored in *Rm*, and set the execution state of the processor (T-bit) based on bit 0 of *Rm* (bit 0 of *Rm* must be 1 because Cortex-M processor only supports Thumb state). |

**Table 5.32** Instructions for Calling a Function

| Instruction | Description |
| --- | --- |
| BL <label> | Branch to a labeled address and save the return address in LR |
| BLX <Rm> | Branch to an address specified by *Rm*, save the return address in LR, and update T-bit in EPSR with LSB of *Rm* |

When these instructions are executed:

- The Program Counter is set to the branch target address.
- The Link Register (LR/R14) is updated to hold the return address, which is the address of the instruction after the executed BL/BLX instruction.
- If the instruction is BLX, the Thumb bit in EPSR will also be updated using the LSB of the register holding the branch target address.

Since the Cortex-M3 and M4 processors only support the Thumb state, the LSB of the register used in a BLX operation must be set to 1. Otherwise, it indicates an attempt to switch to the ARM state and will result in a fault exception.

---

### SAVE THE LR IF YOU NEED TO CALL A SUBROUTINE

The BL instruction will destroy the current content of your LR register. So, if your program code needs the LR register later, you should save your LR before you use BL. The most common method is to push the LR to stack in the beginning of your subroutine. For example:

```
main
     ...
     BL functionA
     ...
functionA
     PUSH {LR} ; Save LR content to stack
     ...
     BL functionB ; Note: return address in LR will be changed
     ...
     POP {PC} ; Use stacked LR content to return to main
functionB
     PUSH {LR}
     ...
     POP {PC} ; Use stacked LR content to return to functionA
```

In addition, if the subroutine you call is a C function, you might also need to save the contents in R0–R3 and R12 if these values will be needed at a later stage. According to *AAPCS* (reference 8), the contents in these registers could be changed by a C function.

---

Note that in traditional ARM processors such as ARM926, there is also a BLX <label> instruction. This instruction always switches the state of the processor. Since the Cortex-M3 and M4 processors only support one state (Thumb), the BLX <label> instruction is not supported.

### *Conditional branches*

Conditional branches are executed conditionally based on the current value in APSR (N, Z, C, and V flags, as shown in Table 5.33).

APSR flags can be affected by the following:

- Most of the 16-bit data processing instructions
- 32-bit (Thumb-2) data processing instructions with the S suffix; for example, ADDS.W

**Table 5.33** Flags (status bits) in APSR, which can be used for Controlling Conditional Branch

| Flag | PSR Bit | Description |
|------|---------|-------------|
| N | 31 | Negative flag (last operation result is a negative value). |
| Z | 30 | Zero (last operation result returns a zero value; for example, compare of two registers with identical values). |
| C | 29 | Carry (last operation results in a carry out or does not result in a borrow; it can also be the last bit shifted out in a shift or rotate operation). |
| V | 28 | Overflow (last operation results in an overflow). |

- Compare (e.g., CMP) and Test (e.g., TST, TEQ)
- Write to APSR/xPSR directly

There is another flag bit at bit[27], called the $Q$ flag. This is for saturated arithmetic operations and is not used for conditional branches.

The required condition for a conditional branch to take place is indicated by a suffix (indicated as <cond> in Table 5.34 below). Conditional branch instructions are available in 16-bit and 32-bit versions, with different branch ranges, as shown in Table 5.34.

The <cond> is one of the 14 possible condition suffixes, as given in Table 5.35.

For example, consider the following operation:

The program flow in Figure 5.8 can be implemented using conditional branch and simple branch instructions:

```
CMP    R0, #1 ; compare R0 to 1
BEQ    p2     ; if Equal, then go to p2
MOVS   R3, #1 ; R3 = 1
B      p3     ; go to p3
p2            ; label p2
  MOVS R3, #2
p3            ; label p3
  ...         ; other subsequence operations
```

**Table 5.34** Instructions for Conditional Branch

| Instruction | Operation |
|-------------|-----------|
| B<cond> <label><br>B<cond>.W <label> | Branch to label if condition is true. E.g.,<br>CMP R0, #1<br>BEQ loop ; Branch to "loop" if R0 equal 1.<br>If a branch range of over ±254Bytes is needed, you might need to specify B.W to use 32-bit version of branch instruction for wider range. |

**Table 5.35** Suffixes for Conditional Branches and Conditional Execution

| Suffix | Branch Condition | Flags (APSR) |
|---|---|---|
| EQ | Equal | Z flag is set |
| NE | Not equal | Z flag is cleared |
| CS/HS | Carry set / unsigned higher or same | C flag is set |
| CC/LO | Carry clear / unsigned lower | C flag is cleared |
| MI | Minus / negative | N flag is set (minus) |
| PL | Plus / positive or zero | N flag is cleared |
| VS | Overflow | V flag is set |
| VC | No overflow | V flag is cleared |
| HI | Unsigned higher | C flag is set and Z is cleared |
| LS | Unsigned lower or same | C flag is cleared or Z is set |
| GE | Signed greater than or equal | N flag is set and V flag is set, or N flag is cleared and V flag is cleared (N == V) |
| LT | Signed less than | N flag is set and V flag is cleared, or N flag is cleared and V flag is set (N != V) |
| GT | Signed greater then | Z flag is cleared, and either both N flag and V flag are set, or both N flag and V flag are cleared (Z == 0 and N == V) |
| LE | Signed less than or equal | Z flag is set, or either N flag set with V flag cleared, or N flag cleared and V flag set (Z == 1 or N != V) |

### Compare and branches

With the ARMv7-M architecture, two new instructions are provided to supply a combined compare-to-zero and conditional branch operation. These two instructions are CBZ (Compare and Branch if Zero) and CBNZ (Compare and Branch if NonZero). They only support forward branches and not backward branches.

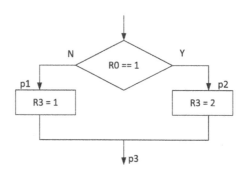

**FIGURE 5.8**

Simple conditional branch

CBZ and CBNZ are very useful in loop structures such as while loops. For example:

```
i = 5;
while (i != 0 ){
  func1();     // call a function
  i--;
}
```

This can be compiled into:

```
      MOV R0, #5        ; Set loop counter
loop1 CBZ R0,loop1exit ; if loop counter = 0 then exit the loop
      BL func1          ; call a function
      SUBS R0, #1       ; loop counter decrement
      B loop1           ; next loop
loop1exit
```

The usage of CBNZ is very similar to CBZ, apart from the fact that the branch is taken if the Z flag is Not set (result is not zero). For example:

```
status = strchr(email_address, '@');
if (status == 0){//status is 0 if @ is not in email_address
    show_error_message();
    exit(1);
    }
```

This can be compiled into:

```
...
BL    strchr
CBNZ R0, email_looks_okay ; Branch if result is not zero
BL    show_error_message
BL    exit
email_looks_okay
...
```

The APSR value is not affected by the CBZ and CBNZ instructions.

### Conditional execution (IF-THEN instruction)

Besides conditional branches, Cortex-M3 and Cortex-M4 processors also support conditional execution. After an IT (IF-THEN) instruction is executed, up to four of the subsequent instructions can be conditionally executed based on the condition specified by the IT instruction and the APSR value.

An IT instruction block consists of an IT instruction, with conditional execution details, followed by one to four conditional execution instructions. The conditional execution instructions can be data processing instructions or memory access instructions. The last conditional execution instruction in the IT block can also be a conditional branch instruction.

The IT instruction statement contains the IT instruction opcode with up to an additional three optional suffixes of "T" (then) and "E" (else), followed by the condition to check against, which is the same as the condition symbol for conditional branches. The "T"/"E" indicates how many subsequence instructions are inside the IT instruction block, and whether they should or should not be executed if the condition is met.

For example, using the same program flow in Figure 5.8, we can write the operation as:

```
CMP   R0, #1 ; Compare R0 to 1
ITE   EQ     ; The next instruction executes if Z is set (EQ),
             ; the one after next executes if Z is cleared(NE)
MOVEQ R3, #2 ; Set R3 to 2 if EQ
MOVNE R3, #1 ; Set R3 to 1 if not EQ (NE)
```

Note that when the "E" suffix is used, the execution condition for the corresponding instruction in the IT block must be the inverse of the condition specified by the IT instruction.

Different combinations of "T" and "E" sequence are possible:

- Just one conditional execution instruction: IT
- Two conditional execution instructions: ITT, ITE
- Three conditional execution instructions: ITTT, ITTE, ITET, ITEE
- Four conditional execution instructions: ITTTT, ITTTE, ITTET, ITTEE, ITETT, ITETE, ITEET, ITEEE

Table 5.36 listed various forms of IT instruction block sequence and examples: where:

**Table 5.36** IT Instruction Block of Various Sizes

|  | IT block (each of <x>, <y> and <z> can either be T (true) or E (else)) | Examples |
|---|---|---|
| Only one conditional instruction | IT  <cond><br>instr1<cond> | IT EQ<br>ADDEQ R0, R0, R1 |
| Two conditional instructions | IT<x>  <cond><br>instr1<cond><br>instr2<cond or ~(cond)> | ITE GE<br>ADDGE R0, R0, R1<br>ADDLT R0, R0, R3 |
| Three conditional instructions | IT<x><y>  <cond><br>instr1<cond><br>instr2<cond or ~(cond)><br>instr3<cond or ~(cond)> | ITET GT<br>ADDGT R0, R0, R1<br>ADDLE R0, R0, R3<br>ADDGT R2, R4, #1 |
| Four conditional instructions | IT<x><y><z> <cond><br>instr1<cond><br>instr2<cond or ~(cond)><br>instr3<cond or ~(cond)><br>instr4<cond or ~(cond)> | ITETT NE<br>ADDNE R0, R0, R1<br>ADDEQ R0, R0, R3<br>ADDNE R2, R4, #1<br>MOVNE R5, R3 |

- $<x>$ specifies the execution condition for the second instruction
- $<y>$ specifies the execution condition for the third instruction
- $<z>$ specifies the execution condition for the fourth instruction
- $<cond>$ specifies the base condition of the instruction block; the first instruction following IT executes if $<cond>$ is true

If "AL" is used as $<cond>$, then you cannot use "E" in the condition control as it implies the instruction will never be executed.

In some assembler tools, it is not necessary to use the IT instruction in the code. Just by adding the condition suffix to a normal instruction, the assembler tool (e.g., ARM Assembler in DS-5™ Professional or Keil™ MDK-ARM) can automatically insert the required IT instruction in front. This helps when porting assembly application code from classic ARM processors (e.g., ARM7TDMI™) to Cortex-M3/M4, as you do not need to manually insert IT instructions.

For example, when using the following assembly code with Keil MDK-ARM, the assembler can automatically insert the required IT instruction, as shown in Table 5.37.

Data processing instructions inside an IT instruction block should not change the APSR value. Note that when some 16-bit data processing instructions are used inside an IT instruction block, the APSR will not be updated. This is different from their usual behavior, which updates the APSR. This enables the use of 16-bit data processing instruction within IT instructions to reduce code size.

In many cases, the IT instruction can help improve the performance of program code significantly because it avoids some of the branch penalty, as well as reducing the number of branch instructions. For example, a short IF-THEN-ELSE program sequence that normally requires one conditional branch and an unconditional branch can be replaced by a single IT instruction.

In some other cases, traditional branch methods can be better than the IT instruction because a conditional failed instruction in an IT instruction sequence will still take a cycle to run. So if you specified ITTTT $<cond>$ and the condition failed due to the APSR value in run-time, it could be quicker to use a conditional branch (three cycles) than using IT instruction block (up to five cycles in this case, including the IT instruction itself).

### Table branches

The Cortex-M3 and Cortex-M4 support two table branch instructions: TBB (Table Branch Byte) and TBH (Table Branch Half-word). These instructions are used

**Table 5.37** Automatic Insertion of IT Instruction in ARM Assembler

| Original Assembler Code | Disassembled Assembly Code from Generated Object File |
|---|---|
| ...<br>CMP   R1, #2<br>ADDEQ R0, R1, #1<br>... | ...<br>CMP   R1, #2<br>IT    EQ<br>ADDEQ R0, R1, #1<br>... |

with branch tables, often used to implement switch statements in C code. Since bit 0 of the program counter value is always zero, a branch table using table branch instructions does not have to store that bit and therefore the branch offset is multiplied by two in the target address calculation.

The TBB is used when all the entries in the branch table are organized as a byte array (offset from base address is less than $2 \times 2^8 = 512$ bytes), and TBH is used when all the entries are organized as a half-word array (offset from base address is less than $2 \times 2^{16} = 128K$ bytes). The base address could be the current program counter (PC) value or from another register. Due to the pipelined nature of the Cortex-M processors, the current PC value is the address of the TBB/TBH instruction plus 4, and this must be factored in during the generation of the branch table. Both TBB and TBH support forward branches only.

The TBB instruction has the syntax:

```
TBB  [Rn, Rm]
```

where *Rn* stores the base address of the branch table and *Rm* is the branch table index. The immediate value for the TBB offset calculation is located in memory[*Rn* + *Rm*]. If R15/PC is used as Rn, we can view the operation as shown in Figure 5.9.

The operation of the TBH instruction is very similar, except that each entry in the branch table is two bytes in size, so that the array indexing is different and the branch offset range is larger. The syntax of TBH is slightly different to reflect the indexing difference:

```
TBH  [Rn, Rm, LSL #1]
```

Again, the following diagram assumes that Rn is set to the PC, as shown in Figure 5.10.

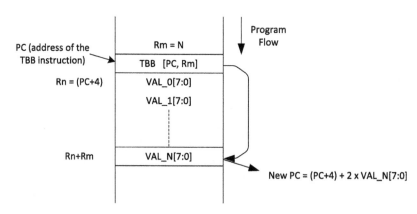

**FIGURE 5.9**

TBB operation

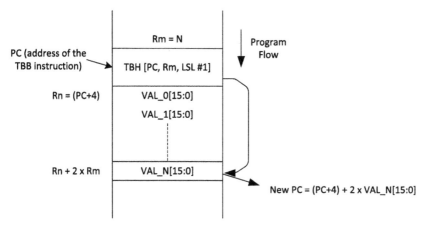

**FIGURE 5.10**

TBH operation

The TBB and TBH instructions are typically used by C compilers in switch (case) statements. Using these instructions directly in assembly programming is less straightforward, as the values in the branch table are relative to the current program counter value. If the branch target address is not in the same assembly program file, the address offset value cannot be determined in the assembly stage.

In ARM assembler (armasm), including the Keil MDK-ARM, the TBB branch table can be created in assembly in the following way:

```
        TBB [pc, r0] ; when executing this instruction, PC equal
                     ; branchtable because size of TBB instruction
                     ; is 32-bit
branchtable      ; Start of the branch table
        DCB ((dest0 — branchtable)/2) ; Note that DCB is used
                                      ; because the values are 8-bit
        DCB ((dest1 — branchtable)/2)
        DCB ((dest2 — branchtable)/2)
        DCB ((dest3 — branchtable)/2)
dest0
        ... ; Execute if r0 = 0
dest1
        ... ; Execute if r0 = 1
dest2
        ... ; Execute if r0 = 2
dest3
        ... ; Execute if r0 = 3
```

In the above example, when the TBB instruction is executed, the current PC value is the address of TBB instruction plus 4 (due to the pipeline structure), which is the same as the start of the *branchtable* because the TBB instruction is 4 bytes in size. (Both TBB and TBH are 32-bit instructions.)

Similar, an example for the TBH instruction can be written as:

```
        TBH [pc, r0, LSL #1]
branchtable ; Start of the branch table
        ; Note that DCI is used
        ; because the values are 16-bit
        DCI ((dest0 — branchtable)/2)
        DCI ((dest1 — branchtable)/2)
        DCI ((dest2 — branchtable)/2)
        DCI ((dest3 — branchtable)/2)
dest0
        ... ; Execute if r0 = 0
dest1
        ... ; Execute if r0 = 1
dest2
        ... ; Execute if r0 = 2
dest3
        ... ; Execute if r0 = 3
```

Please note that the coding syntax to create the branch table could be dependent on the development tools.

### 5.6.10 Saturation operations

The Cortex®-M3 processor supports two instructions that provide saturation adjustment of signed and unsigned data. They are SSAT (for signed data) and USAT (for unsigned data). The Cortex-M4 processor also supports them, and in addition, supports instructions for saturated algorithms. In this section, we will cover the SSAT and USAT instructions first. The instructions for saturated algorithm will be covered in section 5.7.

Saturation is commonly used in signal processing. For example, after certain operations such as amplification, the amplitude of a signal can exceed the maximum allowed output range. If the value is adjusted by simply cutting off the MSB bits, the resulted signal waveform could be completely distorted (Figure 5.11).

The saturation operation reduces the distortion by forcing the value to the maximum allowed value. The distortion still exists, but if the value does not exceed the maximum range by too much it is less noticeable.

The syntax for SSAT and USAT instructions are as follows:

```
SSAT <Rd>, #<immed>, <Rn>, {,<shift>}  ; Saturation for signed value
USAT <Rd>, #<immed>, <Rn>, {,<shift>}  ; Saturation for a signed
                                            value into unsigned value
Where
<Rn>    : input value
```

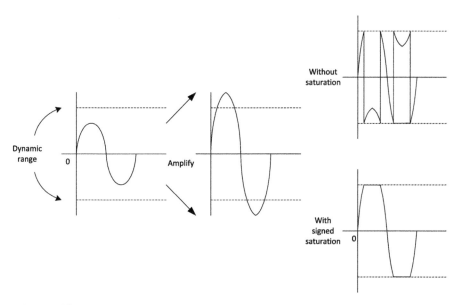

**FIGURE 5.11**

Signed Saturation operation

```
<shift> : optional shift operation for input value before saturation.
It can be #LSL N or #ASR N
<immed> : Bit position where the saturation is carried out
<Rd>    : destination register
```

Besides the destination register, the Q-bit in the APSR can also be affected by the result. The Q flag is set if saturation takes place in the operation and it can be cleared by writing to the APSR (see section 4.3.2). For example, if a 32-bit signed value is to be saturated into a 16-bit signed value, the following instruction can be used:

```
SSAT R1, #16, R0
```

Table 5.38 shows several examples of SSAT operation result.

USAT is slightly different in that the result is an unsigned value. This will provide a saturation operation as shown in Figure 5.12.

For example, you can convert a 32-bit signed value to a 16-bit unsigned value using:

```
USAT R1, #16, R0
```

Table 5.39 shows several examples of USAT operation result.

### 5.6.11 Exception-related instructions

The SuperVisor Call (SVC) instruction is used to generate the SVC exception (exception type 11). Typically, SVC is used with an embedded OS/RealTime OS

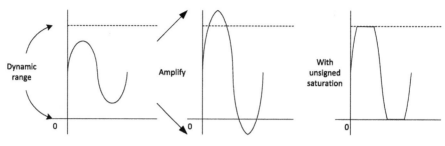

**FIGURE 5.12**

USAT saturation operation

**Table 5.38** Examples of SSAT Results

| Input (R0) | Output (R1) | Q Bit |
|---|---|---|
| 0x00020000 | 0x00007FFF | Set |
| 0x00008000 | 0x00007FFF | Set |
| 0x00007FFF | 0x00007FFF | Unchanged |
| 0x00000000 | 0x00000000 | Unchanged |
| 0xFFFF8000 | 0xFFFF8000 | Unchanged |
| 0xFFFF7FFF | 0xFFFF8000 | Set |
| 0xFFFE0000 | 0xFFFF8000 | Set |

**Table 5.39** Examples of USAT Results

| Input (R0) | Output (R1) | Q Bit |
|---|---|---|
| 0x00020000 | 0x0000FFFF | Set |
| 0x00008000 | 0x00008000 | Unchanged |
| 0x00007FFF | 0x00007FFF | Unchanged |
| 0x00000000 | 0x00000000 | Unchanged |
| 0xFFFF8000 | 0x00000000 | Set |
| 0xFFFF8001 | 0x00000000 | Set |
| 0xFFFFFFFF | 0x00000000 | Set |

(RTOS), where an application task running in an unprivileged execution state can request services from the OS, which runs in the privileged execution state. The SVC exception mechanism provides the transition from unprivileged to privileged.

In addition, the SVC mechanism is useful as a gateway for application tasks to access various services (including OS services, or other API functions) because an application task can request a service without knowing the actual program memory address of the service. It only needs to know the SVC service number, the input parameters, and the returned results.

The SVC instruction requires that the priority of the SVC exception be higher than the current priority level and that the exception is not masked by masking registers like PRIMASK. Otherwise, a fault exception would be triggered instead. As a result, you cannot use SVC in a NMI handler or HardFault handler, because the priority levels of these handlers are always higher than the SVC exception.

The SVC instruction has the following syntax:

```
SVC #<immed>
```

The immediate value is 8-bit. The value itself does not affect the behavior of the SVC exception, but the SVC handler can extract this value in software and use it as an input parameter; for example, to determine what service is requested by the application task.

In traditional ARM® assembly syntax, for the SVC instruction the immediate value does not need the "#" sign. So you can write the instruction as:

```
SVC <immed>
```

You can still use this syntax in most assembler tools, but for new software the use of the "#" sign is recommended.

Another instruction that is related to exceptions is the Change Processor State (CPS) instruction. For the Cortex®-M processors, you can use this instruction to set or clear interrupt masking registers such as PRIMASK and FAULTMASK. Note that these registers can also be accessed using MSR and MRS instructions.

The CPS instruction must be used with one of these suffixes: IE (interrupt enable) or ID (interrupt disable). You must also specify which masking register to set/clear because the Cortex-M3 and Cortex-M4 processors have several interrupt mask registers. Table 5.40 listed various forms of the CPS instructions available for the Cortex-M3 and Cortex-M4 processors.

The switching of PRIMASK or FAULTMASK to disable and enable an interrupt is commonly used to ensure timing critical code can finish quickly without getting interrupted.

**Table 5.40** Instructions for Setting and Clearing PRIMASK and FAULTMASK

| Instruction | Operation |
| --- | --- |
| CPSIE I | Enable interrupts (clear PRIMASK). Same as __enable_irq(); |
| CPSID I | Disable interrupts (set PRIMASK). NMI and HardFault are not affected. Same as __disable_irq(); |
| CPSIE F | Enable interrupt (clear FAULTMASK). Same as __enable_fault_irq(); |
| CPSID F | Disable fault interrupt (set FAULTMASK). NMI is not affected. Same as __disable_fault_irq(); |

### 5.6.12 Sleep mode-related instructions

There are two main instructions for entering sleep modes (note: there is another way to enter sleep mode called Sleep-on-Exit, which allows the processor to enter sleep upon exception exit; see section 9.2.5):

```
WFI  ; Wait for Interrupt (enter sleep)
```

Or in C programming with CMSIS-compliant device driver:

```
__WFI(); // Wait for Interrupt (enter sleep)
```

The WFI (Wait for Interrupt) instruction causes the processor to enter sleep mode immediately. The processor can be woken up from the sleep mode by interrupts, reset, or by debug operation.

Another instruction called WFE (Wait for Event) causes the processor to enter sleep mode conditionally:

```
WFE  ; Wait for Event (enter sleep conditionally)
```

Or in C programming with CMSIS-compliant device-driver:

```
__WFE(); // Wait for Event (enter sleep conditionally)
```

Inside the Cortex®-M3/M4 processor, there is a single bit internal register to record events. If this register is set, the WFE instruction will not enter sleep and will just clear the event register and continue to the next instruction. If this register is clear, the processor will enter sleep, and will wake up upon an event, which can be an interrupt, debug operation, reset, or a pulse signal at an external event input (e.g., event pulse can be generated by another processor or an peripheral).

The interface signals of the Cortex-M processor include an event input and an event output. The event input of a processor can be generated from event outputs from other processors in a multi-processor system. So a processor in WFE sleep (e.g., waiting for a spin lock) can be woken up by other processors. In some cases, these signals are connected to the I/O port pins of a Cortex-M microcontroller. While in some other Cortex-M microcontroller the event input could be tied low and the event output could be unused.

The event output can be triggered using the SEV (Send Event) instruction:

```
SEV  ; Send Event
```

Or in C programming with CMSIS-compliant device driver:

```
__SEV(); // Send Event
```

When SEV is executed, a single cycle pulse is generated at the event output interface. The SEV instruction also sets the event register of the same processor.

More information on WFE and the event register is given in Chapter 9.

### 5.6.13 **Memory barrier instructions**

The ARM® architectures (including the ARMv7-M architecture) allow memory transfers to take place in a completion order different from the program code, provided that it doesn't affect the result of the data processing. This often happens in high-end processors with superscalar or out-of-order-execution capabilities. However, by re-ordering memory accesses, and if the data is shared between multiple processors, the data sequence observed by another processor can be different from the programmed sequence and this can causes error or glitches in the applications.

The memory barrier instructions can be used to:

- Enforce ordering between memory accesses
- Enforce ordering between a memory access and another processor operation
- Ensure effect of a system configuration change taken place before subsequent operations

The Cortex®-M processors support three memory barrier instructions, as shown in Table 5.41.

In C programming, these instructions can be accessed using the following functions if CMSIS-compliant device driver is used:

```
void __DMB(void); // Data Memory Barrier
void __DSB(void); // Data Synchronization Barrier
void __ISB(void); // Instruction Synchronization Barrier
```

Since the Cortex-M processors have relatively simple pipelines, and the AHB Lite bus protocol use does not allow re-ordering of transfers in the memory system, most applications work fine without any memory barrier instructions. However, there are several cases where these barrier instructions should be used, as outlined in Table 5.42.

From an architectural point of view, there can be additional situations where a memory barrier should be used between two operations, although omitting the memory barrier will not cause any error in current Cortex-M processors, as can be seen in Table 5.43.

An application note on the use of memory barrier instructions on the Cortex-M processor is available from ARM called "A Programmer's Guide to the Memory Barrier Instruction for ARM Cortex-M Family Processor" (reference 9, ARM DAI0321A).

**Table 5.41** Memory Barrier Instructions

| Instruction | Description |
| --- | --- |
| DMB | Data Memory Barrier; ensures that all memory accesses are completed before new memory access is committed |
| DSB | Data Synchronization Barrier; ensures that all memory accesses are completed before next instruction is executed |
| ISB | Instruction Synchronization Barrier; flushes the pipeline and ensures that all previous instructions are completed before executing new instructions |

**Table 5.42** Examples of Where Memory Barrier Instruction(s) are Needed

| Scenarios (Required in Current Cortex-M3 & Cortex-M4 Implementations) | Barrier Instruction(s) Needed |
|---|---|
| After updating the CONTROL register with MSR instruction, the ISB instruction should be used to ensure the updated configuration is used for subsequent operations. | ISB |
| If the SLEEPONEXIT bit in the System Control Register is changed inside an exception handler, a DSB should be used before exception return. | DSB |
| If an exception which was pending is enabled, and if you want to ensure the pended exception taken place before subsequent operation. | DSB, followed by an ISB |
| When disabling an interrupt using NVIC clear enable register, and if you want make sure that the effect of interrupt disabling take place immediately before starting the next operation. | DSB, followed by an ISB |
| Self-modifying code (subsequent instructions are already fetched and need to be flushed). | DSB, followed by an ISB |
| Program memory map changed by a control register in a peripheral and the new program memory map is to be used immediately. (Assume the memory map is updated immediately after write completion). | DSB, followed by and ISB |
| Data memory map changed by a control register in a peripheral and the new data memory map is to be used immediately. (Assume the memory map is updated immediately after write completion.) | DSB |
| Update of Memory Protection Unit (MPU) configuration, and then immediately fetch and execute an instruction in the memory region affected by the MPU configuration change. | DSB, followed by an ISB |

### 5.6.14 Other instructions

There are a few other miscellaneous instructions.

The Cortex®-M processors support a NOP instruction. This instruction can be used to produce instruction alignment, or to introduce delay:

```
NOP ; Do nothing
```

Or in C programming:

```
__NOP(); // Do nothing
```

Please note that in general the delay created by NOP instruction is not guaranteed, and can vary between different systems (e.g., memory wait states, processor type). If the timing delay needs to be accurate, a hardware timer should be used.

**Table 5.43** Examples of where Memory Barrier Instruction(s) are Recommended by Architecture Definitions

| Scenarios (Recommendation Based on Architecture) | Barrier Instruction(s) Needed |
|---|---|
| Update of Memory Protection Unit (MPU) configuration, and then perform a data access to the memory region affected by the MPU configuration change. (Data accesses only region, no instruction fetch.) | DSB |
| Before entering sleep (WFI or WFE) | DSB |
| Semaphore operations | DMB or DSB |
| Changing priority level of an exception (e.g., SVC) and then triggering it. | DSB |
| Relocating the vector Table to a new location using Vector Table Offset Register (VTOR), and then trigger an exception with new vector. | DSB |
| Changing a vector entry in the vector table (if it is relocated to SRAM), and then trigger the same exception immediately. | DSB |
| Just before a self-reset (there can be an ongoing data transfer still active). | DSB |

The Breakpoint (BKPT) instruction is used for creating software breakpoints in an application during software development/debugging. If a program is executed from SRAM, usually this instruction is inserted by a debugger, replacing the original instruction. When the breakpoint is hit, the processor would be halted, and the debugger then restores the original instruction and the user can then carry out the debug tasks through the debugger. The BKPT instruction can also be used to generate a debug monitor exception. The BKPT instruction has an 8-bit immediate data, and the debugger or debug monitor exception handler can extract this data and decide what action to carry out based on this information. For example, some special values can be used to indicate semi-hosting requests (this is toolchain dependent).

The syntax of BKPT instruction is:

```
BKPT #<immed> ; Breakpoint
```

Similarly to SVC, you can also omit the "#" sign in most assembler tools:

```
BKPT <immed> ; Breakpoint
```

Or in C programming:

```
__BKPT (immed);
```

In addition to BKPT, the Cortex-M3 and Cortex-M4 processors also support a breakpoint unit, which provides up to eight hardware breakpoints and does not require replacing the original program image.

### 5.6.15 **Unsupported instructions**

The whole Thumb instruction set (including 32-bit instructions covered by the Thumb®-2 Technology) is designed to support a wide range of processor hardware. A couple of these instructions cannot be used in the Cortex®-M processors, as shown in Table 5.44.

Some of the Change Processor State (CPS) instructions defined in the Thumb instruction set are also unsupported. This is because the Program Status Register (PSR) definition of the Cortex-M processors is different from traditional ARM® or Cortex-A/R processors, as shown in Table 5.45.

Even within the ARMv7-M architecture, a number of instructions are defined, but are not supported in the existing Cortex-M3 and Cortex-M4 processor design. For example, both Cortex-M3 and Cortex-M4 processors do not support co-processors, therefore the co-processor instructions given in Table 5.46 will result in a fault exception when executed.

There are a number of Hint instructions defined in the Thumb instruction set. These instructions are executed as NOP on the Cortex-M3 and Cortex-M4 processors, as shown in Table 5.47.

All other undefined instructions, when executed, will cause a fault exception (either HardFault or Usage fault exception) to take place.

**Table 5.44** Unsupported Thumb Instructions

| Unsupported Thumb Instruction | Reasons |
| --- | --- |
| BLX <label> | This is branch with link and exchange state. In a format with immediate data, BLX always changes to ARM state. Since the Cortex-M3 and Cortex-M4 processors do not support the ARM state, instructions like this one that attempt to switch to the ARM state will result in a fault exception called *usage fault*. |
| SETEND | This Thumb instruction, introduced in architecture v6, switches the endian configuration during run time. Since the Cortex-M3 and Cortex-M4 processors do not support dynamic endian, using the SETEND instruction will result in a fault exception. |

**Table 5.45** Unsupported CPS Instructions

| Unsupported CPS Instructions | Reasons |
| --- | --- |
| CPS<IE|ID>.W A | There is no A bit in the Cortex-M3 |
| CPS.W #mode | There is no mode bit in the Cortex-M3 PSR |

**Table 5.46** Unsupported Co-processor Instructions

| Unsupported Co-processor Instructions | Function |
|---|---|
| MCR | Move to co-processor from ARM processor |
| MCR2 | Move to co-processor from ARM processor |
| MCRR | Move to co-processor from two ARM register |
| MRC | Move to ARM register from co-processor |
| MRC2 | Move to ARM register from co-processor |
| MRRC | Move to two ARM registers from co-processor |
| LDC | Load co-processor; load memory data from a sequence of consecutive memory addresses to a co-processor |
| STC | Store co-processor; stores data from a co-processor to a sequence of consecutive memory addresses |

**Table 5.47** Miscellaneous Unsupported Instructions

| Unsupported Instruction | Function |
|---|---|
| DBG | A hint instruction to debug and trace system. The exact effect is dependent on the processor product. In Cortex-M3 and Cortex-M4, this instruction is not used. |
| PLD | Pre-load data. This is a hint instruction typically use for cache memory to accelerate data accesses. However, since there is no cache inside the Cortex-M3 and Cortex-M4 processors, this instruction behaves as NOP. |
| PLI | Pre-load instruction. This is a hint instruction typically use for cache memory system to accelerate instruction accesses by indicating the program code a certain memory region is to be used. However, since there is no cache inside the Cortex-M3 and Cortex-M4 processors, this instruction behaves as NOP. |
| YIELD | A hint instruction to allow an application task in a multithreading system to indicate that it is doing a task that can be swapped out (e.g., stalled or waiting for something to happen). This hint information can be used by processors with hardware multi-threading support to improve overall system performance. Since the Cortex-M3 and Cortex-M4 processors do not have any hardware multi-threading support, this hint instruction is executed as a NOP. |

## 5.7 Cortex®-M4-specific instructions

### 5.7.1 Overview of enhanced DSP extension in Cortex-M4

Compared to the Cortex®-M3 processor, the Cortex-M4 processor supports a number of additional instructions such as:

- Single Instruction, Multiple Data (SIMD)
- Saturating instructions

- Additional multiply and MAC (multiply and accumulate) instructions
- Packing and unpacking instructions
- Optional floating point instructions (if floating unit is available)

These instructions enable the Cortex-M4, a general-purpose processor, to handle real-time digital signal processing (DSP) tasks more efficiently.

First, we'll have a look at how SIMD data is handled in Cortex-M4 processors. Very often, the data needed to be processed are 16-bit or 8-bit in size. For example, most audio is sampled with ADCs (Analog to Digital Converters) with 16-bit resolution or less, and image pixels are often represented with multiple channels of 8-bit data (e.g., RGB color space). Since the data path inside the processor is 32-bit, we can utilize the data path to handle 2x16-bit data or 4x8-bit data. We also need to consider that sometimes data are to be processed as signed values, and some other times as unsigned values. So a 32-bit register can be used for four types of SIMD data, as shown in Figure 5.13.

In most cases, the data inside a SIMD data set would be the same type (i.e., no mixture of signed and unsigned data, and no mixture of 16-bit and 8-bit data inside a SIMD data set). This allows simpler SIMD instruction set design.

In order to handle SIMD data, additional instructions have been added to the ARM® Cortex-M architecture, and this is called Enhanced DSP extension. Sometimes this is referred to as the ARMv7E-M architecture. The Enhanced DSP extension of Cortex-M4 is source-level compatible with the Enhanced DSP extension in ARMv9E architecture, allowing codecs developed for the ARM9E processors (e.g., ARM926, ARM946) to be ported to Cortex-M4 easily. (Note: this is not binary-level compatible.)

Since the SIMD data types are not native data types in C language, normally C compilers cannot generate the required DSP instructions in normal C code. In order to make it easier for software developers, intrinsic functions are added in the header files in CMSIS compliant driver libraries. In this way the software developers can access these instructions easily. To make it even better, ARM also provides a DSP library called CMSIS-DSP, which can be used by software developers free of charge.

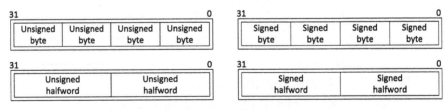

**FIGURE 5.13**

Various possible SIMD data representations in a 32-bit register

**Table 5.48** C99 Data Types Used in CMSIS

| Type | Size (bits) | Equivalent C/data Types |
|------|-------------|-------------------------|
| uint8_t | 8 | unsigned char |
| uint16_t | 16 | unsigned short int |
| uint32_t | 32 | unsigned int |
| int8_t | 8 | signed char |
| int16_t | 16 | signed short int |
| int32_t | 32 | signed int |

Appendix B contains a number of diagrams to illustrate the operations of the DSP instructions. Inside the diagrams, the following C99 data types (Table 5.48) are used to represent the data.

### 5.7.2 SIMD and saturating instructions

There is quite a long list of SIMD and saturating instructions. Some of the saturating instructions also support SIMD. Many of the SIMD instructions contain similar operations, but with different prefixes to tell if the instruction is intended for signed or unsigned data, as shown in Table 5.49.

The descriptions of the base operations are given in Table 5.50.

In addition, the SIMD instructions also included, are given in Table 5.51.

Some of the saturating instructions are not SIMD, as shown in Table 5.52.

The syntaxes of these instructions are given in Table 5.53.

Please note that some of these instructions set the Q-bit if saturation occurred. However, the Q bit does not get cleared by these instructions and the Q bit has to be cleared manually by writing to the APSR. Usually the program code has to examine the value of the Q bit in the APSR to detect if saturation has taken place in any of the steps during a computation flow. Therefore the Q bit does not get cleared unless explicitly specified.

### 5.7.3 Multiply and MAC instructions

There are also a number of various multiply and MAC instructions. In the earlier section of this chapter we covered some of the multiply and MAC instructions that are available in both Cortex®-M3 and Cortex-M4 processors. These are given in Table 5.54.

Note: In the Cortex-M4, the UMULL, UMLAL, SMULL, and SMLAL instructions have a faster execution time than the Cortex-M3 processor.

In addition, the Cortex-M4 processor also supports additional multiply and MAC instructions, and some of them come in multiple forms for selecting lower and upper half-words from input operands, as shown in Table 5.55.

**Table 5.49** SIMD Instructions

| Prefix Operation (see next table) | S[1] Signed | Q[2] Signed Saturating | SH[3] Signed Halving | U[1] Unsigned | UQ[2] Unsigned Saturating | UH[3] Unsigned Halving |
|---|---|---|---|---|---|---|
| ADD8 | SADD8 | QADD8 | SHADD8 | UADD8 | UQADD8 | UHADD8 |
| SUB8 | SSUB8 | QSUB8 | SHSUB8 | USUB8 | UQSUB8 | UHSUB8 |
| ADD16 | SADD16 | QADD16 | SHADD16 | UADD16 | UQADD16 | UHADD16 |
| SUB16 | SSUB16 | QSUB16 | SHSUB16 | USUB16 | UQSUB16 | UHSUB16 |
| ASX | SASX | QASX | SHASX | UASX | UQASX | UHASX |
| SAX | SSAX | QSAX | SHSAX | USAX | UQSAX | UHSAX |

[1] GE bits updates.
[2] Q bit is set when saturation occurs.
[3] Each data in the SIMD operation result is divided by 2 in Signed Halving (SH) and Unsigned Halving (UH) operations.

**Table 5.50** Base Operations for SIMD Instructions

| Operation | Descriptions |
|-----------|--------------|
| ADD8 | Add 4 pairs of 8-bit data |
| SUB8 | Subtract 4 pairs of 8-bit data |
| ADD16 | Add 2 pairs of 16-bit data |
| SUB16 | Subtract 2 pairs of 16-bit data |
| ASX | Exchange half-words of the second operand register, then add top half-words and subtract bottom half-word |
| SAX | Exchange half-words of the second operand register, then subtract top half-words and add bottom half-word |

**Table 5.51** Additional SIMD Instruction

| Operation | Descriptions |
|-----------|--------------|
| USAD8 | Unsigned Sum of Absolute Difference between 4 pairs of 8-bit data |
| USADA8 | Unsigned Sum of Absolute Difference between 4 pairs of 8-bit data and Accumulate |
| USAT16 | Unsigned saturate 2 signed 16-bit values to a selected unsigned range |
| SSAT16 | Signed saturate 2 signed 16-bit values to a selected unsigned range |
| SEL | Select Byte from first or second operand based on GE flags |

**Table 5.52** Additional non-SIMD Instruction

| Operation | Descriptions |
|-----------|--------------|
| SSAT | Signed saturation (supported in Cortex-M3) |
| USAT | Unsigned saturation (supported in Cortex-M3) |
| QADD | Saturating Add two signed 32-bit integer |
| QDADD | Double a 32-bit signed integer and add to another 32-bit signed integer, saturating possible for both operations |
| QSUB | Saturating Subtract two signed 32-bit integer |
| QDSUB | Double a 32-bit signed integer and subtract it from another 32-bit signed integer, saturating possible for both operations |

Please note that some of these instructions set the Q bit when signed overflow occurrs. However, the Q bit does not get cleared by these instructions and has to be cleared manually by writing to the APSR. Usually the program code has to examine the value of the Q bit in the APSR to detect if overflow has taken place

**Table 5.53** Syntax of SIMD and Saturating Instructions

| Mnemonic | Operands | Brief Description | Flags | Figure |
|----------|----------|------------------|-------|--------|
| SADD8 | {Rd,} Rn, Rm | Signed Add 8 | GE[3:0] | B.13 |
| SADD16 | {Rd,} Rn, Rm | Signed Add 16 | GE[3:0] | B.14 |
| SSUB8 | {Rd,} Rn, Rm | Signed Subtract 8 | GE[3:0] | B.17 |
| SSUB16 | {Rd,} Rn, Rm | Signed Subtract 16 | GE[3:0] | B.18 |
| SASX | {Rd,} Rn, Rm | Signed Add and Subtract with Exchange | GE[3:0] | B.21 |
| SSAX | {Rd,} Rn, Rm | Signed Subtract and Add with Exchange | GE[3:0] | B.22 |
| QADD8 | {Rd,} Rn, Rm | Saturating Add 8 | Q | B.5 |
| QADD16 | {Rd,} Rn, Rm | Saturating Add 16 | Q | B.4 |
| QSUB8 | {Rd,} Rn, Rm | Saturating Subtract 8 | Q | B.9 |
| QSUB16 | {Rd,} Rn, Rm | Saturating Subtract 16 | Q | B.8 |
| QASX | {Rd,} Rn, Rm | Saturating Add and Subtract with Exchange | Q | B.10 |
| QSAX | {Rd,} Rn, Rm | Saturating Subtract and Add with Exchange | Q | B.11 |
| SHADD8 | {Rd,} Rn, Rm | Signed Halving Add 8 | | B.15 |
| SHADD16 | {Rd,} Rn, Rm | Signed Halving Add 16 | | B.16 |
| SHSUB8 | {Rd,} Rn, Rm | Signed Halving Subtract 8 | | B.19 |
| SHSUB16 | {Rd,} Rn, Rm | Signed Halving Subtract 16 | | B.20 |
| SHASX | {Rd,} Rn, Rm | Signed Halving Add and Subtract with Exchange | | B.23 |
| SHSAX | {Rd,} Rn, Rm | Signed Halving Subtract and Add with Exchange | | B.24 |
| UADD8 | {Rd,} Rn, Rm | Unsigned Add 8 | GE[3:0] | B.69 |
| UADD16 | {Rd,} Rn, Rm | Unsigned Add 16 | GE[3:0] | B.70 |
| USUB8 | {Rd,} Rn, Rm | Unsigned Subtract 8 | GE[3:0] | B.73 |
| USUB16 | {Rd,} Rn, Rm | Unsigned Subtract 16 | GE[3:0] | B.74 |
| UASX | {Rd,} Rn, Rm | Unsigned Add and Subtract with Exchange | GE[3:0] | B.77 |
| USAX | {Rd,} Rn, Rm | Unsigned Subtract and Add with Exchange | GE[3:0] | B.78 |
| UQADD8 | {Rd,} Rn, Rm | Unsigned Saturating Add 8 | Q | B.85 |
| UQADD16 | {Rd,} Rn, Rm | Unsigned Saturating Add 16 | Q | B.84 |
| UQSUB8 | {Rd,} Rn, Rm | Unsigned Saturating Subtract 8 | Q | B.87 |
| UQSUB16 | {Rd,} Rn, Rm | Unsigned Saturating Subtract 16 | Q | B.86 |
| UQASX | {Rd,} Rn, Rm | Unsigned Saturating Add and Subtract with Exchange | Q | B.88 |

**Table 5.53** Syntax of SIMD and Saturating Instructions—*Cont'd*

| Mnemonic | Operands | Brief Description | Flags | Figure |
|----------|----------|-------------------|-------|--------|
| UQSAX | {Rd,} Rn, Rm | Unsigned Saturating Subtract and Add with Exchange | Q | B.89 |
| UHADD8 | {Rd,} Rn, Rm | Unsigned Halving Add 8 | | B.71 |
| UHADD16 | {Rd,} Rn, Rm | Unsigned Halving Add 16 | | B.72 |
| UHSUB8 | {Rd,} Rn, Rm | Unsigned Halving Subtract 8 | | B.75 |
| UHSUB16 | {Rd,} Rn, Rm | Unsigned Halving Subtract 16 | | B.76 |
| UHASX | {Rd,} Rn, Rm | Unsigned Halving Add and Subtract with Exchange | | B.79 |
| UHSAX | {Rd,} Rn, Rm | Unsigned Halving Subtract and Add with Exchange | | B.80 |
| USAD8 | {Rd,} Rn, Rm | Unsigned Sum of Absolute Differences | | B.81 |
| USADA8 | {Rd,} Rn, Rm, Ra | Unsigned Sum of Absolute Differences and Accumulate | | B.82 |
| USAT16 | Rd, #imm, Rn | Unsigned saturate 2 signed 16-bit values | Q | B.83 |
| SSAT16 | Rd, #imm, Rn | Signed saturate 2 signed 16-bit values | Q | B.62 |
| SEL | {Rd,} Rn, Rm | Select bytes base on GE bits | | B.25 |
| USAT | {Rd,} #imm, Rn {, LSL #n} {Rd,} #imm, Rn {, ASR #n} | Unsigned saturate (optionally shifted) vale | Q | 5.12 |
| SSAT | {Rd,} #imm, Rn {, LSL #n} {Rd,} #imm, Rn {, ASR #n} | Signed saturate (optionally shifted) vale | Q | 5.11 |
| QADD | {Rd,} Rn, Rm | Saturating Add | Q | B.3 |
| QDADD | {Rd,} Rn, Rm | Saturating double and Add | Q | B.6 |
| QSUB | {Rd,} Rn, Rm | Saturating Subtract | Q | B.7 |
| QDSUB | {Rd,} Rn, Rm | Saturating double and Subtract | Q | B.12 |

in any of the steps during a computation flow. Therefore the Q bit does not get cleared unless explicitly specified.

The syntaxes of these instructions are given in Table 5.56.

### 5.7.4 Packing and unpacking

A number of instructions are available to allow easy packing and unpacking of SIMD data (Table 5.57). Some of these instructions support barrel shift or rotate of the second operand. The shift and rotate operation is optional, and in the value

**Table 5.54** Multiply and MAC Instructions Available in both Cortex-M3 and Cortex-M4 Processors

| Instructions | Descriptions (size) | Flags |
|---|---|---|
| MUL / MULS | Unsigned multiply (32b x 32b = 32b) | None, or N and Z |
| UMULL | Unsigned Multiply (32b x 32b = 64b) | None |
| UMLAL | Unsigned MAC ((32b x 32b) + 64b = 64b) | None |
| SMULL | Signed Multiply (32b x 32b = 64b) | None |
| SMLAL | Signed MAC ((32b x 32b) + 64b = 64b) | None |

**Table 5.55** Summary of Multiply and MAC Operations

| Instructions | Descriptions | Flags |
|---|---|---|
| UMAAL | Unsigned MAC ((32b x 32b) + 32b + 32b = 64b) | None |
| SMULxy | Signed multiply (16b x 16b = 32b)<br>SMULBB: lower half-word x lower half-word<br>SMULBT: lower half-word x upper half-word<br>SMULTB: upper half-word x lower half-word<br>SMULTT: upper half-word x upper half-word | |
| SMLAxy | Signed MAC ((16b x 16b) + 32b = 32b)<br>SMLABB: (lower half-word x lower half-word) + word<br>SMLABT: (lower half-word x upper half-word) + word<br>SMLATB: (upper half-word x lower half-word) + word<br>SMLATT: (upper half-word x upper half-word) + word | Q |
| SMULWx | Signed multiply ((32b x 16b = 32b, return upper 32-bit of result, lowest 16-bit ignore)<br>SMULWB: word x lower half-word<br>SMULWT: word x upper half-word | |
| SMLAWx | Signed MAC ((32b x 16b) + 32b<<16 = 32b, return upper 32-bit of result, lowest 16-bit ignored)<br>SMLAWB: (word x lower half-word) + (word<<16)<br>SMLAWT: (word x upper half-word) + (word<<16) | Q |
| SMMUL | Signed multiply (32b x 32b = 32b, return upper 32-bit, lowest 32-bit ignored) | |
| SMMULR | Signed multiply with round (32b x 32b = 32b, round then return upper 32-bit, lowest 32-bit ignored) | |
| SMMLA | Signed MAC ((32b x 32b) + 32b<<32)= 32b, return upper 32-bit, lowest 32-bit ignored) | |
| SMMLAR | Signed MAC with round ((32b x 32b) + 32b<<32 = 32b, round and return upper 32-bit, lowest 32-bit ignored) | |
| SMMLS | Signed multiply and subtract (32b<<32 – (32b x 32b) = 32b, return upper 32-bit, lowest 32-bit ignored) | |
| SMMLSR | Signed multiply and subtract with round (32b<<32 – (32b x 32b) = 32b, round then return upper 32-bit, lowest 32-bit ignored) | |

**Table 5.55** Summary of Multiply and MAC Operations—*Cont'd*

| Instructions | Descriptions | Flags |
|---|---|---|
| SMLALxy | Signed MAC ((16b x 16b) + 64b = 64b)<br>SMLALBB: (lower half-word x lower half-word)<br>+ double-word<br>SMLALBT: (lower half-word x upper half-word)<br>+ double-word<br>SMLALTB: (upper half-word x lower half-word)<br>+ double-word<br>SMLALTT: (upper half-word x upper half-word)<br>+ double-word | |
| SMUAD | Signed dual multiply then add ((16b x 16b)<br>+ (16b x 16b) = 32b) | Q |
| SMUADX | Signed dual multiply with exchange then add ((16b x 16b)<br>+ (16b x 16b) = 32b) | Q |
| SMUSD | Signed dual multiply then subtract ((16b x 16b) -<br>(16b x 16b) = 32b) | |
| SMUSDX | Signed dual multiply with exchange then subtract<br>((16b x 16b) - (16b x 16b) = 32b) | |
| SMLAD | Signed dual multiply then add and accumulate<br>((16b x 16b) + (16b x 16b) + 32b = 32b) | Q |
| SMLADX | Signed dual multiply with exchange then add and<br>accumulate ((16b x 16b) + (16b x 16b) + 32b = 32b) | Q |
| SMLSD | Signed dual multiply then subtract and accumulate<br>((16b x 16b) - (16b x 16b) + 32b = 32b) | Q |
| SMLSDX | Signed dual multiply with exchange then subtract and<br>accumulate ((16b x 16b) - (16b x 16b) + 32b = 32b) | Q |
| SMLALD | Signed dual multiply then add and accumulate<br>((16b x 16b) + (16b x 16b) + 64b = 64b) | |
| SMLALDX | Signed dual multiply with exchange then add and<br>accumulate ((16b x 16b) + (16b x 16b) + 64b = 64b) | |
| SMLSLD | Signed dual multiply then subtract and accumulate<br>((16b x 16b) - (16b x 16b) + 64b = 64b) | |
| SMLSLDX | Signed dual multiply with exchange then subtract and<br>accumulate ((16b x 16b) - (16b x 16b) + 64b = 64b) | |

of "n" for rotate (ROR) in the table below can be 8, 16, or 24. The shift operation in PKHBT and PKHTB can be any number of bits.

### 5.7.5 Floating point instructions

To support floating point operations, the Cortex®-M4 also has a number of instructions for floating point data processing as well as floating point date transfers (Table 5.58). These instructions are not available if the Cortex-M4 device you are using does not have a floating point unit. All the floating point instructions start with the letter V.

**Table 5.56** Syntax of Multiply and MAC Instructions

| Mnemonic | Operands | Brief Description | Flags | Figure |
| --- | --- | --- | --- | --- |
| MUL{S} | {Rd,} Rn, Rm | Unsigned Multiply, 32-bit result | N, Z | |
| SMULL | RdLo, RdHi, Rn, Rm | Signed Multiply, 64-bit result | | B.26 |
| SMLAL | RdLo, RdHi, Rn, Rm | Signed Multiply and Accumulate, 64-bit result | | B.27 |
| UMULL | RdLo, RdHi, Rn, Rm | Unsigned Multiply, 64-bit result | | B.90 |
| UMLAL | RdLo, RdHi, Rn, Rm | Unsigned Multiply and Accumulate, 64-bit result | | B.91 |
| UMAAL | RdLo, RdHi, Rn, Rm | Unsigned Multiply Accumulate | | B.92 |
| SMULBB | {Rd,} Rn, Rm | Signed Multiply (half-words) | | B.28 |
| SMULBT | {Rd,} Rn, Rm | Signed Multiply (half-words) | | B.29 |
| SMULTB | {Rd,} Rn, Rm | Signed Multiply (half-words) | | B.30 |
| SMULTT | {Rd,} Rn, Rm | Signed Multiply (half-words) | | B.31 |
| SMLABB | Rd, Rn, Rm, Ra | Signed Multiply Accumulate (half-words) | Q | B.36 |
| SMLABT | Rd, Rn, Rm, Ra | Signed Multiply Accumulate (half-words) | Q | B.37 |
| SMLATB | Rd, Rn, Rm, Ra | Signed Multiply Accumulate (half-words) | Q | B.38 |
| SMLATT | Rd, Rn, Rm, Ra | Signed Multiply Accumulate (half-words) | Q | B.39 |
| SMULWB | Rd, Rn, Rm, Ra | Signed Multiply (word by half-word) | | B.40 |
| SMULWT | Rd, Rn, Rm, Ra | Signed Multiply (word by half-word) | | B.41 |
| SMLAWB | Rd, Rn, Rm, Ra | Signed Multiply Accumulate (word by half-word) | Q | B.42 |
| SMLAWT | Rd, Rn, Rm, Ra | Signed Multiply Accumulate (word by half-word) | Q | B.43 |
| SMMUL | {Rd,} Rn, Rm | Signed Most significant word Multiply | | B.32 |
| SMMULR | {Rd,} Rn, Rm | Signed Most significant word Multiply with rounded result | | B.33 |
| SMMLA | Rd, Rn, Rm, Ra | Signed Most significant word Multiply Accumulate | | B.34 |
| SMMLAR | Rd, Rn, Rm, Ra | Signed Most significant word Multiply Accumulate with rounded result | | B.35 |
| SMMLS | Rd, Rn, Rm, Ra | Signed Most significant word Multiply Subtract | | B.44 |

**Table 5.56** Syntax of Multiply and MAC Instructions—*Cont'd*

| Mnemonic | Operands | Brief Description | Flags | Figure |
|---|---|---|---|---|
| SMMLSR | Rd, Rn, Rm, Ra | Signed Most significant word Multiply Subtract with rounded result | | B.45 |
| SMLALBB | RdLo, RdHi, Rn, Rm | Signed Multiply Accumulate Long (half-words) | | B.46 |
| SMLALBT | RdLo, RdHi, Rn, Rm | Signed Multiply Accumulate Long (half-words) | | B.47 |
| SMLALTB | RdLo, RdHi, Rn, Rm | Signed Multiply Accumulate Long (half-words) | | B.48 |
| SMLALTT | RdLo, RdHi, Rn, Rm | Signed Multiply Accumulate Long (half-words) | | B.49 |
| SMUAD | {Rd,} Rn, Rm | Signed Dual Multiply Add | Q | B.50 |
| SMUADX | {Rd,} Rn, Rm | Signed Dual Multiply Add with eXchange | Q | B.51 |
| SMUSD | {Rd,} Rn, Rm | Signed Dual Multiply Subtract | | B.56 |
| SMUSDX | {Rd,} Rn, Rm | Signed Dual Multiply Subtract with eXchange | | B.57 |
| SMLAD | Rd, Rn, Rm, Ra | Signed Multiply Accumulate Dual | Q | B.52 |
| SMLADX | Rd, Rn, Rm, Ra | Signed Multiply Accumulate Dual with eXchange | Q | B.53 |
| SMLSD | Rd, Rn, Rm, Ra | Signed Multiply Subtract Dual | Q | B.58 |
| SMLSDX | Rd, Rn, Rm, Ra | Signed Multiply Subtract Dual with eXchange | Q | B.59 |
| SMLALD | RdLo, RdHi, Rn, Rm | Signed Multiply Accumulate Long Dual | | B.54 |
| SMLALDX | RdLo, RdHi, Rn, Rm | Signed Multiply Accumulate Long Dual with eXchange | | B.55 |
| SMLSLD | RdLo, RdHi, Rn, Rm | Signed Multiply Subtract Long Dual | | B.60 |
| SMLSLDX | RdLo, RdHi, Rn, Rm | Signed Multiply Subtract Long Dual with eXchange | | B.61 |

Before using any floating point instruction, you must first enable the floating point unit by setting the CP11, CP10 bit fields in the Co-processor Access Control Register (SCB->CPACR at address 0xE000ED88). This is usually done within the *SystemInit(void)* function in the device initialization code provided by the microcontroller vendors. Also, the "__FPU_PRESENT" directive in the device header file should be set to 1.

In floating point operations, the input operands have to be converted to floating point format. Otherwise we can call this Operand NaN (Not a Number) and some forms of NaN can be used to signal floating point exceptions. See chapter 13 for more details.

**Table 5.57** Syntax of Packing and Unpacking Instructions

| Instructions | Operands | Descriptions | |
|---|---|---|---|
| PKHBT | {Rd,} Rn, Rm {,LSL #imm} | Pack Half-word with lower half from 1st operand and upper half from shifted 2nd operand | B.1 |
| PKHTB | {Rd,} Rn, Rm {,ASR #imm} | Pack Half-word with upper half from 1st operand and lower half from shifted 2nd operand | B.2 |
| SXTB | Rd,Rm {,ROR #n} | Signed eXtend Byte | B.63 |
| SXTH | Rd,Rm {,ROR #n} | Signed eXtend Half-word | B.67 |
| UXTB | Rd,Rm {,ROR #n} | Unsigned eXtend Byte | B.93 |
| UXTH | Rd,Rm {,ROR #n} | Unsigned eXtend Half-word | B.97 |
| SXTB16 | Rd,Rm {,ROR #n} | Signed eXtend two bytes to two half-words | B.64 |
| UXTB16 | Rd,Rm {,ROR #n} | Unsigned eXtend two bytes to two half-words | B.94 |
| SXTAB | {Rd,} Rn, Rm{,ROR #n} | Signed eXtend and Add byte | B.65 |
| SXTAH | {Rd,} Rn, Rm{,ROR #n} | Signed eXtend and Add half-word | B.68 |
| SXTAB16 | {Rd,} Rn, Rm{,ROR #n} | Signed extend two bytes to half-words and dual add | B.66 |
| UXTAB | {Rd,} Rn, Rm{,ROR #n} | Unsigned eXtend and Add byte | B.95 |
| UXTAH | {Rd,} Rn, Rm{,ROR #n} | Unsigned eXtend and Add half-word | B.98 |
| UXTAB16 | {Rd,} Rn, Rm{,ROR #n} | Unsigned extend two bytes to half-words and dual add | B.96 |

## 5.8 Barrel shifter

A number of 32-bit Thumb instructions can make use of the barrel shifter feature in the Cortex®-M3 and Cortex-M4 processors. For example, if the second operand is an ARM® core register ($Rm$), an optional shift can be applied for some of the data processing instructions before the data processing (Figure 5.14).

For example:

$$mnemonic \qquad Rd, Rn, Rm, <shift>$$

where <shift> can be one of the following:

| | |
|---|---|
| ASR #n | Arithmetic shift right $n$ bits, $1 \leq n \leq 32$ |
| LSL #n | Logical shift left $n$ bits, $1 \leq n \leq 31$ |
| LSR #n | Logical shift right $n$ bits, $1 \leq n \leq 32$ |
| ROR #n | Rotate right $n$ bits, $1 \leq n \leq 31$ |
| RRX | Rotate right one bit, with extend (include C flag in rotation) |

**Table 5.58** Floating Point Instructions

| Instructions | Operands | Operations |
|---|---|---|
| VABS.F32 | Sd, Sm | Floating point Absolute value |
| VADD.F32 | {Sd,} Sn, Sm | Floating point Add |
| VCMP{E}.F32 | Sd, Sm | Compare two floating point registers<br>VCMP: Raise Invalid Operation exception if either operand is a signaling NaN<br>VCMPE: Raise Invalid Operation exception if either operand is any types of NaN |
| VCMP{E}.F32 | Sd, #0.0 | Compare a floating point register to zero (#0.0) |
| VCVT.S32.F32 | Sd, Sm | Convert from signed 32-bit integer to floating point (round toward zero rounding mode) |
| VCVTR.S32.F32 | Sd, Sm | Convert from signed 32-bit integer to floating point (use rounding mode specified by FPCSR) |
| VCVT.U32.F32 | Sd, Sm | Convert from unsigned 32-bit integer to floating point (round toward zero rounding mode) |
| VCVTR.U32.F32 | Sd, Sm | Convert from unsigned 32-bit integer to floating point (use rounding mode specified by FPCSR) |
| VCVT.F32.S32 | Sd, Sm | Convert from floating point to 32-bit signed integer |
| VCVT.F32.U32 | Sd, Sm | Convert from floating point to 32-bit unsigned integer |
| VCVT.S16.F32 | Sd, Sd, #fbit | Convert from signed 16-bit fixed point value to floating point.<br>#fbit range from 1 to 16 (fraction bits) |
| VCVT.U16.F32 | Sd, Sd, #fbit | Convert from unsigned 16-bit fixed point value to floating point. #fbit range from 1 to 16 (fraction bits) |
| VCVT.S32.F32 | Sd, Sd, #fbit | Convert from signed 32-bit fixed point value to floating point.<br>#fbit range from 1 to 32 (fraction bits) |
| VCVT.U32.F32 | Sd, Sd, #fbit | Convert from unsigned 32-bit fixed point value to floating point. #fbit range from 1 to 32 (fraction bits) |
| VCVT.F32.S16 | Sd, Sd, #fbit | Convert from floating point to signed 16-bit fixed point value.<br>#fbit range from 1 to 16 (fraction bits) |
| VCVT.F32.U16 | Sd, Sd, #fbit | Convert from floating point to unsigned 16-bit fixed point value. #fbit range from 1 to 16 (fraction bits) |

*(Continued)*

**Table 5.58** Floating Point Instructions—*Cont'd*

| Instructions | Operands | Operations |
|---|---|---|
| VCVT.F32.S32 | Sd, Sd, #fbit | Convert from floating point to signed 32-bit fixed point value.<br>#fbit range from 1 to 32 (fraction bits) |
| VCVT.F32.U32 | Sd, Sd, #fbit | Convert from floating point to unsigned 32-bit fixed point value. #fbit range from 1 to 32 (fraction bits) |
| VCVTB.F32.F16 | Sd, Sm | Convert from Single precision to Half Precision (use bottom 16-bit, upper 16-bit unaffected) |
| VCVTF.F32.F16 | Sd, Sm | Convert from Single precision to Half Precision (use upper 16-bit, bottom 16-bit unaffected) |
| VCVTB.F16.F32 | Sd, Sm | Convert from Half Precision (use bottom 16-bit) to Single precision |
| VCVTF.F16.F32 | Sd, Sm | Convert from Half Precision (use upper 16-bit) to Single precision |
| VDIV.F32 | {Sd,} Sn, Sm | Floating point Divide |
| VFMA.F32 | Sd, Sn, Sm | Floating point Fused Multiply Accumulate Sd=Sd+(Sn*Sm) |
| VFMS.F32 | Sd, Sn, Sm | Floating point Fused Multiply Subtract Sd=Sd-(Sn*Sm) |
| VFNMA.F32 | Sd, Sn, Sm | Floating point Fused Negate Multiply Accumulate Sd=(-Sd)+(Sn*Sm) |
| VFNMS.F32 | Sd, Sn, Sm | Floating point Fused Negate Multiply Subtract Sd=(-Sd)-(Sn*Sm) |
| VLDMIA.32 | Rn{!}, {S_regs} | Floating point Multiple Load Increment After |
| VLDMDB.32 | Rn{!}, {S_regs} | Floating point Multiple Load Decrement Before |
| VLDMIA.64 | Rn{!}, {D_regs} | Floating point Multiple Load Increment After |
| VLDMDB.64 | Rn{!}, {D_regs} | Floating point Multiple Load Decrement Before |
| VLDR.32 | Sd,[Rn{, #imm}] | Load a single-precision data from memory (register + offset) |
| VLDR.32 | Sd, label | Load a single-precision data from memory (literal data) |
| VLDR.32 | Sd, [PC, #imm] | Load a single-precision data from memory (literal data) |
| VLDR.64 | Dd,[Rn{, #imm}] | Load a double-precision data from memory (register + offset) |
| VLDR.64 | Dd, label | Load a double-precision data from memory (literal data) |
| VLDR.64 | Dd, [PC, #imm] | Load a double-precision data from memory (literal data) |

**Table 5.58** Floating Point Instructions—*Cont'd*

| Instructions | Operands | Operations |
|---|---|---|
| VMLA.F32 | Sd, Sn, Sm | Floating point Multiply Accumulate Sd=Sd+(Sn*Sm) |
| VMLS.F32 | Sd, Sn, Sm | Floating point Multiply Subtract Sd= Sd-(Sn*Sm) |
| VMOV{.F32} | Rt, Sm | Copy floating point (scalar) to ARM core register |
| VMOV{.F32} | Sn, Rt | Copy ARM core register to floating point (scalar) |
| VMOV{.F32} | Sd, Sm | Copy floating point register Sm to Sd (single precision) |
| VMOV | Sm, Sm1, Rt, Rt2 | Copy 2 ARM core registers to 2 single-precision register |
| VMOV | Rt, Rt2, Sm, Sm1 | Copy 2 single-precision register to 2 ARM core registers (Alternate syntax: VMOV Rt, Rt2, Dm) |
| VMRS.F32 | Rt, FPCSR | Copy value in FPSCR, a floating point unit system register to Rt |
| VMRS | APSR_nzcv, FPCSR | Copy flags from FPSCR to the flags in APSR |
| VMSR | FPSCR, Rt | Copy Rt to FPSCR, a floating point unit system register |
| VMOV.F32 | Sd, #imm | Move single-precision value into floating point register |
| VMUL.F32 | {Sd,} Sn, Sm | Floating point Multiply |
| VNEG.F32 | Sd, Sm | Floating point Negate |
| VNMUL | {Sd,} Sn, Sm | Floating point Multiply with negation Sd = - (Sn * Sm) |
| VNMLA | Sd, Sn, Sm | Floating point Multiply Accumulate with negation Sd = -(Sd + (Sn * Sm)) |
| VNMLS | Sd, Sn, Sm | Floating point Multiply Accumulate with negation Sd = -(Sd - (Sn * Sm)) |
| VPUSH.32 | {S_regs} | Floating point single-precision register(s) push |
| VPUSH.64 | {D_regs} | Floating point double-precision register(s) push |
| VPOP.32 | {S_regs} | Floating point single-precision register(s) pop |
| VPOP.64 | {D_regs} | Floating point double-precision register(s) pop |
| VSQRT.F32 | Sd, Sm | Floating point Square Root |
| VSTMIA.32 | Rn{!}, <S_regs> | Floating point Multiple Store Increment After |
| VSTMDB.32 | Rn{!}, <S_regs> | Floating point Multiple Store Decrement Before |

*(Continued)*

**Table 5.58** Floating Point Instructions—*Cont'd*

| Instructions | Operands | Operations |
|---|---|---|
| VSTMIA.64 | Rn{!}, <D_regs> | Floating point Multiple Store Increment After |
| VSTMDB.64 | Rn{!}, <D_regs> | Floating point Multiple Store Decrement Before |
| VSTR.32 | Sd,[Rn{, #imm}] | Store a single-precision data to memory (register + offset) |
| VSTR.64 | Dd,[Rn{, #imm}] | Store a double-precision data to memory (register + offset) |
| VSUB.F32 | {Sd,} Sn, Sm | Floating point Subtract |

The shift is optional. So you can write the instruction as:

*mnemonic*          *Rd, Rn, Rm*

if the shift/rotate operation is not required.

The barrel shifter operations can be applied as shown in Table 5.59.

The barrel shifter is also useful in memory access instructions; for example, in the calculation of an address:

```
LDR Rd, [Rn, Rm, LSL #n]
```

This is particularly useful in data array handling where the address equals to array_base + (index * $2^n$).

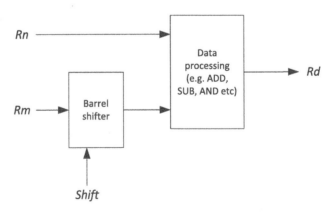

**FIGURE 5.14**

Barrel shifter

**Table 5.59** Data Processing Instructions with Barrel Shifter Support

| Instructions | | Descriptions |
|---|---|---|
| MOV{S} | Rd, Rm, <shift> | Move |
| MVN{S} | Rd, Rm, <shift> | Move Not |
| ADD{S} | Rd, Rm, Rn, <shift> | Add |
| ADC{S} | Rd, Rm, Rn, <shift> | Add with carry |
| SUB{S} | Rd, Rm, Rn, <shift> | Subtract |
| SBC{S} | Rd, Rm, Rn, <shift> | Subtract with borrow |
| RSB{S} | Rd, Rm, Rn, <shift> | Reverse subtract |
| AND{S} | Rd, Rm, Rn, <shift> | Logical And |
| ORR{S} | Rd, Rm, Rn, <shift> | Logical Or |
| EOR{S} | Rd, Rm, Rn, <shift> | Logical Exclusive Or |
| BIC{S} | Rd, Rm, Rn, <shift> | Logical And Not (bit clear) |
| ORN{S} | Rd, Rm, Rn, <shift> | Logical Or Not |
| CMP | Rn, Rm, <shift> | Compare |
| CMN | Rn, Rm, <shift> | Compare Negative |
| TEQ | Rn, Rm, <shift> | Test Equivalent (bitwise XOR) |
| TST | Rn, Rm, <shift> | Test (bitwise AND) |

## 5.9 Accessing special instructions and special registers in programming

### 5.9.1 Overview

Some instructions cannot be generated by the C compiler with normal C statements. For example, instructions that trigger sleep (WFI, WFE) and memory barrier (ISB, DSB, and DMB) are such instructions. There are several ways to get around this:

- Use intrinsic functions provided in the CMSIS (inside CMSIS-Core header files).
- Use compiler-specific intrinsic functions.
- Use inline assembler (or embedded assembler in ARM®/Keil™ toolchains) to insert the required instructions.
- Use compiler specific features such as keywords (e.g., __svc in ARM/Keil toolchains can be used to generate SVC instruction) or idiom recognitions.

In some cases we also need to access special registers inside the processor. There are again several options:

- Use processor access functions provided in CMSIS-Core (Appendix E.4).
- Use compiler specific feature such as Name Register variable feature in ARM C compiler (section 5.9.5).
- Insert assembly code using inline assembler or embedded assembler.

In general, CMSIS-Core functions are preferred because they are portable (compiler independent).

## 5.9.2 Intrinsic functions

There are two types of intrinsic functions, discussed in the following section.

### CMSIS-core intrinsic functions

Header files in the CMSIS-Core define a set of intrinsic functions for accessing special instructions. These functions can be found in the CMSIS-Core file "core_cmInstr.h" and "core_cm4_simd.h" (for Cortex®-M4 SIMD instructions).

A list of CMSIS intrinsic functions is provided in Appendix E.

### Compiler-specific intrinsic functions

Using a compiler-specific intrinsic function is just like using a C function, although the definition of the function is built-in inside the C compiler. Very often this method provides the most optimized code, however, the function definitions are tool-dependent and therefore the application code will not be portable across different toolchains. Examples can be found in Chapter 20 (section 20.6).

In some toolchains, the CMSIS-Core intrinsic function is directly supported by the compiler. In some other compilers (e.g., ARM® C Compiler) the implementation of the two types of intrinsic are separated and it is possible that the code generated from CMSIS-Core intrinsic is less optimized.

Be aware that sometimes the compiler-specific intrinsic functions can have function names which are very similar to those in CMSIS-Core (e.g., void __wfi(void) in the ARM C compiler, and __WFE(void) in CMSIS-Core). Also, the parameter required could be different.

Chapter 20 has additional information on this subject (section 20.6).

## 5.9.3 Inline assembler and embedded assembler

In some cases, you might need to insert assembly instructions in C code using inline assembler; for example, when you want to insert a SVC instruction in gcc. This is also useful for creating optimized code, as it gives you good control of what instruction sequence is generated. However, an application created using inline assembler is toolchain-dependent (less portable). The ARM® C compiler (including Keil™ MDK-ARM and ARM DS-5™ Professional) also supports a feature called embedded assembler. In embedded assembler, you can create an assembly function inside a C program file.

For ARM C Compilers prior to version 5.0.1 or older versions of Keil MDK-ARM, inline assembler only works for ARM instructions and not Thumb instructions. So you need to use embedded assembler to insert assembly code for the old versions of ARM toolchains. Inline assembler is supported since ARM C Compiler 5.01 or Keil MDK-ARM 4.60.

More information on inline assembler and embedded assembler can be found in Chapter 20, in sections 20.5.3 and 20.5.4.

### 5.9.4 Using other compiler-specific features

Most C compilers come with various features that are handy for generating special instructions. For example, in the ARM® C Compiler or Keil™ MDK-ARM, you can insert SVC instruction using the __svc keyword (see the example in Chapter 10).

Another compiler-specific feature is called idiom recognition. By writing a data operation in a C statement of certain form, the C compiler can recognize the function and replace the operation with a simple instruction. This method is toolchain-specific. More information on idiom recognition is covered in section 20.7.

### 5.9.5 Access to special registers

CMSIS-Core provides a number of functions for accessing special registers in the Cortex®-M3/M4 processors. Details of these functions are available in Appendix E.4.

**Table 5.60** Processor Registers that can be Accessed Using "Named Register Variables" Feature

| Register | Character String in __asm |
|---|---|
| APSR | "apsr" |
| BASEPRI | "basepri" |
| BASEPRI_MAX (see section 7.10.3) | "basepri_max" |
| CONTROL | "control" |
| EAPSR (EPSR + APSR) | "eapsr" |
| EPSR | "epsr" |
| FAULTMASK | "faultmask" |
| IAPSR (IPSR + APSR) | "iapsr" |
| IEPSR (IPSR + EPSR) | "iepsr" |
| IPSR | "ipsr" |
| MSP | "msp" |
| PRIMASK | "primask" |
| PSP | "psp" |
| PSR | "psr" |
| r0 to r12 | "r0" to "r12" |
| r13 | "r13" or "sp" |
| r14 | "r14" or "lr" |
| r15 | "r15" or "pc" |
| XPSR | "xpsr" |

If you are using the ARM® C compiler (including Keil™ MDK-ARM, ARM DS-5™ Professional), you can also use the "Named register variables" feature to access to special registers. The syntax is:

```
register type var-name __arm(reg);
```

where:

| | |
|---|---|
| *type* | - data type of the named register variable. |
| *var-name* | - the name of the named register variable. |
| *reg* | - a character string to indicate which register to use. |

For example, you can declare a register name:

```
register unsigned int reg_apsr __asm("apsr");
reg_apsr = reg_apsr & 0xF7FFFFFFUL; // Clear Q bit in APSR
```

You can use the Named register variables feature to access the registers listed in Table 5.60.

# Memory System

6

## CHAPTER OUTLINE

## 6.1 Overview of memory system features

The Cortex®-M processors have 32-bit memory addressing and therefore have 4GB memory space. The memory space is unified, which means instructions and data share the same address space. The 4GB memory space is architecturally divided into a number of regions, which will be explained in Section 6.2 memory maps. In addition, the memory system of the Cortex-M3 and Cortex-M4 processors supports a number of features:

- Multiple bus interfaces to allow concurrent instructions and data accesses (Harvard bus architecture)
- Bus interface designs based on AMBA® (Advanced Microcontroller Bus Architecture), a de facto on-chip bus standard: AHB (AMBA High-performance Bus) Lite protocol for pipelined operations in memory and system bus, and APB (Advanced Peripheral Bus) protocol for communication to debug components
- Support both little endian and big endian memory systems
- Support for unaligned data transfers
- Support exclusive accesses (for semaphore operations in systems with an embedded OS or RTOS)

- Bit addressable memory spaces (bit-band)
- Memory attributes and access permissions for different memory regions
- An optional Memory Protection Unit (MPU). Memory attributes and access permission configurations can be programmed at runtime if the MPU is available.

The processor and the architecture allow very flexible memory configurations, so you can find Cortex-M3 and Cortex-M4 based microcontrollers with many different memory sizes and memory maps. You can also find device-specific memory system features in some Cortex-M microcontroller products, such as memory address remapping/alias.

## 6.2 Memory map

In the 4GB addressable memory space, some parts are allocated to internal peripherals within the processor such as the NVIC and debug components. The memory locations of these internal components are fixed. In addition, the memory space is architecturally divided into a number of memory regions as shown in Figure 6.1. This arrangement allows:

- The processor design to support different types of memories and devices out of the box.
- Optimized arrangement for higher performance.

Although the pre-defined memory map is fixed, the architecture has high flexibility to allow silicon designers to design their products with different memories and peripherals for better product differentiation.

First let us look at the memory region definitions, which are shown on the left-hand side of Figure 6.1. The description of the memory region definitions are listed in Table 6.1.

While it is possible to store and execute program code in SRAM and RAM regions, the processor design is not optimized for such operation and requires one extra clock cycle per instruction for each instruction fetch. As a result, the performance is slightly slower when executing program codes though the system bus.

Program execution from Peripherals, Devices, and System memory regions is not allowed.

Figure 6.1 shows that there are a number of built-in components in the memory map. They are described in Table 6.2.

The memory space for the NVIC, MPU, SCB, and various system peripherals is called the System Control Space (SCS). More information about these components will be covered in various chapters of this book.

## 6.3 Connecting the processor to memory and peripherals

The Cortex®-M processors provide generic bus interfaces based on AMBA (Advanced Microcontroller Bus Architecture). The AMBA® specification supports

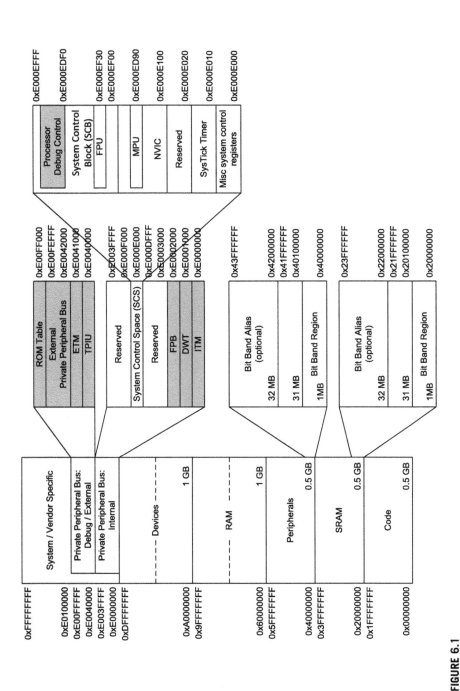

**FIGURE 6.1**

Pre-defined memory map of the Cortex®-M3 and Cortex-M4 processors (shaded areas are components for debug purpose)

**Table 6.1** Memory Regions

| Region | Address Range |
|---|---|
| Code | 0x00000000 to 0x1FFFFFFF |
| | A 512MB memory space primarily for program code, including the default vector table that is a part of the program memory. This region also allow data accesses. |
| SRAM | 0x20000000 to 0x3FFFFFFF |
| | The SRAM region is located in the next 512MB of memory space. It is primarily for connecting SRAM, mostly on-chip SRAM, but there is no limitation of exact memory type. The first 1MB of the SRAM region is bit addressable if the optional bit-band feature is included. You can also execute program code from this region. |
| Peripherals | 0x40000000 to 0x5FFFFFFF |
| | The Peripheral memory region also has the size of 512MB, and is use mostly for on-chip peripherals. Similar to SRAM region, the first 1MB of the peripheral region is bit addressable if the optional bit-band feature is included. |
| RAM | 0x60000000 to 0x9FFFFFFF |
| | The RAM region contains two slots of 512MB memory space (total 1GB) for other RAM such as off-chip memories. The RAM region can be used for program code as well as data. |
| Devices | 0xA0000000 to 0xDFFFFFFF |
| | The Device region contains two slots of 512MB memory space (total 1GB) for other peripherals such as off-chip peripherals. |
| System | 0xE0000000 to 0xFFFFFFFF |
| | The System region contains several parts: |
| | Internal Private Peripheral Bus (PPB), 0xE0040000 to 0xE00FFFFF: |
| | The internal Private Peripheral Bus (PPB) is used to access system components such as the NVIC, SysTick, MPU, as well as debug components inside the Cortex-M3/M4 processors. In most cases this memory space can only be accessed by program code running in privileged state. |
| | External Private Peripheral Bus (PPB), 0xE0040000 to 0xE00FFFFF |
| | An addition PPB region is available for additional optional debug components and so allow silicon vendors to add their own debug or vendor-specific components. This memory space can only be accessed by program code running in privileged state. Note that the base address of debug components on this bus can potentially be changed by silicon designers. |
| | Vendor-specific area, 0xE0100000 to 0xFFFFFFFF |
| | The remaining memory space is reserved for vendor-specific components and in most cases this is not used. |

**Table 6.2** Various Built-in Components in the Cortex-M3 and Cortex-M4 Memory Map

| Component | Descriptions |
|---|---|
| NVIC | Nested Vector Interrupt Controller<br>A built-in interrupt controller for exceptions (including interrupts) handling. |
| MPU | Memory Protection Unit<br>An optional programmable unit to set up memory access permissions and memory access attributes (characteristics or behaviors) for various memory regions. Some Cortex-M3 and Cortex-M4 microcontrollers might not have MPU. |
| SysTick | System Tick timer<br>A 24-bit timer designed mainly for generating regular OS interrupts. It can also be used by application code if OS is not in use. |
| SCB | System Control Block<br>A set of registers that can be used to control the behavior of the processor and provide status information. |
| FPU | Floating Point Unit<br>A number of registers are placed here for controlling the behavior of the floating point unit and to provide status information. This is available only for Cortex-M4, and only if the floating point unit is present. |
| FPB | Flash Patch and Breakpoint Unit<br>For debug operations. It contains up to eight comparators and each can be configured to generate hardware breakpoint events; for example, when an instruction at a breakpoint address is executed. It can also be used to replace instructions and therefore can be used for implementing a patch mechanism for fixed program code. |
| DWT | Data Watchpoint and Trace Unit<br>For debug and trace operations. It contains up to four comparators and each can be configured to generate data watchpoint events; for example, when a certain memory address range is accessed by software. It can also be used for generating data trace packets to allow accesses to monitored memory locations to be observed using a debugger. |
| ITM | Instrumentation Trace Macrocell<br>A component for debug and trace. It is to allow software to generate data trace stimulus, which can be captured via trace interface. It also provides time stamp package generation in the trace system. |
| ETM | Embedded Trace Macrocell<br>A component for generating instruction trace for debugging software. |
| TPIU | Trace Point Interface Unit<br>A component for converting trace packets from trace sources to trace interface protocol so that the trace data can be captured easily with a minimal set of pins. |
| ROM Table | ROM Table<br>A simple look-up table for debug tools to indicate addresses of debug and trace components, so that debug tools can identify the debug components available in the system. It also provides ID registers that could be used for system identifications. |

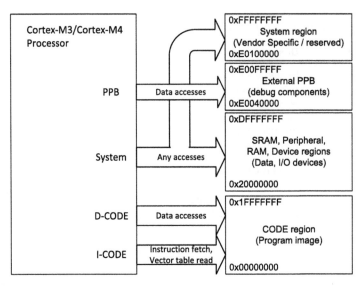

**FIGURE 6.2**

Multiple bus interface for different memory regions

several bus protocols. In the Cortex-M3 and Cortex-M4 processors, the AHB (AMBA High-performance Bus) Lite protocol is used for the main bus interfaces, and the APB protocol is used for the Private Peripheral Bus (PPB), which is mainly used for debug components. Additional bus segments based on APB can be added onto the system bus by using additional bus bridge components.

In order to provide better performance, the CODE memory region has separated the bus interfaces from the system bus, as shown in Figure 6.2. In this way, data accesses and instruction fetches can be carried out in parallel.

The separate bus arrangement also helps to improve the interrupt responsiveness because during the interrupt handling sequence, stack accesses and reading the vector table in the program image can be carried out at the same time.

Wait states can be inserted on the bus interfaces of the processor. Therefore a Cortex-M processor running at high speed can access a slower memory or a slower peripheral. The bus interfaces also support error responses. If an error condition occurs, for example, when the processor accesses a memory location that is not within valid memory space, the bus system can return an error response to the processor, which triggers a fault exception and then the software running on the processor can then report the issue or deal with it.

In a simple microcontroller design, typically the program memory is connected to the I-CODE and D-CODE bus, and the SRAM and peripherals are connected to the system bus. See Figure 6.3 for a simple arrangement of a Cortex-M3 or Cortex-M4 design.

Normally most peripherals will be connected on separate peripheral bus segments, and in a number of existing Cortex-M3 and Cortex-M4 products you

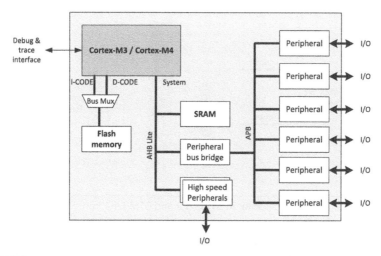

**FIGURE 6.3**

A simple Cortex-M3 or Cortex-M4 processor based system

can find multiple peripheral bus segments in the design. This allows each bus to be operated at different speeds for best power optimization. In some cases, this allows higher system bandwidth (e.g., when the peripheral system needs to support DMA accesses for Ethernet or high speed USB).

Peripheral interfaces are usually based on the APB protocol. However, for high-performance peripherals, AHB Lite could be used instead for higher bandwidth and operation speed.

Note that the Private Peripheral Bus (PPB) is not used for normal peripherals. This is because:

- PPB is privileged access only
- Only 32-bit accesses are allowed
- PPB write accesses need more clock cycles as there is no write buffer (Strongly Ordered device accesses; More details of memory attribute can be found in section 6.9, Table 6.11)
- Peripherals in PPB cannot make use of the bit-band feature (section 6.7)

In addition, there are other restrictions on PPB usage:

- Little endian only, even if the processor is configured to be big endian.
- It is accessible by the processor and debugger, but no other bus master (e.g., in a multiprocessor environment).

Two bus interfaces (I-CODE and D-CODE) are provided for the accesses to program memory. In simple designs, the two buses can be merged together using a simple bus multiplexer component provided by ARM®. Microcontroller vendors can also utilize these two interfaces to develop customized flash access accelerators

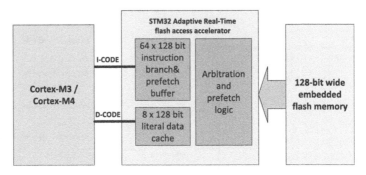

**FIGURE 6.4**

Concept of flash access accelerator in STM32F2 and STM32F4 from ST Microelectronics

to allow the processor to run much faster than to the access speed of the flash memory. For example, the STM32F2 (Cortex-M3-based microcontroller) and STM32F4 (Cortex-M4-based microcontroller) from ST Microelectronics have implemented a flash access accelerator that utilizes the two bus interface (Figure 6.4). Two set of buffers are connected to the two buses, and the buffers are optimized for the access types on of each of the buses.

The concepts and details of flash pre-fetch and branch buffers are beyond the scope of this text and will not be covered. For users of microcontrollers, such an arrangement means that microcontroller vendors can design a product that can provide high processor clock frequency, but without much performance loss resulting from the wait states required for flash memory accesses. Compared to traditional approaches using full feature cache memory systems, this approach requires smaller silicon area and requires lower power consumption.

In a number of Cortex-M3 and Cortex-M4 microcontroller products, you can also find multiple bus masters on the internal bus system such as DMA, Ethernet, and USB controllers. Often in the documentation of these products, the terms "bus matrix" or "multi-layer AHB" are mentioned. What this means is that the internal bus system contains an AHB interconnect which allows simultaneous transfers from multiple bus masters to be carried out to different memory or peripherals. For example, in LPC1700 devices from NXP, the bus matrix arrangement as shown in Figure 6.5 is used.

In the system shown in Figure 6.5, each AHB slave bus segment (which might contain multiple AHB slaves) connected to the AHB interconnect (or bus matrix) can be accessed by multiple bus masters. If two bus masters are trying to access the same bus slave segment at the same time, the internal arbiter in the AHB interconnect will delay the transfer from the lower priority master to allow the transfer from higher priority bus master to get through first.

Since there are multiple SRAM blocks in separate AHB slave bus segments, transfers between SRAM blocks and different bus masters can be carried out at the same time. For example, the processor can be using one of the SRAMs for

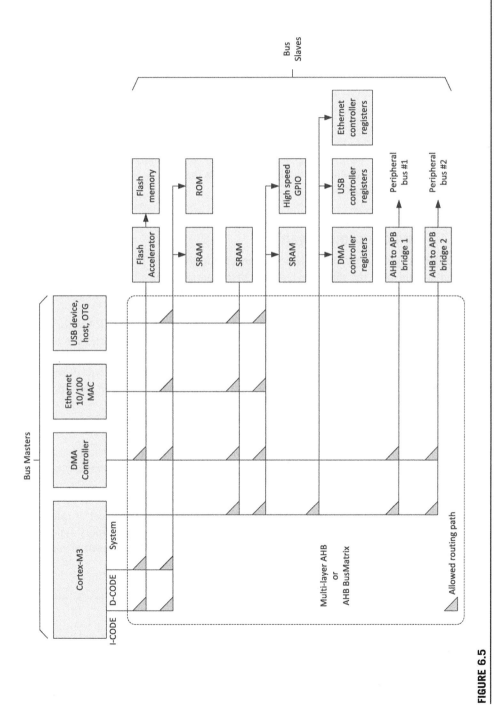

**FIGURE 6.5**

Multi-layer AHB example (NXP LPC1700)

data processing, while the DMA controller could be transferring data between a peripheral and a second SRAM block, and the Ethernet or USB controller could also be using the third SRAM block, all at the same time. This allows applications requiring high-data bandwidth (e.g., hi-speed USB) to work on low-cost microcontrollers, which normally have a relatively low clock frequency.

Many high-performance Cortex-M3 or Cortex-M4 microcontrollers from various vendors support similar approaches, although the detailed bus arrangement and the number of SRAM blocks can all be different, and the graphical representation of the bus architecture in their documentation might use different terminologies and can look completely different. In simpler applications, the embedded software developers do not need to know the details of the AHB operations. With sufficient information about the memory map and the programmer's model, developers can create high-performance embedded applications due to the performance of the Cortex-M3 or Cortex-M4 processors.

## 6.4 Memory requirements

Different types of memories can be connected to the AHB Lite bus interfaces with suitable memory interface logic. Although the bus size is 32-bit, it is possible to connect memories with other width sizes (e.g., 8 bits, 16 bits, 64 bits, 128 bits, etc.) if appropriate conversion hardware is used.

Note that although the architecture uses names like SRAM and RAM for memory region names, there is no real limitation on what types of memories can be connected to the processors. For example, some memories connected could be PSRAM, SDRAM, DDR DRAM, etc. Also, there is no real limitation on what type of program memories have to be used. For example, the program code could be in flash memory, EPROM, OTP ROM, etc.

The memory size is also flexible. Some of the low-cost Cortex®-M microcontrollers might have only 8KB of Flash and 4KB of on-chip SRAM.

The only real requirement for data memory (e.g., SRAM) is that the memory must be byte addressable and the memory interface needs to support byte, halfword, and word transfers.

Some Cortex-M3 and Cortex-M4 microcontrollers support external memories. For these microcontroller products, you need to refer to the datasheet or reference manual from the microcontroller vendors regarding supported memory types and speed because these details depend on the specification of the external memory controller used by the microcontroller design.

## 6.5 Memory endianness

The Cortex®-M3 and Cortex-M4 processors support both little endian and big endian memory systems. In normal cases the memory systems are designed to be either little endian only, or big endian only. The Cortex-M3 and Cortex-M4 processors determine the endianness of the memory system at a system reset. Once it is set, the endianness of the memory system cannot be changed until the next system reset.

In a few cases some peripheral registers can contain data of a different endianness. In that case, the applications accessing such peripheral registers need to convert the data to the correct endianness inside the program code.

Most of the existing Cortex-M microcontrollers are little endian. They have little endian memory system and peripherals. Please refer to the datasheet or reference materials from the microcontroller vendor to confirm the endianness of a given product. With a little endian memory system, the first byte of word-size data is stored in the least significant byte of the 32-bit memory location (Table 6.3).

It is possible to design a Cortex-M3 or Cortex-M4 microcontroller with a big endian memory system. In such case, the first byte of word-size data is stored in the most significant byte of the 32-bit address memory location (Table 6.4).

In the Cortex-M3 and Cortex-M4 processors, the big endian scheme is called *byte-invariant big endian*, also referred to as BE-8. The byte-invariant big endian scheme is supported on ARM® architecture v6, v6-M, v7, and v7-M. The design of BE-8 systems is different from the big endian system built on tradition ARM processors such as ARM7TDMI™. In the classic ARM processors, the big endian scheme is called *word-invariant big endian*, or BE-32. The memory view of both schemes is the same, but the byte lane usages on the bus interface during data transfers are different. Table 6.5 shows the AHB byte lane usages for BE-8 and Table 6.6 shows the AHB byte lane usages for BE-32.

**Table 6.3** Little Endian Memory View

| Address | Bits 31 – 24 | Bits 23 – 16 | Bits 15 – 8 | Bits 7 – 0 |
| --- | --- | --- | --- | --- |
| 0x0003 – 0x0000 | Byte – 0x3 | Byte – 0x2 | Byte – 0x1 | Byte – 0x0 |
| ... | | | | |
| 0x1003 – 0x1000 | Byte – 0x1003 | Byte – 0x1002 | Byte – 0x1001 | Byte – 0x1000 |
| 0x1007 – 0x1004 | Byte – 0x1007 | Byte – 0x1006 | Byte – 0x1005 | Byte – 0x1004 |
| ... | | | | |
| ... | Byte – 4xN+3 | Byte – 4xN+2 | Byte – 4xN+1 | Byte – 4xN |

**Table 6.4** Big Endian Memory View

| Address | Bits 31 – 24 | Bits 23 – 16 | Bits 15 – 8 | Bits 7 – 0 |
| --- | --- | --- | --- | --- |
| 0x0003 – 0x0000 | Byte – 0x0 | Byte – 0x1 | Byte – 0x2 | Byte – 0x3 |
| ... | | | | |
| 0x1003 – 0x1000 | Byte – 0x1000 | Byte – 0x1001 | Byte – 0x1002 | Byte – 0x1003 |
| 0x1007 – 0x1004 | Byte – 0x1004 | Byte – 0x1005 | Byte – 0x1006 | Byte – 0x1007 |
| ... | | | | |
| ... | Byte – 4xN | Byte – 4xN+1 | Byte – 4xN+2 | Byte – 4xN+3 |

**Table 6.5** The Cortex-M3 and Cortex-M4 (byte-invariant big-endian, BE-8) – Data on the AHB Bus

| Address, Size | Bits 31 – 24 | Bits 23 – 16 | Bits 15 – 8 | Bits 7 – 0 |
| --- | --- | --- | --- | --- |
| 0x1000, word | Data bit[7:0] | Data bit [15:8] | Data bit [23:16] | Data bit [31:24] |
| 0x1000, half word | - | - | Data bit [7:0] | Data bit [15:8] |
| 0x1002, half word | Data bit [7:0] | Data bit [15:8] | - | - |
| 0x1000, byte | - | - | - | Data bit [7:0] |
| 0x1001, byte | - | - | Data bit [7:0] | - |
| 0x1002, byte | - | Data bit [7:0] | - | - |
| 0x1003, byte | Data bit [7:0] | - | - | - |

**Table 6.6** ARM7TDMI (word-invariant big-endian, BE-32) – Data on the AHB Bus

| Address, Size | Bits 31 – 24 | Bits 23 – 16 | Bits 15 – 8 | Bits 7 – 0 |
| --- | --- | --- | --- | --- |
| 0x1000, word | Data bit [7:0] | Data bit [15:8] | Data bit [23:16] | Data bit [31:24] |
| 0x1000, half word | Data bit [7:0] | Data bit [15:8] | - | - |
| 0x1002, half word | - | - | Data bit [7:0] | Data bit [15:8] |
| 0x1000, byte | Data bit [7:0] | - | - | - |
| 0x1001, byte | - | Data bit [7:0] | - | - |
| 0x1002, byte | - | - | Data bit [7:0] | - |
| 0x1003, byte | - | - | - | Data bit [7:0] |

**Table 6.7** Little Endian – Data on the AHB Bus

| Address, Size | Bits 31 – 24 | Bits 23 – 16 | Bits 15 – 8 | Bits 7 – 0 |
| --- | --- | --- | --- | --- |
| 0x1000, word | Data bit [31:24] | Data bit [23:16] | Data bit [15:8] | Data bit [7:0] |
| 0x1000, half word | - | - | Data bit [15:8] | Data bit [7:0] |
| 0x1002, half word | Data bit [15:8] | Data bit [7:0] | - | - |
| 0x1000, byte | - | - | - | Data bit [7:0] |
| 0x1001, byte | - | - | Data bit [7:0] | - |
| 0x1002, byte | - | Data bit [7:0] | - | - |
| 0x1003, byte | Data bit [7:0] | - | - | - |

With little endian systems, the bus lane usages of Cortex-M3, Cortex-M4, and classic ARM processors are the same, which is shown in Table 6.7.

In the Cortex-M processors:

- Instruction fetches are always in little endian.
- Access to 0xE0000000 to 0xE00FFFFF including System Control Space (SCS), debug components, and Private Peripheral Bus (PPB) are always little endian.

If your software application needs to process big endian data and the microcontroller you are using is little endian, you can convert the data between little endian and big endian using instructions such as REV, REVSH, and REV16.

## 6.6 Data alignment and unaligned data access support

Since the memory system is 32-bit (at least from a programmer's model point of view), a data access that is 32-bit (4 bytes, or word) or 16-bit (2 bytes, or half-word) in size can either be aligned or unaligned. Aligned transfers means the address value is a multiple of the size (in bytes). For example, a word-size aligned transfer can be carried out to address 0x00000000, 0x00000004 … 0x00001000, 0x00001004, …, and so on. Similarly, a half-word size aligned transfer can be carried out to 0x00000000, 0x00000002 … 0x00001000, 0x00001002, …, and so on.

An example of an aligned and unaligned data transfer is shown in Figure 6.6.

Traditionally, most classic ARM® processors (such as the ARM7™/ARM9™/ ARM10) allow only aligned transfers. That means that in accessing memory, a

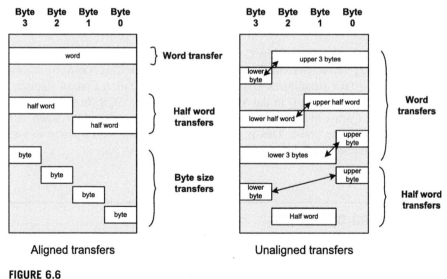

**FIGURE 6.6**

Example of aligned and unaligned data transfers in little endian memory system

word transfer must have address bit[1] and bit[0] equal to 0, and a half-word transfer must have an address bit[0] equal to 0. For example, word data can be located at 0x1000 or 0x1004, but it cannot be located at 0x1001, 0x1002, or 0x1003. For half-word data, the address can be 0x1000 or 0x1002, but it cannot be 0x1001. All byte size transfers are aligned.

The Cortex®-M3 and Cortex-M4 processors support unaligned data transfers in normal memory accesses (e.g., LDR, LDRH, STR, STRH instructions).

There are a number of limitations:

- Unaligned transfers are not supported in Load/Store multiple instructions
- Stack operations (PUSH/POP) must be aligned
- Exclusive accesses (such as LDREX or STREX) must be aligned; otherwise, a fault exception (usage fault) will be triggered
- Unaligned transfers are not supported in bit-band operations. Results will be unpredictable if you attempt to do so

When unaligned transfers are issued by the processor, they are actually converted into multiple aligned transfers by the processor's bus interface unit. This conversion is transparent, so application programmers do not have to worry about it. However, when an unaligned transfer takes place, it is broken into several separate aligned transfers and as a result it takes more clock cycles for a single data access and might not be good in situations in which high performance is required. To get the best performance, it's worth making sure that data are aligned properly.

In most cases, C compilers do not generate unaligned data accesses. It can only happen in:

- Direct manipulation of pointers
- Accessing data structures with "__packed" attributes that contain unaligned data
- Inline/Embedded Assembly code

It is also possible to set up the Cortex-M3 or Cortex-M4 processor so that an exception is triggered when an unaligned transfer takes place. This is done by setting the UNALIGN_TRP (Unaligned Trap) bit in the Configuration Control Register, CCR (address 0xE000ED14) in the System Control Block, SCB. In this way, the Cortex-M3 or Cortex-M4 processor generates usage fault exceptions when unaligned transfers take place. This is useful during software development to test whether an application produces unaligned transfers.

## 6.7 Bit-band operations

### 6.7.1 Overview

Bit-band operation support allows a single load/store operation to access (read/write) to a single data bit. In the Cortex®-M3 and Cortex-M4 processors, this is supported in two pre-defined memory regions called *bit-band regions*. One of them is

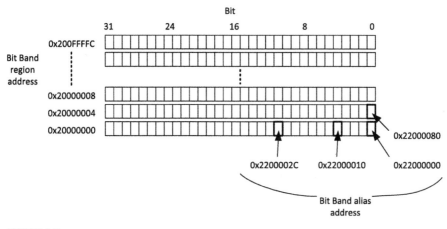

**FIGURE 6.7**

Bit accesses to bit-band region via the bit-band alias (SRAM region)

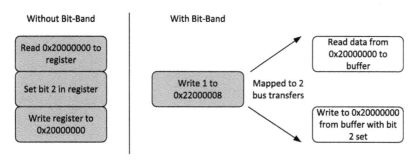

**FIGURE 6.8**

Write to bit-band alias

located in the first 1MB of the SRAM region, and the other is located in the first 1MB of the peripheral region. These two memory regions can be accessed like normal memory, but they can also be accessed via a separate memory region called the *bit-band alias*. When the bit-band alias address is used, each individual bit can be accessed separately in the least significant bit (LSB) of each word-aligned address (Figure 6.7).

For example, to set bit 2 in word data in address 0x20000000, instead of using three instructions to read the data, set the bit, and then write back the result, this task can be carried out by a single instruction (Figure 6.8).

The assembler sequence for these two cases could be like the one shown in Figure 6.9.

| Without Bit-Band | With Bit-Band |
|---|---|
| ```
LDR    R0,=0x20000000 ; Setup address
LDR    R1, [R0]        ; Read
ORR.W  R1, #0x4        ; Modify bit
STR    R1, [R0] ; Write back result
``` | ```
LDR    R0,=0x22000008 ; Setup address
MOV    R1, #1          ; Setup data
STR    R1, [R0]        ; Write
``` |

**FIGURE 6.9**

Example assembler sequence to write a bit with and without bit-band

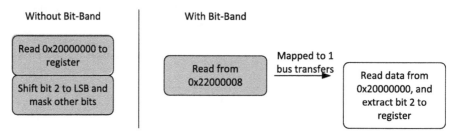

**FIGURE 6.10**

Read from the bit-band alias

| Without Bit-Band | With Bit-Band |
|---|---|
| ```
LDR      R0,=0x20000000 ; Setup address
LDR      R1, [R0]        ; Read
UBFX.W  R1, R1, #2, #1 ; Extract bit[2]
``` | ```
LDR    R0,=0x22000008 ; Setup address
LDR    R1, [R0]        ; Read
``` |

**FIGURE 6.11**

Read from the bit-band alias

Similarly, bit-band support can simplify application code if we need to read a bit in a memory location. For example, if we need to determine bit 2 of address 0x20000000, we use the steps outlined in Figure 6.10.

The assembler sequence for these two cases could be like the one shown in Figure 6.11.

Bit-band operation is not a new idea; in fact, a similar feature has existed for more than 30 years on 8-bit microcontrollers such as the 8051. In these 8-bit processors, the bit addressable data have special data types and need special instructions for bit data accesses. Although the Cortex-M3 and Cortex-M4 processors do not have special instructions for bit operation, special memory regions are defined so that data accesses to these regions are automatically converted into bit-band operations.

Note that the Cortex-M3 and Cortex-M4 processors use the following terms for the bit-band memory addresses:

- Bit-band region: This is a memory address region that supports bit-band operation.
- Bit-band alias: Access to the bit-band alias will cause an access (a bit-band operation) to the bit-band region. (Note: A memory remapping is performed.)

Within the bit-band region, each word is represented by an LSB of 32 words in the bit-band alias address range. What actually happens is that, when the bit-band alias address is accessed, the address is remapped into a bit-band address. For read operations, the word is read and the chosen bit location is shifted to the LSB of the read return data. For write operations, the written bit data is shifted to the required bit position, and a READ-MODIFY-WRITE is performed.

There are two regions of memory for bit-band operations:

- 0x20000000−0x200FFFFF (SRAM, 1MB)
- 0x40000000−0x400FFFFF (Peripherals, 1MB)

For the SRAM memory region, the remapping of the bit-band alias is shown in Table 6.8.

Similarly, the bit-band region of the peripheral memory region can be accessed via bit-band aliased addresses, as shown in Table 6.9.

Here is a simple example:

1. Set address 0x20000000 to a value of 0x3355AACC.
2. Read address 0x22000008. This read access is remapped into read access to 0x20000000. The return value is 1 (bit[2] of 0x3355AACC).
3. Write 0x0 to 0x22000008. This write access is remapped into a READ-MODIFY-WRITE to 0x20000000. The value 0x3355AACC is read from memory, bit 2 is cleared, and a result of 0x3355AAC8 is written back to address 0x20000000.

**Table 6.8** Remapping of Bit-Band Addresses in the SRAM Region

| Bit-Band Region | Aliased Equivalent |
|---|---|
| 0x20000000 bit [0] | 0x22000000 bit [0] |
| 0x20000000 bit [1] | 0x22000004 bit [0] |
| 0x20000000 bit [2] | 0x22000008 bit [0] |
| ... | ... |
| 0x20000000 bit [31] | 0x2200007C bit [0] |
| 0x20000004 bit [0] | 0x22000080 bit [0] |
| ... | ... |
| 0x20000004 bit [31] | 0x220000FC bit [0] |
| ... | ... |
| 0x200FFFFC bit [31] | 0x23FFFFFC bit [0] |

**Table 6.9** Remapping of Bit-Band Addresses in the Peripherals Region

| Bit-Band Region | Aliased Equivalent |
| --- | --- |
| 0x40000000 bit [0] | 0x42000000 bit [0] |
| 0x40000000 bit [1] | 0x42000004 bit [0] |
| 0x40000000 bit [2] | 0x42000008 bit [0] |
| ... | ... |
| 0x40000000 bit [31] | 0x4200007C bit [0] |
| 0x40000004 bit [0] | 0x42000080 bit [0] |
| ... | ... |
| 0x40000004 bit [31] | 0x420000FC bit [0] |
| ... | ... |
| 0x400FFFFC bit [31] | 0x43FFFFFC bit [0] |

**4.** Now read 0x20000000. That gives you a return value of 0x3355AAC8 (bit[2] cleared).

When you access bit-band alias addresses, only the LSB (bit[0]) in the data is used. In addition, accesses to the bit-band alias region should not be unaligned. If an unaligned access is carried out in the bit-band alias address range, the result is unpredictable.

The bit-band feature is optional on Cortex-M3 since version r2p1, and is optional on all current releases of Cortex-M4 processor. It is possible that the Cortex-M3 or Cortex-M4 microcontroller you are using does not support bit-band. Please refer to documentation from your microcontroller vendor for details.

## 6.7.2 Advantages of bit-band operations

So, what are the uses of bit-band operations? We can use them to, for example, implement serial data transfers in general-purpose input/output (GPIO) ports to serial devices. The application code can be implemented easily because access to serial data and clock signals can be separated.

---

**Bit-Band vs. Bit-Bang**

In the Cortex®-M3 and Cortex-M4 processors, we use the term bit-band to indicate that the feature is a special memory band (region) that provides bit accesses. Bit-bang commonly refers to driving I/O pins under software control to provide serial communication functions. The bit-band feature in the Cortex-M3 and Cortex-M4 processors can be used for bit-banging implementations, but the definitions of these two terms are different.

---

Bit-band operations can also be used to simplify branch decisions. For example, if a branch should be carried out based on one single bit in a status register in a peripheral, instead of:

- Reading the whole register
- Masking the unwanted bits
- Comparing and branching

You can simplify the operations to:

- Reading the status bit via the bit-band alias (get 0 or 1)
- Comparing and branching

Besides providing faster bit operations with fewer instructions, the bit-band feature in the Cortex®-M3 and Cortex-M4 processors is also essential for situations in which resources (e.g., various pins in I/O port) are being shared by more than one process. One of the most important advantages or properties of bit-band operation is that it is atomic. In other words, the READ-MODIFY-WRITE sequence cannot be interrupted by other bus activities. Without this behavior in, for example, using a software READ-MODIFY-WRITE sequence, the following problem can occur: Consider a simple output port with bit 0 used by a main program and bit 1 used by an interrupt handler. A software based READ-MODIFY-WRITE operation can cause data conflicts, as shown in Figure 6.12.

With the bit-band feature, this kind of race condition can be avoided because the READ-MODIFY-WRITE is carried out at the hardware level and is atomic (the two transfers cannot be pulled apart) and interrupts cannot take place between them (Figure 6.13).

Similar issues can be found in multitasking systems. For example, if bit 0 of the output port is used by Process A and bit 1 is used by Process B, a data conflict can occur in a software-based READ-MODIFY-WRITE (Figure 6.14).

Again, the bit-band feature can ensure that bit accesses from each task are separated so that no data conflicts occur (Figure 6.15).

Besides I/O functions, the bit-band feature can be used for storing and handling Boolean data in the SRAM region. For example, multiple Boolean variables can be packed into one single memory location to save memory space, whereas access to each bit is still completely separate when the access is carried out via the bit-band alias address range.

For SoC designers designing a bit-band-capable device, the device's memory address should be located within the bit-band memory, and the lock (HMAST-LOCK) signal from the AHB interface must be checked to make sure that during a locked transfer operation, the writable register contents will only be modifiable by the bus and not be changed by the peripheral's hardware operations.

### 6.7.3 Bit-band operation of different data sizes

Bit-band operation is not limited to word transfers. It can be carried out as byte transfers or half-word transfers as well. For example, when a byte access instruction (LDRB/STRB) is used to access a bit-band alias address range, the accesses generated to the bit-band region will be in byte size. Similarly, half-word transfers

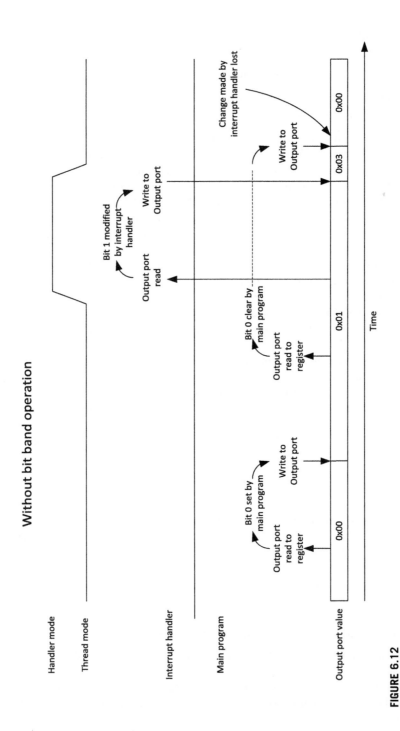

**FIGURE 6.12**

Data are lost when an exception handler modifies a shared memory location

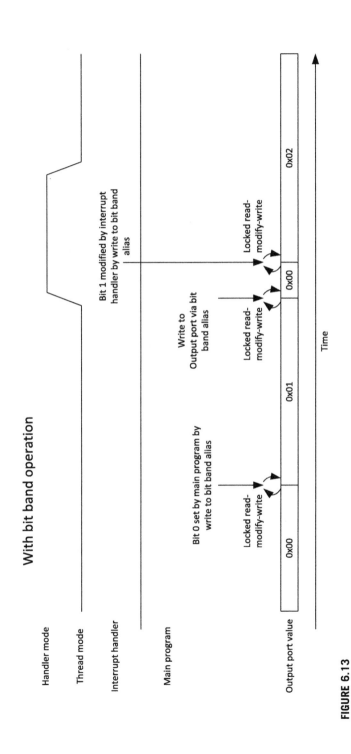

**FIGURE 6.13**

Data loss prevention with locked transfers using the bit-band feature

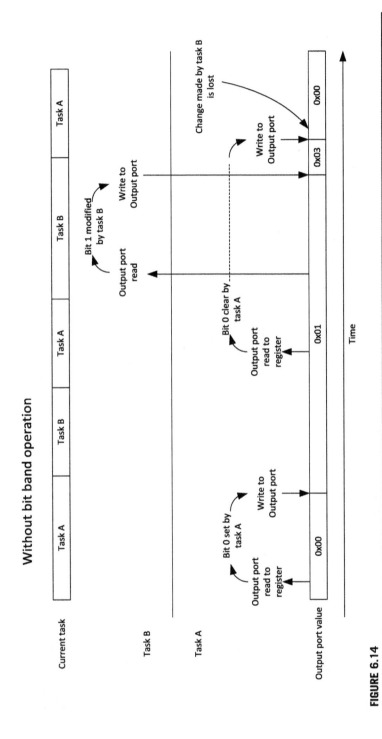

**FIGURE 6.14**

Data are lost when a different task modifies a shared memory location

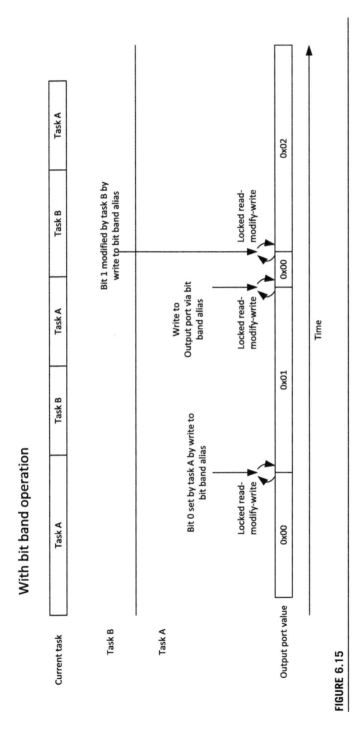

**FIGURE 6.15**

Data loss prevention with locked transfers using the bit-band feature

(LDRH/STRH) carried out to bit-band alias are remapped into half-word size transfer to bit-band region.

When you use non-word transfers to bit-band alias addresses, the address value should still be word aligned.

## 6.7.4 Bit-band operations in C programs

There is no native support of bit-band operations in C/C++ languages. For example, C compilers do not understand that the same memory can be accessed using two different addresses, and they do not know that accesses to the bit-band alias will only access the LSB of the memory location. To use the bit-band feature in C, the simplest solution is to separately declare the address and the bit-band alias of a memory location. For example:

```
#define DEVICE_REG0      *((volatile unsigned long *) (0x40000000))
#define DEVICE_REG0_BIT0 *((volatile unsigned long *) (0x42000000))
#define DEVICE_REG0_BIT1 *((volatile unsigned long *) (0x42000004))
. . .
DEVICE_REG0 = 0xAB; // Accessing the hardware register by normal
                    // address
. . .
DEVICE_REG0 = DEVICE_REG0 | 0x2; // Setting bit 1 without using
                                 // bit-band feature
. . .
DEVICE_REG0_BIT1 = 0x1; // Setting bit 1 using bit-band feature
                        // via the bit-band alias address
```

It is also possible to develop C macros to make accessing the bit-band alias easier. For example, we could set up one macro to convert the bit-band address and the bit number into the bit-band alias address and set up another macro to access the memory location by taking the address value as a pointer:

```
// Convert bit-band address and bit number into
// bit-band alias address
#define BIT BAND(addr,bitnum) ((addr & 0xF0000000)+0x2000000+((addr &
  0xFFFFF)<<5)+(bitnum <<2))

// Convert the address as a pointer
#define MEM_ADDR(addr) *((volatile unsigned long *) (addr))
```

Based on the previous example, we rewrite the code as follows:

```
#define DEVICE_REG0 0x40000000
#define BIT BAND(addr,bitnum) ((addr & 0xF0000000)+0x02000000+((addr &
  0xFFFFF)<<5)1(bitnum<<2))
#define MEM_ADDR(addr) *((volatile unsigned long *) (addr))

   ...
   MEM_ADDR(DEVICE_REG0) = 0xAB; // Accessing the hardware
                                 // register by normal address

   ...
   // Setting bit 1 without using bit-band feature
```

```
MEM_ADDR(DEVICE_REG0) = MEM_ADDR(DEVICE_REG0) | 0x2;
...
// Setting bit 1 with using bit-band feature
MEM_ADDR(BIT BAND(DEVICE_REG0,1)) = 0x1;
```

Note that when the bit-band feature is used, the variables being accessed might need to be declared as *volatile*. The C compilers do not know that the same data could be accessed in two different addresses, so the volatile property is used to ensure that each time a variable is accessed, the memory location is accessed instead of a local copy of the data inside the processor.

Starting from ARM® RealView® Development Suite version 4.0 and Keil™ MDK-ARM 3.80, and in all versions of ARM Development Studio 5 (DS-5™), bit-band support is provided by __attribute__((bit band)) language extension and __bit band command line option (see reference 12).

You can find further examples of bit-band accesses with C macros using ARM C Compiler Tools in ARM Application Note 179 (reference 10).

## 6.8 Default memory access permissions

The Cortex®-M3 and Cortex-M4 memory map has a default configuration for memory access permissions. This prevents user programs (non-privileged) from accessing system control memory spaces such as the NVIC. The default memory access permission is used when either no MPU is present or the MPU is present but disabled.

If the MPU is present and enabled, additional access permission rules defined by the MPU setup will also determine whether user accesses are allowed for other memory regions.

The default memory access permissions are shown in Table 6.10.

When an unprivileged access is blocked, the fault exception takes place immediately. The fault exception can either be a HardFault or BusFault exception, depending on whether the BusFault exception is enabled and the priority level.

## 6.9 Memory access attributes

The memory map shows what is included in each memory region. Aside from decoding which memory block or device is accessed, the memory map also defines the memory attributes of the access. The memory attributes you can find in the Cortex®-M3 and Cortex-M4 processors include the following:

*Bufferable:* Write to memory can be carried out by a write buffer while the processor continues on to next instruction execution.

*Cacheable:* Data obtained from memory read can be copied to a memory cache so that next time it is accessed the value can be obtained from the cache to speed up program execution.

*Executable:* The processor can fetch and execute program code from this memory region.

**Table 6.10** Default Memory Access Permissions

| Memory Region | Address | Access in Unprivileged (User) Program |
|---|---|---|
| Vendor specific | 0xE0100000–0xFFFFFFFF | Full access |
| ROM Table | 0xE00FF000–0xE00FFFFF | Blocked; unprivileged access results in bus fault |
| External PPB | 0xE0042000–0xE00FEFFF | Blocked; unprivileged access results in bus fault |
| ETM | 0xE0041000–0xE0041FFF | Blocked; unprivileged access results in bus fault |
| TPIU | 0xE0040000–0xE0040FFF | Blocked; unprivileged access results in bus fault |
| Internal PPB | 0xE000F000–0xE003FFFF | Blocked; unprivileged access results in bus fault |
| NVIC | 0xE000E000–0xE000EFFF | Blocked; unprivileged access results in bus fault, except Software Trigger Interrupt Register that can be programmed to allow user accesses |
| FPB | 0xE0002000–0xE0003FFF | Blocked; unprivileged access results in bus fault |
| DWT | 0xE0001000–0xE0001FFF | Blocked; unprivileged access results in bus fault |
| ITM | 0xE0000000–0xE0000FFF | Read allows; write ignored except for stimulus ports with unprivileged access enabled (run-time configurable) |
| External Device | 0xA0000000–0xDFFFFFFF | Full access |
| External RAM | 0x60000000–0x9FFFFFFF | Full access |
| Peripheral | 0x40000000–0x5FFFFFFF | Full access |
| SRAM | 0x20000000–0x3FFFFFFF | Full access |
| Code | 0x00000000–0x1FFFFFFF | Full access |

*Sharable:* Data in this memory region could be shared by multiple bus masters. The memory system needs to ensure coherency of data between different bus masters in the shareable memory region.

The processor bus interfaces output the memory access attribute information to the memory system for each instruction and data transfer. The default memory attribute settings can be overridden if the MPU is present and the MPU region configurations are programmed differently from the default.

In most existing Cortex-M3 and Cortex-M4 microcontrollers, only the Executable and Bufferable attributes affect the operation of the applications. The Cacheable and

**Table 6.11** Memory Attributes in Relation to Memory Types

| Bufferable | Cacheable | Memory Type |
|---|---|---|
| 0 | 0 | Strongly Ordered – The Cortex-M3 and Cortex-M4 processors wait until the transfer is completed on the bus interface before continuing to the next operation. Architecturally the processor could proceed to the next operation but cannot start another access to Strongly Ordered or Devices type memory. |
| 1 | 0 | Devices – The Cortex-M3 and Cortex-M4 processors can use a write buffer to handle the transfer while continuing to the next instructions unless the next instruction is also a memory access. Architecturally the processor can proceed to the next operation but cannot start another access to Strongly Ordered or Devices type memory. |
| 0 | 1 | Normal memory with Write Through (WT) cache. |
| 1 | 1 | Normal memory with Write Back (WB) cache. |

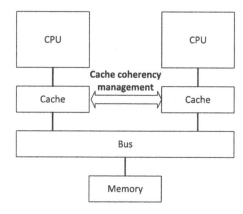

**FIGURE 6.16**

Cache coherency in multi-processor system needs Sharable attribute

Sharable attributes are usually used by a cache controller, which specifies memory types and caching scheme, as shown in Table 6.11.

The Sharable memory attribute is needed in systems with multiple processors and multiple cache units with cache coherency control (Figure 6.16). When a data access is indicated as Sharable, the cache controller needs to ensure the value is coherent with other cache units as it could have been cached and modified by another processor.

Though the Cortex-M3 and Cortex-M4 processors do not have a cache memory or cache controller, a cache unit can be added on the microcontroller, which can use the memory attribute information to define the memory access behaviors. In addition, the cache attributes might also affect the operation of memory controllers

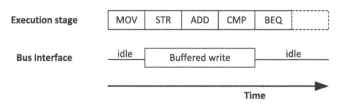

**FIGURE 6.17**

Buffered write operation

for on-chip memory and off-chip memory, depending on the memory controllers used by the chip manufacturers.

The Bufferable attribute is used inside the processor. In order to provide better performance, the Cortex-M3 and Cortex-M4 processors support a single entry write buffer on the bus interface. A data write to a bufferable memory region can be carried out in a single clock cycle and continue to the next instruction execution, even if the actual transfer needs several clock cycles to be completed on the bus interface (Figure 6.17).

The default memory access attributes for each memory region are shown in Table 6.12. The "XN" in the table is eXecute Never, which means program execution in this region is prohibited.

Note that from Revision 1 of the Cortex-M3 and all releases of Cortex-M4 processors, the CODE region memory attribute signals on the processor's I-CODE and D-CODE bus interface are hardwired to indicate transfers as cacheable and non-bufferable. This cannot be overridden by MPU configuration. This only affects cache memory systems outside the processor (e.g., level 2 cache and certain types of memory controllers with cache features). Within the processor, the internal write buffer can still be used for write transfers accessing the CODE region.

## 6.10 Exclusive accesses

You might have noticed that the Cortex®-M3 and Cortex-M4 processors has no SWP instruction (swap), which was used for semaphore operations in traditional ARM® processors like ARM7TDMI™. This is now being replaced by exclusive access operations. Exclusive accesses were first supported in architecture v6 (e.g., in the ARM1136).

Semaphores are commonly used for allocating shared resources to applications. When a shared resource can only service one client or application processor, we also call it Mutual Exclusion (MUTEX). In such cases, when a resource is being used by one process, it is locked to that process and cannot serve another process until the lock is released. To set up a MUTEX semaphore, a memory location is defined as the lock flag to indicate whether a shared resource is locked by a process. When a process or application wants to use the resource, it needs to check whether the resource has been locked first. If it is not being used, it can set the lock flag to indicate

**Table 6.12** Default Memory Attributes

| Region | Memory/Device type | XN | Cache | Note |
| --- | --- | --- | --- | --- |
| CODE memory region (0x00000000–0x1FFFFFFF) | Normal | - | WT | Internal write buffer enabled, exported memory attribute is always cacheable, Non-Bufferable |
| SRAM memory region (0x20000000–0x3FFFFFFF) | Normal | - | WB-WA | Write Back, Write Allocate |
| Peripheral region (0x40000000–0x5FFFFFFF) | Devices | Y | - | Bufferable , Non-cacheable |
| RAM region (0x60000000–0x7FFFFFFF) | Normal | - | WB-WA | Write Back, Write Allocate |
| RAM region (0x80000000–0x9FFFFFFF) | Normal | - | WT | Write Through |
| Devices (0xA0000000–0xBFFFFFFF) | Devices | Y | - | Bufferable, Non-cacheable |
| Devices (0xC0000000–0xDFFFFFFF) | Devices | Y | - | Bufferable, Non-cacheable |
| System - PPB (0xE0000000–0xE00FFFFF) | Strongly Order | Y | - | Non-bufferable, Non-cacheable |
| System – Vendor Specific (0xE0100000 –0xFFFFFFFF) | Devices | Y | - | Bufferable, Non-cacheable |

that the resource is now locked. In traditional ARM processors, the access to the lock flag is carried out by the SWP instruction. It allows the lock flag read and write to be atomic, preventing the resource from being locked by two processes at the same time.

In newer ARM processors, the read/write access can be carried out on separate buses. In such situations, the SWP instructions can no longer be used to make the memory access atomic because the read and write in a locked transfer sequence must be on the same bus. Therefore, the locked transfers are replaced by exclusive accesses. The concept of exclusive access operation is quite simple but different from SWP; it allows the possibility that the memory location for a semaphore can be accessed by another bus master or another process running on the same processor (Figure 6.18).

The exclusive write (e.g., STREX) can fail if one of the following conditions takes place:

- A CLREX instruction has been executed
- A context switch has occurred (e.g., an interrupt)
- There wasn't a LDREX executed beforehand
- An external hardware return exclusive fail status to the processor via a side band signal on the bus interface

If the exclusive store gets a failed status, the actual write will not take place in the memory, as it will be either blocked by the processor core or by an external hardware.

To allow exclusive access to work properly in a multiple processor environment, an additional hardware unit called the "exclusive access monitor" is required. This monitor checks the transfers toward shared address locations and replies to the

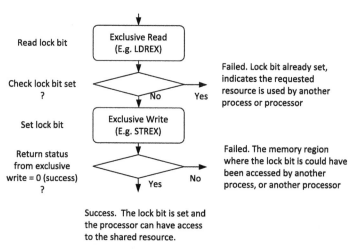

**FIGURE 6.18**

Using exclusive access in MUTEX semaphore

processor if an exclusive access is successful. The processor bus interface also provides additional control signals[1] to this monitor to indicate if the transfer is an exclusive access.

If the memory device has been accessed by another bus master between the
exclusive read and the exclusive write, the exclusive access monitor will flag an
exclusive fail through the bus system when the processor attempts the exclusive
write. This will cause the return status of the exclusive write to be 1. In the case
of a failed exclusive write, the exclusive access monitor also blocks the write transfer
from getting to the exclusive access address.

Exclusive access instructions in the Cortex-M3 and Cortex-M4 processors
include LDREX (word), LDREXB (byte), LDREXH (half word), STREX (word),
STREXB (byte), and STREXH (half word). A simple example of the syntax is as
follows:

```
LDREX <Rxf>, [Rn, #offset]
STREX <Rd>, <Rxf>,[Rn, #offset]
```

where Rd is the return status of the exclusive write (0 = success and 1 =
failure).

Example code for exclusive accesses can be found in section 9.2.9. You can
access exclusive access instructions in C using intrinsic functions provided in
CMSIS-compliant device driver libraries from microcontroller vendors: __LDREX,
__LEDEXB, __LDREXB, __STREX, __STREXH, __STREXB. A quick reference
for these functions is included in Appendix E.

When exclusive accesses are used, the internal write buffers in the processor's
bus interface will be bypassed, even when the MPU defines the region as bufferable.
This ensures that semaphore information on the physical memory is always up to
date and coherent between bus masters. SoC designers using Cortex-M3 or
Cortex-M4 on multiprocessor systems should ensure that the memory system
enforces data coherency when exclusive transfers occur.

## 6.11 Memory barriers

In Chapter 5 we covered the memory barrier instructions (ISB, DSB, DMB) and
their uses. In most applications running in Cortex®-M3 and Cortex-M4 microcontrollers, omission of memory barrier instructions does not cause any issues because:

- The Cortex-M3 and Cortex-M4 processors do not re-order any memory transfers
  or instruction execution (this can happen in some superscalar processors or high-
  performance processors with out-of-order-execution capability).

---

[1]Exclusive access signals are available on the system bus and the D-CODE bus of the Cortex-M3/M4
processor. They are EXREQD and EXRESPD for the D-CODE bus and EXREQS and EXRESPS for
the system bus. The I-CODE bus that is used for instruction fetch cannot generate exclusive accesses.

- The simple nature of the AHB Lite and APB protocol does not allow a transfer to start before the previous one finishes.

However, due to the fact that the processor has a write buffer (as mentioned in section 6.9), a data write that takes multiple clock cycles might happen in parallel with execution of subsequent operations. Most of the times it does not matter and this behavior is favorable because it enhances the performance of the system. In some cases you might want to make sure that the next operation does not start until the write is completed. For example, some microcontrollers have a memory-remapping feature that is controlled by a register in a system controller peripheral. After writing to the remap control register for switching of the memory mapping, a DSB instruction can be used to ensure that all subsequen operations (e.g., data access) do not start until the buffered write is indeed completed. You can also add an ISB instruction after the DSB to re-fetch all subsequent instructions. Other scenarios where the memory barrier instructions are needed are covered in section 5.6.13.

Note that in some cases additional write buffers could exist in the system bus infrastructure. In those cases the barrier instructions cannot be used to guarantee the completion of a buffered write. Some of these write buffers can be "drained" using a dummy read operation. Please consult with your microcontroller vendor or distributor if you require this level of device-specific detail.

## 6.12 Memory system in a microcontroller

In many microcontroller devices, the designs integrate additional memory system features such as:

- Boot loader
- Memory remapping
- Memory alias

These features are not a part of the processor, and are implemented differently in microcontrollers from different vendors. There features can be used together to provide a more flexible memory-map arrangement.

In many cases, aside from the program memory (e.g., flash) that you put your program code into, the microcontrollers might have a separate ROM, which could be flash memory, or could be a mask ROM that cannot be changed. Very often this separate program memory contains a boot loader, which is a program that executes before your own application starts.

There are many different reasons why chip designers put a boot loader into the system. For example, to:

- Provide a flash programming utility, so that you can program the flash using a simple UART interface, or even program some parts of the flash memory inside your application when your program is running

- Provide additional firmware such as a communication protocol stack that can be used by software developers via API calls
- Provide Built-In Self Test (BIST) for the chip

For chips with a boot loader ROM, the boot loader is executed when the system is started, so it has to be located in address 0 when the system starts at power up. However, the next time the system starts, it might not need to execute the boot loader again and can run the application in the flash directly, so the memory map needs to be changed.

In order to do this, the address decoder needs to be programmable. A hardware register (e.g., a peripheral register in a system control unit) can be used, as shown in Figure 6.19.

The operation to switch the memory map is called "Memory Remap." This operation is done by the boot loader. However, you cannot switch the memory map and branch to the new location of the boot loader at the same time. So a method called alias is used. With memory address alias, the boot loader ROM is accessible from two different memory regions, as shown in Figure 6.20.

Normally the boot loader is accessible from address 0 using a memory address alias, and the alias can be turned off. There are many possible memory system

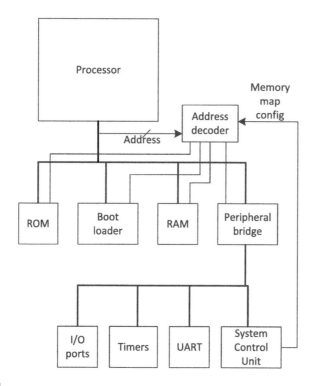

**FIGURE 6.19**

A simplified memory system with configurable memory map

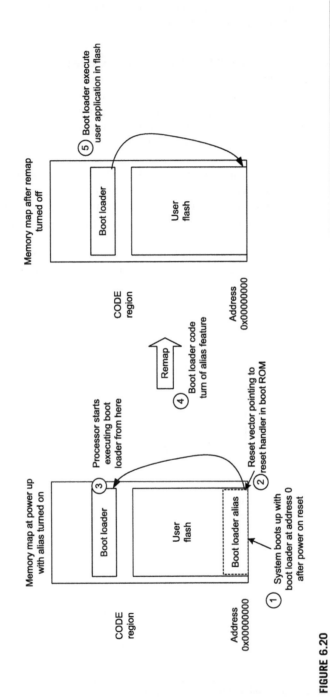

**FIGURE 6.20**

An example of a memory remap implementation with boot loader configurations. What is shown in Figure 6.20 is just one of the possibilities. In some microcontrollers, memory remapping is not needed as the boot loaders present in address 0 are executed every time the system starts up. The vector table is then relocated using the vector table relocation feature provided by the processor, so there is no need to use any remap to handle vector fetches.

# Exceptions and Interrupts

# 7

The Definitive Guide to ARM® Cortex®-M3 and Cortex-M4 Processors. http://dx.doi.org/10.1016/B978-0-12-408082-9.00007-5

## 7.1 Overview of exceptions and interrupts

Interrupts are a common feature available in almost all microcontrollers. Interrupts are events typically generated by hardware (e.g., peripherals or external input pins) that cause changes in program flow control outside a normal programmed sequence (e.g., to provide service to a peripheral). When a peripheral or hardware needs service from the processor, typically the following sequence would occur:

**1.** The peripheral asserts an interrupt request to the processor
**2.** The processor suspends the currently executing task
**3.** The processor executes an Interrupt Service Routine (ISR) to service the peripheral, and optionally clear the interrupt request by software if needed
**4.** The processor resumes the previously suspended task

All Cortex®-M processors provide a Nested Vectored Interrupt Controller (NVIC) for interrupt handling. In addition to interrupt requests, there are other events that need servicing and we called them "exceptions." In ARM® terminology, an interrupt is one type of exception. Other exceptions in Cortex-M processors included fault exceptions and other system exceptions to support the OS (e.g., SVC instruction). The pieces of program code that handle exceptions are often called exception handlers. They are part of the compiled program image.

In a typical Cortex-M microcontroller, the NVIC receives interrupt requests from various sources, as shown in Figure 7.1.

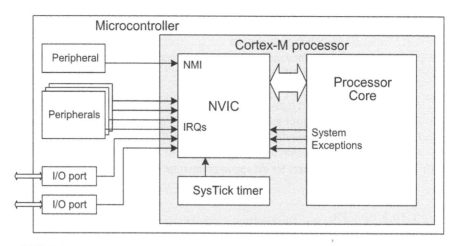

**FIGURE 7.1**

Various sources of exceptions in a typical microcontroller

The Cortex-M3 and Cortex-M4 NVIC supports up to 240 IRQs (Interrupt Requests), a Non-Maskable Interrupt (NMI), a SysTick (System Tick) timer interrupt, and a number of system exceptions. Most of the IRQs are generated by peripherals such as timers, I/O ports, and communication interfaces (e.g., UART, I2C). The NMI is usually generated from peripherals like a watchdog timer or Brown-Out Detector (BOD). The rest of the exceptions are from the processor core. Interrupts can also be generated using software.

In order to resume the interrupted program, the exception sequence needs some way to store the status of the interrupted program so that this can be restored after the exception handler is completed. In general this can either be done by a hardware mechanism or a mixture of hardware and software operations. In the Cortex-M processors, some of the registers are saved onto the stack automatically when an exception is accepted, and are also automatically restored in an exception return sequence. This mechanism allows the exception handlers to be written as normal C functions without any additional software overhead.

## 7.2 Exception types

The Cortex®-M processors provide a feature-packed exception architecture that supports a number of system exceptions and external interrupts. Exceptions are numbered 1−15 for system exceptions and 16 and above for interrupt inputs (inputs to the processor, but not necessarily accessible on the I/O pins of the package). Most of the exceptions, including all interrupts, have programmable priorities, and a few system exceptions have fixed priority.

Different Cortex-M3 or Cortex-M4 microcontrollers can have different numbers of interrupt sources (from 1−240) and different numbers of priority levels. This is because chip designers can configure the Cortex-M3 or Cortex-M4 design source code for different application requirements.

Exception types 1−15 are system exceptions (there is no exception type 0), as outlined in Table 7.1. Exceptions of type 16 or above are external interrupt inputs (see Table 7.2).

Note that here the interrupt number (e.g., Interrupt #0) refers to the interrupt inputs to the NVIC on the Cortex-M processor. In actual microcontroller products or System-on-Chips (SoCs), the external interrupt input pin number might not match the interrupt input number on the NVIC. For example, some of the first few interrupt inputs might be assigned to internal peripherals, and external interrupt pins could be assigned to the next couple of interrupt inputs. Therefore, you need to check the chip manufacturer's datasheets to determine the numbering of the interrupts.

The exception number is used as the identification for each exception and is used in various places in the ARMv7-M architecture. For example, the value of the currently running exception is indicated by the special register Interrupt Program Status Register (IPSR), or by one of the registers in the NVIC called the Interrupt Control State Register (the VECTACTIVE field).

In general programming with CMSIS-Core, the interrupt identification is handled by an interrupt enumeration, starting with value 0 for Interrupt #0.

**Table 7.1** List of System Exceptions

| Exception Number | Exception Type | Priority | Descriptions |
|---|---|---|---|
| 1 | Reset | −3 (Highest) | Reset |
| 2 | NMI | −2 | Non-Maskable Interrupt (NMI), can be generated from on chip peripherals or from external sources. |
| 3 | Hard Fault | −1 | All fault conditions, if the corresponding fault handler is not enabled |
| 4 | MemManage Fault | Programmable | Memory management fault; MPU violation or program execution from address locations with XN (eXecute Never) memory attribute. |
| 5 | Bus Fault | Programmable | Bus error; usually occurs when AHB interface receives an error response from a bus slave (also called *prefetch abort* if it is an instruction fetch or *data abort* if it is a data access). Can also be caused by other illegal accesses. |
| 6 | Usage Fault | Programmable | Exceptions due to program error or trying to access co-processor (the Cortex-M3 and Cortex-M4 processor do not support co-processors). |
| 7–10 | Reserved | NA | – |
| 11 | SVC | Programmable | SuperVisor Call; usually used in OS environment to allow application tasks to access system services. |
| 12 | Debug Monitor | Programmable | Debug monitor; exception for debug events like breakpoints, watchpoints when software based debug solution is used. |
| 13 | Reserved | NA | – |
| 14 | PendSV | Programmable | Pendable service call; An exception usually used by an OS in processes like context switching. |
| 15 | SYSTICK | Programmable | System Tick Timer; Exception generates by a timer peripheral which is included in the processor. This can be used by an OS or can be used as a simple timer peripheral. |

**Table 7.2** List of Interrupts

| Exception Number | Exception Type | Priority | Descriptions |
|---|---|---|---|
| 16 | Interrupt #0 | Programmable | It can be generated from on chip peripherals or from external sources. |
| 17 | Interrupt #1 | Programmable | |
| ... | ... | ... | |
| 255 | Interrupt #239 | Programmable | |

The system exceptions use negative values in the enumeration, as shown in Table 7.3. The CMSIS-Core also defines the names of the system exception handlers.

The reason for using a different number system in CMSIS-Core access functions is to allow slightly better efficiency in some of these API functions (e.g., when setting up priority levels). The interrupt number and enumeration definitions of interrupts are device-specific, and are defined in device-specific header files provided by microcontroller vendors, in a "typedef" section called "IRQn." The enumeration definitions are used by various NVIC access functions in the CMSIS-Core.

## 7.3 Overview of interrupt management

The Cortex®-M processors have a number of programmable registers for managing interrupts and exceptions. Most of these registers are inside the NVIC and System Control Block (SCB). (Physically the SCB is implemented as part of the NVIC but the CMSIS-Core defines the registers in separated data structures.) There are also special registers inside the processor core for interrupt masking (e.g., PRIMASK, FAULTMASK, and BASEPRI). To make it easier to manage interrupts and exceptions, the CMSIS-Core provides a number of access functions.

The NVIC and SCB are located inside the System Control Space (SCS) address range from 0xE000E000, with a size of 4KB. The SCS also contains registers for the SysTick timer, Memory Protection Unit (MPU), debug registers, etc. Almost all of the registers in this address range can only be accessed by code running in privileged access level. The only exception is a register called the Software Trigger Interrupt Register (STIR), which can be set up to be accessible in unprivileged mode.

For general application programming, the best practice is to use the CMSIS-Core access functions. For example, the most commonly used interrupt control functions are shown in Table 7.4.

You can also directly access registers in NVIC or SCB if needed. But such practice might limit the software portability when porting the code from one Cortex-M processor to another with a different processor type (e.g., switching between ARMv6-M and ARM7™-M architecture).

**Table 7.3** CMSIS-Core Exception Definitions

| Exception Number | Exception Type | CMSIS-Core Enumeration (IRQn) | CMSIS-Core Enumeration Value | Exception Handler Name |
|---|---|---|---|---|
| 1 | Reset | - | - | Reset_Handler |
| 2 | NMI | NonMaskableInt_IRQn | −14 | NMI_Handler |
| 3 | Hard Fault | HardFault_IRQn | −13 | HardFault_Handler |
| 4 | MemManage Fault | MemoryManagement_IRQn | −12 | MemManage_Handler |
| 5 | Bus Fault | BusFault_IRQn | −11 | BusFault_Handler |
| 6 | Usage Fault | UsageFault_IRQn | −10 | UsageFault_Handler |
| 11 | SVC | SVCall_IRQn | −5 | SVC_Handler |
| 12 | Debug Monitor | DebugMonitor_IRQn | −4 | DebugMon_Handler |
| 14 | PendSV | PendSV_IRQn | −2 | PendSV_Handler |
| 15 | SYSTICK | SysTick_IRQn | −1 | SysTick_Handler |
| 16 | Interrupt #0 | (device-specific) | 0 | (device-specific) |
| 17 | Interrupt #1 - #239 | (device-specific) | 1 to 239 | (device-specific) |

| Table 7.4 Commonly Used CMSIS-Core Functions for Basic Interrupt Control | |
|---|---|
| **Functions** | **Usage** |
| **void NVIC_EnableIRQ (IRQn_Type IRQn)** | Enable an external interrupt |
| **void NVIC_DisableIRQ (IRQn_Type IRQn)** | Disable an external interrupt |
| **void NVIC_SetPriority (IRQn_Type IRQn, uint32_t priority)** | Set the priority of an interrupt |
| **void __enable_irq(void)** | Clear PRIMASK to enable interrupts |
| **void __disable_irq(void)** | Set PRIMASK to disable all interrupts |
| **void NVIC_SetPriorityGrouping(uint32_t PriorityGroup)** | Set priority grouping configuration |

After reset, all interrupts are disabled and given a priority-level value of 0. Before using any interrupts, you need to:

- Set up the priority level of the required interrupt (this step is optional)
- Enable the interrupt generation control in the peripheral that triggers the interrupt
- Enable the interrupt in the NVIC

In most typical applications, this is all you need to do. When the interrupt triggers, the corresponding Interrupt Service Routine (ISR) will execute (you might need to clear the interrupt request from the peripheral within the handler). The name of the ISR can be found inside the vector table inside the startup code, which is also provided by the microcontroller vendor. The name of the ISR needs to match the name used in the vector table so that the linker can place the starting address of the ISR into the vector table correctly.

## 7.4 Definitions of priority

In the Cortex®-M processors (both ARMv6-M and ARMv7-M), whether and when an exception can be accepted by the processor and get its handler executed can be dependent on the priority of the exception and the current priority of the processor. A higher-priority (smaller number in priority level) exception can pre-empt a lower-priority (larger number in priority level) exception; this is the nested exception/interrupt scenario. Some of the exceptions (reset, NMI, and HardFault) have fixed priority levels. Their priority levels are represented with negative numbers to indicate that they are of higher priority than other exceptions. Other exceptions have programmable priority levels, which range from 0 to 255.

The design of the Cortex-M3 and Cortex-M4 processors support three fixed highest-priority levels and up to 256 levels of programmable priority (with a maximum of 128 levels of pre-emption). The actual number of available programmable priority levels is decided by silicon chip designers. Most Cortex-M3 or Cortex-M4 chips have fewer supported levels — for example, 8, 16, 32, and so on.

This is because having large number of priority levels can increases the complexity of the NVIC and can increase power consumption and reduce the speed of the design. In most cases, applications only require a small number of programmable priority levels. Therefore silicon chip designers need to customize their processor design based on the number of priority levels in the targeted applications. This reduction in levels is implemented by cutting out the least significant bit (LSB) part of the priority configuration registers.

Interrupt-priority levels are controlled by priority-level registers, with width of 3 bits to 8 bits. For example, if only 3 bits of priority level are implemented in the design, a priority-level configuration register will look like Figure 7.2.

For the example in Figure 7.2, because bit 4 to bit 0 are not implemented, they are always read as zero, and writes to these bits will be ignored. With this setup, we have possible priority levels of 0x00 (high priority), 0x20, 0x40, 0x60, 0x80, 0xA0, 0xC0, and 0xE0 (the lowest) (Figure 7.4).

Similarly, if 4 bits of priority level are implemented in the design, a priority-level configuration register will look like Figure 7.3. This allows 16 levels of programmable priority level (also illustrated in Figure 7.4).

The more bits are implemented, the more priority levels will be available. However, more priority bits can also increase gate counts and hence the power consumption of the silicon designs. For ARMv7-M architecture, the minimum number of implemented priority register widths is 3 bits (eight levels). In Cortex-M3 and Cortex-M4 processors, all the priority-level registers have a reset value of 0.

The reason for removing the least significant bit (LSB) of the priority-level registers instead of the most significant bit (MSB) is to make it easier to port software from one Cortex-M device to another. In this way, a program written for devices with 4-bit priority configuration registers is likely to be able to run on devices with 3-bit

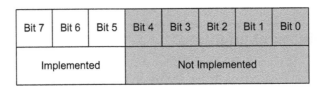

**FIGURE 7.2**

A priority-level register with 3 bits implemented (8 programmable priority levels)

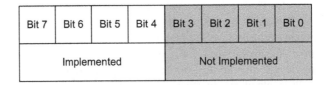

**FIGURE 7.3**

A priority-level register with 4 bits implemented (16 programmable priority levels)

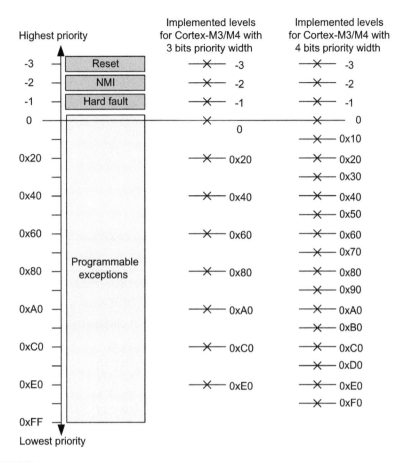

**FIGURE 7.4**

Available priority levels with 3-bit or 4-bit priority width

priority configuration registers. If the MSB is removed instead of the LSB, you might get an inversion of the priority arrangement when porting an application from one Cortex-M chip to another. For example, if an application uses priority level 0x05 for IRQ #0 and level 0x03 for IRQ #1, IRQ #1 should have higher priority. But when MSB bit 2 is removed, IRQ #0 will become level 0x01 and have a higher priority than IRQ #1.

Examples of available exception priority levels for devices with 3-bit, 5-bit, and 8-bit priority registers are shown in Table 7.5.

Some readers might wonder why, if the priority-level configuration registers are 8-bits wide, there are only 128 pre-emption levels? This is because the 8-bit register is further divided into two parts: *group priority* and *sub-priority*. (Note: Group priority was also called Pre-emption Priority in older versions of Technical Reference Manuals and previous edition of this book).

**Table 7.5** Available Priority Levels for DEVICES with 3-Bit, 5-Bit and 8-Bit Priority-Level Registers

| Priority Level | Exception Type | Devices with 3-Bit Priority Configuration Registers | Devices with 5-Bit Priority Configuration Registers | Devices with 8-Bit Priority Configuration Registers |
|---|---|---|---|---|
| −3 (Highest) | Reset | −3 | −3 | −3 |
| −2 | NMI | −2 | −2 | −2 |
| −1 | Hard fault | −1 | −1 | −1 |
| 0, 1, ... 0xFF | Exceptions with programmable priority level | 0x00 0x20 ... 0xE0 | 0x00 0x08 ... 0xF8 | 0x00, 0x01 0x02, 0x03 ... 0xFE, 0xFE |

**Table 7.6** Definition of Pre-empt Priority Field and Sub-priority Field in a Priority-level Register in Different Priority Group Settings

| Priority Group | Pre-empt Priority Field | Sub-priority Field |
|---|---|---|
| 0 (default) | Bit [7:1] | Bit [0] |
| 1 | Bit [7:2] | Bit [1:0] |
| 2 | Bit [7:3] | Bit [2:0] |
| 3 | Bit [7:4] | Bit [3:0] |
| 4 | Bit [7:5] | Bit [4:0] |
| 5 | Bit [7:6] | Bit [5:0] |
| 6 | Bit [7] | Bit [6:0] |
| 7 | None | Bit [7:0] |

Using a configuration register in the System Control Block (SCB) called Priority Group (a part of the Application Interrupt and Reset Control register in the SCB, section 7.9.4), the priority-level configuration registers for each exception with programmable priority levels is divided into two halves. The upper half (left bits) is the group (pre-empt) priority, and the lower half (right bits) is the sub-priority (Table 7.7).

The *group priority level* defines whether an interrupt can take place when the processor is already running another interrupt handler. The *sub-priority level* value is used only when two exceptions with same group-priority level occur at the same time. In this case, the exception with higher *sub-priority* (lower value) will be handled first.

You can access the priority group setting, and encode/decode priority information using CMSIS-Core functions as shown in Table 7.7.

Please note that the results for NVIC_DecodePriority are returned by modified values pointed by the pointers (`uint32_t *pPre emptPriority`, `uint32_t *pSub priority`).

As a result of the priority grouping, the maximum width of group (pre-empt) priority is 7, so there can be 128 levels. When the priority group is set to 7, all exceptions with a programmable priority level will be in the same level, and no pre-emption between these exceptions will take place, except that hard fault, NMI, and reset, which have priority of $-1$, $-2$, and $-3$, respectively, can pre-empt these exceptions.

When deciding the effective group-priority level and sub-priority level, you must take these factors into account:

- Implemented priority-level configuration registers
- Priority group setting

For example, if the width of the configuration registers is 3 (bit 7 to bit 5 are available) and priority group is set to 5, you can have four levels of group/pre-empt priority levels (bit 7 to bit 6), and inside each group/pre-empt level there are two levels of sub-priority (bit 5) (Figure 7.5).

With settings as shown in Figure 7.5, the available priority levels are illustrated in Figure 7.6.

**Table 7.7** CMSIS-Core Functions for Managing Priority Grouping

|  | **Functions** | **Usage** |
|---|---|---|
| void | **NVIC_SetPriorityGrouping (uint32_t PriorityGroup)** | Set priority grouping value |
| uint32_t | **NVIC_GetPriorityGrouping (uint32_t PriorityGroup)** | Get priority grouping value |
| uint32_t | **NVIC_EncodePriority (uint32_t PriorityGroup, uint32_t Pre emptPriority, uint32_t Sub priority)** | Generate encoded priority value based on priority grouping, group priority and sub-priority |
| void | **NVIC_DecodePriority (uint32_t Priority, uint32_t PriorityGroup, uint32_t *pPre emptPriority, uint32_t *pSub priority)** | Extract, group priority and sub-priority from a priority value |

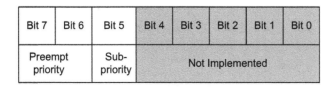

| Bit 7 | Bit 6 | Bit 5 | Bit 4 | Bit 3 | Bit 2 | Bit 1 | Bit 0 |
|---|---|---|---|---|---|---|---|
| Preempt priority | | Sub-priority | Not Implemented | | | | |

**FIGURE 7.5**

Definition of priority fields in a 3-bit priority-level register with priority group set to 5

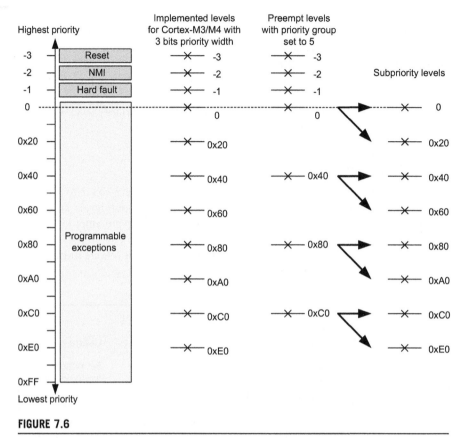

**FIGURE 7.6**

Available priority levels with 3-bit priority width and priority group set to 5

For the same design, if the priority group is set to 0x1, there can be only eight group priority levels and no further sub-priority levels inside each pre-empt level. (Bit [1:0] of priority level register is always 0.) The definition of the priority level configuration registers is shown in Figure 7.7, and the available priority levels are illustrated in Figure 7.8.

| Bit 7 | Bit 6 | Bit 5 | Bit 4 | Bit 3 | Bit 2 | Bit 1 | Bit 0 |
|---|---|---|---|---|---|---|---|
| Preempt priority[5:3] | | | Preempt priority[2:0] (always 0) | | | Sub-priority [1:0] (always 0) | |

**FIGURE 7.7**

Definition of priority fields in a 3-bit priority-level register with priority group set to 1

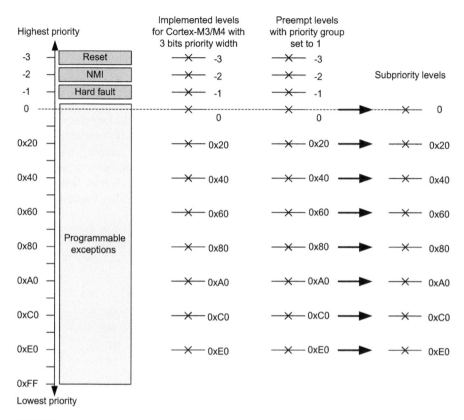

**FIGURE 7.8**

Available priority levels with 3-bit priority width and priority group set to 1

If a Cortex-M3/M4 device has implemented all 8 bits in the priority-level config-uration registers, the maximum number of pre-emption levels it can have is only 128, using a priority group setting of 0. The priority fields definition is shown in Figure 7.9.

When two interrupts are asserted at the same time with exactly the same group/pre-empt priority level and sub-priority level, the interrupt with the smaller exception number has higher priority. (IRQ #0 has higher priority than IRQ #1.)

To avoid unexpected changes of priority levels for interrupts, be careful when writing to the Application Interrupt and Reset Control Register (address

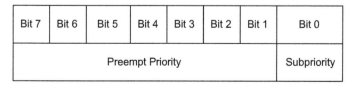

| Bit 7 | Bit 6 | Bit 5 | Bit 4 | Bit 3 | Bit 2 | Bit 1 | Bit 0 |
|-------|-------|-------|-------|-------|-------|-------|-------|
| Preempt Priority | | | | | | | Subpriority |

**FIGURE 7.9**

Definition of priority fields in an 8-bit priority-level register with priority group set to 0

0xE000ED0C). In most cases, after the priority group is configured, there is no need to use this register except to generate a reset (see Table 7.20 for AIRCR descriptions, and section 9.6 for self-reset descriptions).

## 7.5 Vector table and vector table relocation

When the Cortex®-M processor accepts an exception request, the processor needs to determine the starting address of the exception handler (or ISR if the exception is an interrupt). This information is stored in the vector table in the memory. By default, the vector table starts at memory address 0, and the vector address is arranged according to the exception number times four (Figure 7.10). The vector table is normally defined in the startup codes provided by the microcontroller vendors.

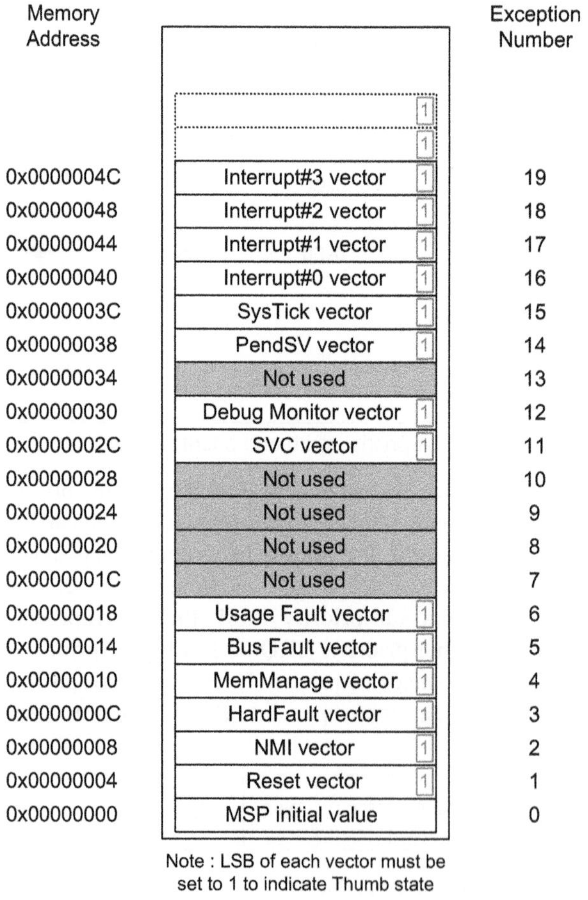

| Memory Address | | Exception Number |
|---|---|---|
| | 1 | |
| | 1 | |
| 0x0000004C | Interrupt#3 vector 1 | 19 |
| 0x00000048 | Interrupt#2 vector 1 | 18 |
| 0x00000044 | Interrupt#1 vector 1 | 17 |
| 0x00000040 | Interrupt#0 vector 1 | 16 |
| 0x0000003C | SysTick vector 1 | 15 |
| 0x00000038 | PendSV vector 1 | 14 |
| 0x00000034 | Not used | 13 |
| 0x00000030 | Debug Monitor vector 1 | 12 |
| 0x0000002C | SVC vector 1 | 11 |
| 0x00000028 | Not used | 10 |
| 0x00000024 | Not used | 9 |
| 0x00000020 | Not used | 8 |
| 0x0000001C | Not used | 7 |
| 0x00000018 | Usage Fault vector 1 | 6 |
| 0x00000014 | Bus Fault vector 1 | 5 |
| 0x00000010 | MemManage vector 1 | 4 |
| 0x0000000C | HardFault vector 1 | 3 |
| 0x00000008 | NMI vector 1 | 2 |
| 0x00000004 | Reset vector 1 | 1 |
| 0x00000000 | MSP initial value | 0 |

Note : LSB of each vector must be set to 1 to indicate Thumb state

**FIGURE 7.10**

Vector table

The vector table used in startup also contains the initial value of the main stack pointer (MSP). It is needed because some exception such as NMI could happen as the processor just came out from reset and before any other initialization steps are executed.

Note that the vector tables in the Cortex-M processors are different from the vector tables in traditional ARM® processors such as the ARM7TDMI™. In traditional ARM processors, the vector tables contain instructions such as branch instructions to branch to appropriate handlers, whereas in Cortex-M, the vector table contains the starting addresses of exception handlers.

Usually, the starting address (0x00000000) should be boot memory, and it will usually be either flash memory or ROM devices, and the value cannot be changed at run-time. However, in some applications it is useful to be able to modify or define exception vectors at run-time. In order to handle this, the Cortex-M3 and Cortex-M4 processors support a feature called Vector Table Relocation.

The Vector Table Relocation feature provides a programmable register called the Vector Table Offset Register (VTOR). This register defines the starting address of the memory being used as the vector table (Figure 7.11). Please note that this register is slightly different between Cortex-M3 revision r2p0 and revision r2p1. In Cortex-M3 r2p0 or older versions, the vector table can only be in the CODE region or the SRAM region. This restriction is removed from Cortex-M3 r2p1 and Cortex-M4.

The VTOR register has a reset value of zero, and in application programming with a CMSIS-compliant device driver, this register can be accessed as "SCB->VTOR." To relocate the vector table to the beginning of the SRAM region, you can use the following code:

```
Example code to copy vector table to beginning of SRAM (0x20000000)

// Note that the use of memory barrier instructions shown below are
// based on architecture recommendation but not strictly required
// for Cortex-M3 and Cortex-M4 processors.
```

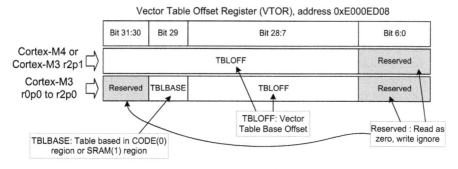

**FIGURE 7.11**

Vector table offset register

```
// Macros for word access
#define HW32_REG(ADDRESS) (*((volatile unsigned long *)(ADDRESS)))

#define VTOR_NEW_ADDR  0x20000000

 int i; // loop counter

// Copy original vector table to SRAM first before programming VTOR
for (i=0;i<48;i++){ // Assume maximum number of exception is 48
  // Copy each vector table entry from flash to SRAM
  HW32_REG((VTOR_NEW_ADDR + (i<<2))) = HW32_REG((i<<2));
  }

__DMB(); // Data Memory Barrier
        // to ensure write to memory is completed

 SCB->VTOR = VTOR_NEW_ADDR; // Set VTOR to the new vector table
//location

__DSB(); // Data Synchronization Barrier to ensure all
        // subsequence instructions use the new configuation
```

When using VTOR, the base address of the new vector table must be aligned to the size of the vector table extended to the next larger power of 2.

### Example 1: 32 interrupt sources in the microcontroller

The vector table size is (32 (for interrupts) +16 (for system exception space)) x 4 (bytes for each vector) = 192 (0xC0). Extending it to the next power of two makes it 256 bytes. So the vector table base address can be programmed as 0x00000000, 0x00000100, 0x00000200, and so on.

### Example 2: 75 interrupt sources in the microcontroller

The vector table size is (75 (for interrupts) +16 (for system exception space)) x 4 (bytes for each vector) = 364 (0x16C). Extending it to the next power of two makes it 512 bytes. So the vector table base address can be programmed as 0x00000000, 0x00000200, 0x00000400, and so on.

Since the minimum number of interrupt is 1, the minimum vector table alignment is 128 bytes. Therefore the lowest 7 bits of the VTOR are reserved and forced to zero.

The vector table relocation feature can be useful in a number of cases:

### 1. Devices with boot loader (Figure 7.12)

In some microcontrollers there are multiple program memories: boot ROM and user flash memory. The boot loaders are often pre-programmed in the boot ROM by the microcontroller manufacturer. When the microcontrollers start, they first execute the boot loader code in the boot ROM, and before branching to the user application

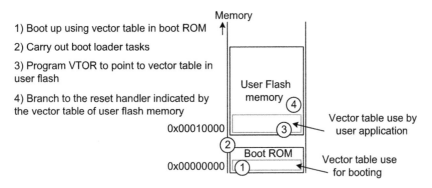

1) Boot up using vector table in boot ROM

2) Carry out boot loader tasks

3) Program VTOR to point to vector table in user flash

4) Branch to the reset handler indicated by the vector table of user flash memory

**FIGURE 7.12**

Vector table relocation in devices with boot ROM and user flash memory

in the user flash, the VTOR is programmed to point to the starting point of the user flash memory so that the vector table in user flash will be used.

## 2. Applications load into RAM (Figure 7.13)

In some situations, the application could be loaded from an external source to the RAM and then get executed. It could have been stored on an SD card, or even need to be transferred through a network. In this case, a program stored in on-chip memory for booting will need to initialize some hardware, copy the externally stored application into RAM, update the VTOR and then execute the externally stored application.

1) Boot up using vector table in flash memory

2) Initialize hardware and copy externally stored application to RAM

3) Program VTOR to point to vector table in the application in SRAM

4) Branch to the reset handler indicated by the vector table in SRAM and start application

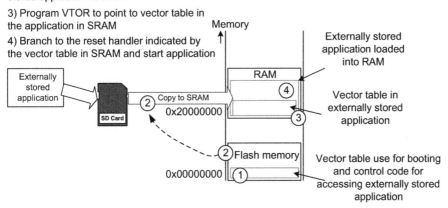

**FIGURE 7.13**

Vector table relocation for applications that are transferred from external storage

### 3. Dynamic changing of vector

In some cases, you might want to have multiple implementations of an interrupt handler in ROM and want to switch between them at different stages of the application. In this case, you can copy the vector table from the program memory to SRAM, and program the VTOR to point to the vector table in SRAM. Since the contents in SRAM can be modified at any time, you can then modify the interrupt vector easily at different stages of the application.

In a minimal setup, the vector table needs to provide the initial MSP value and the reset vector for the system to boot up. In addition, depending on your application, you might also need to include the NMI vector, as some devices could have NMI triggered as soon as it started, and the HardFault vector for error handling.

## 7.6 Interrupt inputs and pending behaviors

There are various status attributes applicable to each interrupt:

- Each interrupt can either be disabled (default) or enabled
- Each interrupt can either be pending (a request is waiting to be served) or not pending
- Each interrupt can either be in an active (being served) or inactive state

To support this, the NVIC contains programmable registers for interrupt enable control, pending status, and read-only active status bits.

Different combinations of these status attributes are possible. For example, while you are serving an interrupt (active), you can disable it, and then a new request for the same interrupt arrives again before the interrupt exits, causing the interrupt to be disabled while active and with a pending status. An interrupt request can be accepted by the processor if:

- The pending status is set,
- The interrupt is enabled, and
- The priority of the interrupt is higher than the current level (including interrupt masking register configuration).

The NVIC is designed to support peripherals that generate pulsed interrupt requests as well as peripherals with high level interrupt request. There is no need to configure any NVIC register to select either interrupt type. For pulsed interrupt requests, the pulse must be at least one clock cycle long. For level triggered interrupts, the peripheral requesting service asserts the request signal until it is cleared by an operation inside the ISR (e.g., write to a register to clear the interrupt request). The request signals received by the NVIC are active high, although the external interrupt request at the I/O pin level could be active low.

The pending status of the interrupts are stored in programmable registers in the NVIC. When an interrupt input of the NVIC is asserted, it causes the pending status of the interrupt to be asserted. The pending status remains high even if the interrupt request is de-asserted. In this way, the NVIC can work with pulsed interrupt requests.

The pending status means it is put into a state of waiting for the processor to serve the interrupt. In some cases, the processor serves the request as soon as an interrupt becomes pending. However, if the processor is already serving another interrupt of higher or equal priority, or if the interrupt is masked by one of the interrupt masking registers, the pended request will remain until the other interrupt handler is finished, or when the interrupt masking is cleared.

This is different from traditional ARM® processors. Previously, the devices that generate interrupts, such as interrupt request (IRQ)/fast interrupt request (FIQ), must hold the request until they are served. Now, with the pending status registers in the NVIC holding the requests, an occurred interrupt will be handled even if the source requesting the interrupt deserts its request signal.

When the processor starts to process an interrupt request, the pending status of the interrupt is cleared automatically, as shown in Figure 7.14.

When the interrupt is being served, it is in the active state. Please note that in the interrupt entry sequence, a number of registers are pushed onto the stack automatically. This is called stacking. Meanwhile, the starting address of the ISR is fetched from the vector table.

In many microcontroller designs, the peripherals operate with level-triggered interrupts and therefore the ISR will have to clear the interrupt request manually; for example, by writing to a register in the peripheral. After the interrupt service is completed, the processor carries out an exception return (covered in section 7.7.4). The registers that were automatically stacked are restored and the interrupted program is resumed. The active status of the interrupt is also cleared automatically.

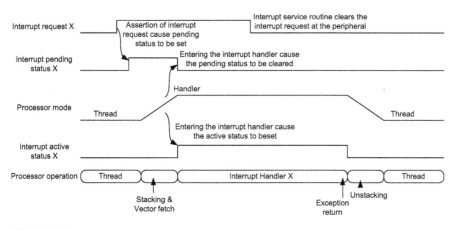

**FIGURE 7.14**

A simple case of interrupt pending and activation behavior

When an interrupt is active, you cannot accept the same interrupt request again until it has completed and terminated with an exception return (sometimes called an exception exit).

The pending status of interrupts are stored in interrupt pending status registers, which are accessible from software code. So you can clear the pending status of an interrupt or set it manually. If an interrupt request arrives when the processor is serving another higher-priority interrupt and the pending status is cleared before the processor starts responding to the pending request, the request is cancelled and will not be served (Figure 7.15).

If a peripheral continuously asserts the interrupt request, and the software attempts to clear the pending status, the pending status will be set again (Figure 7.16).

If an interrupt source continues to assert its interrupt request after it has been serviced, the interrupt will be in the pending state again and will get serviced by the processor again. This is shown in Figure 7.17.

For pulsed interrupt requests, if an interrupt request signal is pulsed several times before the processor starts processing, the request will be treated as one single interrupt request, as illustrated in Figure 7.18.

The pending status of an interrupt can be set again when it is being served. For example, in Figure 7.19 a new interrupt request arrived while the previous

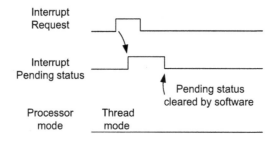

**FIGURE 7.15**

Interrupt pending status cleared before processor serving the interrupt

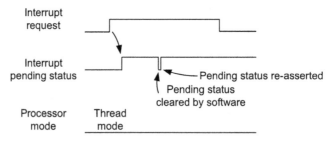

**FIGURE 7.16**

Interrupt pending status cleared and re-asserted due to continuous interrupt request

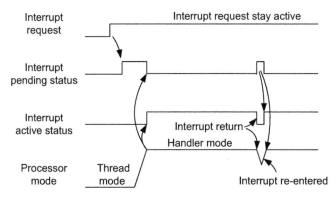

**FIGURE 7.17**

Interrupt gets its pending status set again if the request is still asserted after exception exit

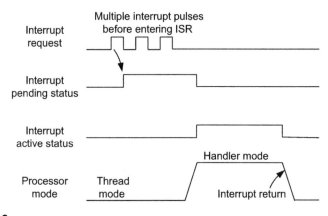

**FIGURE 7.18**

Interrupt pending status set only once with several interrupt request pulses

request was still being served, and this caused a new pending status and therefore the processor needs to serve this interrupt again after the first ISR is completed.

Please note that the pending status of an interrupt can be set even when the interrupt is disabled. In this case, when the interrupt is enabled later, it can be triggered and get served. In some cases this might not be desirable, so in this case you will have to clear the pending status manually before enabling the interrupt in the NVIC.

In general the NMI request behavior is the same as interrupts. Unless an NMI handler is already running, or the processor is halted or in a locked up state, a NMI request will be executed almost immediately because it has the highest priority and cannot be disabled.

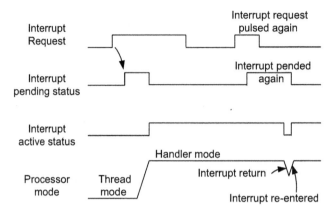

**FIGURE 7.19**

Interrupt pending occurs again during the execution of ISR

## 7.7 Exception sequence overview

### 7.7.1 Acceptance of exception request

The processor accepts an exception if the following conditions are met:

- The processor is running (not halted or in reset state)
- The exception is enabled (with special cases for NMI and HardFault exceptions, which are always enabled)
- The exception has higher priority than the current priority level
- The exception is not blocked by an exception masking register (e.g., PRIMASK)

Note that for the SVC exception, if the SVC instruction is accidentally used in an exception handler that has the same or higher priority than the SVC exception itself, it will cause the HardFault exception handler to execute.

### 7.7.2 Exception entrance sequence

An exception entrance sequence contains several operations:

- Stacking of a number of registers, including return address to the currently selected stack. This enables an exception handler to be written as a normal C function. If the processor was in Thread mode and was using the Process Stack Pointer (PSP), the stack area pointed to by the PSP will be used for this stacking. Otherwise the stack area pointed to by the Main Stack Pointer (MSP) will be used.
- Fetching the exception vector (starting address of the exception handler/ISR). This can happen in parallel to the stacking operation to reduce latency.
- Fetching the instructions for the exception handler to be executed. After the starting address of the exception handler is determined, the instructions can be fetched.

- Update of various NVIC registers and core registers. This includes the pending status and active status of the exception, and registers in the processor core including the Program Status Register (PSR), Link Register (LR), Program Counter (PC), and Stack Pointer (SP).

Depending on which stack was used for stacking, either the MSP or PSP value would be adjusted accordingly just before the exception handler starts. The PC is also updated to the starting address of the exception handler and the Link Register (LR) is updated with a special value called EXC_RE-TURN. This value is 32-bit, and the upper 27 bits are set to 1. Some of the lower 5 bits are used to hold status information about the exception sequence (e.g., which stack was used for stacking). This value is to be used in the exception return.

### 7.7.3 Exception handler execution

Within the exception handler, you can carry out services for the peripheral that requires service. The processor is in Handler mode when executing an exception handler. In Handler mode:

- The Main Stack Pointer (MSP) is used for stack operations
- The processor is executing in privileged access level

If a higher-priority exception arrives during this stage, the new interrupt will be accepted, and the currently executing handler will be suspended and pre-empted by the higher-priority handler. This is called a nested exception.

If another exception with the same or lower priority arrives during this stage, the newly arrived exception will stay in the pending state and will be serviced when the current exception handler is completed.

At the end of the exception handler, the program code executes a return that causes the EXC_RETURN value to be loaded into the Program Counter (PC). This triggers the exception return mechanism.

### 7.7.4 Exception return

In some processor architectures, a special instruction is used for exception return. However, this means that the exception handlers cannot be written and compiled as normal C code. In ARM® Cortex®-M processors, the exception return mechanism is triggered using a special return address called EXC_RETURN. This value is generated at exception entrance and is stored in the Link Register (LR). When this value is written to the PC with one of the allowed exception return instructions, it triggers the exception return sequence.

The exception return can be generated by the instructions shown in Table 7.8. When the exception return mechanism is triggered, the processor accesses the previously stacked register values in the stack memory during exception entrance and restores them back to the register bank. This is called unstacking. In addition, a

**Table 7.8** Instructions that can be Used for Triggering Exception Return

| Return Instruction | Description |
|---|---|
| BX <reg> | If the EXC_RETURN value is still in the LR when the exception handler ends, we can use the "BX LR" instruction to perform the exception return. |
| POP {PC}, or POP {....., PC} | Very often the value of LR is pushed to the stack after entering the exception handler. We can use the POP instruction, either a single register POP or a POP operation with multiple registers including the PC, to put the EXC_RETURN value to the Program Counter (PC). This will cause the processor to perform the exception return. |
| Load (LDR) or Load multiple (LDM) | It is possible to produce an exception return using the LDR or LDM instructions with PC as the destination register. |

number of NVIC registers (e.g., active status) and registers in the processor core (e.g., PSR, SP, CONTROL) will be updated.

In parallel to the unstacking operation, the processor can start fetching the instructions of the previously interrupted program to allow the program to resume operation as soon as possible.

The use of the EXC_RETURN value for triggering exception returns allows exception handlers (including Interrupt Service Routines) to be written as a normal C function/subroutine. In code generation, the C compiler handles the EXC_RETURN value in LR as a normal return address. Due to the EXC_RETURN mechanism, it is impossible to have a normal function return to address 0xF0000000 to 0xFFFFFFFF. However, since the architecture specified that this address range cannot be used for program code (has the Execute Never (XN) memory attribute), it does not create any confusion.

## 7.8 Details of NVIC registers for interrupt control
### 7.8.1 Summary

There are a number of registers in the NVIC for interrupt control (exception type 16 up to 255). These registers are located in the System Control Space (SCS) address range. Table 7.9 shows a summary of these registers.

All of these registers, with the exception of the Software Trigger Interrupt Register (STIR), can only be accessed in privileged access level. By default STIR can only be accessed in privileged access level, but can be configured to be accessed in unprivileged access level.

By default, after a system reset:

- All interrupts are disabled (enable bit = 0)
- All interrupts have priority level of 0 (highest programmable level)
- All interrupt pending statuses are cleared

**Table 7.9** Summary of the Registers in NVIC for Interrupt Control

| Address | Register | CMSIS-Core Symbol | Function |
|---------|----------|-------------------|----------|
| 0xE000E100 to 0xE000E11C | Interrupt Set Enable Registers | NVIC->ISER [0] to NVIC->ISER [7] | Write 1 to set enable |
| 0xE000E180 to 0xE000E19C | Interrupt Clear Enable Registers | NVIC->ICER [0] to NVIC->ICER [7] | Write 1 to clear enable |
| 0xE000E200 to 0xE000E21C | Interrupt Set Pending Registers | NVIC->ISPR [0] to NVIC->ISPR [7] | Write 1 to set pending status |
| 0xE000E280 to 0xE000E29C | Interrupt Clear Pending Registers | NVIC->ICPR [0] to NVIC->ICPR [7] | Write 1 to clear pending status |
| 0xE000E300 to 0xE000E31C | Interrupt Active Bit Registers | NVIC->IABR [0] to NVIC->IABR [7] | Active status bit. Read only. |
| 0xE000E400 to 0xE000E4EF | Interrupt-Priority Registers | NVIC->IP [0] to NVIC->IR [239] | Interrupt-Priority Level (8-bit wide) for each interrupt |
| 0xE000EF00 | Software Trigger Interrupt Register | NVIC->STIR | Write an interrupt number to set its pending status |

## 7.8.2 Interrupt enable registers

The Interrupt Enable register is programmed through two addresses. To set the enable bit, you need to write to the NVIC->ISER[n] register address; to clear the enable bit, you need to write to the NVIC->ICER[n] register address. In this way, enabling or disabling an interrupt will not affect other interrupt enable states. The ISER/ICER registers are 32-bits wide; each bit represents one interrupt input.

As there could be more than 32 external interrupts in the Cortex®-M3 or Cortex-M4 processors, you often find more than one ISER and ICER register — for example, NVIC->ISER[0], NVIC->ISER[1], and so on (Table 7.10). Only the enable bits for interrupts that exist are implemented. So, if you have only 32 interrupt inputs, you will only have ISER and ICER. Although the CMSIS-Core header file defines the ISER and ICER as words (32-bit), these registers can be accessed as word, half word, or byte. As the first 16 exception types are system exceptions, external Interrupt #0 has a start exception number of 16 (see Table 7.1).

The CMSIS-Core provides the following functions for accessing Interrupt Enable registers:

```
void NVIC_EnableIRQ (IRQn_Type IRQn); // Enable an interrupt
void NVIC_DisableIRQ (IRQn_Type IRQn); // Disable an interrupt
```

## 7.8.3 Interrupt set pending and clear pending

If an interrupt takes place but cannot be executed immediately (for instance, if another higher-priority interrupt handler is running), it will be pended.

**Table 7.10** Interrupt Set Enable Registers and Interrupt Clear Enable Registers (0xE000E100-0xE000E11C, 0xE000E180-0xE000E19C)

| Address | Name | Type | Reset Value | Description |
|---|---|---|---|---|
| 0xE000E100 | NVIC->ISER [0] | R/W | 0 | Enable for external interrupt #0–31 bit [0] for interrupt #0 (exception #16) bit [1] for interrupt #1 (exception #17) … bit [31] for interrupt #31 (exception #47) Write 1 to set bit to 1; write 0 has no effect Read value indicates the current status |
| 0xE000E104 | NVIC->ISER [1] | R/W | 0 | Enable for external interrupt #32–63 Write 1 to set bit to 1; write 0 has no effect Read value indicates the current status |
| 0xE000E108 | NVIC->ISER [2] | R/W | 0 | Enable for external interrupt #64–95 Write 1 to set bit to 1; write 0 has no effect Read value indicates the current status |
| … | … | … | … | … |
| 0xE000E180 | NVIC->ICER [0] | R/W | 0 | Clear enable for external interrupt #0–31 bit [0] for interrupt #0 bit [1] for interrupt #1 … bit [31] for interrupt #31 Write 1 to clear bit to 0; write 0 has no effect Read value indicates the current enable status |
| 0xE000E184 | NVIC->ICER [1] | R/W | 0 | Clear Enable for external interrupt #32–63 Write 1 to clear bit to 0; write 0 has no effect Read value indicates the current enable status |
| 0xE000E188 | NVIC->ICER [2] | R/W | 0 | Clear enable for external interrupt #64–95 Write 1 to clear bit to 0; write 0 has no effect Read value indicates the current enable status |
| … | … | … | … | … |

The interrupt-pending status can be accessed through the Interrupt Set Pending (NVIC->ISPR[n]) and Interrupt Clear Pending (NVIC->ICPR[n]) registers. Similarly to the enable registers, the pending status controls might contain more than one register if there are more than 32 external interrupt inputs.

The values of the pending status registers can be changed by software, so you can cancel a current pended exception through the NVIC->ICPR[n] register, or generate software interrupts through the NVIC->ISPR[n] register (Table 7.11).

The CMSIS-Core provides the following functions for accessing Interrupt Pending registers:

```
    void NVIC_SetPendingIRQ(IRQn_Type IRQn); // Set the pending status
of an interrupt
    void NVIC_ClearPendingIRQ(IRQn_Type IRQn); // Clear the pending
status of an interrupt
    uint32_t NVIC_GetPendingIRQ(IRQn_Type IRQn); // Get the pending
status of an interrupt
```

### 7.8.4 Active status

Each external interrupt has an active status bit. When the processor starts the interrupt handler, that bit is set to 1 and cleared when the interrupt return is executed. However, during an Interrupt Service Routine (ISR) execution, a higher-priority interrupt might occur and cause pre-emption. During this period, although the processor is executing another interrupt handler, the previous interrupt is still defined as active. Although the IPSR (Figure 4.5) indicates the currently executing exception services, it cannot tell you whether an exception is active when there is a nested exception. The Interrupt Active Status registers are 32 bits, but can also be accessed using half-word or byte-size transfers. If there are more than 32 external interrupts, there will be more than one active register. The active status registers for external interrupts are read-only (Table 7.12).

The CMSIS-Core provides the following functions for accessing Interrupt Active status registers:

```
    uint32_t NVIC_GetActive(IRQn_Type IRQn); // Get the active status of
an interrupt
```

### 7.8.5 Priority level

Each interrupt has an associated priority-level register, which has a maximum width of 8 bits and a minimum width of 3 bits. As described in section 7.4, each register can be further divided into group priority level and sub-priority level based on priority group settings. The priority-level registers can be accessed as byte, half-word, or word. The number of priority-level registers depends on how many external interrupts the chip contains (Table 7.13).

**Table 7.11** Interrupt Set Pending Registers and Interrupt Clear Pending Registers (0xE000E200–0xE000E21C, 0xE000E280–0xE000E29C)

| Address | Name | Type | Reset Value | Description |
|---|---|---|---|---|
| 0xE000E200 | NVIC->ISPR [0] | R/W | 0 | Pending for external interrupt #0–31 bit [0] for interrupt #0 (exception #16) bit [1] for interrupt #1 (exception #17) … bit [31] for interrupt #31 (exception #47) Write 1 to set bit to 1; write 0 has no effect Read value indicates the current status |
| 0xE000E204 | NVIC->ISPR [1] | R/W | 0 | Pending for external interrupt #32–63 Write 1 to set bit to 1; write 0 has no effect Read value indicates the current status |
| 0xE000E208 | NVIC->ISPR [2] | R/W | 0 | Pending for external interrupt #64–95 Write 1 to set bit to 1; write 0 has no effect Read value indicates the current status |
| … | … | … | … | … |
| 0xE000E280 | NVIC->ICPR [0] | R/W | 0 | Clear pending for external interrupt #0–31 bit [0] for interrupt #0 (exception #16) bit [1] for interrupt #1 (exception #17) … bit [31] for interrupt #31 (exception #47) Write 1 to clear bit to 0; write 0 has no effect Read value indicates the current pending status |
| 0xE000E284 | NVIC->ICPR [1] | R/W | 0 | Clear pending for external interrupt #32–63 Write 1 to clear bit to 0; write 0 has no effect Read value indicates the current pending status |
| 0xE000E288 | NVIC->ICPR [2] | R/W | 0 | Clear pending for external interrupt #64–95 Write 1 to clear bit to 1; write 0 has no effect Read value indicates the current pending status |
| … | … | … | … | … |

**Table 7.12** Interrupt Active Status Registers (0xE000E300-0xE000E31C)

| Address | Name | Type | Reset Value | Description |
|---|---|---|---|---|
| 0xE000E300 | NVIC-> IABR [0] | R | 0 | Active status for external interrupt #0–31 bit[0] for interrupt #0 bit[1] for interrupt #1 ... bit[31] for interrupt #31 |
| 0xE000E304 | NVIC-> IABR [1] | R | 0 | Active status for external interrupt #32–63 |
| ... | – | – | – | – |

The CMSIS-Core provides the following functions for accessing Interrupt Priority Level registers:

```
void NVIC_SetPriority(IRQn_Type IRQn, uint32_t priority); // Set the
priority level of an interrupt or exception
    uint32_t NVIC_GetPriority(IRQn_Type IRQn); // Get the priority level
of an interrupt or exception
```

Additional functions for handling priority grouping are listed in Table 7.7.

If you need to determine the number of priority levels available in the NVIC, you can use the "__NVIC_PRIO_BITS" directive provided in the CMSIS-Core header file provided by your microcontroller supplier. Alternatively you can write 0xFF to one of the Interrupt Priority-Level registers and read back to see how many bits are set. In the case that the device implemented eight levels of interrupt priority levels (3-bits), the read back value would be 0xE0.

### 7.8.6 Software trigger interrupt register

Besides using NVIC-ISPR[n] registers, you can also use a Software Trigger Interrupt Register (NVIC->STIR, see Table 7.14) to trigger an interrupt using software.

**Table 7.13** Interrupt Priority-Level Registers (0xE000E400-0xE000E4EF)

| Address | Name | Type | Reset Value | Description |
|---|---|---|---|---|
| 0xE000E400 | NVIC->IP [0] | R/W | 0 (8-bit) | Priority-level external interrupt #0 |
| 0xE000E401 | NVIC->IP [1] | R/W | 0 (8-bit) | Priority-level external interrupt #1 |
| ... | – | – | – | – |
| 0xE000E41F | NVIC->IP [31] | R/W | 0 (8-bit) | Priority-level external interrupt #31 |
| ... | – | – | – | – |

**Table 7.14** Software Trigger Interrupt Register (0xE000EF00)

| Bits | Name | Type | Reset Value | Description |
|------|------|------|-------------|-------------|
| 8:0 | NVIC->STIR | W | – | Writing the interrupt number sets the pending bit of the interrupt; for example, write 0 to pend external interrupt #0 |

For example, you can generate interrupt #3 by writing the following code in C:

```
NVIC->STIR = 3;
```

This has the same function as using the following CMSIS-Core function call which uses NVIC-ISPR[$n$]:

```
NVIC_SetPendingIRQ(Timer0_IRQn); // Assume Timer0_IRQn equal 3
// Timer0_IRQn is an enumeration defined in device-specific header
```

Unlike NVIC->ISPR[$n$], which can only be accessed in privileged access level, you can enable unprivileged program code to trigger a software interrupt by setting bit 1 (USERSETMPEND) of the Configuration Control Register (address 0xE000ED14). By default the USERSETMPEND is cleared, which means only privileged code can use NVIC->STIR.

Similar to NVIC->ISPR[$n$], NVIC->STIR cannot be used to trigger system exceptions like NMI, SysTick, etc. Additional registers in the System Control Block (SCB) data structure are available for system exception management.

### 7.8.7 Interrupt controller type register

The NVIC also have an Interrupt Controller Type Register in address 0xE000E004. This read-only register gives the number of interrupt inputs supported by the NVIC in granularities of 32 (Table 7.15).

In the CMSIS device-driver library, you can access this read-only register using SCnSCB->ICTR. (SCnSCB refers to "System Control Registers not in SCB"). While the Interrupt Controller Type register can give you an approximate number of interrupts available, you can obtain the exact number of interrupts available by writing to interrupt control registers such as interrupt enable/pending registers while the

**Table 7.15** Interrupt Controller Type Register (SCnSCB->ICTR, 0xE000E004)

| Bits | Name | Type | Reset Value | Description |
|------|------|------|-------------|-------------|
| 4:0 | INTLINESNUM | R | – | Number of interrupt inputs in step of 32<br>0 = 1 to 32<br>1 = 33 to 64<br>... |

PRIMASK register is set (to disable the interrupt from taking place), and read back to see exactly how many bits are implemented in the interrupt enable/pending registers.

## 7.9 Details of SCB registers for exception and interrupt control

### 7.9.1 Summary of the SCB registers

Besides the NVIC data structure in CMSIS-Core, the System Control Block (SCB) data structure also contains some registers that are commonly used for interrupt control. Table 7.16 shows the list of registers in the SCB data structure. Only some of these registers are related to interrupts or exception control and they will be covered in this section.

### 7.9.2 Interrupt control and state register (ICSR)

The ICSR register can be used by application code to:

- Set and clear the pending status of system exceptions including SysTick, PendSV, and NMI.
- Determine the currently executing exception/interrupt number by reading VECTACTIVE.

In addition, it can also be used by the debugger to determine the interrupt status. The VECTACTIVE field is equivalent to the IPSR, as can be seen from Table 7.17.

In this register, quite a number of the bit fields are for the debugger to determine the system exception status. In most cases, only the pending bits would be useful for application development.

### 7.9.3 Vector table offset register (VTOR)

The VTOR was introduced in section 7.5. Please note that different versions of Cortex®-M3 and Cortex-M4 processors have slightly different VTOR definitions. In both cases the address of the VTOR register is located in 0xE000ED0C, and can be accessed using the CMSIS-Core symbol SCB->VTOR.

In the Cortex-M4 processor or Cortex-M3 r2p1 (or after), the definition of the VTOR is shown in Table 7.18.

In the Cortex-M3 rp0p to r2p0, the VTOR definition is shown in Table 7.19.

The version of the Cortex-M processor can be determined using the CPUID register in SCB (section 9.7).

### 7.9.4 Application interrupt and reset control register (AIRCR)

The AIRCR register (Table 7.20) is used for:

- Controlling priority grouping in exception/interrupt priority management
- Providing information about the endianness of the system (can be used by software as well as the debugger)
- Providing a self-reset feature

**Table 7.16** Summary of the Registers in SCB

| Address | Register | CMSIS-Core symbol | Function |
|---------|----------|-------------------|----------|
| 0xE000ED00 | CPU ID | SCB->CPUID | An ID code to allow identification of processor type and revision |
| 0xE000ED04 | Interrupt Control and State Register | SCB->ICSR | Control and status of system exceptions |
| 0xE000ED08 | Vector Table Offset Register | SCB->VTOR | Enable the vector table to be relocated to other address location |
| 0xE000ED0C | Application Interrupt / Reset Control Register | SCB->AIRCR | Configuration for priority grouping, and self-reset control |
| 0xE000ED10 | System Control Register | SCB->SCR | Configuration for sleep modes and low power features |
| 0xE000ED14 | Configuration Control Register | SCB->CCR | Configuration for advanced features |
| 0xE000ED18 to 0xE000ED23 | System Handler Priority Registers | SCB->SHP [0] to SCB->SHP [11] | Exception priority setting for system exceptions |
| 0xE000ED24 | System Handle Control and State Register | SCB->SHCSR | Enable control of fault exceptions, and status of system exceptions |
| 0xE000ED28 | Configurable Fault Status Register | SCB->CFSR | Hint information for causes of fault exceptions |
| 0xE000ED2C | HardFault Status Register | SCB->HFSR | Hint information for causes of HardFault exception |
| 0xE000ED30 | Debug Fault Status Register | SCB->DFSR | Hint information for causes of debug events |
| 0xE000ED34 | MemManage Fault Address Register | SCB->MMFAR | Address Value of Memory Management Fault |
| 0xE000ED38 | Bus Fault Address Register | SCB->BFAR | Address Value of Bus Fault |
| 0xE000ED3C | Auxiliary Fault Status Register | SCB->AFSR | Information for device-specific fault status |
| 0xE000ED40 to 0xE000ED44 | Processor Feature Registers | SCB->PFR [0] to SCB->PFR [1] | Read only information on available processor features |
| 0xE000ED48 | Debug Feature Register | SCB->DFR | Read only information on available debug features |
| 0xE000ED4C | Auxiliary Feature Register | SCB->AFR | Read only information on available auxiliary features |

**Table 7.16** Summary of the Registers in SCB—*Cont'd*

| Address | Register | CMSIS-Core symbol | Function |
|---|---|---|---|
| 0xE000ED50 to 0xE000ED5C | Memory Model Feature Registers | SCB->MMFR [0] to SCB->MMFR [3] | Read only information on available memory model features |
| 0xE000ED60 to 0xE000ED70 | Instruction Set Attributes Register | SCB->ISAR [0] to SCB->ISAR [4] | Read only information on instruction set features |
| 0xE000ED88 | Co-processor Access Control Register | SCB->CPACR | Register to enable floating point unit feature; available on Cortex®-M4 with floating point unit only |

**Table 7.17** Interrupt Control and State Register (SCB->ICSR, 0xE000ED04)

| Bits | Name | Type | Reset Value | Description |
|---|---|---|---|---|
| 31 | NMIPENDSET | R/W | 0 | Write 1 to pend NMI Read vale indicates NMI pending status |
| 28 | PENDSVSET | R/W | 0 | Write 1 to pend system call Read value indicates pending status |
| 27 | PENDSVCLR | W | 0 | Write 1 to clear PendSV pending status |
| 26 | PENDSTSET | R/W | 0 | Write 1 to pend SYSTICK exception Read value indicates pending status |
| 25 | PENDSTCLR | W | 0 | Write 1 to clear SYSTICK pending status |
| 23 | ISRPRE-EMPT | R | 0 | Indicates that a pending interrupt is going to be active in the next step (for debug) |
| 22 | ISRPENDING | R | 0 | External interrupt pending (excluding system exceptions such as NMI for fault) |
| 21:12 | VECTPENDING | R | 0 | Pending ISR number |
| 11 | RETTOBASE | R | 0 | Set to 1 when the processor is running an exception handler; will return to Thread level if interrupt return and no other exceptions pending |
| 9:0 | VECTACTIVE | R | 0 | Current running interrupt service routine |

**Table 7.18** Vector Table Offset Register in Cortex-M4 or Cortex-M3 r2p1

| Bits | Name | Type | Reset Value | Description |
|------|------|------|-------------|-------------|
| 31:7 | TBLOFF | R/W | 0 | Vector Table offset value |

**Table 7.19** Vector Table Offset Register in Cortex-M3 r2p0 or Earlier Versions

| Bits | Name | Type | Reset Value | Description |
|------|------|------|-------------|-------------|
| **31:30** | **Reserved** | – | – | **Not implemented, tied to 0** |
| 29 | TBLBASE | R/W | 0 | Table base in Code (0) or RAM (1) |
| 28:7 | TBLOFF | R/W | 0 | Table offset value from Code region or RAM region |

**Table 7.20** Application Interrupt and Reset Control Register (SCB->AIRCR, Address 0xE000ED0C)

| Bits | Name | Type | Reset Value | Description |
|------|------|------|-------------|-------------|
| 31:16 | VECTKEY | R/W | – | Access key; 0x05FA must be written to this field to write to this register, otherwise the write will be ignored; the read-back value of the upper half-word is 0xFA05 |
| 15 | ENDIANNESS | R | – | Indicates endianness for data: 1 for big endian (BE8) and 0 for little endian; this can only change after a reset |
| 10:8 | PRIGROUP | R/W | 0 | Priority group |
| 2 | SYSRESETREQ | W | – | Requests chip control logic to generate a reset |
| 1 | VECTCLRACTIVE | W | – | Clears all active state information for exceptions; typically used in debug or OS to allow system to recover from system error (reset is safer). |
| 0 | VECTRESET | W | – | Resets the Cortex®-M3/M4 processor (except debug logic), but this will not reset circuits outside the processor. This is intended for debug operations. Do not use this at the same time as SYSRESETREQ. |

The priority grouping feature is covered in section 7.4. In most cases, the PRIGROUP field can be accessed by CMSIS-Core functions "NVIC_SetPriority Grouping" and "NVIC_GetPriorityGrouping," as shown in Table 7.20.

The VECTRESET and VECTCLRACTIVE bit fields are intended to be used by debuggers. Although software can trigger a processor reset using VECTRESET, it is less desirable in most applications because it does not reset the rest of the system such as peripherals. If you want to generate a system reset, in most cases (depending on the chip design as well as the application reset requirements) the SYSRESETREQ should be used (see section 9.6 for additional information).

Please note that the SYSRESETREQ and VECTRESET fields should not be set at the same time. Doing so could result in a glitch in the reset circuitry in some Cortex®-M3/M4 devices, as the VECTRESET signal resets the SYSRESETREQ.

Depending on the design of the reset circuitry in the microcontroller, after writing 1 to SYSRESETREQ, the processor could continue to execute a number of instructions before the reset actually takes place. Therefore it is common to add an endless loop after the system reset request.

## 7.9.5 System handler priority registers (SCB->SHP[0 to 11])

The bit field definitions of the SCB->SHP[0] to SCB->SHP[11] are the same as the interrupt priority registers, apart from the difference that they are for system exceptions. Only some of these registers are implemented. You can use the "NVIC_Set-Priority" and "NVIC_GetPriority" functions from CMSIS-Core to adjust or access the priority level of system exceptions (Table 7.21).

**Table 7.21** System Handler Priority Registers (SCB->SHP[0 to 11])

| Address | Name | Type | Reset Value | Description |
|---|---|---|---|---|
| 0xE000ED18 | SCB->SHP [0] | R/W | 0 (8-bit) | MemManage Fault Priority Level |
| 0xE000ED19 | SCB->SHP [1] | R/W | 0 (8-bit) | Bus Fault Priority Level |
| 0xE000ED1A | SCB->SHP [2] | R/W | 0 (8-bit) | Usage Fault Priority Level |
| 0xE000ED1B | SCB->SHP [3] | – | – | – (not implemented) |
| 0xE000ED1C | SCB->SHP [4] | – | – | – (not implemented) |
| 0xE000ED1D | SCB->SHP [5] | – | – | – (not implemented) |
| 0xE000ED1E | SCB->SHP [6] | – | – | – (not implemented) |
| 0xE000ED1F | SCB->SHP [7] | R/W | 0 (8-bit) | SVC Priority Level |
| 0xE000ED20 | SCB->SHP [8] | R/W | 0 (8-bit) | Debug Monitor Priority Level |
| 0xE000ED21 | SCB->SHP [9] | – | – | – (not implemented) |
| 0xE000ED22 | SCB->SHP[10] | R/W | 0 (8-bit) | PendSV Priority Level |
| 0xE000ED23 | SCB->SHP[11] | R/W | 0 (8-bit) | SysTick Priority Level |

### 7.9.6 System handler control and state register (SCB->SHCSR)

Usage faults, memory management (MemManage) faults, and bus fault exceptions are enabled by the System Handler Control and State Register (0xE000ED24). The pending status of faults and active status of most system exceptions are also available from this register (Table 7.22).

In most cases, this register is used by application code only for enabling configurable fault handlers (MemManage Fault, Bus Fault, and Usage Fault).

**Table 7.22** System Handler Control and State Register (SCB->SHCSR, Address 0xE000ED24)

| Bits | Name | Type | Reset Value | Description |
|------|------|------|------|-------------|
| 18 | USGFAULTENA | R/W | 0 | Usage fault handler enable |
| 17 | BUSFAULTENA | R/W | 0 | Bus fault handler enable |
| 16 | MEMFAULTENA | R/W | 0 | Memory management fault enable |
| 15 | SVCALLPENDED | R/W | 0 | SVC pended; SVCall was started but was replaced by a higher-priority exception |
| 14 | BUSFAULTPENDED | R/W | 0 | Bus fault pended; bus fault handler was started but was replaced by a higher-priority exception |
| 13 | MEMFAULTPENDED | R/W | 0 | Memory management fault pended; memory management fault started but was replaced by a higher-priority exception |
| 12 | USGFAULTPENDED | R/W | 0 | Usage fault pended; usage fault started but was replaced by a higher-priority exception |
| 11 | SYSTICKACT | R/W | 0 | Read as 1 if SYSTICK exception is active |
| 10 | PENDSVACT | R/W | 0 | Read as 1 if PendSV exception is active |
| 8 | MONITORACT | R/W | 0 | Read as 1 if debug monitor exception is active |
| 7 | SVCALLACT | R/W | 0 | Read as 1 if SVCall exception is active |
| 3 | USGFAULTACT | R/W | 0 | Read as 1 if usage fault exception is active |
| 1 | BUSFAULTACT | R/W | 0 | Read as 1 if bus fault exception is active |
| 0 | MEMFAULTACT | R/W | 0 | Read as 1 if memory management fault is active |

Be cautious when writing to this register; make sure that the active status bits of system exceptions are not changed accidentally. For example, to enable the Bus Fault exception, you should use a read-modify-write operation:

```
SCB->SHCSR |= 1<<17; // Enable Bus Fault exception
```

Otherwise, if an activated system exception has its active state cleared by accident, a fault exception will be generated when the system exception handler generates an exception exit.

## 7.10 Details of special registers for exception or interrupt masking

### 7.10.1 PRIMASK

In many applications you might need to temporarily disable all interrupts to carry out some timing critical tasks. You can use the PRIMASK register for this purpose. The PRIMASK register can only be accessed in privileged state.

The PRIMASK register is used to disable all exceptions except NMI and Hard-Fault. It effectively changes the current priority level to 0 (highest programmable level). In C programming, you can use the functions provided in CMSIS-Core to set and clear PRIMASK:

```
void __enable_irq(); // Clear PRIMASK
void __disable_irq(); // Set PRIMASK
void __set_PRIMASK(uint32_t priMask); // Set PRIMASK to value
uint32_t __get_PRIMASK(void); // Read the PRIMASK value
```

In assembly language programming, you can change the value of PRIMARK register using CPS (Change Processor State) instructions:

```
CPSIE I ; Clear PRIMASK (Enable interrupts)
CPSID I ; Set PRIMASK (Disable interrupts)
```

The PRIMASK register can also be accessed using the MRS and MSR instructions. For example:

```
MOVS R0, #1
MSR PRIMASK, R0 ; Write 1 to PRIMASK to disable all
                ; interrupts
```

and:

```
MOVS R0, #0
MSR PRIMASK, R0 ; Write 0 to PRIMASK to allow interrupts
```

When the PRIMASK is set, all fault events will trigger the HardFault exception even if the corresponding configurable fault exception (i.e., MemManage, Bus Fault, Usage Fault) is enabled.

## 7.10.2 FAULTMASK

In terms of behavior, the FAULTMASK is very similar to PRIMASK except that it changes the effective current priority level to −1, so that even the HardFault handler is blocked. Only the NMI exception handler can be executed when FAULTMASK is set.

In terms of usage, FAULTMASK is intended for configurable fault handlers (i.e., MemManage, Bus Fault, Usage Fault) to raise its priority to −1 so that these handlers can have access to some special features for HardFault exceptions including:

- Bypass MPU (see HFNMIENA in MPU Control Register, Table 11.3)
- Ignore data Bus Fault (see BFHFMIGN bit in Configuration Control Register, section 9.8.3) for device/memory probing

By raising the current priority level to −1, the FAULTMASK also allows configurable fault handlers to prevent other exceptions or interrupt handlers executing while an issue is being addressed. More information about fault handling can be found in Chapter 12.

The FAULTMASK register can only be accessed in privileged state, but cannot be set within NMI and HardFault handlers. In C programming with CMSIS-compliant driver libraries, you can use the following CMSIS-Core functions to set and clear the FAULTMASK as follows:

```
void __enable_fault_irq(void); // Clear FAULTMASK
void __disable_fault_irq(void); // Set FAULTMASK to disable
interrupts
void __set_FAULTMASK(uint32_t faultMask);
uint32_t __get_FAULTMASK(void);
```

For assembly language users, you can change the current status of the FAULT-MASK using CPS instructions as follows:

```
CPSIE F ; Clear FAULTMASK
CPSID F ; Set FAULTMASK
```

You can also access the FAULTMASK register using MRS and MSR instructions:

```
MOVS R0, #1
MSR FAULTMASK, R0 ; Write 1 to FAULTMASK to disable all
                  ; interrupts
```

and:

```
MOVS R0, #0
MSR FAULTMASK, R0 ; Write 0 to FAULTMASK to allow interrupts
```

FAULTMASK is cleared automatically upon exiting the exception handler except return from NMI handler. This characteristic provides an interesting usage for FAULTMASK: if in a lower-priority exception handler we want to trigger a higher-priority handler (except NMI), but want this higher-priority handler to start AFTER the lower-priority handler is completed, we can:

- Set the FAULTMASK to disable all interrupts and exceptions (apart from the NMI exception)
- Set the pending status of the higher-priority interrupt or exception
- Exit the handler

Because the pending higher-priority exception handler cannot start while the FAULT-MASK is set, the higher-priority exception stays in the pending state until FAULTMASK is cleared, which happens when the lower-priority handler finishes. As a result, you can force the higher-priority handler to start after the lower-priority handler ends.

### 7.10.3 BASEPRI

In some cases, you might only want to disable interrupts with priority lower than a certain level. In this case, you could use the BASEPRI register. To do this, simply write the required masking priority level to the BASEPRI register. For example, if you want to block all exceptions with priority level equal to or lower than 0x60, you can write the value to BASEPRI:

```
__set_BASEPRI(0x60); // Disable interrupts with priority
                     // 0x60-0xFF using CMSIS-Core function
```

For users of assembly language, the same operation can be written as:

```
MOVS R0, #0x60
MSR BASEPRI, R0 ; Disable interrupts with priority
                ; 0x60-0xFF
```

You can also read back the value of BASEPRI:

```
x = __get_BASEPRI(void); // Read value of BASEPRI
```

or in assembly language:

```
MRS R0, BASEPRI
```

To cancel the masking, just write 0 to the BASEPRI register:

```
__set_BASEPRI(0x0); // Turn off BASEPRI masking
```

Or in assembly language:

```
MOVS R0, #0x0
MSR BASEPRI, R0 ; Turn off BASEPRI masking
```

The BASEPRI register can also be accessed using the BASEPRI_MAX register name. It is actually the same register, but when you use it with this name, it will give you a conditional write operation. (As far as hardware is concerned, BASEPRI and BASEPRI_MAX are the same register, but in the assembler code they use different register name coding.) When you use BASEPRI_MAX as a register, the processor hardware automatically compares the current value and the new value and only allows the update if it is to be changed to a higher-priority level; it cannot be changed to lower-priority levels. For example, consider the following instruction sequence:

```
MOVS R0, #0x60
MSR BASEPRI_MAX, R0 ; Disable interrupts with priority
                    ; 0x60, 0x61,...., etc
MOVS R0, #0xF0
MSR BASEPRI_MAX, R0 ; This write will be ignored because
                    ; it is lower level than 0x60
MOVS R0, #0x40
MSR BASEPRI_MAX, R0 ; This write is allowed and change the
                    ; masking level to 0x40
```

To change to a lower masking level or disable the masking, the BASEPRI register name should be used. The BASEPRI/ BASEPRI_MAX register cannot be set in the unprivileged state.

As with other priority-level registers, the formatting of the BASEPRI register is affected by the number of implemented priority register widths. For example, if only 3 bits are implemented for priority-level registers, BASEPRI can be programmed as 0x00, 0x20, 0x40 … 0xC0, and 0xE0.

## 7.11 Example procedures in setting up interrupts
### 7.11.1 Simple cases

In most applications, the program code including the vector table is stored in read-only memory (ROM) such as flash memory and there is no need to modify the vector table during operation. In this case we can rely on the vector table stored in the ROM and no vector table relocation is needed. The only steps that you would need to set up an interrupt will be as follows:

1. Set up the priority group setting. This step is optional. By default the priority group setting is 0 (only bit 0 of the priority-level registers is use for the subpriority).
2. Set up the priority level of the interrupt. This step is also optional. By default the priority levels of interrupts are 0 (highest programmable level).
3. Enable the interrupt at the NVIC and possibly at the peripheral that generates the interrupt.

Here is a simple example procedure for setting up an interrupt:

```
// Set priority group to 5
NVIC_SetPriorityGrouping(5);
// Set Timer0_IRQn priority level to 0xC0 (4 bit priority)
NVIC_SetPriority(Timer0_IRQn, 0xC); //Shift to 0xC0 by CMSIS function
```

```
// Enable Timer 0 interrupt at NVIC
NVIC_EnableIRQ(Timer0_IRQn);
```

By default the vector table is located at the beginning of the memory (address 0x0). However, some microcontrollers have a boot loader and the boot loader code could have relocated the vector table to the beginning of the flash memory (the vector table defined by the software developer) before the program stored in the flash is executed. In this case there is also no need to reprogram VTOR and relocate the vector table as the boot loader will handle this for you.

Besides enabling the interrupts, you should also make sure that you have enough stack memory if you allow a large number of nested interrupts. Because in Handler mode the handlers always use the Main Stack Pointer (MSP), the main stack memory should contain enough stack space for the worst case (maximum number of nested interrupt/exceptions). The stack space calculation should include the stack space used by the handler, and the space used by the stack frames for each level (see Chapter 8 for details on stack frames).

### 7.11.2 **With vector table relocation**

If the vector table needs to be relocated, for example, to SRAM so that some of the exception vectors can be updated at different stages of the application, a couple more steps are required:

- When the system boots up, the priority group setting might need to be set up. This step is optional. By default the priority group setting is 0 (bit 0 of the priority-level registers is use for the sub-priority, bit 7 down to 1 are used for the pre-emption priority).
- If the vector table needs to be relocated to SRAM, copy the current vector table to its new location in SRAM.
- Set up the Vector Table Offset Register (VTOR) to point to the new vector table.
- Update the exception vector(s) if required.
- Set up the priority level of the required interrupt.
- Enable the interrupt.

An example of vector table relocation is shown in section 7.5. Once the vector table is copied to SRAM, you can update an exception vector using the following code:

```
// Macros for word access
#define HW32_REG(ADDRESS) (*((volatile unsigned long *)(ADDRESS)))
void new_timer0_handler(void); // New Timer 0 Interrupt Handler
unsigned int vect_addr;
// Calculate the address of the exception vector
// Assumed the interrupt to have the vector replaced is identified by
// Timer0_IRQn
```

```
vect_addr = SCB->VTOR + ((((int) Timer0_IRQn) + 16) << 2);
// Update vector to the address of new_timer0_handler()
HW32_REG(vect_addr) = (unsigned int) new_timer0_handler;
```

## 7.12 Software interrupts

You can trigger exceptions or interrupts using software code. The most common reason for doing this is to allow an application task running in unprivileged state in a multi-tasking environment (e.g., a RTOS) to access some system services that need to be carried out in privileged state. Depending on what type of exception or interrupts you want to trigger, different methods are available.

If you want to trigger an interrupt (exception type 16 or above), the easiest way is to use a CMSIS-Core function called "NVIC_SetPendingIRQ":

```
NVIC_SetPendingIRQ(Timer0_IRQn);
// Timer0_IRQn is an enumeration defined in device-specific header
```

This sets the pending status of the interrupt, and triggers the interrupt if it is enabled and if the priority of this interrupt is higher than the current level. Please note that there could be a few clock cycles delay in the starting of the interrupt handler even if the priority is higher than current level. If you want to have the interrupt handler executed before the next operation, you will need to insert memory barrier instructions:

```
NVIC_SetPendingIRQ(Timer0_IRQn);
__DSB(); // Ensure transfer is completed
__ISB(); // Ensure side effect of the write is visible
```

More details on this topic can be found in ARM® Application Note 321, "ARM Cortex®-M Programming Guide to Memory Barrier Instructions" (reference 9).

Instead of using the CMSIS-Core function, you can set the pending status of the interrupt by writing to Interrupt Set Pending Register (see section 7.8.3) or Software Trigger Interrupt Register (see section 7.8.6). But this code could become less portable and might need modification to be used on other Cortex-M processors.

If you want to trigger a SVC exception, you need to execute the SVC instruction. The way to do it in C language is compiler-specific, which is covered in section 10.3, or use an inline/embedded assembly method, covered in section 20.5. This is likely to be used in an OS environment.

For some other system exceptions like NMI, PendSV, and SysTick, you can trigger them by setting the pending status in Interrupt Control and State Register (see ICSR in section 7.9.2). Similar to interrupts, setting the pending status of these exceptions does not always guarantee that they will be executed immediately.

The remaining system exceptions are fault exceptions and should not be used as software interrupts.

## 7.13 **Tips and hints**

If you are developing an application that is going to be used on both ARMv7-M (e.g., Cortex®-M3, Cortex-M4) and ARMv6-M (e.g., Cortex-M0, Cortex-M0+), there are a few features that you need to be aware of:

- The System Control Space (SCS) registers, including NVIC and SCB are word accesses only on ARMv6-M, whereas in ARMv7-M, these registers can be accessed as word, half-word or bytes. As a result, the definitions of the interrupt priority registers NVIC->IP are different between the two architectures. To ensure software portability, you should use the CMSIS-Core functions to handle the interrupt configurations.
- There is no Software Trigger Interrupt Register (NVIC->STIR) in ARMv6-M. To set the pending status of an interrupt, you need to use Interrupt Set Pending Register (NVIC->ISPR).
- There is no vector table relocation feature in Cortex-M0. This feature is available on Cortex-M3, Cortex-M4, and is optional on the Cortex-M0+ processor.

**Table 7.23** Summary of NVIC Feature Differences in Various Cortex-M Processors

|  | Cortex-M0 | Cortex-M0+ | Cortex-M1 | Cortex-M3 | Cortex-M4 |
|---|---|---|---|---|---|
| Number of Interrupts | 1 to 32 | 1 to 32 | 1, 8, 16, 32 | 1 to 240 | 1 to 240 |
| NMI | Y | Y | Y | Y | Y |
| Width of Priority Registers | 2 | 2 | 2 | 3 to 8 | 3 to 8 |
| Access | Word | Word | Word | Word, Half-word, Byte | Word, Half-word, Byte |
| PRIMASK | Y | Y | Y | Y | Y |
| FAULTMASK | N | N | N | Y | Y |
| BASEPRI | N | N | N | Y | Y |
| Vector Table Offset Register | N | Y (optional) | N | Y | Y |
| Dynamic Priority Change | N | N | N | Y | Y |
| Interrupt Active Status | N | N | N | Y | Y |
| Fault Handling | HardFault | HardFault | HardFault | HardFault + 3 other fault exceptions | HardFault + 3 other fault exceptions |
| Debug Monitor Exception | N | N | N | Y | Y |

- There are no interrupt active status registers in ARMv6-M, so the NVIC->IABR register and the associated CMSIS-Core function "NVIC_GetActive" are not available on ARMv6-M.
- There is no priority grouping in ARMv6-M. Therefore CMSIS-Core functions "NVIC_EncodePriority"and "NVIC_DecodePriority" are not available.

In ARMv7-M the priority of an interrupt can be changed dynamically at run-time. In ARMv6-M the priority of an interrupt should only be changed when it is disabled.

The FAULTMASK and BASEPRI features are not available on ARMv6-M.

Table 7.23 shows a comparison of the NVIC features available in various Cortex-M processors.

# Exception Handling in Detail

## 8.1 Introduction

### 8.1.1 About this chapter

In the last chapter, we covered a wide range of information about exceptions and interrupts, including the types of exceptions available, an overview of the exception handling sequence, and how to set up the NVIC.

In this chapter, we look into the details of the exception handling sequences, including the stacking and unstacking operations, details of the EXC_RETURN code, which is used for exception returns, and additional programmable registers in the System Control Block (SCB) that are related to these behaviors.

Many of the topics in this chapter contain more advanced information, and many software developers do not need to know about them. However, they are very useful in debugging issues related to exceptions and interrupts, and most of these topics are essential for programmers who work on RealTime Operating System (RTOS) design.

The Definitive Guide to ARM® Cortex®-M3 and Cortex-M4 Processors. http://dx.doi.org/10.1016/B978-0-12-408082-9.00008-7

## 8.1.2 Exception handlers in C

We mentioned that with the Cortex®-M processor, you can program your exception handlers or Interrupt Service Routines (ISR) as normal C routines/functions. In order to understand the exact mechanism to support this, we first look at how C functions work on ARM® architecture.

C compilers for ARM architecture follow a specification from ARM called the AAPCS, Procedure Call Standard for ARM Architecture (reference 13). According to this standard, a C function can modify R0 to R3, R12, R14 (LR), and PSR. If the C function needs to use R4 to R11, it should save these registers on to the stack memory and restore them before the end of the function (see Figure 8.1).

R0−R3, R12, LR, and PSR are called "caller saved registers." The program code that calls a subroutine needs to save these register contents into memory (e.g., stack) before the function call if these values will still be needed after the function call. Register values that are not required after the function call don't have to be saved.

R4−R11 are called "callee-saved registers." The subroutine or function being called needs to make sure the contents of these registers are unaltered at the end of the function (same value as when the function is entered). The values of these registers could change in the middle of the function execution, but need to be restored to their original values before function exit.

Similar requirements apply to the registers in the floating point unit if the processor used is a Cortex-M4 with floating point support:

S0−S15 are "caller saved registers."

S16−S31 are "callee-saved registers."

Typically, a function call uses R0 to R3 as input parameters, and R0 as the return result. If the return value is 64 bits, R1 will also be used as the return result (see Figure 8.1).

In order to allow a C function to be used as an exception handler, the exception mechanism needs to save R0 to R3, R12, LR, and PSR at exception entrance automatically, and restore them at exception exit under the control of the processor's hardware. In this way when returned to the interrupted program, all the registers would have the same value as when the interrupt entry sequence started. In addition, since the value of the return address (PC) is not stored in LR as in normal C function calls (the exception mechanism puts an EXC_RETURN code in LR at exception entry, which is used in exception return), the value of the return address also needs to be saved by the exception sequence. So in total eight registers need to be saved during the exception handling sequence on the Cortex-M3 or Cortex-M4 processors without a floating point unit.

For the Cortex-M4 processor with floating point unit, the exception mechanism also needs to save S0−S15 and FPSCR if the floating point unit is used. This is indicated by a bit in the CONTROL register called FPCA (Floating Point Context Active).

## 8.1.3 Stack frames

The block of data that are pushed to the stack memory at exception entrance is called a stack frame. For the Cortex®-M3 processor or Cortex-M4 processor without the

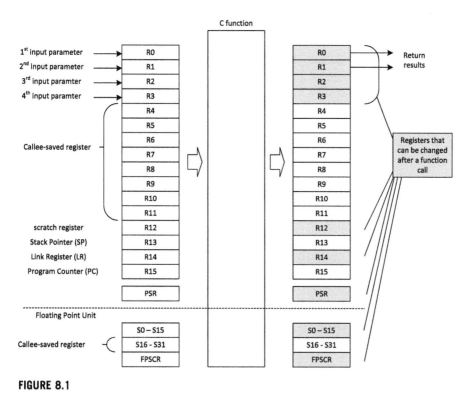

**FIGURE 8.1**

Register usage in a function call in AAPCS

floating point unit, the stack frames are always eight words (Figure 8.2 and Figure 8.3). For the Cortex-M4 with floating point unit, the stack frame can either be 8 or 26 words.

Another requirement of the AAPCS is that the stack pointer value should be double-word aligned at function entry or exit boundary. As a result, the Cortex-M3 and Cortex-M4 processors can insert an additional word of padding space in the stack automatically if the stack pointer was not aligned to double-word location when the interrupt happened. In this way, we can guarantee that the stack pointer will be at the beginning of the exception handler. This "double-word stack alignment" feature is programmable, and can be turned off if the exception handlers do not need full AAPCS compliance.

The bit 9 of the stacked xPSR is used to indicate if the value of the stack pointer has been adjusted. In Figure 8.2, the stack pointer was aligned to double-word address location, so no padding was inserted and bit 9 of the stack xPSR is set to 0. The same stack frame behavior can also be found when the double-word stack alignment feature is turned off, even if the value of stack pointer wasn't aligned to double-word boundary.

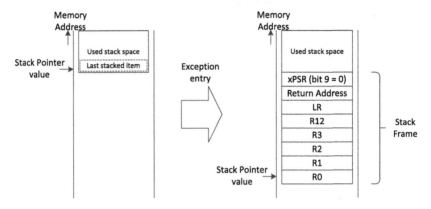

**FIGURE 8.2**

Exception Stack Frame of the Cortex-M3 or Cortex-M4 processor (without floating point context), when double-word stack alignment adjustment is not required or disabled

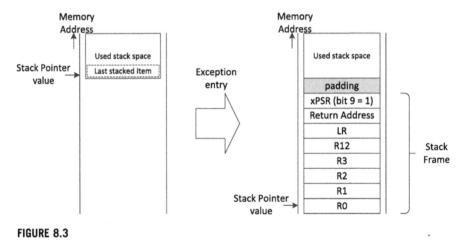

**FIGURE 8.3**

Exception Stack Frame of the Cortex-M3 or Cortex-M4 processor (without floating point context), when double-word stack alignment adjustment is enabled and required

If the double-word stack alignment feature was enabled, and the value of the stack pointer was not aligned to double-word boundary, a padding spaces is inserted to the stack to force the stack pointer to be aligned to double-word location, and bit 9 of the stacked xPSR is set to 1, to indicate the present of the padding space, as shown in Figure 8.3.

The bit 9 of the stacked xPSR is used in the exception exit sequence to decide whether the value of SP has to be adjusted to remove the padding.

For Cortex-M4 with floating point unit, if the floating point unit has been enabled and used, the stack frame will include S0 to S15 of the registers in the floating point unit register bank, as shown in Figure 8.4.

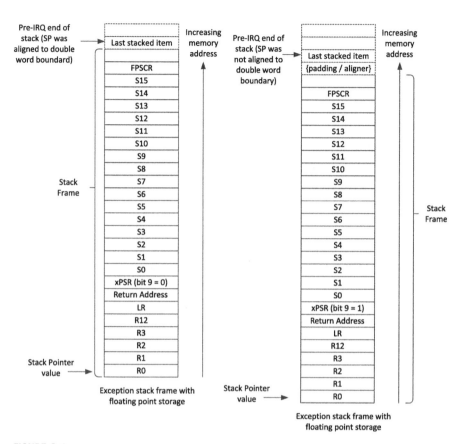

**FIGURE 8.4**

Stack frame format with floating point context

The values of general-purposed registers R0-R3 are located at the bottom of the stack frame. They can be accessed easily using SP-related addressing. In some cases these stacked registers can be used to pass information to software triggered interrupt handlers or SVC services.

The double workd stack alignment feature is enabled by a control bit in the Configuration Control Register (CCR, address 0xE000ED14) in the System Control Block (SCB). You can enable this feature during the initialization stage by the following C program code:

```
SCB->CCR |= SCB_CCR_STKALIGN_Msk; // Set STKALIGN bit (bit 9) of CCR
```

The double-word stack alignment is:

- Enabled by default in the Cortex-M4 processor
- Enabled by default in Cortex-M3 r2p0 and newer versions

- Disabled by default in Cortex-M3 r1p0 and r1p1
- Not available in Cortex-M3 r0p0

It is highly recommended to enable this feature, although it can result in slightly larger stack memory space usage. The bit should not be changed inside exception handlers.

### 8.1.4 EXC_RETURN

As the processor enters the exception handler or Interrupt Service Routine (ISR), the value of the Link Register (LR) is updated to a code called EXC_RETURN. The value of this code is used to trigger the exception return mechanism when it is loaded into the Program Counter (PC) using BX, POP, or memory load instructions (LDR or LDM).

Some bits of the EXC_RETURN code are used to provide additional information about the exception sequence. The definition of the EXC_RETURN value is shown in Table 8.1. The Valid values of the EXC_RETURN are shown in Table 8.2.

As a result of the EXC_RETURN number format, you cannot perform an interrupt return to an address in the 0xF0000000 to 0xFFFFFFFF memory range. However, since this address range is in the system region that is defined as non-executable in the architecture, it is not a problem.

**Table 8.1** Bit Fields of the EXC_RETURN

| Bits | Descriptions | Values |
|---|---|---|
| 31:28 | EXC_RETURN indicator | 0xF |
| 27:5 | Reserved (all 1) | 0xEFFFFF (23 bits of 1) |
| 4 | Stack Frame type | 1 (8 words) or 0 (26 words). Always 1 when the floating unit is unavailable. This value is set to the inverted value of FPCA bit in the CONTROL register when entering an exception handler. |
| 3 | Return mode | 1 (Return to Thread) or 0 (Return to Handler) |
| 2 | Return stack | 1 (Return with Process Stack) or 0 (Return with Main Stack) |
| 1 | Reserved | 0 |
| 0 | Reserved | 1 |

**Table 8.2** Valid Values for EXC_RETURN

| | Floating Point Unit was used before Interrupt (FPCA = 1) | Floating Point Unit was not used before Interrupt (FPCA = 0) |
|---|---|---|
| Return to Handler mode (always use Main Stack) | 0xFFFFFFE1 | 0xFFFFFFF1 |
| Return to Thread mode and use the Main Stack for return | 0xFFFFFFE9 | 0xFFFFFFF9 |
| Return to Thread mode and use the Process Stack for return | 0xFFFFFFED | 0xFFFFFFFD |

## 8.2 Exception sequences

### 8.2.1 Exception entrance and stacking

When an exception occurs and is accepted by the processor, the stacking sequence starts to push the registers into the stack, forming the stack frame, as shown in Figure 8.5.

The Cortex®-M3 and Cortex-M4 processors have multiple bus interfaces (as shown in Figure 6.2). In parallel to the stacking operation (usually on the System bus), the processor can also start the vector fetch (typically through the I-CODE bus), and then start the instruction fetch. As a result, the Harvard bus architecture allows the interrupt latency to be reduced because the stacking operation and flash memory accesses (vector fetch, instruction fetches) can take place in parallel. If the

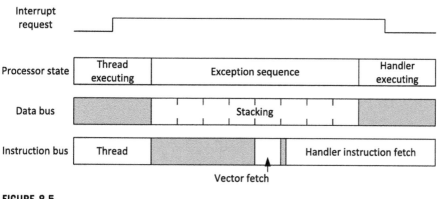

**FIGURE 8.5**

Stacking and vector fetch

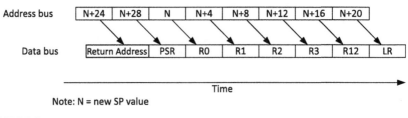

Note: N = new SP value

**FIGURE 8.6**

Stacking sequence in the Cortex-M3 processor on AHB lite interface

vector table is relocated to SRAM or if the exception handlers are also stored in SRAM, then this can increase the interrupt latency slightly.

Please note that the exact order of stack accesses during stacking is NOT the same order as the registers in the stack frame. For example, the Cortex-M3 processor stacks the PC and xPSR first, as shown in Figure 8.6, before other registers in the register banks so that the PC can be updated sooner with the vector fetch. Due to the pipeline nature of the AHB Lite interface, the data transfer lags behind the address by at least one clock cycle.

The stack being used in the stacking operations can either be the Main stack (using Main Stack Pointer, MSP) or the Process stack (using Process Stack Pointer, PSP).

If the processor was in Thread mode, and was using MSP (bit 1 of the CONTROL register is 0, as in the default setting), the stacking operation is carried out in the main stack with MSP (as shown in Figure 8.7).

If the processor was in Thread mode, and was using the process stack (bit 1 of the CONTROL register is 1), then the stacking operation is carried out in the process stack with PSP. After entering Handler mode, the processor must be using the MSP, so the stacking operation of all nested interrupts are carried out with the main stack with MSP (as shown in Figure 8.8).

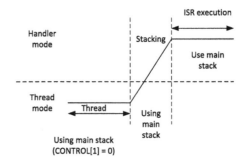

**FIGURE 8.7**

Exception stacking in thread mode using the main stack

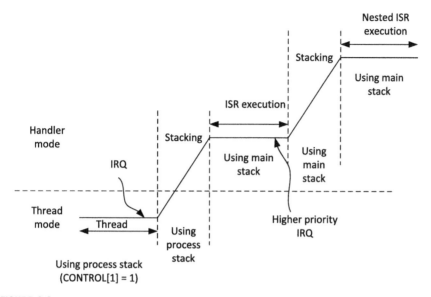

**FIGURE 8.8**

Exception stacking in thread mode using the process stack, and nested interrupt stacking using the main stack

## 8.2.2 Exception return and unstacking

At the end of an exception handler, the bit 2 of the EXC_RETURN value generated at the exception entrance is used to decide which stack pointer should be used to extract the stack frame. If bit 2 is 0, the processor knows that the main stack was used for stacking, as shown in Figure 8.9.

If bit 2 is 1, the processor knows that the process stack was used for stacking, as shown in the second unstacking operation in Figure 8.10.

At the end of each unstacking operation, the processor also checks the bit 9 of the unstacked xPSR value, and adjusts the stack pointer accordingly to remove the padding space if it was inserted during stacking, as shown in Figure 8.11.

To reduce the time required for the unstacking operation, the return address (stacked PC) value is accessed first, so that instruction fetch can start in parallel with the rest of the unstacking operation.

## 8.3 Interrupt latency and exception handling optimization
### 8.3.1 What is interrupt latency?

The term *interrupt latency* refers to the delay from the start of the interrupt request to the start of the interrupt handler execution. In the Cortex®-M3 and Cortex-M4

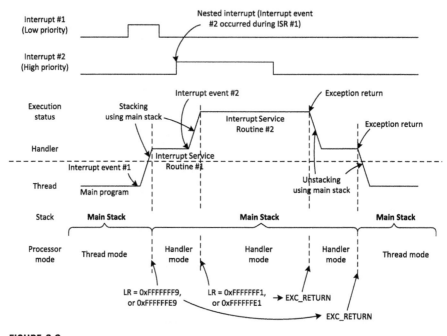

**FIGURE 8.9**

LR set to EXC_RETURN at exception (Main stack used in thread mode)

processors, if the memory system has zero latency, and provided that the bus system design allows vector fetch and stacking to happen at the same time, the interrupt latency is only 12 clock cycles. This includes stacking the registers, vector fetch, and fetching instructions for the interrupt handler. However, in many cases the latency can be higher due to wait states in the memory system. If the processor is carrying out a memory transfer, including buffered write operations, the outstanding transfer has to be completed before the exception sequence starts. The duration of the execution sequence also depends on memory access speed.

Besides wait states generated by the memory devices or peripherals, there can be other situations that can increase the interrupt latency:

- The processor was serving another exception at the same or higher priority.
- Debugger accesses to the memory system.
- The processor was carrying out an unaligned transfer. From the processor point of view, this might be a single access, but at the bus level it takes multiple cycles, as the bus interface needs to convert the unaligned transfer into multiple aligned transfers.
- The processor was carrying out a write to bit-band alias. The internal bus system converts this into a read-modify-write sequence, which takes at least two cycles.

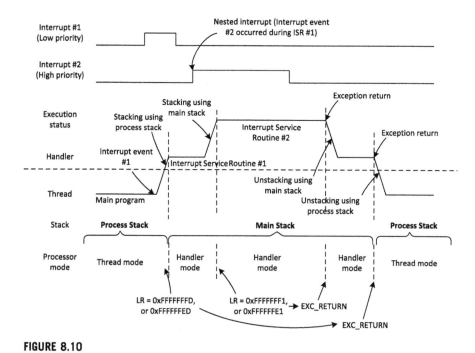

**FIGURE 8.10**

LR set to EXC_RETURN at exception (Process stack used in thread mode)

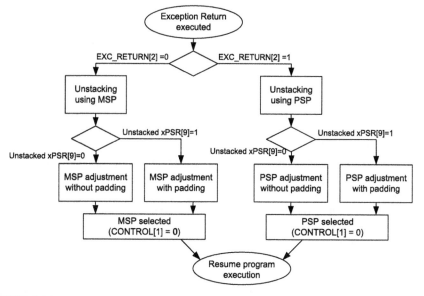

**FIGURE 8.11**

Unstacking operation

- The Cortex-M3 and Cortex-M4 processors use a number of methods to reduce the latency of servicing interrupts. For instance, most of the operations such as nested interrupt handling are automatically handled by the processor hardware. Also you don't need to use software code to determine which interrupt to service, or use software code to locate the starting addresses of ISRs.

### 8.3.2 Interrupts at multiple-cycle instructions

Some of the instructions take multiple clock cycles to execute. If an interrupt request arrives when the processor is executing a multiple cycle instruction, such as an integer divide, the instruction could be abandoned and restarted after the interrupt handler completes. This behavior also applies to load double-word (LDRD) and store double-word (STRD) instructions.

In addition, the Cortex®-M3 and Cortex-M4 processors allow exceptions to be taken in the middle of Multiple Load and Store instructions (LDM/STM) and stack push/pop instructions. If one of these LDM/STM/PUSH/POP instructions is executing when the interrupt request arrives, the current memory accesses will be completed, and the next register number will be saved in the stacked xPSR (Interrupt-Continuable Instruction [ICI] bits). After the exception handler completes, the multiple load/store/push/pop will resume from the point at which the transfer stopped. The same approach applies to floating point memory access instructions (i.e. VLDM, VSTM, VPUSH and VPOP) for Cortex-M4 processor with floating point unit. There is a corner case: If the multiple load/store/push/pop instruction being interrupted is part of an IF-THEN (IT) instruction block, the instruction will be canceled and restarted when the interrupt is completed. This is because the ICI bits and IT execution status bits share the same space in the Execution Program Status Register (EPSR).

For Cortex-M4 processor with floating point unit, if the interrupt request arrive when the processor is executing VSQRT (floating point square root) or VDIV (floating point divide), the floating point instruction execution continues in parallel with the stacking operation.

### 8.3.3 Tail chaining

When an exception takes place but the processor is handling another exception of the same or higher priority, the exception will enter the pending state. When the processor finishes executing the current exception handler, it can then proceed to process the pending exception/interrupt request. Instead of restoring the registers back from the stack (unstacking) and then pushing them on to the stack again (stacking), the processor skips the unstacking and stacking steps and enters the exception handler of the pended exception as soon as possible (Figure 8.12). In this way, the timing gap between the two exception handlers is considerably reduced. For a memory system with no-wait state, the tail-chain latency is only six clock cycles.

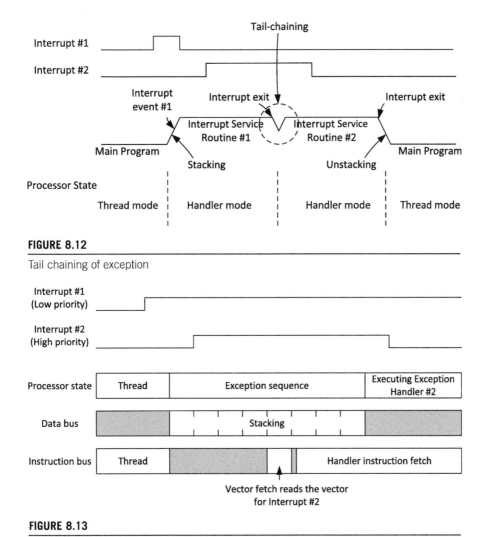

**FIGURE 8.12**

Tail chaining of exception

**FIGURE 8.13**

Late arrival exception behavior

The tail chaining optimization also makes the system more energy efficient because it reduces the amount of stack memory accesses, and each memory transfer consumes energy.

### 8.3.4 Late arrival

When an exception takes place, the processor accepts the exception request and starts the stacking operation. If during this stacking operation another exception of higher priority takes place, the higher priority late arrival exception will be serviced first.

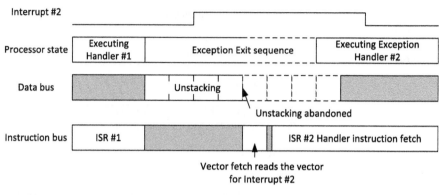

**FIGURE 8.14**

Pop pre-emption behavior

For example, if Exception #1 (lower priority) takes place a few cycles before Exception #2 (higher priority), the processor will behave as shown in Figure 8.13, such that Handler #2 is executed as soon as the stacking completes.

### 8.3.5 Pop preemption

If an exception request arrives during the unstacking process of another exception handler that has just finished, the unstacking operation would be abandoned and the vector fetch and instruction fetch for the next exception service begins. This optimization is called pop pre-emption. (Figure 8.14)

### 8.3.6 Lazy stacking

Lazy stacking is a feature related to stacking of the registers in the floating point unit. Therefore it is only relevant to Cortex®-M4 devices with a floating point unit. It is not needed for Cortex-M3 devices and Cortex-M4 devices that don't have the floating point unit.

If the floating point unit is available and enabled, and if it has been used, the registers in the register bank of the floating point unit will contain data that might need to be saved. As can be seen in Figure 8.4, if we need to stack the required floating point registers for each exception, we will need to carry out an additional 17 memory pushes each time, which will increase the interrupt latency from 12 to 29 cycles.

In order to reduce the interrupt latency, the Cortex-M4 processor has a feature called *lazy stacking*. By default this feature is enabled. When an exception arrives with the floating point unit enabled and used (indicated by bit 2 of the CONTROL register called FPCA), the longer stack frame format is used (as shown in Figure 8.4). However, the values of these floating point registers are not actually written into the stack frame. The lazy stacking mechanism only reserves the stack space for these registers, but only the R0-R3, R12, LR, Return Address, and xPSR are stacked. In this way, the interrupt latency remains at 12 clock cycles. When the lazy stacking happens, an internal register called LSPACT (Lazy Stacking Preservation Active) is set and another 32-bit register called Floating Point Context Address Register (FPCAR) stores the address of the reserved stack space for the floating point register.

If the exception handler does not require any floating point operation, the floating point registers remain unchanged throughout the operation of exception handler, and are not restored at exception exit. If the exception handler does need floating point operations, the processor detects the conflict and stalls the processor, pushes the floating point registers into the reserved stack space and clears LSPACT. After that, the exception handler resumes. In this way, the floating point registers are stacked only if they are necessary.

Lazy stacking operations may be interrupted. When an interrupt request arrives during lazy stacking, the lazy stacking operation will be stopped and normal exception stacking starts. Because the floating point instruction which triggered the lazy stacking has not yet been executed, the PC value stacked for the interrupt will point to the floating point instruction. When the interrupt service completed, the exception return will return to this floating point instruction and reattempting this instruction will again trigger the lazy stacking operation.

If the current executing context (either Thread or Handler) do not use the floating point unit, as indicated by a zero value in FPCA (bit 2 of the CONTROL register), the shorter stack frame format would be used.

Details of lazy stacking are covered in Chapter 13 (section 13.3).

# Low Power and System Control Features

## CHAPTER OUTLINE

The Definitive Guide to ARM® Cortex®-M3 and Cortex-M4 Processors. http://dx.doi.org/10.1016/B978-0-12-408082-9.00009-9

## 9.1 Low power designs

### 9.1.1 What does low power mean in microcontrollers?

Many embedded system products require low power microcontrollers, especially portable products that run on batteries. In addition, low power characteristics can benefit product designs in many ways, including:

- Smaller battery size (smaller product size and lower cost) or longer battery life
- Lower electromagnetic interference (EMI), which allows better wireless communication quality
- Simpler power supply design, avoiding heat dissipation issues
- In some cases it even allows the system to be powered using alternate energy sources (solar panel, energy harvesting from the environment)

One of the major benefits of the Cortex®-M microcontrollers are their energy efficiency and low power characteristics. Energy efficiency is typically measured by how much work can be done with a limited amount of energy, for example, in the form of DMIPS/µW or CoreMark/µW, whereas low power measurement

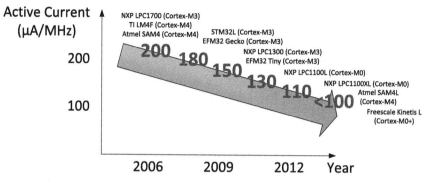

**FIGURE 9.1**

Trend of active power in low power Cortex-M microcontrollers and examples

traditionally only focuses on the active current and sleep mode current, measured in μA/MHz for active current and μA for sleep current (clock is stopped).

A number of years ago, low power microcontrollers were mostly 8-bit and 16-bit designs because many of them had very low active current and sleep currents. However, the low power microcontroller market has moved on and now, just having low active current and sleep current is often inadequate for low power designs. For example, many of these microcontrollers might need long execution times to complete a processing task, which results in higher overall power consumption. Therefore energy efficiency is equally important. In recent years, the Cortex-M microcontrollers have narrowed the gap between the active current and sleep current (Figure 9.1) and at the same time provide much better performance than many 8-bit and 16-bit microcontrollers. Therefore the Cortex-M microcontrollers are becoming much more attractive for many low power system designers.

Today we see many low power Cortex-M microcontrollers with very sophisticated system features that enable longer battery life. For example:

- Various run modes and sleep modes are available
- Ultra low power Real-Time Clock (RTC), watchdog and Brown-Out Detector (BOD)
- Smart peripherals that can operate while the processor remains in sleep mode
- Flexible clock system control features to allow clock signals for inactive parts of the design to be turned off

While we won't be able to cover the details of all the low power features in individual microcontroller devices here, in this chapter we will provide an overview of what is available with the Cortex-M3 and Cortex-M4 processors. Since different microcontrollers have different low power features, if you want to fully utilize their low power capability you will need to check out the relevant details in reference materials or the examples available from the microcontroller vendors. In many cases, example code will be available for download from the manufacturer's website.

**Table 9.1** Typical Low Power Requirements and Related Considerations

| Requirements | Typical Measurements and Consideration |
|---|---|
| Active current | Usually measured in μA/MHz. The active current is mostly caused by dynamic power needed by the memories, peripherals and the processor. To simplify the calculation, very often we assume that the power consumption of the microcontroller is directly proportional to the clock frequency (not strictly true). Also, the actual program code being used can also affect the results. |
| Sleep mode current | Usually measured in μA, as in most cases all clock signals are stopped for the lowest power sleep mode. This is generally composed of leakage current in the transistors and the current consumed by some of the analog circuits and I/O pads. Typically most peripherals would be turned off for such measurement. However, in real applications you would likely to have some peripherals remain active. |
| Energy efficiency | The measurement is typically based on popular benchmarks such as Dhrystone (DMIPS/μW) or CoreMark (CoreMark/μW). However, these benchmarks might be very different from the actual data processing activities in your application. |
| Wake-up latency | Usually measured in number of clock cycles, or sometime μsec. Typically this is the time from a hardware request (e.g., peripheral interrupt) to the time the processor resume program execution. If measured in μsec, the clock frequency will affect the result directly. In some designs, you can use a very low power sleep mode, but might take longer to wake-up (e.g., clock circuitry like the PLL could be turned off and it will take longer to resume normal clock output). Product designers need to decide which sleep mode should be used for their applications. |

### 9.1.2 Low power system requirements

Different low power systems have different requirements. Typically, we can summarize these requirements into the categories outlined in Table 9.1.

In a given application some of these factors could be more important than others. For example, in some battery powered products energy efficiency is the most important factor, while in some industrial control applications the wake-up latency can be critical.

There are also different approaches to designing low power systems. Nowadays a lot of embedded systems are designed to be interrupt driven (Figure 9.2). This means that the system stays in sleep mode when there is no request to be processed. When an interrupt request arrives, the processor wakes up and processes the requests, and goes back into sleep mode when the processing is done.

Alternatively, if the data processing request is periodic and has a constant duration, and if the data processing latency is not an issue, you could run the system at the slowest possibly clock speed to reduce the power. There is no clear answer to which approach is better, as the choice will be dependent on the data processing requirements of the application, the microcontroller being used, and other factors like the type of power source available.

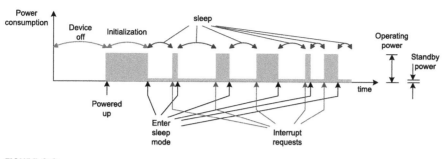

**FIGURE 9.2**

Activities in an interrupt-driven application

### 9.1.3 Low power characteristics of the Cortex®-M3 and Cortex-M4 processors

There are various reasons why Cortex®-M3 and Cortex-M4 are successful in the low power microcontroller market, as discussed in the following sections.

#### Low power design

The Cortex-M3 and Cortex-M4 processors have a relatively small silicon area compared to most other 32-bit processors. While the silicon area is larger than 8-bit processors and some 16-bit processors, the actual power consumption is not much larger because of the various low power optimizations that are used in the design of the Cortex-M3 and Cortex-M4 processors.

#### High performance

Since the Cortex-M3 and Cortex-M4 processors offer high performance, you can run the microcontroller at a lower clock frequency, or have the processor finish the processing task quicker and stay in sleep mode longer.

#### High code density

Since the Thumb instruction set offers excellent code density, for the same application task you can use a microcontroller with a smaller flash memory to reduce power consumption and cost. This also allows some of the microcontroller chips to be produced in smaller chip packages. This is important for some applications that need small device sizes, such as sensors and medical implants.

## 9.2 Low power features
### 9.2.1 Sleep modes

Sleep modes are common features in most microcontroller designs. In the Cortex®-M processors, the processors support two sleep modes: Sleep and Deep Sleep (Figure 9.3). These sleep modes can be further extended using device-specific power

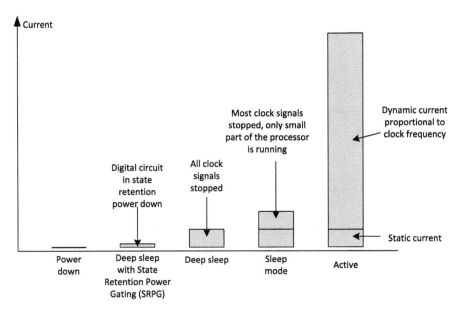

**FIGURE 9.3**

Various power modes including sleep modes

management features, and in some cases the Deep Sleep mode can be used with advanced chip design technologies such as State Retention Power Gating (SRPG) to further reduce the power.

What exactly happens during sleep modes depends on the chip design. In most cases some of the clock signals can be stopped to reduce the power consumption. However, the chip can also be designed such that part of it shuts down to further reduce power. In some cases it is also possible to power down the whole chip completely, and the only way to wake up the system from such power down mode is via a system reset.

### 9.2.2 System control register (SCR)

The Cortex®-M processor provides a register called the System Control Register (SCR) to allow you to select between the Sleep mode and Deep Sleep mode. This register is at address 0xE000ED10, and can be accessed in C programming using the "SCB->SCR" symbol. The details of the SCR bit fields are listed in Table 9.2. Just like most other registers in the System Control Block (SCB), the SCR can only be accessed in privileged state.

The SLEEPDEEP bit (bit 2) can be set to enable the Deep Sleep mode. This register can also be used to control other lower power features like Sleep-On-Exit and SEV-On-Pend. We will cover these features later in this chapter (in sections 9.2.5 and 9.2.6).

**Table 9.2** System Control Register (SCB->SCR, 0xE000ED10)

| Bits | Name | Type | Reset Value | Description |
|------|------|------|-------------|-------------|
| 4 | SEVONPEND | R/W | 0 | Send Event on Pending; when this is set to 1, the processor wakes up from WFE if a new interrupt is pended, regardless of whether the interrupt has priority higher than the current level and whether it was enabled |
| 3 | Reserved | – | – | – |
| 2 | SLEEPDEEP | R/W | 0 | When set to 1, the Deep Sleep mode is selected. Otherwise the sleep mode is selected. |
| 1 | SLEEPONEXIT | R/W | 0 | When this bit is set to 1, it enables the Sleep-On-Exit feature, which cause the processor to enter sleep mode automatically when exiting an exception handler and is returning to Thread. |
| 0 | Reserved | – | – | – |

### 9.2.3 Entering sleep modes

The processor provides two instructions for entering sleep modes: WFI and WFE (see Table 9.3).

Both WFI sleep and WFE sleep can be woken up by interrupt requests (depending on the priority of the interrupt, current priority level, and interrupt mask settings; see section 9.2.4).

WFE can be woken up by events. This includes a pulse from an event input signal (called RXEV on the processor), and events that happened in the past. Inside the

**Table 9.3** Instructions for Entering Sleep Mode

| Instruction | CMSIS-Core | Descriptions |
|-------------|------------|--------------|
| WFI | void __WFI(void); | Wait for Interrupt<br>Enter sleep mode. The processor can wake-up by interrupt request, debug request or reset. |
| WFE | void __WFE(void); | Wait for Event<br>Enter sleep mode conditionally. If the internal event register is clear, the processor enters sleep mode. Otherwise the internal event register is cleared and the processor continues. The processor can wake-up by interrupt request, event input, debug request or reset. |

processor, there is a single bit event register that can indicate that an event had occurred previously. This event register can be set by:

- Exception entrance and exit
- When SEV-On-Pend feature is enabled, the event register can be set when an interrupt pending status is changed from 0 to 1
- An External Event Signal (RXEV) from on-chip hardware (this is device-specific)
- Execution of the SEV (Send Event) instruction
- Debug event (e.g., halting request)

Similar to WFI sleep, during WFE sleep the processor can be woken up by an interrupt request if the interrupt has higher priority than the current level, including the priority of the interrupt masking such as BASEPRI, regardless of the SEV-On-Pend feature setting.

## 9.2.4 Wake-up conditions

In most cases, interrupts (including NMI and SysTick timer interrupts) can be used to wake up the Cortex®-M3 or Cortex-M4 microcontrollers from sleep modes. However, you also need to check the microcontroller's reference manual carefully because some of the sleep modes might turn off clock signals to the NVIC or peripherals, which would then prevent the interrupts (or some of them) from waking up the processor.

If the sleep mode is entered using WFI or Sleep-On-Exit, the interrupt request needs to be enabled, and have a higher priority level than the current level for the wake-up to occur (Table 9.4). For example, if the processor enters sleep mode while

**Table 9.4** Wake-up Conditions for WFI or Sleep-On-Exit

| IRQ Priority Condition | PRIMASK | Wake-up | IRQ Execution |
|---|---|---|---|
| Incoming IRQ higher than current priority level:<br>(IRQ priority > Current priority) AND (IRQ priority > BASEPRI) | 0 | Y | Y |
| Incoming IRQ same or lower than current priority level:<br>((IRQ priority =< Current priority) OR (IRQ priority =< BASEPRI)) | 0 | N | N |
| Incoming IRQ higher than current priority level:<br>(IRQ priority > Current priority) AND (IRQ priority > BASEPRI) | 1 | Y | N |
| Incoming IRQ same or lower than current priority level:<br>(IRQ priority =< Current priority) OR (IRQ priority =< BASEPRI) | 1 | N | N |

running an exception handler, or if the BASEPRI register was set before entering sleep mode, then the priority of the incoming interrupt will need to be higher than the current level to wake-up the processor.

The PRIMASK wake-up condition is a special feature to allow software code to be used to restore certain system resources between waking up and execution of ISRs. For example, a microcontroller could allow its clocking of Phase Locked Loops (PLL) to be turned off during sleep mode to reduce power, and restore it before executing the ISRs:

**(i)** Before entering sleep mode, the PRIMASK is set, the clock source switched to crystal clock, and then the PLL is turned off.

**(ii)** The microcontroller enters sleep mode with the PLL turned off to save power.

**(iii)** An interrupt request arrives, wakes up the microcontroller, and resumes program execution from the point after the WFI instruction.

**(iv)** The software code re-enables the PLL and then switches back to using the PLL clock, then clears the PRIMASK and services the interrupt request.

More information on using such an arrangement is given in sections 9.4.3 and Figure 9.14.

If the sleep mode is entered using a WFE instruction, the wake-up conditions are slightly different (Table 9.5). A feature called SEVONPEND can be used to generate a wake-up event when an interrupt request arrives and set the pending status, even if the interrupt was disabled or had the same or lower priority than the current level.

Note that the SEVONPEND feature generates the wake-up event only when a pending status switches from 0 to 1. If the pending status of the incoming interrupt was already set, it will not generate a wake-up event.

### 9.2.5 Sleep-on-Exit feature

The Sleep-on-Exit feature is very useful for interrupt-driven applications where all operations (apart from the initialization) are carried out inside interrupt handlers. This is a programmable feature, and can be enabled or disabled using bit 1 of the System Control Register (SCR; see section 9.2.2). When enabled, the Cortex®-M processor automatically enters sleep mode (with WFI behavior) when exiting from an exception handler and returning to Thread mode (i.e., when no other exception request is waiting to be processed).

For example, a program utilizing the Sleep-on-Exit might have a program flow as shown in Figure 9.4. The activities of such a system are shown in Figure 9.5. Unlike normal interrupt handling sequences, the stacking and unstacking processes are minimized to save power in the processor as well as memory (apart from the first stacking, which is still required).

Please note that the "loop" in Figure 9.4 is required because the processor could still be woken up by debug requests when a debugger is attached.

**Table 9.5** Wake-up Conditions for WFE

| IRQ Priority Condition | PRIMASK | SEVONPEND | Wake-up | IRQ Execution |
|---|---|---|---|---|
| Incoming IRQ higher than current priority level: (IRQ priority > Current priority) AND (IRQ priority > BASEPRI) | 0 | 0 | Y | Y |
| Incoming IRQ same or lower than current priority level: (IRQ priority =< Current priority) OR (IRQ priority =< BASEPRI) | 0 | 0 | N | N |
| Incoming IRQ higher than current priority level: (IRQ priority > Current priority) AND (IRQ priority > BASEPRI) | 1 | 0 | N | N |
| Incoming IRQ same or lower than current priority level: (IRQ priority =< Current priority) OR (IRQ priority =< BASEPRI) | 1 | 0 | N | N |
| Incoming IRQ higher than current priority level: (IRQ priority > Current priority) AND (IRQ priority > BASEPRI) | 0 | 1 | Y | Y |
| Incoming IRQ same or lower than current priority level: (IRQ priority =< Current priority) OR (IRQ priority =< BASEPRI) | 0 | 1 | Y | N |
| Incoming IRQ higher than current priority level: (IRQ priority > Current priority) AND (IRQ priority > BASEPRI) | 1 | 1 | Y | N |
| Incoming IRQ same or lower than current priority level: (IRQ priority =< Current priority) OR (IRQ priority =< BASEPRI) | 1 | 1 | Y | N |

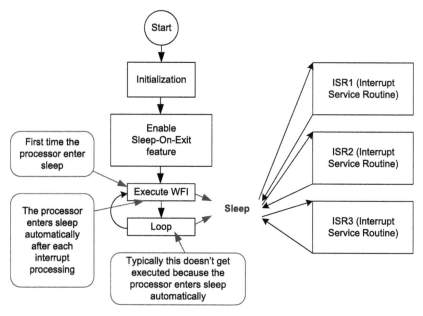

**FIGURE 9.4**

Sleep-on-Exit program flow

The Sleep-On-Exit feature should be enabled at the end of the initialization stage. Otherwise, if an interrupt event happened during the initialization stage and if the Sleep-on-Exit feature was already enabled, the processor will enter sleep even if the initialization stage was not yet completed.

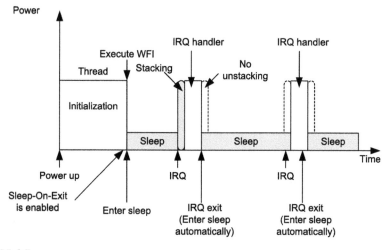

**FIGURE 9.5**

Sleep-on-Exit operations

### 9.2.6 Send event on pend (SEVONPEND)

One of the programmable control bits in the System Control Register (SCR) is the SEVONPEND. This feature is used with the WFE sleep operation. When this bit is set to 1, a new interrupt pending status set triggers an event and can wake up the processor. The interrupt does not have to be enabled, but the pending status before entering WFE needs to be 0 to trigger the wake-up event.

### 9.2.7 Sleep extension/wake-up delay

On some microcontrollers, some of the low power sleep modes might reduce power consumption aggressively by, for example, reducing the voltage supply to SRAM and turning off the power to the flash memory. In some cases, these hardware circuits might take a bit longer to get ready to run again after an interrupt request arrives. The Cortex®-M3 and Cortex-M4 processors provide a set of handshaking signals to allow the waking up to be delayed so that the rest of the system can get ready. This feature is only visible to silicon designers and is completely transparent to software. However, the microcontroller users might observe a longer interrupt latency when this feature is used.

### 9.2.8 Wake-up interrupt controller (WIC)

During Deep Sleep, when all the clock signals to the processor are stopped, the NVIC cannot detect incoming interrupt requests. In order to allow the microcontroller to be woken up by interrupt requests even when clock signals are unavailable, a feature called the Wake-up Interrupt Controller (WIC) was introduced in the Cortex-M3 revision r2p0.

The WIC is a small, optional interrupt detection circuit that is coupled with the NVIC in the Cortex®-M processor via a special interface, and also linked to the device-specific power control system such as a Power Management Unit (PMU) (Figure 9.6). The WIC does not contain any programmable registers, and the

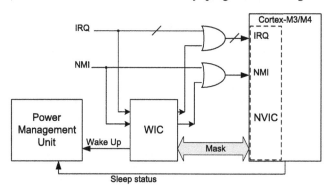

**FIGURE 9.6**

WIC mirrors the interrupt detection function when the clock signals to the processor stop interrupt masking information is transferred from the NVIC to the WIC just before entering Deep Sleep mode.

The interrupt detection logic of the WIC can be customized by silicon designers to support asynchronous operation. This means that the WIC can operate without any clock signal. When an interrupt request arrives, the WIC detects the request and informs the PMU to restore the clock, then the processor can wake up, resume operations, and service the interrupt request.

In some designs, advanced power-saving techniques called State Retention Power Gating (SRPG) can be used to reduce the leakage current of the chip by a wide margin. In SRPG designs, the registers (often called flip-flops in IC design terminology) have a separate power supply for state retention elements inside the registers (Figure 9.7). When the system is in Deep Sleep mode, the normal power supply can be turned off, leaving only the power to the state retention elements on. The leakage of this type of design is greatly reduced because the combinatorial logic, clock buffers, and most parts of the registers are powered down.

While the SRPG power down state can help to reduce the sleep mode current of the microcontroller significantly, the processor that is powered down cannot detect interrupt requests. Therefore the WIC is needed when the microcontroller implements SRPG technology. Figure 9.8 illustrates Deep Sleep operation when SRPG is used.

In SRPG designs, because the states of the processor are retained, it can resume operation from the point where the program was suspended and is therefore able to service the interrupt request almost immediately, just like normal sleep mode. In practice, the power-up sequence does take time to complete and therefore can increase the interrupt latency. The exact latency depends on semiconductor technology, memories, clocking arrangements, power system design (e.g., how long it takes for the voltage to be stabilized), etc.

For Cortex-M3 and Cortex-M4 processors, the WIC is only used in Deep Sleep mode (when the SLEEPDEEP bit in the System Control Register is set). In normal

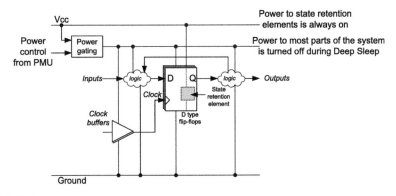

**FIGURE 9.7**

SRPG technology allows most parts of a digital system to be powered down without state loss

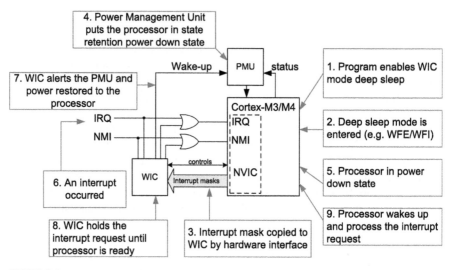

**FIGURE 9.8**

Illustration of the WIC mode Deep Sleep operations

Sleep mode, it does not enable the WIC operations and should not trigger SRPG power-down operations. Depending on the microcontroller you might also need to program additional control registers in the device-specific Power Management Unit (PMU) to enable the WIC feature.

Not all Cortex-M3 and Cortex-M4 microcontrollers support the WIC feature. Early Cortex-M3 products (revision r1p1 and earlier versions) do not support the WIC feature at all. Please note that if all the clock signals are stopped in the Deep Sleep mode, the SysTick timer inside the Cortex-M3 or Cortex-M4 processors will also be stopped and therefore cannot generate SysTick exceptions.

Also note that when a debugger is attached to the system, it might disable some of the low power capability. For example, the microcontroller could be designed in a way such that the clock continues to run in Deep Sleep mode when a debugger is attached, so that the debugger can continue to examine the system status even though the application code attempts to use the Deep Sleep mode.

## 9.2.9 Event communication interface

We mention that the WFE instruction can be woken up by an input signal on the Cortex®-M processor called RXEV (Receive Event). This signal is part of the event communication interface feature. The processor also has an output signal called TXEV (Transmit Event). The TXEV outputs a single cycle pulse when executing the SEV (Send Event) instruction.

The primary goal of the event communication interface is to allow the processor to stay in sleep mode until a certain event has occurred. The event communication interface can be used in a number of ways. For example, it can:

- Allow event communication between a peripheral and the processor
- Allow event communication between multiple processors

For example, in a single processor design with a DMA controller, the processor can set up a memory copy operation using the DMA controller for higher performance. However, if a polling loop is used on the processor to detect when the DMA is complete, it will waste energy and could reduce the performance because some of the bus bandwidth is used by the processor. If the DMA Controller has an output signal that generates a pulse when a DMA operation is done, we can put the processor into sleep mode using WFE instruction, and use the output signal pulse from the DMA controller to trigger a wake-up event when the DMA operation is completed.

Of course, you can also connect the DMA_Done output signal from the DMA to the NVIC and use interrupt mechanism to wake up the processor. However, it would mean the next step after the polling loop would have to wait longer for the interrupt entrance and exit sequence to be completed. The polling loop in the example in Figure 9.9 is necessary because the processor could be woken up by other interrupts, or by debug events.

Event communication is also important in multi-processor designs. For example, if processor A is waiting for processor B to finish a task by polling a variable in a shared memory space, then processor A might need to wait for a long time and this wastes energy (Figure 9.10).

In order to reduce power, we connect the event communication interface of the two Cortex-M processors together as shown in Figure 9.11. The connection arrangement can be extended to support more processors.

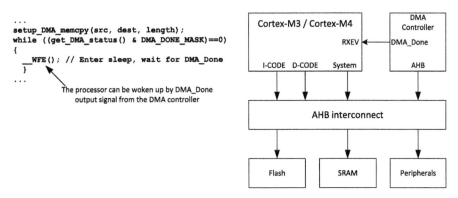

**FIGURE 9.9**

Example use of event input

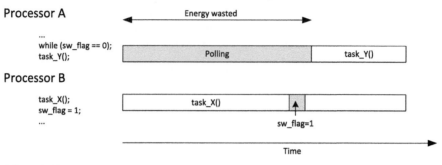

**FIGURE 9.10**

Multi-processor communication wastes energy in simple polling loops

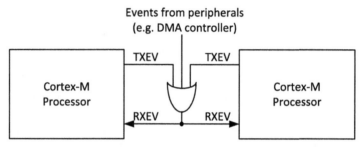

**FIGURE 9.11**

Event communication connections in a dual processor system

With this arrangement, we can use the SEV instruction to send an event to another processor to wake up the other processor from WFE sleep. As a result, the polling loop can be changed to include the sleep operation, as shown in Figure 9.12.

Again, the polling loop is still required because the processor can be woken up by other interrupts or debug events.

Using this method, the event communication interface feature can reduce the power consumption in various multiple processor communications such as task synchronizations, semaphores, and so on.

In a task synchronization situation, an example flow might be like the one shown in Figure 9.13, where multiple processors all stay in sleep mode and wait for an event from a "master controller," which generates the event pulse. When the event signal is triggered, the processors can start executing the next task together.

Please note that such task synchronization does not guarantee the processors in the system will start the task at exactly the same time. Because the processor could be woken up by other events, a check of the task status is needed and this could cause variation in the timing in each processor. Also, other factors such as the event propagation path in the chip design and memory system can also affect the execution timing of the synchronized tasks.

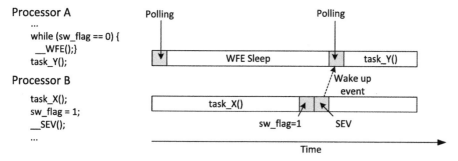

**FIGURE 9.12**

WFE sleep mode added to event communication routine to reduce power consumption

Another use for the event communication interface can be semaphores and MUTEX (mutual exclusive, which is one type of semaphore operation). For example, in semaphore operations which do not use the event communication feature, a processor might have to use a polling loop to detect when a locking variable is free, which can waste a lot of power:

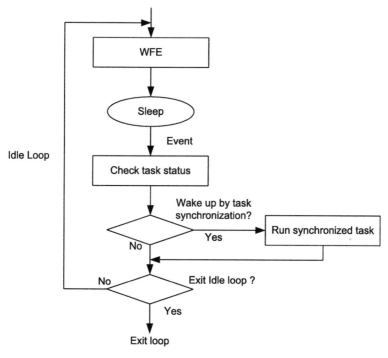

**FIGURE 9.13**

Example code of using WFE for task synchronization in a multiple core system

```
// Function to gain a lock in MUTEX (mutual exclusive)/semaphore
void get_lock(volatile int * Lock_Variable)
{ // Note: __LDREXW and __STREXW are functions in CMSIS-Core
 int status;
 do {
 while ( __LDREXW(&Lock_Variable) != 0); // Polling: Wait until lock
                                         // variable is free
 status = __STREXW(1, &Lock_Variable);   // Try set Lock_Variable
                                         // to 1 using STREX
 } while (status != 0);                  // retry until lock
                                            successfully
 __DMB();                                // Data memory Barrier
 return;
}
```

On the other hand, a process using the resource should unlock the resource when it is no longer required:

```
void free_lock(volatile int * Lock_Variable)
{
 __DMB();             // Data memory Barrier
 Lock_Variable = 0; // Free the lock
 return;
}
```

The polling loop (sometimes also called spin lock) can result in high power consumption, and can reduce the system performance by occupying memory bus bandwidth. As a result, we can add the WFE and SEV instructions into the semaphore operations so that the processor waiting for the lock can enter sleep and wake up when the lock is released:

```
// Function to gain a lock in MUTEX (mutual exclusive)/semaphore
void get_lock_with_WFE(volatile int * Lock_Variable)
{// Note: __LDREXW and __STREXW are functions in CMSIS-Core
 int status;
 do {
 while ( __LDREXW(&Lock_Variable) != 0){ // Wait until lock
 __WFE();} // variable is free, if not, enter sleep until event
 status = __STREXW(1, &Lock_Variable); // Try set Lock_Variable
                                       // to 1 using STREX
 } while (status != 0);                // retry until lock successfully
 __DMB();                              // Data memory Barrier
 return;
}
```

And the function for releasing the lock becomes:

```
void free_lock(volatile int * Lock_Variable)
```

```
{
  __DMB();              // Data memory Barrier
  Lock_Variable = 0; // Free the lock
  __SEV(); // Send Event to wake-up other processors
  return;
}
```

These examples are just illustrations of how event communication can help reduce power in simple semaphore operations. In systems with an embedded OS, the semaphore operations can be very different, because the OS can suspend a task while waiting for a semaphore and execute some other tasks instead.

## 9.3 Using WFI and WFE instructions in programming

As mentioned early in this chapter, just using the WFI and WFE instructions will not help you to fully utilize the low power features/optimizations in the Cortex®-M3 or Cortex-M4 microcontrollers. In most cases, microcontroller vendors provide example code or device-driver libraries to help you to get the most out of their products. But first, it is important to make sure that the WFI and WFE instructions are used correctly.

### 9.3.1 When to use WFI

The WFI instruction triggers sleep mode unconditionally. This is typically used in interrupt-driven applications. For example, an interrupt-driven application might have a program flow as shown below:

```
int main(void)
{

  setup_Io();
  setup_NVIC();
  ...
  SCB->SCR |= 1<< 1; // Enable Sleep-on-exit feature
  while(1) {
  __WFI(); // Keep in sleep mode
  }
}
```

In other cases, WFI may or may not be suitable, depending on if the expected timing of the interrupt events, which can wake up the processor. The following code demonstrates one such scenario:

```
setup_timer0(); // Setup a timer to trigger a timer interrupt
NVIC_EnableIRQ(Timer0_IRQn); // Enable Timer0 interrupt at NVIC
__WFI(); // Enter sleep and wait for timer #0 interrupt
Toggle_LED();
```

If the timer interrupt takes a long time to trigger, we can be sure that the processor will enter sleep mode well before the timer interrupt fires, then this code will toggle an LED after a bit of a time delay. However, if the timer is set to fire within a few cycles, or if another interrupt occurs after the timer is configured, and the execution time of the other interrupt handler is long enough, then the timer interrupt will fire before the WFI is executed. So we could end up executing the WFI after the timer interrupt handler is completed and the processor will possibly wait forever and be stuck.

Even if we changed the code to the following:

```
volatile int timer0irq_flag; // Set to 1 by timer0 ISR

...
setup_timer0(); // Setup a timer to trigger a timer interrupt
NVIC_EnableIRQ(Timer0_IRQn); // Enable Timer0 interrupt at NVIC
if (timer0irq_flag==0) { // timer0irq_flag is set in timer0 ISR
  __WFI(); // Enter sleep and wait for timer #0 interrupt
  }
Toggle_LED();
```

this is still not 100% safe. If the timer 0 interrupt takes place just after the compare of the software flag and before the WFI, the processor will still enter sleep mode and wait for a timer interrupt event that it has missed.

If we are not certainly about the timing, then we need to make the sleep operation conditional by using WFE.

## 9.3.2 Using WFE

WFE instruction is commonly used in idle loops, including idle task in RTOS design. Since the WFE is conditional, it might not enter sleep so you cannot change the code simply by replacing WFI with WFE. Taking the last example of toggling an LED, we can modify the code to:

```
volatile int timer0irq_flag;

...
timer0irq_flag = 0; // Clear flag
set_timer0();
NVIC_EnableIRQ(Timer0_IRQn);
while (timer0irq_flag==0) {
  __WFE(); // Enter sleep and wait for timer #0 interrupt
  };
Toggle_LED();
```

Here we change the sleep operation to an idle loop. If the WFE didn't enter sleep the first time due to previous events such as interrupts, the loop will execute again with the event latch cleared after the first WFE. So it will enter sleep mode only if the timer0 interrupt has not been triggered.

If the timer0 interrupt has been triggered before entering the while loop, the while loop will be skipped because the software flag is set.

If the timer0 interrupt is triggered just between the compare and the WFE, the interrupt sets the internal event register, and the WFE will be skipped. As a result the loop is repeated and the condition is checked again causing the loop to exit and toggle the LED.

Please note that in the Cortex®-M3 p0p0 to r2p0, there is a defect that affects the setting of the internal event register at run-time. To solve this issue you can insert a SEV instruction (__SEV();) inside the interrupt handlers to ensure that the interrupts will set the event register correctly.

The status of the internal event register cannot be read directly in software code. However, you can set it to 1 by executing the SEV instruction. If you want to clear the event register, you can execute SEV, then WFE:

```
__SEV(); // Set the internal event register
__WFE(); // Since the event register was set,
         // this WFE does not trigger sleep and
         // just clear the event register.
```

In most cases, you could get the processor to enter sleep mode using WFE using a sequence of:

```
__SEV(); // Set the internal event register
__WFE(); // Clear event register
__WFE(); // Enter sleep
```

However, if an interrupt occurs just after the first WFE instruction, the second WFE will not enter sleep because the event register is set by the interrupt event.

The WFE instruction should also be used if the SEVONPEND feature is needed.

## 9.4 Developing low power applications

Almost all Cortex®-M microcontrollers come with various low power operations and sleep modes to help product designers to reduce the power consumption of the products and get longer battery life. Since every microcontroller product is different, it is essential for designers to spend time learning about the low power features of the microcontrollers they are using. It is important to program the low power control features of a microcontroller correctly if you want to get the best energy efficiency from the resultant product.

It is impossible to cover all low power design methods for all the different microcontroller types. Here we will only cover some general considerations in designing low power embedded systems.

## 9.4.1 Reducing the active power

### Choose the right microcontroller device

Obviously the choice of the microcontroller device plays a significant role in getting low power consumption. Besides the typical device's electrical characteristics, you should also consider the size of memory required for your projects. If the microcontroller you use has much larger flash or SRAM than you need, you could be wasting power because of the power consumed by the memory.

### Run at the right clock frequency

Most applications do not need a high clock frequency, so you could potentially reduce the power consumption of the system by reducing the clock frequency. However, pushing the clock frequency too low risks reducing the system responsiveness or even failing timing requirements of the application tasks.

In most cases, you should run the microcontroller at a decent clock speed to ensure that the responsiveness of the system is good, and then put the system in sleep mode when there is no processing to do. Sometimes you might have to do some benchmarking to decide if you should run the system faster and then enter sleep, or run it slower to keep the active current down.

### Select the right clock source

Some microcontrollers provide multiple clock sources with different capabilities in terms of frequency as well as accuracy. Depending on your application, you might be better off using an internal clock source to save power, as an oscillator for external crystal could consume a fair amount of power. Alternatively you can also switch between different clock sources in different work-load conditions.

### Turn off unused clock signals

Many modern microcontrollers allow you to turn off the clock signals to unused peripherals, or in many cases the clock signals need to be turned on before the peripherals are used. In additional, some devices also allow you to power down some of these unused peripherals to save power.

### Utilize clock system features

Some microcontrollers provide various clock dividers for different parts of the system. You can use them to reduce the speed of peripherals, peripheral buses, etc.

### Power supply design

A good power supply design is another key factor of getting high energy efficiency. For example, if you use a voltage source with higher voltage than required, you need to reduce the voltage and the conversion often wastes power.

### Running the program from SRAM

It sounds a bit strange, but in some microcontrollers you can run the application entirely from SRAM, and you can turn off the power to the internal flash memory to save power. To achieve this, the microcontroller starts up with the program code in flash, and the reset handler can copy the program image to SRAM, execute from there, and then turn off the flash memory to save power.

However, many microcontrollers have only got a small SRAM space and it is impossible to copy the whole program here. In this case, it is still possible to copy just some of the frequently used parts of the program to SRAM, and turn on the flash memory only when the remaining parts of the program are needed.

### Using the right I/O port configurations

Some microcontrollers have configurable I/O port configurations for different drive currents and skew rates. Depending on the devices connected to the I/O pins, you could reduce the power consumption of the I/O interface logic by having a lower drive strength or slower skew rate configuration.

## 9.4.2 Reduction of active cycles

### Utilizing sleep modes

The most obvious point is to utilize the sleep mode features of the microcontroller as much as possible, even if the idle time only lasts for a short period. Also, features like sleep-on-exit can also help reduce the active cycles.

### Reducing run-time

If you have spare space in the flash memory, you can optimize the program (or some parts of it) for execution speed. In this way, the tasks can complete more quickly, and the system stays in sleep mode for longer.

## 9.4.3 Sleep mode current reduction

### Using the right sleep mode

Some microcontrollers provide various sleep modes and some peripherals can operate in some of these sleep modes without waking up the processor. By using the right sleep mode for your application, you can reduce the power consumption significantly. However, in some cases this can affect the wake-up latency.

### Utilizing power control features

Some microcontrollers allow you to fine-tune the power management in different active and sleep modes. For example, you could turn off the PLL, some peripherals, or flash memory during some sleep modes. However, in some cases this can affect the wake-up latency.

### Power off flash memory during sleep

In many Cortex®-M microcontrollers, the flash memory can be switched off automatically during some of the sleep modes, as well as being able to be switched off manually. This allows a significant reduction of sleep mode current, and allows the user to run applications in SRAM only to save power.

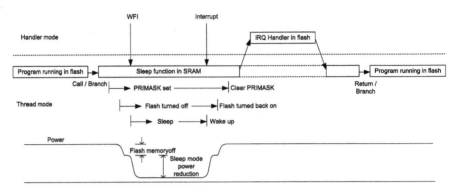

**FIGURE 9.14**

You can turn off flash memory in sleep mode and use PRIMASK to delay ISR execution

However, if the sleep mode you want to use does not support the automatically switching off of flash memory, and the SRAM size is not big enough to hold the entire application, you could just copy a function that handles sleep entry and sleep exit into the SRAM and execute the function from there. In this way the flash memory can be turned off manually during sleep.

In section 9.2.4 we mentioned that by setting PRIMASK before executing WFI, the processor can wake up from interrupts but not execute them. We can utilize this characteristic as shown in Figure 9.14.

## 9.5 The SysTick timer

### 9.5.1 Why have a SysTick timer?

The Cortex®-M processors have a small integrated timer called the SysTick (System Tick) timer. It is integrated as a part of the NVIC and can generate the SysTick exception (exception type #15). The SysTick timer is a simple decrement 24-bit timer, and can run on processor clock frequency or from a reference clock frequency (normally an on-chip clock source).

In modern operating systems, a periodic interrupt is needed to ensure that the OS kernel can invoke regularly; for example, for task management and context switching. This enables a processor to handle different tasks in different time slots. The processor design also ensures that the application tasks running at unprivileged level cannot disable this timer; otherwise, these tasks could disable the SysTick timer and lock out the whole system.

The reason for having the timer inside the processor is to help software portability. Since all the Cortex-M processors have the same SysTick timer, an OS written for one Cortex-M3/M4 microcontroller can be reused on other Cortex-M3/M4 microcontrollers.

If you do not need an embedded OS in your application, the SysTick timer can be used as a simple timer peripheral for periodic interrupt generation, delay generation, or timing measurement.

### 9.5.2 Operations of the SysTick timer

The SysTick timer contains four registers, as shown in Table 9.6. A data structure called SysTick is defined in the CMSIS-Core header file to allow these registers to be accessed easily.

The counter inside the SysTick is a 24-bit decrement counter (Figure 9.15). It can decrement using the processor's clock, or using a reference clock signal (called STCLK in Cortex®-M3 or Cortex-M4 Technical Reference Manual from ARM®). The exact implementation of the reference clock signal depends on the microcontroller design. In some cases it might not be available. The reference clock must be at least two times slower than the processor clock because of the sampling logic used for rising edge detection.

When the counter is enabled by setting bit 0 of the Control and Status register, the current value register decrements at every processor clock cycle or every rising edge of the reference clock. If it reaches zero, it will then load the value from the reload value register and continue.

An additional register called SysTick Calibration Register is available to allow the on-chip hardware to provide calibration information for the software. In CMSIS-Core, the use of SysTick Calibration Register is not required because the CMSIS-Core provides a software variable called "SystemCoreClock" (from CMSIS 1.2 and later; CMSIS 1.1 or prior versions use the variable "SystemFrequency"). This variable is set up in the system initialization function "SystemInit()" and is also updated every time the system clock configuration is changed. This software approach is much more flexible then the hardware approach using the SysTick Calibration Register.

Details of the SysTick registers are shown in Tables 9.7 to 9.10.

**Table 9.6** Summary of the SysTick Registers

| Address | CMSIS-Core Symbol | Register |
| --- | --- | --- |
| 0xE000E010 | SysTick->CTRL | SysTick Control and Status Register |
| 0xE000E014 | SysTick->LOAD | SysTick Reload Value Register |
| 0xE000E018 | SysTick->VAL | SysTick Current Value Register |
| 0xE000E01C | SysTick->CALIB | SysTick Calibration Register |

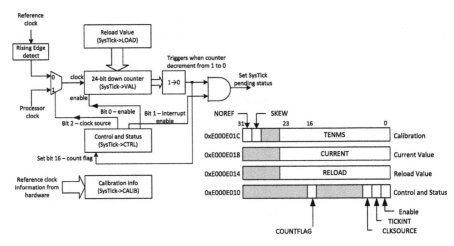

**FIGURE 9.15**

A simplified block diagram of SysTick timer

### 9.5.3 Using the SysTick timer

If you only want to generate a periodic SysTick interrupt, the easiest way is to use a CMSIS-Core function called "SysTick_Config":

```
uint32_t SysTick_Config(uint32_t ticks);
```

This function sets the SysTick interrupt interval to "ticks," enables the counter using the processor clock, and enables the SysTick exception with the lowest exception priority.

For example, if you have a clock frequency of 30MHz and you want to trigger a SysTick exception of 1KHz, you can use:

```
SysTick_Config(SystemCoreClock / 1000);
```

The variable "SystemCoreClock" should hold the correct clock frequency value of $30 \times 10^6$. Alternatively, you can just use:

```
SysTick_Config(30000); // 30MHz / 1000 = 30000
```

The "SysTick_Handler(void)" will then be triggered at a rate of 1 kHz.

If the input parameter of SysTick_Config function cannot be fit into the 24-bit reload value register (larger than 0xFFFFFF), the SysTick_Config function returns 1; otherwise, it returns 0.

In many cases you might not want to use the SysTick_Config function because you might want to use the reference clock or you might not want to enable the SysTick interrupt. In these cases you need to program the SysTick registers directly, and the following sequence is recommended:

**Table 9.7** SYSTICK Control and Status Register (0xE000E010)

| Bits | Name | Type | Reset Value | Description |
|------|------|------|-------------|-------------|
| 16 | COUNTFLAG | R | 0 | Read as 1 if counter reaches 0 since last time this register is read; clear to 0 automatically when read or when current counter value is cleared |
| 2 | CLKSOURCE | R/W | 0 | 0 = External reference clock (STCLK) 1 = Use core clock |
| 1 | TICKINT | R/W | 0 | 1 = Enable SYSTICK interrupt generation when SYSTICK timer reaches 0 0 = Do not generate interrupt |
| 0 | ENABLE | R/W | 0 | SYSTICK timer enable |

**Table 9.8** SYSTICK Reload Value Register (0xE000E014)

| Bits | Name | Type | Reset Value | Description |
|------|------|------|-------------|-------------|
| 23:0 | RELOAD | R/W | 0 | Reload value when timer reaches 0 |

**Table 9.9** SYSTICK Current Value Register (0xE000E018)

| Bits | Name | Type | Reset Value | Description |
|------|------|------|-------------|-------------|
| 23:0 | CURRENT | R/Wc | 0 | Read to return current value of the timer. Write to clear counter to 0. Clearing of current value also clears COUNTFLAG in SYSTICK Control and Status Register |

1. Disable the SysTick timer by writing 0 to SysTick->CTRL. This step is optional. It is recommended for reusable code because the SysTick could have been enabled previously.
2. Write the new reload value to SysTick->LOAD. The reload value should be the interval value $-1$.
3. Write to the SysTick Current Value register SysTick->VAL with any value to clear the current value to 0.

**Table 9.10** SYSTICK Calibration Value Register (0xE000E01C)

| Bits | Name | Type | Reset Value | Description |
|------|------|------|-------------|-------------|
| 31 | NOREF | R | – | 1 = No external reference clock (STCLK not available)<br>0 = External reference clock available |
| 30 | SKEW | R | – | 1 = Calibration value is not exactly 10 ms<br>0 = Calibration value is accurate |
| 23:0 | TENMS | R | – | Calibration value for 10 ms.; chip designer should provide this value via Cortex-M3/M4 input signals. If this value is read as 0, calibration value is not available |

**4.** Write to the SysTick Control and Status register SysTick->CTRL to start the SysTick timer.

Since the SysTick timer counts down to 0, if you want to set the SysTick interval to 1000, you should set the reload value (SysTick->LOAD) to 999.

If you want to use the SysTick timer in polling mode, you can use the count flag in the SysTick Control and Status Register (SysTick->CTRL) to determine when the timer reaches zero. For example, you can create a timed delay by setting the SysTick timer to a certain value and waiting until it reaches zero:

```
SysTick->CTRL = 0;     // Disable SysTick
SysTick->LOAD = 0xFF;  // Count from 255 to 0 (256 cycles)
SysTick->VAL = 0;      // Clear current value as well as count flag
SysTick->CTRL = 5;     // Enable SysTick timer with processor clock
while ((SysTick->CTRL & 0x00010000)==0);// Wait until count flag is set
SysTick->CTRL = 0; // Disable SysTick
```

If you want to schedule the SysTick interrupt for one-shot operation, which triggers in a certain time, you can reduce the reload value by 12 cycles to compensate for the interrupt latency. For example, if we want to have the SysTick handler to execute in 300 clock cycle time:

```
volatile int SysTickFired; // A global software flag to
                           // indicate SysTickAlarm executed
...
SysTick->CTRL = 0;         // Disable SysTick
SysTick->LOAD = (300-12);  // Set Reload value
                           // Minus 12 because of exception latency
SysTick->VAL = 0;          // Clear current value to 0
SysTickFired = 0;          // Setup software flag to zero
SysTick->CTRL = 0x7;       // Enable SysTick, enable SysTick
```

```
                              // exception and use processor clock
while (SysTickFired == 0);   // Wait until software flag is set by
                             // SYSTICK handler
```

Inside the one-shot SysTick Handler, we need to disable the SysTick so that the SysTick exception only triggers once. We might also need to clear the SysTick pending status in case the pending status has been set again when the required processing task takes some time:

```
void SysTick_Handler(void) // SYSTICK exception handler
{
SysTick->CTRL = 0x0;       // Disable SysTick
...;                       // Execute required processing task
SCB->ICSR |= 1<<25;        // Clear SYSTICK pend bit
                           // in case it has been pended again
SysTickFired++;            // Update software flag so that the
                           // main program know that SysTick alarm
                           // task has been carried out

return;
}
```

If there is another exception happening at the same time, the SysTick exception could be delayed.

The SysTick timer can be used for timing measurement. For example, you can measure the duration of a short function using the following code:

```
unsigned int start_time, stop_time, cycle_count;
SysTick->CTRL = 0;            // Disable SysTick
SysTick->LOAD = 0xFFFFFFFF;   // Set Reload value to maximum
SysTick->VAL = 0;            // Clear current value to 0
SysTick->CTRL = 0x5;         // Enable SysTick, use processor clock
while(SysTick->VAL != 0);    // Wait until SysTick reloaded
start_time = SysTick->VAL;   // Get start time
function();                  // Execute function to be measured
stop_time = SysTick->VAL;    // Get stop time
cycle_count = start_time - stop_time;
```

Since the SysTick is a decrement counter, the value of start_time is larger than stop_time. You might want to include a check for the count_flag at the end of the timing measurement. If the count_flag is set, the duration being measured is longer than 0xFFFFFF clock cycles. In that case you will have to enable the SysTick exception and use the SysTick Handler to count how many times the SysTick counter underflows. The total number of clock cycles will then also include the SysTick exceptions.

The SysTick timer provides a register to provide a calibration value. If this information is available, the lowest 24 bits of the SysTick->CALIB register provide the reload value required to get 10 msec SysTick intervals. However, many microcontrollers do not have this information and the TENMS bit field would read as zero. The CMSIS-Core approach of providing a software variable (SystemCore-Clock) for clock frequency information is more flexible and supported by most microcontroller vendors.

You can use the bit 31 of the SysTick Calibration register to determine if a reference clock is available.

### 9.5.4 Other considerations

There are a number of considerations when using the SysTick timer:

- The registers in the SysTick timer can only be accessed when in privileged state.
- The reference clock might not be available in some microcontroller designs.
- When you are using an embedded OS in your application, the SysTick timer will be used by the OS and therefore should not be used by the application tasks.
- The SysTick Timer stops counting when the processor is halted during debugging.
- Depending on the design of the microcontroller, the SysTick timer may stop in certain sleep modes.

## 9.6 Self-reset

The Cortex®-M processor provides a mechanism for triggering self-reset in software. In section 7.9.4, we discussed the Application Interrupt and Reset Control Register (AIRCR). This register has two control bits for reset.

The SYSRESETREQ bit (bit 2) generates a system reset request to the microcontroller's system reset control logic. Because the system reset control logic is not part of the processor design, the exact timing of the reset is device-specific (e.g., How many clock cycles delay before the system actually goes into the reset state?). In normal cases, this should not reset the debug logic.

To use the SYSRESETREQ feature (or any access to the AIRCR), the program must be running in privileged state. The easiest way is to use a function provided in the CMSIS-Core header file called "NVIC_SystemReset(void)."

Instead of using CMSIS-Core, you can access the AIRCR register directly:

```
// Use DMB to wait until all outstanding
// memory accesses are completed
__DMB();
// Read back PRIGROUP and merge with SYSRESETREQ
SCB->AIRCR = 0x05FA0004 | (SCB->AIRCR & 0x700);
while(1); // Wait until reset happen
```

The Data Memory Barrier (DMB) instruction is needed to make sure previous memory accesses are completed before the reset happens. When writing to AIRCR, the upper 16 bits of the write value should be set to 0x05FA, which is key to preventing accidentally resetting the system.

The second reset feature is the VECTRESET control bit (bit 0). This is indented for use by debuggers. Writing 1 to this bit resets the Cortex-M3/Cortex-M4 processor excluding the debug logic. This reset feature does not reset the peripherals in the microcontroller. In some cases, this feature can be useful for systems with multiple processor cores because you might want to reset just one processor, but not the rest of the system.

Writing to VECTREST is similar to setting SYSRESETREQ:

```
// Use DMB to wait until all outstanding
// memory accesses are completed
__DMB();
// Read back PRIGROUP and merge with SYSRESETREQ
SCB->AIRCR = 0x05FA0001 | (SCB->AIRCR & 0x700);
while(1); // Wait until reset happen
```

Unlike SYSRESETREQ, the VECTRESET happens almost immediately because the reset path is not dependent on other logic circuits in the microcontroller. However, because the VECTRESET does not reset peripherals, it is not recommended for general use in application programming.

Do not set SYSRESETREQ and VECTRESET simultaneously. In some chip designs this can result in a glitch in the reset system (SYSRESETREQ gets asserted for a short period and then gets cleared by the reset as a result of the VECTRESET path). The result can be unpredictable.

In some cases, you might want to set PRIMASK to disable processing before starting the self-reset operation. Otherwise, if the system reset takes some time to trigger, an interrupt could occur during the delay and the system reset could happen in the middle of the interrupt handler. In most cases this is not a problem, but in some applications this needs to be avoided.

## 9.7 **CPU ID base register**

Inside the System Control Block (SCB) there is a register called the CPU ID Base Register. This is a read-only register that shows the processor type and the revision number. The address of this register is 0xE000ED00 (privileged accesses only). In C language programming you can access to this register using "SCB->CPUID". For reference, the CPU IDs of all existing Cortex®-M processors are shown in Table 9.11.

Individual debug components inside the Cortex-M processors also carry their own ID registers, and their revision fields might also be different between different revisions.

**Table 9.11** CPU ID Base Register (SCB->CPUID, 0xE000ED00)

| Processor and Revisions | Implementer Bit [31:24] | Variant Bit [23:20] | Constant Bit [19:16] | PartNo Bit [15:4] | Revision Bit [3:0] |
|---|---|---|---|---|---|
| Cortex-M0 - r0p0 | 0x41 | 0x0 | 0xC | 0xC20 | 0x0 |
| Cortex-M0+ - r0p0 | 0x41 | 0x0 | 0xC | 0xC60 | 0x0 |
| Cortex-M1 - r0p1 | 0x41 | 0x0 | 0xC | 0xC21 | 0x0 |
| Cortex-M1 - r0p1 | 0x41 | 0x0 | 0xC | 0xC21 | 0x1 |
| Cortex-M1 - r1p0 | 0x41 | 0x1 | 0xC | 0xC21 | 0x0 |
| Cortex-M3 - r0p0 | 0x41 | 0x0 | 0xF | 0xC23 | 0x0 |
| Cortex-M3 - r1p0 | 0x41 | 0x0 | 0xF | 0xC23 | 0x1 |
| Cortex-M3 - r1p1 | 0x41 | 0x1 | 0xF | 0xC23 | 0x1 |
| Cortex-M3 - r2p0 | 0x41 | 0x2 | 0xF | 0xC23 | 0x0 |
| Cortex-M3 - r2p1 | 0x41 | 0x2 | 0xF | 0xC23 | 0x1 |
| Cortex-M4 - r0p0 | 0x41 | 0x0 | 0xF | 0xC24 | 0x0 |
| Cortex-M4 - r0p1 | 0x41 | 0x0 | 0xF | 0xC24 | 0x1 |

## 9.8 Configuration control register

### 9.8.1 Overview of CCR

There is a register in the System Control Block (SCB) called the Configuration Control Register (CCR). This can be used to adjust some of the behaviors in the processor and for controlling advanced features. The address of this register is 0xE000ED14 (privileged accesses only). The details of the CCR bit fields are listed in Table 9.12. In C programming you can access this register using SCB->CCR.

### 9.8.2 STKALIGN bit

When the STKALIGN bit is set to 1, it forces the stack frame to be placed in double-word aligned memory locations. If the stack pointer was not pointing to a double-word aligned address when an interrupt occurred, a padding word will be added

**Table 9.12** Configuration Control Register (SCB->CCR, 0xE000ED14)

| Bits | Name | Type | Reset Value | Descriptions |
|------|------|------|-------------|--------------|
| 9 | STKALIGN | R/W | 0 or 1 | Force exception stacking start in double word aligned address. This bit is reset as zero on Cortex®-M3 revision r1p0 and r1p1, and is reset as one on revision 2. Revision r0p0 does not have this feature. In Cortex-M4 this bit is reset as 1. |
| 8 | BFHFNMIGN | R/W | 0 | Ignore data bus fault during HardFault and NMI handlers. |
| 7:5 | Reserved | – | – | Reserved |
| 4 | DIV_0_TRP | R/W | 0 | Trap on divide by 0 |
| 3 | UNALIGN_TRP | R/W | 0 | Trap on unaligned accesses. |
| 2 | Reserved | – | – | Reserved |
| 1 | USERSETMPEND | R/W | 0 | If set to 1, allow user code to write to Software Trigger Interrupt Register. |
| 0 | NONBASETHRDENA | R/W | 0 | Non-base thread enable. If set to 1, allows exception handler to return to thread state at any level by controlling EXC_RETURN value. |

during stacking and bit 9 of the stacked xPSR will be set to 1 to indicate that the stack has been adjusted, and the adjustment will be reversed at unstacking.

If you are using Cortex®-M3 r1p0 or r1p1, it is highly recommended that you enable the double-word stack alignment feature at the start of the program to ensure that the interrupt handling mechanism fully conforms to AAPCS requirement (see section 8.1.3). This can be done by adding the following code:

```
SCB->CCR |= SCB_CCR_STKALIGN_Msk; /* Set STKALIGN */
```

If this feature is not enabled, the stack frame will be aligned to word (4 bytes) address boundaries. In some applications, problems can occur if the C compiler or run-time library functions make assumptions that the stack pointer is double-word aligned; for example, when handling pointers computation.

### 9.8.3 BFHFNMIGN bit

When this bit is set, handlers with priority of −1 (e.g., HardFault) or −2 (e.g., NMI) ignore data bus faults caused by load and store instructions. This can also be used when configurable fault exception handlers (i.e., BusFault, Usage Fault, or Mem-Menage fault) are executing with the FAULTMASK bit set.

If this bit is not set, a data bus fault in NMI or HardFault handler causes the system to enter a lock-up state (see section 12.7 in chapter 12).

The bit is typically used in fault handlers that need to probe various memory locations to detect system buses or memory controller issues.

### 9.8.4 DIV_0_TRP bit

When this bit is set, a Usage Fault exception is triggered when a divide by zero occurs in SDIV (signed divide) or UDIV (unsigned divide) instructions. Otherwise, the operation will complete with a quotient of 0.

If the Usage Fault handler is not enabled, the HardFault exception would be triggered (see chapter 12, section 12.1 and Figure 12.1).

### 9.8.5 UNALIGN_TRP bit

The Cortex®-M3 and Cortex-M4 processors support unaligned data transfers (see section 6.6). However, in some cases occurrence of unaligned transfers might indicate incorrect program code (e.g., used of incorrect data type), and could result in slower performance because each unaligned transfer needs multiple clock cycles to be carried out. Therefore a trap exception mechanism is implemented to detect the presence of unaligned transfers.

If the UNALIGN_TRP bit is set to 1, the Usage Fault exception is triggered when an unaligned transfer occurs. Otherwise (UNALIGN_TRP set to 0, the default value), unaligned transfers are allowed for single-load and store instructions LDR, LDRT, LDRH, LDRSH, LDRHT, LDRSHT, STR, STRH, STRT, and STRHT.

Multiple transfer instructions such as LDM, STM, LDRD, and STRD always trigger faults if the address is unaligned regardless of the UNALIGN_TRP value.

Byte size transfers are always aligned.

### 9.8.6 USERSETMPEND bit

By default, the Software Trigger Interrupt Register (NVIC->STIR) can only be accessed in privileged state. If the USERSETMPEND is set to 1, unprivileged accesses are allowed on this register (but do not allow unprivileged accesses other NVIC or SCB registers).

Setting USERSETMPEND can lead to another problem. After it is set, unprivileged tasks can trigger any software interrupt apart from system exceptions. As a result, if the USERSETMPEND is used and the system contains untrusted user tasks, the interrupt handlers need to check whether the exception handling should be carried out because it could have been triggered from untrusted programs.

### 9.8.7 **NONBASETHRDENA bit**

By default, the processor executing an exception handler can only return to Thread mode if there is no other active exception being served. Otherwise, the Cortex®-M3 or Cortex-M4 processors will trigger a Usage Fault that indicates there has been an error. The NONBASETHRDENA bit enables returning to Thread mode even when exiting a nested exception. More details of using NONBASETHRDENA are covered in section 23.5.

This feature is rarely used in application software development. In most cases it should be disabled, as the integrity check of NVIC status is useful in detecting issues such as stack corruption in exception handling.

## 9.9 **Auxiliary control register**

The Cortex®-M3 and Cortex-M4 processors have an additional control register called the Auxiliary Control Register that controls other processor-specific behavior. This is used for debugging, and in normal application programming this register is unlikely to be used. This register was added to the Cortex-M3 at revision r2p0. Older versions of the Cortex-M3 processor do not have this register.

The address of this register is 0xE000E008 (privileged accesses only). In C programming you can access this register using the "SCnSCB->ACTLR" symbol.

In the Cortex-M3 or Cortex-M4 processors without a floating point unit, the definition of the Auxiliary Control Register is shown in Table 9.13.

In the Cortex-M4 processor with a floating point unit, the Auxiliary Control Register has additional bit fields, as shown in Table 9.14.

### *DISFOLD (bit 2)*

In some situations, the processor can start executing the first instruction in an IT block while it is still executing the IT instruction. This behavior is called IT folding,

**Table 9.13** Auxiliary Control Register (SCnSCB -> ACTLR, 0xE000E008) in the Cortex-M3 Processor

| Bits | Name | Type | Reset Value | Description |
|------|------|------|-------------|-------------|
| 2 | DISFOLD | R/W | 0 | Disable IT folding (Prevent overlap of IT instruction execution phase with following instruction) |
| 1 | DISDEFWBUF | R/W | 0 | Disable write buffer for default memory map (memory accesses in MPU mapped regions are not affected) |
| 0 | DISMCYCINT | R/W | 0 | Disable interruption of multiple cycle instructions like LDM, STM, 64-bit multiply and divide instructions |

**Table 9.14** Auxiliary Control Register (SCnSCB -> ACTLR, 0xE000E008) in the Cortex-M4 Processor with Floating Point Unit

| Bits | Name | Type | Reset Value | Description |
|------|------|------|-------------|-------------|
| 9 | DISOOFP | R/W | 0 | Disable floating point instructions completing out of order with respect to integer instructions |
| 8 | DISFPCA | R/W | 0 | Disable automatic update of FPCA bit in the CONTROL register |
| 7:3 | – | – | – | Reserved – Not used in current design |
| 2 | DISFOLD | R/W | 0 | Disable IT folding (Prevent overlap of IT instruction execution phase with following instruction) |
| 1 | DISDEFWBUF | R/W | 0 | Disable write buffer for default memory map (memory accesses in MPU mapped regions are not affected) |
| 0 | DISMCYCINT | R/W | 0 | Disable interruption of multiple cycle instructions like LDM, STM, 64-bit multiply and divide instructions |

and improves performance by overlapping the execution cycles. However, IT folding can cause jitter in looping. If a task must avoid jitter, set the DISFOLD bit to 1 before executing the task to disable IT folding.

### DISDEFWBUF (bit 1)

The Cortex-M3 and Cortex-M4 processors have a write buffer feature that when a write is carried out to a bufferable memory region, the processor can proceed to the next instruction before the transfer is completed. This is great for performance, but can cause some complexity in debugging imprecise bus faults.

For example, consider the program flow given in Figure 9.16. If an error takes place at a bus write, causing the bus fault, but the processor has already proceeded by a number of instructions including branches, it will not be easy to locate the faulting instruction because the imprecise bus fault could have been caused by any one of the three store instructions. Unless you have a debugger that supports ETM instruction trace, it could be difficult to tell which one caused the bus fault.

The DISDEFWBUF bit disables the write buffer in the processor interface (unless the processor has MPU and the MPU region setting overrides this). In this way the processor will not continue executing the next instruction until the write operation is completed, and so you can see the bus fault immediately at the STR

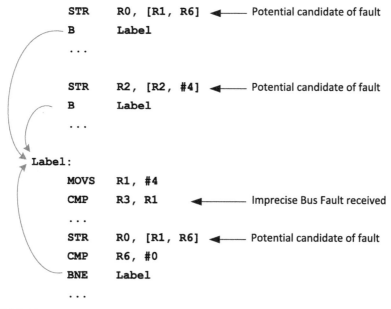

```
        STR     R0, [R1, R6]    ◄──── Potential candidate of fault
        B       Label
        . . .

        STR     R2, [R2, #4]    ◄──── Potential candidate of fault
        B       Label
        . . .

Label:
        MOVS    R1, #4
        CMP     R3, R1          ◄──────── Imprecise Bus Fault received
        . . .

        STR     R0, [R1, R6]    ◄──── Potential candidate of fault
        CMP     R6, #0
        BNE     Label
        . . .
```

**FIGURE 9.16**

The source of an imprecise bus fault can be difficult to locate

instruction (a precise bus fault) and find out which store instruction caused the fault pointed from the stacked return address (stacked program counter) easily. It is typically used during the debugging of bus fault situations (see section 12.3.2 for details).

### DISMCYCINT (bit 0)

When set to 1, this disables interruption of load multiple and store multiple instructions. This increases the interrupt latency of the processor because any LDM or STM must complete before the processor can stack the current state and enter the interrupt handler.

Both DISOOFP and DISFPCA are reserved for ARM® processor design team testing. The use of these bits must be avoided in normal application programming.

## 9.10 Co-processor access control register

The Co-processor Access Control Register is available in Cortex®-M4 with floating point unit for enabling the floating point unit. This register is located in address 0xE000ED88 (privileged accesses only). In C language programming you can access this register using the "SCB->CPACR" symbol. By default the floating point unit is turned off to reduce the power consumption.

**Table 9.15** Co-processor Access Control Register (SCB->CPACR, 0xE000ED88)

| Bits | Name | Type | Reset Value | Descriptions |
|------|------|------|-------------|--------------|
| 31:24 | Reserved | – | – | Reserved. Read as Zero. Write ignore |
| 23:22 | CP11 | R/W | 0 | Access for floating point unit |
| 21:20 | CP10 | R/W | 0 | Access for floating point unit |
| 19:0 | Reserved | – | – | Reserved. Read as Zero. Write ignore |

**Table 9.16** CP10 and CP11 Settings

| Bits | Setting |
|------|---------|
| 00 | Access denied. Any attempted access generate a Usage Fault (type NOCP – No Co-processor) |
| 01 | Privileged Access only. Unprivileged access generate a Usage Fault |
| 10 | Reserved – result unpredictable |
| 11 | Full access |

The encoding for CP10 and CP11 is shown in Table 9.16, and the value 01 or 11 must be set to use the floating point unit.

The settings for CP10 and CP11 must be identical. Usually, when the floating point unit is needed, you can enable the floating point unit using the following code:

```
SCB->CPACR|= 0x00F00000; // Enable the floating point unit for full
access
```

This step is typically carried out inside the SystemInit() function provided in the device-specific software package file. It is executed by the reset handler.

# OS Support Features

# 10

## CHAPTER OUTLINE

## 10.1 Overview of OS support features

The Cortex®-M processors are designed with OS support in mind. Currently there are over 30 different embedded OSs (including many RealTime OS, or RTOS) available for Cortex-M microcontrollers, and the number is still growing. A number of features are implemented in the architecture to make OS implementation easier and more efficient. For example:

- Shadowed stack pointer: Two stack pointers are available. The MSP is used for the OS Kernel and interrupt handlers. The PSP is used by application tasks.
- SysTick timer: A simple timer included inside the processor. This enables an embedded OS to be used on the wide range of Cortex-M microcontrollers available. Details of the SysTick timer are covered in section 9.5 of this book.
- SVC and PendSV exceptions: These two exception types are essential for the operations of embedded OSs such as the implementation of context switching.
- Unprivileged execution level: This allows a basic security model that restricts the access rights of some application tasks. The privileged and unprivileged separation can also be used in conjunction with the Memory Protection Unit (MPU), thus further enhancing the robustness of embedded systems.
- Exclusive accesses: The exclusive load and store instructions are useful for semaphore and mutual exclusive (MUTEX) operations in the OS.

In addition, the low interrupt latency nature and various features in the instruction set also help embedded OSs to work efficiently. For example, the context-switching

The Definitive Guide to ARM® Cortex®-M3 and Cortex-M4 Processors. http://dx.doi.org/10.1016/B978-0-12-408082-9.00010-5

overhead is low due to low interrupt latency. Also, one of the debug features, called Instrumentation Trace Macrocell (ITM), is used in many debug tools for OS-aware debugging.

## 10.2 Shadowed stack pointer

In Chapter 4 we said that there are two stack pointers in the Cortex®-M processors:

- Main Stack Pointer (MSP) is the default stack pointer. It is used in the Thread mode when the CONTROL bit[1] (SPSEL) is 0, and it is always used in Handler mode.
- Processor Stack Pointer (PSP) is used in Thread mode when the CONTROL bit [1] (SPSEL) is set to 1.

Stack operations like PUSH and POP instructions, and most instructions that use SP (R13) use the currently selected stack pointer. You can also access the MSP and PSP directly using MRS and MSR instructions. In simple applications without an embedded OS or RTOS, you can just use the MSP for all operations and ignore the PSP.

In systems with an embedded OS or RTOS, the exception handlers (including part of the OS kernel) use the MSP, while the application tasks use the PSP. Each application task has its own stack space (Figure 10.1), and the context-switching code in the OS updates the PSP each time the context is switched.

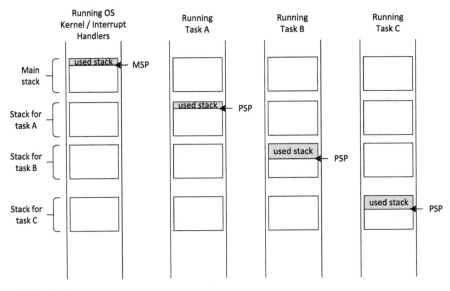

**FIGURE 10.1**

The stack for each task is separated from the others

This arrangement has several benefits:

- If an application task encounters a problem that leads to a stack corruption, the stack used by the OS kernel and other tasks is still likely to be intact, thus helping to improve system reliability.
- The stack space for each task only needs to cover the maximum stack usage plus one level of stack frame (maximum 9 words including padding in Cortex-M3 or Cortex-M4 without floating point unit, or maximum 27 words for Cortex-M4 with floating point unit). Stack space needed for the ISR and nested interrupt handling is allocated in the main stack only.
- It makes it easy to create an efficient OS for the Cortex-M processors.
- An OS can also utilize the Memory Protection Unit (MPU) to define the stack region which an application task can use. If an application task has a stack overflow problem, the MPU can trigger a MemManage fault exception and prevent the task from overwriting memory regions outside the allocated stack space for this task. After power up, the MSP is initialized from the vector table as a part of the processor's reset sequence. The C startup code added by the toolchain can also carry out another stage of stack initialization for the main stack. It is then possible to start using PSP by initializing it using the MSR instruction and then write to the CONTROL register to set SPSEL, but it is uncommon to do so.

    The simplest way to initialize and start using PSP (not suitable for most OS):

```
LDR    R0,=PSP_TOP   ; PSP_TOP is a constant defines the top address
                          of stack
MSR    PSP, R0       ; Set PSP to the top of a process stack
MRS    R0, CONTROL   ; Read current CONTROL
ORRS   R0, R0, #0x2  ; Set SPSEL
MSR    CONTROL, R0   ; write to CONTROL
ISB    ; Execute and ISB after updating CONTROL,
       ; this is an architectural recommendation
```

Typically, to use the process stack, put OS in Handler mode, and program the PSP directly, then use an exception return sequence to "jump" into the application task.

For example, when an OS first starts in Thread mode, it can use the SVC exception to enter the Handler mode (Figure 10.2). Then it can create a stack frame in the process stack, and trigger an exception return that uses the PSP. When the stack frame is loaded, the application task is started.

In OS designs, we need to switch between different tasks. This is typically called context switching. Context switching is usually carried out in the PendSV exception handler, which can be triggered by the periodic SysTick exception. Inside the context-switching operation, we need to:

- Save the current status of the registers in the current task
- Save the current PSP value

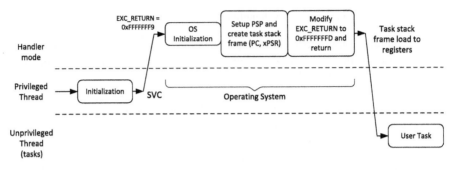

**FIGURE 10.2**

Initialization of a task in a simple OS

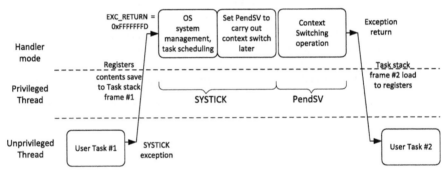

**FIGURE 10.3**

Concept of context switching

- Set the PSP to the last SP value for the next task
- Restore the last values for the next tasks
- Use exception return to switch to the task

For example, in Figure 10.3, a simplified context-switching operation is shown. Note that the context switching is carried out in PendSV, which is typically programmed to the lowest priority level. This prevents context switching from happening in the middle of an interrupt handler. This is explained in detail in section 10.4.

## 10.3 SVC exception

The SVC (Supervisor Call) and PendSV (Pendable Service Call) exceptions are important to OS designs. SVC is exception type 11, and has a programmable priority level.

The SVC exception is triggered by the SVC instruction. Although it is possible to trigger an interrupt using software by writing to NVIC (e.g., Software Trigger Interrupt Register, NVIC->STIR), the behavior is a bit different: Interrupts are imprecise. It means that a number of instructions could be executed after setting the pending status but before the interrupt actually takes place. On the other hand, SVC is precise. The SVC handler must execute after the SVC instruction, except when another higher-priority exception arrives at the same time.

In many systems, the SVC mechanism can be used as an API to allow application tasks to access system resources, as shown in Figure 10.4.

In systems with high-reliability requirements, the application tasks can be running in unprivileged access level, and some of the hardware resources can be set up to be privileged accessed only (using MPU). The only way an application task can access these protected hardware resources is via services from the OS. In this way, an embedded system can be more robust and secure, because the application tasks cannot gain unauthorized access to critical hardware.

In some cases this also makes the programming of the application tasks easier because the application tasks do not need to know the programming details of the underlying hardware if the OS services provide what the task needs.

SVC also allows application tasks to be developed independently of the OS because the application tasks do not need to know the exact address of the OS service functions. The application tasks only need to know the SVC service number and the parameters that the OS services requires. The actual hardware-level programming is handled by device drivers (Figure 10.4).

The SVC exception is generated using the SVC instruction. An immediate value is required for this instruction, which works as a parameter-passing method. The SVC exception handler can then extract the parameter and determine what action it needs to perform. For example:

```
SVC #0x3 ; Call SVC function 3
```

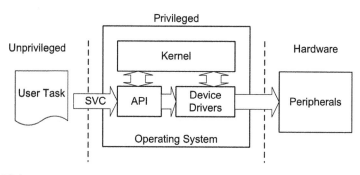

**FIGURE 10.4**

SVC as a gateway for OS services

The traditional syntax for SVC is also acceptable (without the "#") in ARM®
toolchains:

```
SVC 0x3 ; Call SVC function 3
```

For C language development with ARM toolchains (KEIL™ Microcontroller
Development Kit for ARM, or ARM Development Studio 5), the SVC instruction
can be generated using the __svc function. In gcc and some other toolchains, this
can be generated using inline assembly.

When the SVC handler is executed, you can determine the immediate data value
in the SVC instruction by reading the stacked Program Counter (PC) value, then
reading the instruction from that address and masking out the unneeded bits. However, the program that executed the SVC could have either been using the main stack
or the process stack. So we need to find out which stack was used for the stacking
process before extracting the stacked PC value. This can be determined from the
link register value when the handler is entered (Figure 10.5).

For assembly programming, we can find out which stack was used, and extract
the SVC service number with the following code:

```
SVC_Handler
    TST    LR, #4         ; Test bit 2 of EXC_RETURN
    ITE    EQ
    MRSEQ  R0, MSP         ; if 0, stacking used MSP, copy to R0
    MRSNE  R0, PSP         ; if 1, stacking used PSP, copy to R0
    LDR    R0, [R0, #24]   ; Get stacked PC from the stack frame
    ; (stacked PC = address of instruction after SVC)
    LDRB   R0, [R0, #-2]   ; Get first byte of the SVC instruction
    ; now the SVC number is in R0
    ...
```

For the C programming environment, we need to separate the SVC handler into
two parts:

- The first part extracts the starting address of the stack frame, and passes it to the
  second part as an input parameter. This needs to be done in assembly because

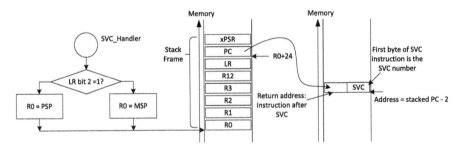

**FIGURE 10.5**

Extraction of the SVC service number in assembly

we need to check the value of the LR (EXC_RETURN), which cannot be done in C.
- The second part extracts the stacked PC from the stack frame, and then the SVC number from the program code. It can optionally extract other information like stacked register values.

Assuming that you are using the Keil MDK-ARM toolchain, you can create the first part of the handler as follows:

```
__asm void SVC_Handler(void)
{
    TST    LR, #4    ; Test bit 2 of EXC_RETURN
    ITE    EQ
    MRSEQ  R0, MSP   ; if 0, stacking used MSP, copy to R0
    MRSNE  R0, PSP   ; if 1, stacking used PSP, copy to R0
    B      __cpp(SVC_Handler_C)
    ALIGN 4
}
```

The function name "SVC_Handler" is standardized in CMSIS-Core. After getting the starting address of the stack frame, it is passed on to the "SVC_Handler_C," a C program part of the SVC handler:

```
void SVC_Handler_C(unsigned int * svc_args)
{
    uint8_t  svc_number;
    uint32_t stacked_r0, stacked_r1, stacked_r2, stacked_r3;

    svc_number = ((char *) svc_args[6])[-2]; //Memory[(Stacked PC)-2]
    stacked_r0 = svc_args[0];
    stacked_r1 = svc_args[1];
    stacked_r2 = svc_args[2];
    stacked_r3 = svc_args[3];
    // ... other processing
    ...
    // Return result (e.g. sum of first two arguments)
    svc_args[0] = stacked_r0 + stacked_r1;
    return;
}
```

The advantage of passing the address of the stack frame is that it allows the C handler to extract any information in the stack frame, including the stacked registers. This is essential if you want to pass parameters to a SVC service and get a return value for the SVC service. Due to the fact that exception handlers can be normal C functions, if a SVC service is called, and a higher-priority interrupt arrives at the same time, the higher-priority ISR will be executed first and this can change the values in R0-R3, R12, etc. To ensure your SVC handler gets the correct input parameters, you need to get the parameter values from the stack frame.

If your SVC service needs to return a value, you need to return the value using the stack frame. Otherwise, the return value stored in the register bank would be overwritten during the unstacking operation of an exception return. An example of a SVC service that passes parameters and returns a value is shown below.

```
/* SVC service example with parameter passing and return value –
based on Keil MDK-ARM */
#include <stdio.h>

// Define SVC functions
int __svc(0x00) svc_service_add(int x, int y); // Service #0 : Add
int __svc(0x01) svc_service_sub(int x, int y); // Service #1 : Sub
int __svc(0x02) svc_service_incr(int x);       // Service #2 : Incr
void SVC_Handler_main(unsigned int * svc_args);

// Function declarations
int main(void)
{
  int x, y, z;

  x = 3; y = 5;
  z = svc_service_add(x, y);
  printf ("3+5 = %d \n", z);

  x = 9; y = 2;
  z = svc_service_sub(x, y);
  printf ("9-2 = %d \n", z);

  x = 3;
  z = svc_service_incr(x);
  printf ("3++ = %d \n", z);

  while(1);
}

// SVC handler - Assembly wrapper to extract
//                stack frame starting address
__asm void SVC_Handler(void)
{
  TST   LR, #4    ; Test bit 2 of EXC_RETURN
  ITE   EQ
```

```
   MRSEQ R0, MSP   ; if 0, stacking used MSP, copy to R0
   MRSNE R0, PSP   ; if 1, stacking used PSP, copy to R0
   B      __cpp(SVC_Handler_C)
   ALIGN 4
}


// SVC handler - main code to handle processing
// Input parameter is stack frame starting address
// obtained from assembly wrapper.
void SVC_Handler_main(unsigned int * svc_args)
{
  // Stack frame contains:
  // r0, r1, r2, r3, r12, r14, the return address and xPSR
  // - Stacked R0  = svc_args[0]
  // - Stacked R1  = svc_args[1]
  // - Stacked R2  = svc_args[2]
  // - Stacked R3  = svc_args[3]
  // - Stacked R12 = svc_args[4]
  // - Stacked LR  = svc_args[5]
  // - Stacked PC  = svc_args[6]
  // - Stacked xPSR= svc_args[7]

  unsigned int svc_number;
  svc_number = ((char *)svc_args[6])[-2];
  switch(svc_number)
    {
    case 0: svc_args[0] = svc_args[0] + svc_args[1];
            break;
    case 1: svc_args[0] = svc_args[0] - svc_args[1];
            break;
    case 2: svc_args[0] = svc_args[0] + 1;
            break;
    default: // Unknown SVC request
            break;
    }
  return;
}
```

Because of the exception priority model, you cannot use SVC inside an SVC handler (because the priority is the same as the current priority). Doing so will result in a usage fault exception. For the same reason, you cannot use SVC in the NMI handler or the HardFault handler.

If you have used traditional ARM processors such as the ARM7TDMI™, you might know that these processors have a software interrupt instruction called

SWI. The SVC has a similar functionality, and in fact the binary encoding of SVC and SWI instructions are the same. In recent architectures the SWI has been changed to SVC. However, the SVC handler code for ARM7TDMI is different from Cortex®-M because there are many differences in the exception models.

## 10.4 PendSV exception

PendSV (Pended Service Call) is another exception type that is important for supporting OS operations. It is exception type 14 and has a programmable priority level. The PendSV Exception is triggered by setting its pending status by writing to the Interrupt Control and State Register (ICSR) (see section 7.9.2). Unlike the SVC exception, it is not precise. So its pending status can be set inside a higher priority exception handler and executed when the higher-priority handler finishes.

Using this characteristic, we can schedule the PendSV exception handler to be executed after all other interrupt processing tasks are done, by making sure that the PendSV has the lowest exception priority level. This is very useful for a context-switching operation, which is a key operation in various OS designs.

First, let us look at some basic concepts of context switching. In a typical system with an embedded OS, the processing time is divided into a number of time slots. For a system with only two tasks, the two tasks are executed alternatively, as shown in Figure 10.6.

The execution of an OS kernel can be triggered by:

- Execution of SVC instruction from application tasks. For example, when an application task is stalled because it is waiting for some data or event, it can call a system service to swap in another task.
- Periodic SysTick exception.

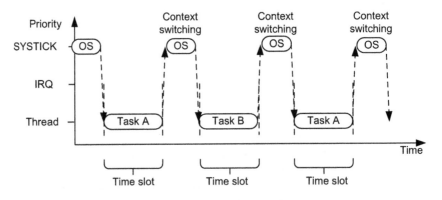

**FIGURE 10.6**

A simple example of context switching

Inside the OS code, the task scheduler can decide if context switching should be carried out. The operations shown in Figure 10.6 assume that the OS kernel execution is triggered by a SysTick exception, and each time it decides to switch to a different task.

If an interrupt request (IRQ) takes place before the SysTick exception, the SysTick exception might preempt the IRQ handler. In this case, the OS should not carry out the context switching. Otherwise, the IRQ handler process will be delayed (Figure 10.7). And for the Cortex®-M3 and Cortex-M4 processors, by default the design does not allow return to Thread mode when there is an active interrupt service (but there is an exception — see Non-base Thread Enable description in section 23.5). If the OS attempts to return to the Thread mode with an active interrupt service running, it triggers a Usage fault exception.

In some OS designs, this problem is solved by not carrying out context switching if an interrupt service is running. This can easily be done by checking the stacked xPSR from the stack frame, or checking the interrupt active status registers in the NVIC (see section 7.8.4). However, this might affect the performance of the system, especially when an interrupt source keeps generating requests around the SysTick triggering time, which can prevent context switching from happening.

The PendSV exception solves the problem by delaying the context-switching request until all other IRQ handlers have completed their processing. To do this, the PendSV is programmed as the lowest-priority exception. If the OS decides that context switching is needed, it sets the pending status of the PendSV, and carries out the context switching within the PendSV exception, as shown in Figure 10.8 show an example sequence of context switching using PendSV with the following event sequence.

Figure 10.8 show an example sequence of context switching using PendSV with the following event sequence:

1. Task A calls SVC for task switching (for example, waiting for some work to complete).

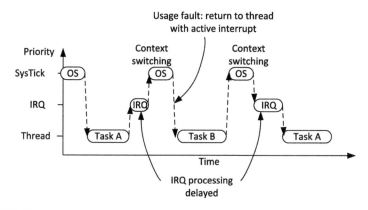

**FIGURE 10.7**

Context switching during ISR execution can delay interrupt service

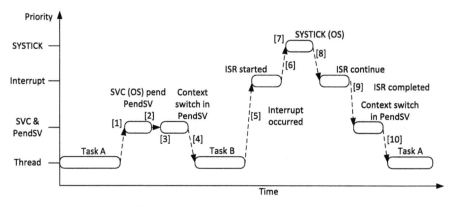

**FIGURE 10.8**

Example context switching with PendSV

2. The OS receives the request, prepares for context switching, and pends the PendSV exception.
3. When the CPU exits SVC, it enters PendSV immediately and does the context switch.
4. When PendSV finishes and returns to Thread level, it executes Task B.
5. An interrupt occurs and the interrupt handler is entered.
6. While running the interrupt handler routine, a SYSTICK exception (for OS tick) takes place.
7. The OS carries out the essential operation, then pends the PendSV exception and gets ready for the context switch.
8. When the SYSTICK exception exits, it returns to the interrupt service routine.
9. When the interrupt service routine completes, the PendSV starts and does the actual context-switch operations.
10. When PendSV is complete, the program returns to Thread level; this time it returns to Task A and continues the processing.

Besides context switching in an OS environment, PendSV can also be used in systems without an OS. For example, an interrupt service can need a fair amount of time to process. The first portion of the processing might need a high priority, but if the whole ISR is executed with a high priority level, other interrupt services would be blocked out for a long time. In such cases, we can partition the interrupt service processing into two halves (Figure 10.9):

- The first half is the time-critical part that needs to be executed quickly with high priority. It is put inside the normal ISR. At the end of the ISR, it sets the pending status of the PendSV.
- The second half contains the remaining processing work needed for the interrupt service. It is placed inside the PendSV handler and is executed with low exception priority.

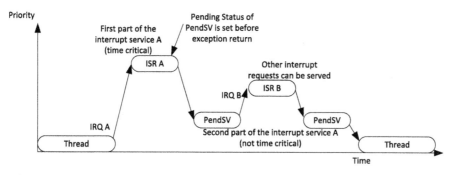

**FIGURE 10.9**

Using PendSV to partition an interrupt service into two sections

## 10.5 **Context switching in action**

To demonstrate context-switching operation in a real example, we use a simple task scheduler that switches between four tasks in a round-robin arrangement. The first example assumes that there are no floating point registers, and therefore can be used on the Cortex®-M3 processor as well as the Cortex-M4 processor without the floating point unit. This example code is developed with the STM32F4 Discovery Kit, which provides four LEDs. Each task toggles one of the LEDs at different speeds.

The context-switching operation is carried out by the PendSV exception handler. Since the exception sequence already saved registers R0-R3, R12, LR, return address (PC), and xPSR, the PendSV only needs to store R4-R11 to the process stack (Figure 10.10).

The code for the project can be as simple as:

```
Example code for a multi-tasking system with four tasks
#include "stm32f4xx.h"
#define LED0 (1<<12)
#define LED1 (1<<13)
#define LED2 (1<<14)
#define LED3 (1<<15)
/* Macros for word accesses */
#define HW32_REG(ADDRESS)
    (*((volatile unsigned long *)(ADDRESS)))
/* Use Breakpoint to stop when error is detected
    (KEIL MDK specific intrinsic) */
/* it can be changed to while(1) XXif needed */
#define stop_cpu __breakpoint(0)
void LED_initialize(void); // Initialize LED
void task0(void);       // Toggle LED0
void task1(void);       // Toggle LED1
```

```
void task2(void);        // Toggle LED2
void task3(void);        // Toggle LED3

// Event to tasks
volatile uint32_t systick_count=0;
// Stack for each task (8Kbytes each - 1024 x 8 bytes)
long long task0_stack[1024], task1_stack[1024],
  task2_stack[1024], task3_stack[1024];
// Data use by OS
uint32_t curr_task=0;  // Current task
uint32_t next_task=1;  // Next task
uint32_t PSP_array[4]; // Process Stack Pointer for each task


// ------------------------------------------------------------
// Start of main program
int main(void)
{
SCB->CCR |= SCB_CCR_STKALIGN_Msk; // Enable double word stack
  alignment
//(recommended in Cortex-M3 r1p1, default in Cortex-M3 r2px and
  Cortex-M4)

LED_initialize();
// Starting the task scheduler
// Create stack frame for task0
PSP_array[0] = ((unsigned int) task0_stack)
  + (sizeof task0_stack) - 16*4;
HW32_REG((PSP_array[0] + (14<<2))) = (unsigned long) task0;
  // initial Program Counter
HW32_REG((PSP_array[0] + (15<<2))) = 0x01000000;   // initial xPSR

// Create stack frame for task1
PSP_array[1] = ((unsigned int) task1_stack)
  + (sizeof task1_stack) - 16*4;
HW32_REG((PSP_array[1] + (14<<2))) = (unsigned long) task1;
  // initial Program Counter
HW32_REG((PSP_array[1] + (15<<2))) = 0x01000000;   // initial xPSR

// Create stack frame for task2
PSP_array[2] = ((unsigned int) task2_stack)
  + (sizeof task2_stack) - 16*4;
HW32_REG((PSP_array[2] + (14<<2))) = (unsigned long) task2;
  // initial Program Counter
HW32_REG((PSP_array[2] + (15<<2))) = 0x01000000;   // initial xPSR
```

```
// Create stack frame for task3
PSP_array[3] = ((unsigned int) task3_stack)
 + (sizeof task3_stack) - 16*4;
HW32_REG((PSP_array[3] + (14<<2))) = (unsigned long) task3;
 // initial Program Counter
HW32_REG((PSP_array[3] + (15<<2))) = 0x01000000;   // initial xPSR

curr_task = 0; // Switch to task #0 (Current task)
__set_PSP((PSP_array[curr_task] + 16*4)); // Set PSP to top
                                   of task 0 stack
NVIC_SetPriority(PendSV_IRQn, 0xFF); // Set PendSV to lowest
                              possible priority
SysTick_Config(168000);              // 1000 Hz SysTick interrupt
                              on 168MHz core clock
__set_CONTROL(0x3); // Switch to use Process Stack, unprivileged
                    state
__ISB();  // Execute ISB after changing CONTROL (architectural
          recommendation)
task0();  // Start task 0

while(1){
      stop_cpu;// Should not be here
};
}
// ------------------------------------------------------------
void task0(void) // Toggle LED #0
{ while (1) {
  if (systick_count & 0x80) {GPIOD->BSRRL = LED0;} // Set   LED 0
  else                      {GPIOD->BSRRH = LED0;} // Clear LED 0
  };
}
// ------------------------------------------------------------
void task1(void) // Toggle LED #1
{   while (1) {
  if (systick_count & 0x100){GPIOD->BSRRL = LED1;} // Set   LED 1
  else                      {GPIOD->BSRRH = LED1;} // Clear LED 1
  };
}
// ------------------------------------------------------------
void task2(void) // Toggle LED #2
{   while (1) {
  if (systick_count & 0x200){GPIOD->BSRRL = LED2;} // Set   LED 2
  else                      {GPIOD->BSRRH = LED2;} // Clear LED 2
  };
}
```

```
// -------------------------------------------------------------
void task3(void) // Toggle LED #3
{  while (1) {
   if (systick_count & 0x400){GPIOD->BSRRL = LED3;} // Set   LED 3
   else                      {GPIOD->BSRRH = LED3;} // Clear LED 3
   };
}
// -------------------------------------------------------------
__asm void PendSV_Handler(void)
{ // Context switching code
 // Simple version - assume No floating point support
 // ------------------------
 // Save current context
 MRS     R0, PSP        // Get current process stack pointer value
 STMDB   R0!,{R4-R11}   // Save R4 to R11 in task stack (8 regs)
 LDR     R1,=__cpp(&curr_task)
 LDR     R2,[R1]        // Get current task ID
 LDR     R3,=__cpp(&PSP_array)
 STR     R0,[R3, R2, LSL #2]  // Save PSP value into PSP_array
 // ------------------------
 // Load next context
 LDR     R4,=__cpp(&next_task)
 LDR     R4,[R4]        // Get next task ID
 STR     R4,[R1]        // Set curr_task = next_task
 LDR     R0,[R3, R4, LSL #2] // Load PSP value from PSP_array
 LDMIA   R0!,{R4-R11}        // Load R4 to R11 from task
                             //              stack (8 regs)
 MSR     PSP, R0        // Set PSP to next task
 BX      LR             // Return
 ALIGN   4
}
// -------------------------------------------------------------
void SysTick_Handler(void) // 1KHz
{ // Increment systick counter for LED blinking
 systick_count++;
 // Simple task round robin scheduler
 switch(curr_task) {
   case(0): next_task=1; break;
   case(1): next_task=2; break;
   case(2): next_task=3; break;
   case(3): next_task=0; break;
   default: next_task=0;
     stop_cpu;
     break; // Should not be here
   }
```

```
  if (curr_task!=next_task){ // Context switching needed
    SCB->ICSR |= SCB_ICSR_PENDSVSET_Msk; // Set PendSV to pending
    }
  return;
}
// -----------------------------------------------------------
void LED_initialize(void)
{
 // Configure LED outputs
 RCC->AHB1ENR |= RCC_AHB1ENR_GPIODEN; // Enable Port D clock
 // Set pin 12, 13, 14, 15 as general purpose output mode
   (pull-push)
 GPIOD->MODER |=   (GPIO_MODER_MODER12_0 |
                    GPIO_MODER_MODER13_0 |
                    GPIO_MODER_MODER14_0 |
                    GPIO_MODER_MODER15_0 ) ;
 // GPIOD->OTYPER |= 0; // No need to change - use pull-push output
 GPIOD->OSPEEDR |=  (GPIO_OSPEEDER_OSPEEDR12 | // 100MHz operations
                    GPIO_OSPEEDER_OSPEEDR13 |
                    GPIO_OSPEEDER_OSPEEDR14 |
                    GPIO_OSPEEDER_OSPEEDR15 );
 GPIOD->PUPDR = 0; // No pull up , no pull down
 return;
}
// -----------------------------------------------------------
```

The example also shows a simple method to start the first task:

```
curr_task = 0; // Switch to task #0 (Current task)
__set_PSP((PSP_array[curr_task] + 16*4)); // Set PSP to top of task
                                              0 stack
...
__set_CONTROL(0x3); // Switch to use Process Stack,
                       unprivileged state
__ISB();  // Execute ISB after changing CONTROL (architectural
              recommendation)
task0();  // Start task 0
```

Using this method, the PSP is set up to task 0 before task 0 is executed. It is not strictly required to initialize the stack frame for task 0 (at the stack setup steps after LED_initialize()), but we included the task 0 initialization code there so that all the tasks set up have the same look and feel.

With this simple design, you can either run all tasks in unprivileged state by setting CONTROL to 3, or run all tasks in privileged state by setting CONTROL

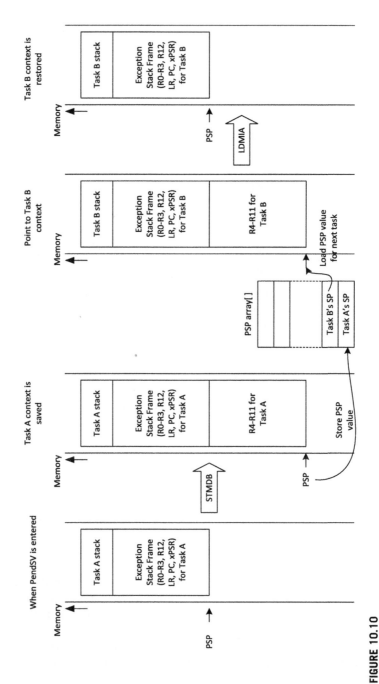

**FIGURE 10.10**

Context switching

to 2. The execution of ISB is a recommendation in the architecture. For existing Cortex-M3 and Cortex-M4 processors, omitting the ISB does not cause any error.

This example can be further extended to support context saving of registers in the Cortex-M4 floating point unit. To do this, we need to extend the registers being stacked in the context-switching sequence to include EXC_RETURN, because bit 4 of this value indicates when the exception stack frame contains floating point registers.

To enhance the design further, we also include the CONTROL register for each task so that you can run some of the tasks in privileged state and some in unprivileged state. The stacking of floating point registers is conditional, and based on whether the task is a floating point operation, as indicated by bit 4 of the EXC_RETURN value (Figure 10.11).

In this example, we also demonstrate a different way to start the first task. Instead of running the OS initialization step in Thread, we can run this in the SVC handler and use an exception return to switch into the first task. The SVC mechanism shown in the example is using __svc keyword, a feature in the ARM® C compiler toolchain.

/* Example code for a multi-tasking system with four tasks, enhanced to support floating point registers, privileged and unprivileged task execution level, and use SVC for OS initialization */

```
#include "stm32f4xx.h"
#include "stdio.h"
#define LED0 (1<<12)
#define LED1 (1<<13)
#define LED2 (1<<14)
#define LED3 (1<<15)
/* Macros for word accesses */
#define HW32_REG(ADDRESS) (*((volatile unsigned long *)(ADDRESS)))
/* Use Breakpoint to stop when error is detected (KEIL MDK specific
intrinsic) */
/* it can be changed to while(1) if needed */
#define stop_cpu __breakpoint(0)
void LED_initialize(void); // Initialize LED
void task0(void);   // Toggle LED0
void task1(void);   // Toggle LED1
void task2(void);   // Toggle LED2
void task3(void);   // Toggle LED3
void __svc(0x00) os_start(void); // OS initialization
void SVC_Handler_C(unsigned int * svc_args);

// Event to tasks
volatile uint32_t systick_count=0;
// Stack for each task (8Kbytes each - 1024 x 8 bytes)
long long task0_stack[1024], task1_stack[1024], task2_stack[1024],
task3_stack[1024];
// Data use by OS
```

```
uint32_t curr_task=0;    // Current task
uint32_t next_task=1;    // Next task
uint32_t PSP_array[4];   // Process Stack Pointer for each task
uint32_t svc_exc_return; // EXC_RETURN use by SVC


// ----------------------------------------------------------------
// Start of main program
int main(void)
{
   SCB->CCR |= SCB_CCR_STKALIGN_Msk; // Enable double word stack
                                     alignment
   //(recommended in Cortex-M3 r1p1, default in Cortex-M3 r2px and
     Cortex-M4)

   puts("Simple context switching demo");
   LED_initialize();
   os_start();

   while(1){
     stop_cpu;// Should not be here
     };
}
// ----------------------------------------------------------------
void task0(void) // Toggle LED #0
{  while (1) {
   if (systick_count & 0x80) {GPIOD->BSRRL = LED0;} // Set   LED 0
   else                      {GPIOD->BSRRH = LED0;} // Clear LED 0
   };
}
// ----------------------------------------------------------------
void task1(void) // Toggle LED #1
{  while (1) {
   if (systick_count & 0x100){GPIOD->BSRRL = LED1;} // Set   LED 1
   else                      {GPIOD->BSRRH = LED1;} // Clear LED 1
   };
}
// ----------------------------------------------------------------
void task2(void) // Toggle LED #2
{  while (1) {
   if (systick_count & 0x200){GPIOD->BSRRL = LED2;} // Set   LED 2
   else                      {GPIOD->BSRRH = LED2;} // Clear LED 2
   };
}
// ----------------------------------------------------------------
```

```
void task3(void) // Toggle LED #3
{  while (1) {
   if (systick_count & 0x400){GPIOD->BSRRL = LED3;} // Set   LED 3
   else                      {GPIOD->BSRRH = LED3;} // Clear LED 3
   };
}
// -----------------------------------------------------------
__asm void SVC_Handler(void)
{
   TST     LR, #4 // Extract stack frame location
   ITE     EQ
   MRSEQ   R0, MSP
   MRSNE   R0, PSP
   LDR     R1,=__cpp(&svc_exc_return) // Save current EXC_RETURN
   STR     LR,[R1]
   BL      __cpp(SVC_Handler_C)       // Run C part of SVC_Handler
   LDR     R1,=__cpp(&svc_exc_return) // Load new EXC_RETURN
   LDR     LR,[R1]
   BX      LR
   ALIGN 4
}
void SVC_Handler_C(unsigned int * svc_args)
{
   uint8_t svc_number;
   svc_number = ((char *) svc_args[6])[-2]; // Memory[(Stacked PC)-2]
   switch(svc_number) {
     case (0): // OS init
       // Starting the task scheduler
       // Create stack frame for task0
        PSP_array[0] = ((unsigned int) task0_stack)
         + (sizeof task0_stack) - 18*4;
       HW32_REG((PSP_array[0] + (16<<2))) = (unsigned long) task0;
         // initial Program Counter
       HW32_REG((PSP_array[0] + (17<<2))) = 0x01000000;
         // initial xPSR
       HW32_REG((PSP_array[0]            )) = 0xFFFFFFFDUL;
         // initial EXC_RETURN
       HW32_REG((PSP_array[0] + ( 1<<2))) = 0x3;// initial CONTROL :
      unprivileged, PSP, no FP

        // Create stack frame for task1
        PSP_array[1] = ((unsigned int) task1_stack)
         + (sizeof task1_stack) - 18*4;
```

```
HW32_REG((PSP_array[1] + (16<<2))) = (unsigned long) task1;
  // initial Program Counter
HW32_REG((PSP_array[1] + (17<<2))) = 0x01000000;
  // initial xPSR
HW32_REG((PSP_array[1]              )) = 0xFFFFFFFDUL;
  // initial EXC_RETURN
HW32_REG((PSP_array[1] + ( 1<<2))) = 0x3;// initial CONTROL :
  unprivileged, PSP, no FP

// Create stack frame for task2
PSP_array[2] = ((unsigned int) task2_stack)
  + (sizeof task2_stack) - 18*4;
HW32_REG((PSP_array[2] + (16<<2))) = (unsigned long) task2;
  // initial Program Counter
HW32_REG((PSP_array[2] + (17<<2))) = 0x01000000;
  // initial xPSR
HW32_REG((PSP_array[2]              )) = 0xFFFFFFFDUL;
  // initial EXC_RETURN
HW32_REG((PSP_array[2] + ( 1<<2))) = 0x3;// initial CONTROL :
  unprivileged, PSP, no FP

// Create stack frame for task3
PSP_array[3] = ((unsigned int) task3_stack)
  + (sizeof task3_stack) - 18*4;
HW32_REG((PSP_array[3] + (16<<2))) = (unsigned long) task3;
  // initial Program Counter
HW32_REG((PSP_array[3] + (17<<2))) = 0x01000000;
  // initial xPSR
HW32_REG((PSP_array[3]              )) = 0xFFFFFFFDUL;
  // initial EXC_RETURN
HW32_REG((PSP_array[3] + ( 1<<2))) = 0x3;// initial CONTROL :
  unprivileged, PSP, no FP

curr_task = 0; // Switch to task #0 (Current task)
svc_exc_return = HW32_REG((PSP_array[curr_task]));
  // Return to thread with PSP
__set_PSP((PSP_array[curr_task] + 10*4));
  // Set PSP to @R0 in task 0 stack frame
NVIC_SetPriority(PendSV_IRQn, 0xFF); // Set PendSV to lowest
                                     //        possible priority
SysTick_Config(168000);              // 1000 Hz SysTick
                                     //   interrupt on 168MHz
                                     //   core clock

__set_CONTROL(0x3);                  // Switch to use Process
                                     //   Stack, unprivileged
                                     //   state
```

```
        __ISB();  // Execute ISB after changing CONTROL (architectural
                   recommendation)
      break;
    default:
      puts ("ERROR: Unknown SVC service number");
      printf("- SVC number 0x%x\n", svc_number);
      stop_cpu;
      break;
  } // end switch
}


// ----------------------------------------------------------------
__asm void PendSV_Handler(void)
{ // Context switching code
  // -------------------------
  // Save current context
  MRS      R0, PSP   // Get current process stack pointer value
  TST      LR, #0x10 // Test bit 4. If zero, need to stack floating
           point regs
  IT       EQ
  VSTMDBEQ R0!, {S16-S31} // Save floating point registers
  MOV      R2, LR
  MRS      R3, CONTROL
  STMDB    R0!,{R2-R11}// Save LR,CONTROL and R4 to R11 in task
           stack (10 regs)
  LDR      R1,=__cpp(&curr_task)
  LDR      R2,[R1]     // Get current task ID
  LDR      R3,=__cpp(&PSP_array)
  STR      R0,[R3, R2, LSL #2] // Save PSP value into PSP_array
  // -------------------------
  // Load next context
  LDR      R4,=__cpp(&next_task)
  LDR      R4,[R4]     // Get next task ID
  STR      R4,[R1]     // Set curr_task = next_task
  LDR      R0,[R3, R4, LSL #2] // Load PSP value from PSP_array
  LDMIA    R0!,{R2-R11}// Load LR, CONTROL and R4 to R11 from task
           stack (10 regs)
  MOV      LR, R2
  MSR      CONTROL, R3
  ISB
  TST      LR, #0x10     // Test bit 4. If zero, need to unstack
                         floating point regs
  IT       EQ
  VLDMIAEQ R0!, {S16-S31} // Load floating point registers
```

```
  MSR      PSP, R0         // Set PSP to next task
  BX       LR              // Return
  ALIGN 4
}
// -----------------------------------------------------------------
void SysTick_Handler(void) // 1KHz
{ // Increment systick counter for LED blinking
  systick_count++;
  // Simple task round robin scheduler
  switch(curr_task) {
    case(0): next_task=1; break;
    case(1): next_task=2; break;
    case(2): next_task=3; break;
    case(3): next_task=0; break;
    default: next_task=0;
        printf("ERROR:curr_task = %x\n", curr_task);
        stop_cpu;
        break; // Should not be here
      }
    if (curr_task!=next_task){ // Context switching needed
       SCB->ICSR |= SCB_ICSR_PENDSVSET_Msk; // Set PendSV to pending
       }
    return;
}
// -----------------------------------------------------------------
void LED_initialize(void)
{
  // Configure LED outputs
  RCC->AHB1ENR |= RCC_AHB1ENR_GPIODEN; // Enable Port D clock
  // Set pin 12, 13, 14, 15 as general purpose output mode (pull-push)
  GPIOD->MODER |= (GPIO_MODER_MODER12_0 |
                   GPIO_MODER_MODER13_0 |
                   GPIO_MODER_MODER14_0 |
                   GPIO_MODER_MODER15_0 ) ;
  // GPIOD->OTYPER |= 0; // No need to change - use pull-push output
  GPIOD->OSPEEDR |= (GPIO_OSPEEDER_OSPEEDR12 |
                     GPIO_OSPEEDER_OSPEEDR13 |
                     GPIO_OSPEEDER_OSPEEDR14 |
                     GPIO_OSPEEDER_OSPEEDR15 );
  GPIOD->PUPDR = 0; // No pull up , no pull down
  return;
}
// -----------------------------------------------------------------
```

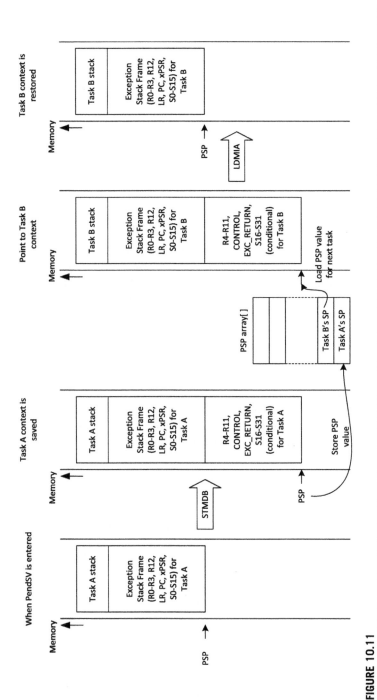

The enhanced example has more registers saved to the process stack, and also the stack frame initialization has additional values to setup. However, the context switching is still very straightforward to implement.

The use of SVC for starting this little OS example added a small complexity because the value of EXC_RETURN needs to be controllable. That is why we created the `svc_exc_return` variable. By setting this variable to 0xFFFFFFFD at the end of the SVC handler, the SVC handler returns to Thread mode using PSP. As a result, the stack frame information that we write to task 0 was used and this caused task 0 to start. Note: return from SVC to task 0 does not go through PendSV, so only eight words are unstacked. As a result the PSP is not set to the lowest word of the stack frame created.

## 10.6 Exclusive accesses and embedded OS

In a multi-tasking system, it is common that a number of tasks need to share a limited number of resources. For example, there might be just one console display output and a number of different tasks might need to display information through this. As a result, almost all OSs have some sort of built-in mechanism to allow a task to "lock" a resource and "free" it when the task no longer needs it. The locking mechanism is usually based on software variables. If a lock variable is set, other tasks see that it is "locked" and have to wait.

This feature is usually called a semaphore. For the case when only one resource is available, it is also called Mutual Exclusive (MUTEX), which is one type of semaphore. In general, semaphores can support multiple tokens. For example, a communication stack might support up to four channels, and the semaphore software variable can be implemented as a "token counter," with a starting value of 4 (Figure 10.12). When a task needs to access a channel, it decrements the counter

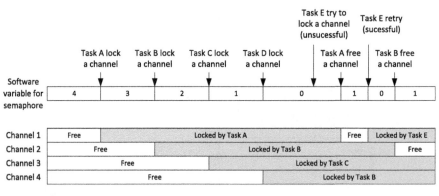

**FIGURE 10.12**

Semaphore operation example

using a semaphore operation. When the counter reaches zero, all the channels are used and any task that needs a communication channel will have to wait until one of the tasks that gained access initially has finished using the channel and releases the token by incrementing the counter.

However, the decrement of the counter variable is not atomic because you need:

- One instruction to read the variable
- One instruction to decrement it
- Another instruction to write it back to the memory

If a context switching happened just between the read and write, another task can read the same value and then both tasks might think that they have got the last token! (Figure 10.13).

There are several ways to prevent this issue. The simplest way is to disable context switching (e.g., by disabling exceptions) when handling semaphores. This can increase interrupt latency, and can only work for single processor designs. In multi-processor designs, there is a chance that two tasks running on two different processors try to decrement the semaphore variable at the same time. Stopping exception handling does not solve this issue.

In order to allow semaphores to work in both single processor and multi-processor environments, the Cortex®-M3 and Cortex-M4 processors support a feature called exclusive access. The semaphore variables are read and written using exclusive load and exclusive store. If during the store operation it was found that the access cannot be guaranteed to be exclusive, the exclusive store fails and the write will not take place. The processor should then retry the exclusive access sequence.

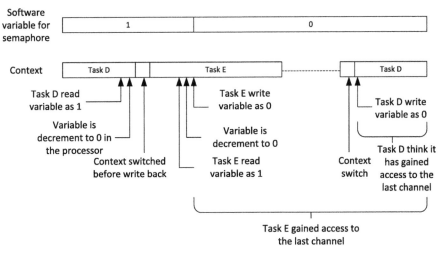

**FIGURE 10.13**

Context switching at the middle of the read-modify-write sequence

Inside the processor, there is a small hardware unit called the local monitor. In a normal situation it is in the Open Access state. After executing an exclusive load instruction, it switches to the Exclusive Access state. An exclusive store can only be carried out if the local monitor is in the Exclusive Access state and if the bus system does not response with an exclusive fail response.

The exclusive write (e.g., STREX) can fail if one of the following operations has taken place:

- A CLREX instruction has been executed, switching the local monitor to the Open Access state.
- A context switch has occurred (e.g., an interrupt).
- There wasn't a LDREX executed beforehand.
- An external hardware returns an exclusive fail status to the processor via a side-band signal on the bus interface.

If the exclusive store gets a failed status, the actual write will not take place in the memory, since it will either be blocked by the processor core or by external hardware.

In multi-processor environments, a component called the global exclusive access monitor is needed in the system bus infrastructure to monitor accesses from various processors, and to generate an exclusive fail status signal to the processor running the exclusive store if a conflict is detected.

More information on exclusive load and store is covered in sections 5.6.2, 6.10, and 9.2.9.

Please note that in ARMv7-M architecture (including Cortex-M3 and Cortex-M4), a context switch (or any exception sequence) automatically clears the exclusive state in the local monitor. This is different in ARMv7-A/R, where a context-switching code has to execute the CLREX instruction (or use a dummy STREX) to ensure the local monitor is switched to the Open Access state. In general, CLREX is not strictly required for Cortex-M3 and Cortex-M4 processors, but is included in the architecture for consistency and easier software porting.

# Memory Protection Unit (MPU)

# 11

## CHAPTER OUTLINE

## 11.1 Overview of the MPU

### 11.1.1 About the MPU

The Cortex®-M3 and Cortex-M4 processors support an optional feature called the Memory Protection Unit (MPU). Some of the Cortex-M3 and Cortex-M4 microcontrollers have this feature, and some do not.

The MPU is a programmable device that can be used to define memory access permissions (e.g., privileged access only or full access) and memory attributes (e.g., bufferable, cacheable) for different memory regions. The MPU on Cortex-M3 and Cortex-M4 processors can support up to eight programmable memory regions, each with their own programmable starting addresses, sizes, and settings. It also supports a background region feature.

The Definitive Guide to ARM® Cortex®-M3 and Cortex-M4 Processors. http://dx.doi.org/10.1016/B978-0-12-408082-9.00011-7

The MPU can be used to make an embedded system more robust, and in some cases it can make the system more secure by:

- Preventing application tasks from corrupting stack or data memory used by other tasks and the OS kernel
- Preventing unprivileged tasks from accessing certain peripherals that can be critical to the reliability or security of the system
- Defining SRAM or RAM space as non-executable (eXecute Never, XN) to prevent code injection attacks

You can also use the MPU to define other memory attributes such as cacheability, which can be exported to the system-level cache unit or memory controllers.

If a memory access violates the access permissions defined by the MPU, or accesses a memory location that is not defined in the programmed MPU regions, the transfer would be blocked and a fault exception would be triggered. The fault exception handler triggered could either be the MemManage (Memory Management) fault or the HardFault exception, depending on whether the MemManage fault is enabled, and the current priority levels. The exception handler can then decide if the system should be reset or just terminate the offending task in an OS environment.

The MPU needs to be programmed and enabled before use. If the MPU is not enabled, the processor no MPU is present. MPU regions can be overlapped. If a memory location falls in two programmed MPU regions, the memory access attributes and permission will be based on the highest-numbered region. For example, if a transfer address is within the address range defined for region 1 and region 4, the region 4 settings will be used.

### 11.1.2 Using the MPU

The MPU can be set up in a number of ways.

In systems without an embedded OS, the MPU can be programmed to have a static configuration. The configuration can be used for functions like:

- Setting a RAM/SRAM region to be read-only to protect important data from accidental corruption
- Making a portion of RAM/SRAM space at the bottom of the stack inaccessible to detect stack overflow
- Setting a RAM/SRAM region to be XN to prevent code injection attacks
- Defining memory attribute settings that can be used by system level cache (level 2) or the memory controllers

In systems with an embedded OS, the MPU can be programmed at each context switch so that each application task can have a different MPU configuration. In this way, you can:

- Define memory access permissions so that stack operations of an application task can only access their own allocated stack space, thus preventing stack corruptions of other stacks in the case of a stack leak

- Define memory access permissions so that an application task can only have access to a limited set of peripherals
- Define memory access permissions so that an application task can only access its own data, or access its own program data (these are much trickier to set up because in most cases the OS and the program code are compiled together, so the data could be mixed together in the memory map)

Systems with an embedded OS can also use a static configuration if preferred.

## 11.2 MPU registers

The MPU contains a number of registers. These registers are located in the System Control Space (SCS). The CMSIS-Core header file has defined a data structure for MPU registers to allow them to be accessed easily. A summary of these registers is shown in Table 11.1.

### 11.2.1 MPU type register

The first one is the MPU Type register. The MPU Type register can be used to determine whether the MPU is fitted. If the DREGION field is read as 0, the MPU is not implemented (Table 11.2).

### 11.2.2 MPU control register

The MPU is controlled by a number of registers. The first one is the MPU Control register (Table 11.3). This register has three control bits. After reset, the value of this register is zero, which disables the MPU. To enable the MPU, the software should first set up the settings for each MPU region, and then set the ENABLE bit in the MPU Control register.

The PRIVDEFENA bit in the MPU Control Register is used to enable the background region (region −1). By using PRIVDEFENA and if no other regions are set up, privileged programs will be able to access all memory locations, and only unprivileged programs will be blocked. However, if other MPU regions are programmed and enabled, they can override the background region. For example, for two systems with similar region setups but only one with PRIVDEFENA set to 1 (the right-hand side in Figure 11.1), the one with PRIVDEFENA set to 1 will allow privileged access to background regions.

The HFNMIENA is used to define the behavior of the MPU during execution of NMI, HardFault handlers, or when FAULTMASK is set. By default, the MPU is bypassed (disabled) in these cases. This allows the HardFault handler and the NMI handler to execute even if the MPU was set up incorrectly.

Setting the enable bit in the MPU Control register is usually the last step in the MPU setup code. Otherwise, the MPU might generate faults accidentally before the region configuration is done. In many cases, especially in embedded OSs with

**Table 11.1** Summary of the MPU Registers

| Address | Register | CMSIS-Core Symbol | Function |
|---|---|---|---|
| 0xE000ED90 | MPU Type Register | MPU->TYPE | Provides information about the MPU |
| 0xE000ED94 | MPU Control Register | MPU->CTRL | MPU enable/disable and background region control |
| 0xE000ED98 | MPU Region Number Register | MPU->RNR | Select which MPU region to be configured |
| 0xE000ED9C | MPU Region Base Address Register | MPU->RBAR | Defines base address of a MPU region |
| 0xE000EDA0 | MPU Region Base Attribute and Size Register | MPU->RASR | Defines size and attributes of a MPU region |
| 0xE000EDA4 | MPU Alias 1 Region Base Address Register | MPU->RBAR_A1 | Alias of MPU->RBAR |
| 0xE000EDA8 | MPU Alias 1 Region Base Attribute and Size Register | MPU->RASR_A1 | Alias of MPU->RASR |
| 0xE000EDAC | MPU Alias 2 Region Base Address Register | MPU->RBAR_A2 | Alias of MPU->RBAR |
| 0xE000EDB0 | MPU Alias 2 Region Base Attribute and Size Register | MPU->RASR_A2 | Alias of MPU->RASR |
| 0xE000EDB4 | MPU Alias 3 Region Base Address Register | MPU->RBAR_A3 | Alias of MPU->RBAR |
| 0xE000EDB8 | MPU Alias 3 Region Base Attribute and Size Register | MPU->RASR_A3 | Alias of MPU->RASR |

**Table 11.2** MPU Type Register (MPU->TYPE, 0xE000ED90)

| Bits | Name | Type | Reset Value | Description |
|---|---|---|---|---|
| 23:16 | IREGION | R | 0 | Number of instruction regions supported by this MPU; because ARMv7-M architecture uses a unified MPU, this is always 0. |
| 15:8 | DREGION | R | 0 or 8 | Number of regions supported by this MPU; in the Cortex®-M3 and Cortex-M4 processors this is either 0 (MPU not present) or 8 (MPU present). |
| 0 | SEPARATE | R | 0 | This is always 0 as the MPU is unified. |

**Table 11.3** MPU Control Register (MPU->CTRL, 0xE000ED94)

| Bits | Name | Type | Reset Value | Description |
|------|------|------|-------------|-------------|
| 2 | PRIVDEFENA | R/W | 0 | Privileged default memory map enable. When set to 1 and if the MPU is enabled, the default memory map will be used for privileged accesses as a background region. If this bit is not set, the background region is disabled and any access not covered by any enabled region will cause a fault. |
| 1 | HFNMIENA | R/W | 0 | If set to 1, it enables the MPU during the HardFault handler and NMI handler; otherwise, the MPU is not enabled for the HardFault handler and NMI. |
| 0 | ENABLE | R/W | 0 | Enables the MPU if set to 1. |

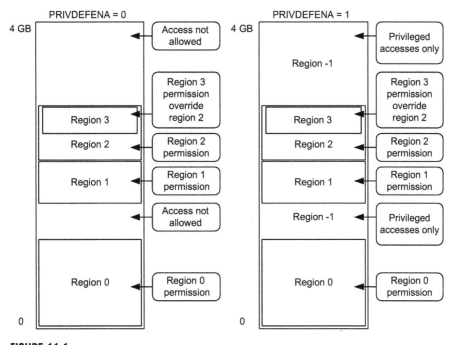

**FIGURE 11.1**

The effect of the PRIVDEFENA bit (background region)

dynamic MPU configurations, the MPU should be disabled at the start of the MPU configuration routine to make sure that the MemManage fault won't be triggered by accident during configuration of MPU regions.

### 11.2.3 MPU region number register

The next MPU control register is the MPU Region Number register (Table 11.4). Before each region is set up, write to this register to select the region to be programmed.

### 11.2.4 MPU region base address register

The starting address of each region is defined by the MPU Region Base Address register (Table 11.5). Using the VALID and REGION fields in this register, we can skip the step of programming the MPU Region Number register. This can reduce the complexity of the program code, especially if the whole MPU setup is defined in a lookup table.

**Table 11.4** MPU Region Number Register (MPU->RNR, 0xE000ED98)

| Bits | Name | Type | Reset Value | Description |
|------|------|------|-------------|-------------|
| 7:0 | REGION | R/W | — | Select the region that is being programmed. Since eight regions are supported in the MPU, only bit[2:0] of this register is implemented. |

**Table 11.5** MPU Region Base Address Register (MPU->RBAR, 0xE000ED9C)

| Bits | Name | Type | Reset Value | Description |
|------|------|------|-------------|-------------|
| 31:N | ADDR | R/W | — | Base address of the region; N is dependent on the region size – for example, a 64 k size region will have a base address field of [31:16]. |
| 4 | VALID | R/W | — | If this is 1, the REGION defined in bit[3:0] will be used in this programming step; otherwise, the region selected by the MPU Region Number register is used. |
| 3:0 | REGION | R/W | — | This field overrides the MPU Region Number register if VALID is 1; otherwise, it is ignored. Since eight regions are supported in the Cortex®-M3 and Cortex-M4 MPU, the region number override is ignored if the value of the REGION field is larger than 7. |

### 11.2.5 **MPU region base attribute and size register**

We also need to define the properties of each region. This is controlled by the MPU Region Base Attribute and Size register (Table 11.6).

The REGION SIZE field (5 bits) in the MPU Region Base Attribute and Size register determines the size of the region (Table 11.7).

The sub-region disable field (bit [15:8] of the MPU Region Base Attribute and Size register) is used to divide a region into eight equal sub-regions and then to define each as enabled or disabled. If a sub-region is disabled and overlaps another region, the access rules for the other region are applied. If the sub-region is disabled and does not overlap any other region, access to this memory range will result in a MemManage fault. Sub-regions cannot be used if the region size is 128 bytes or less.

The data Access Permission (AP) field (bit [26:24]) defines the AP of the region (Table 11.8).

The XN (Execute Never) field (bit [28]) decides whether an instruction fetch from this region is allowed. When this field is set to 1, all instructions fetched from this region will generate a MemManage fault when they enter the execution stage.

The TEX (Type extension), S (Shareable), B (Bufferable), and C (Cacheable) fields (bit [21:16]) are more complex. These memory attributes are exported to the bus system together with each instruction and data access, and the information can be used by the bus system such as write buffers or cache units, as shown in Figure 11.2.

**Table 11.6** MPU Region Base Attribute and Size Register (MPU->RASR, 0xE000EDA0)

| Bits | Name | Type | Reset Value | Description |
|------|------|------|-------------|-------------|
| 31:29 | Reserved | — | — | — |
| 28 | XN | R/W | — | Instruction Access Disable (1 = Disable instruction fetch from this region; an attempt to do so will result in a memory management fault). |
| 27 | Reserved | — | — | — |
| 26:24 | AP | R/W | — | Data Access Permission field |
| 23:22 | Reserved | — | — | — |
| 21:19 | TEX | R/W | — | Type Extension field |
| 18 | S | R/W | — | Shareable |
| 17 | C | R/W | — | Cacheable |
| 16 | B | R/W | — | Bufferable |
| 15:8 | SRD | R/W | — | Sub-region disable |
| 7:6 | Reserved | — | — | |
| 5:1 | REGION SIZE | R/W | — | MPU Protection Region size |
| 0 | ENABLE | R/W | — | Region enable |

**Table 11.7** Encoding of REGION SIZE field for Different Memory Region Sizes

| REGION Size | Size | | REGION Size | Size |
|---|---|---|---|---|
| b00000 | Reserved | | b10000 | 128 KB |
| b00001 | Reserved | | b10001 | 256 KB |
| b00010 | Reserved | | b10010 | 512 KB |
| b00011 | Reserved | | b10011 | 1 MB |
| b00100 | 32 Byte | | b10100 | 2 MB |
| b00101 | 64 Byte | | b10101 | 4 MB |
| b00110 | 128 Byte | | b10110 | 8 MB |
| b00111 | 256 Byte | | b10111 | 16 MB |
| b01000 | 512 Byte | | b11000 | 32 MB |
| b01001 | 1 KB | | b11001 | 64 MB |
| b01010 | 2 KB | | b11010 | 128 MB |
| b01011 | 4 KB | | b11011 | 256 MB |
| b01100 | 8 KB | | b11100 | 512 MB |
| b01101 | 16 KB | | b11101 | 1 GB |
| b01110 | 32 KB | | b11110 | 2 GB |
| b01111 | 64 KB | | b11111 | 4 GB |

Although the Cortex®-M3 and Cortex-M4 processors do not include cache controllers, their implementations follow the ARMv7-M architecture, which can support external cache controllers on the system bus level, including advanced memory systems with caching capabilities. In addition, there is a write buffer in the processor's internal bus system, which is affected by the bufferable attribute. Therefore, the region access properties TEX, S, B, and C fields should to be programmed correctly to support different types of memory or devices. The definitions of these bit fields are shown in Table 11.9. However, in many microcontrollers, these

**Table 11.8** Encoding of AP Field for Various Access Permission Configurations

| AP Value | Privileged Access | User Access | Description |
|---|---|---|---|
| 000 | No access | No access | No access |
| 001 | Read/Write | No access | Privileged access only |
| 010 | Read/Write | Read-only | Write in a user program generates a fault |
| 011 | Read/Write | Read/Write | Full access |
| 100 | Unpredictable | Unpredictable | Unpredictable |
| 101 | Read-only | No access | Privileged read only |
| 110 | Read-only | Read-only | Read-only |
| 111 | Read-only | Read-only | Read-only |

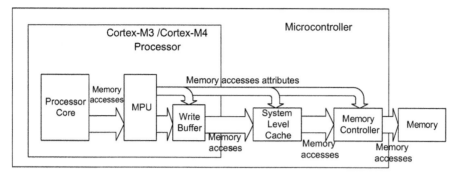

**FIGURE 11.2**

Memory attributes can be used inside the processor or by components on the bus system

memory attributes are not used by the bus system and only the B (Bufferable) attribute affects the write buffer in the processor.

The memory attribute settings can support two cache levels: inner cache and outer cache. They can have different caching policies. If a system-level cache is implemented, it can either be using the inner cache attribute or the outer cache attribute. This is device-specific and therefore you need to refer to the documentation from silicon vendors regarding suitable settings. In most cases, the memory attributes can be configured as shown in Table 11.10, with the same attributes on both levels.

**Table 11.9** ARMv7-M Memory Attributes

| TEX | C | B | Description | Region Shareability |
|------|---|---|-------------|---------------------|
| b000 | 0 | 0 | Strongly ordered (transfers carry out and complete in programmed order) | Shareable |
| b000 | 0 | 1 | Shared device (write can be buffered) | Shareable |
| b000 | 1 | 0 | Outer and inner write-through; no write allocate | [S] |
| b000 | 1 | 1 | Outer and inner write-back; no write allocate | [S] |
| b001 | 0 | 0 | Outer and inner non-cacheable | [S] |
| b001 | 0 | 1 | Reserved | Reserved |
| b001 | 1 | 0 | Implementation defined | — |
| b001 | 1 | 1 | Outer and inner write-back; write and read allocate | [S] |
| b010 | 0 | 0 | Non-shared device | Not shared |
| b010 | 0 | 1 | Reserved | Reserved |
| b010 | 1 | X | Reserved | Reserved |
| b1BB | A | A | Cached memory; BB = outer policy, AA = inner policy | [S] |

*Note: [S] Indicates that shareability is determined by the S bit field (shared by multiple processors)*

**Table 11.10** Commonly Used Memory Attributes in Microcontrollers

| Type | Memory Type | Commonly Used Memory Attributes |
|------|-------------|---------------------------------|
| ROM, flash (program memories) | Normal memory | Non-shareable, write through C = 1, B = 0, TEX = 0, S = 0 |
| Internal SRAM | Normal memory | Shareable, write through C = 1, B = 0, TEX = 0, S = 1 |
| External RAM | Normal memory | Shareable, write back C = 1, B = 1, TEX = 0, S = 1 |
| Peripherals | Device | Shareable devices C = 0, B = 1, TEX = 0, S = 1 |

**Table 11.11** Encoding of Inner and Outer Cache Policy When Most Significant Bit of TEX is set to 1

| Memory Attribute Encoding (AA and BB) | Cache Policy |
|----------------------------------------|--------------|
| 00 | Non-cacheable |
| 01 | Write back, write and read allocate |
| 10 | Write through, no write allocate |
| 11 | Write back, no write allocate |

In the cases where you need to have different inner and outer cache policies, you need to set the bit 2 of TEX to 1. In this case, the definition of TEX[1:0] will become the outer policy (indicated as BB in Table 11.9), and the C and B bits will become the inner policy (indicated as AA in Table 11.9). The definitions of the cache policy setting (AA and BB) are shown in Table 11.11.

If you are using a microcontroller with cache memory, and if you are using the MPU to define access permissions in your application, then you should also make sure that the memory attribute settings match the memory type and the cache policy you want to use (e.g., cache disable, write-through cache, or write-back cache).

The shareable attribute is important for a multi-processor system with caches. In these systems, if a transfer is marked as shareable, then the cache system might need to do extra work to ensure data coherency between the different processors (Figure 6.16). In single processor systems, the shareable attribute is normally not used.

## 11.2.6 MPU alias registers

The remaining registers (MPU->RBAR_Ax and MPU->RASR_Ax) are alias addresses locations. When these addresses are accessed they are just accessing MPU->RBAR or MPU->RASR. The reason for having these register aliases is to allow multiple MPU regions to be programmed in one go; for example, by using the store-multiple (STM) instruction.

## 11.3 Setting up the MPU

Most simple applications do not require an MPU. By default the MPU is disabled and the system works as if it is not present. Before using the MPU, you need to work out what memory regions the program or application tasks need to (and are allowed to) access:

- Program code for privileged applications including handlers and OS kernel, typically privileged accesses only
- Data memory including stack for privileged applications including handlers and OS kernels, typically privileged accesses only
- Program code for unprivileged applications (application tasks), full access
- Data memory including stack for unprivileged applications (application tasks), full accesses
- Peripherals that are for privileged applications including handlers and OS kernel, privileged accesses only
- Peripherals that can be used by unprivileged applications (application tasks), full accesses

When defining the address and size of the memory region, be aware that the base address of a region must be aligned to an integer multiple value of the region size. For example, if the region size is 4KB (0x1000), the starting address must be "$N$ x 0x1000" where $N$ is an integer (Figure 11.3).

If the goal for using the MPU is to prevent unprivileged tasks from accessing certain memory regions, the background region feature is very useful as it reduces the setup required. You only need to set up the region setting for unprivileged tasks, and privileged tasks and handlers have full access to other memory spaces using the background region.

**FIGURE 11.3**

MPU region addresses must be aligned to integer multiples of the region sizes

There is no need to set up memory regions for Private Peripheral Bus (PPB) address ranges (including System Control Space, SCS) and the vector table. Accesses to PPB (including MPU, NVIC, SysTick, ITM) are always allowed in privileged state, and vector fetches are always permitted by the MPU.

You also need to define the fault handler for either the HardFault or MemManage (Memory Management) fault. By default the MemManage exception is disabled, and you can enable this by setting the MEMFAULTENA bit in the System Handler Control and State Register (SCB->SHCSR):

```
SCB->SHCSR |= SCB_SHCSR_MEMFAULTENA_Msk; // Set bit 16
```

The HardFault handler (void HardFault_Handler(void)) should always be defined even if you have enabled the MemManage exception. The MemManage handler (void MemManage_Handler(void)) should also be defined if the MemManage exception is enabled. By default, the vector table in the startup code should contain the exception vector definition for these handlers. If you are using the vector table relocation feature, you might need to ensure that the vector table is set up accordingly. More information about using fault handlers is covered in Chapter 12.

To help set the MPU, we define a number of constant values:

```
#define MPU_DEFS_RASR_SIZE_32B (0x04 << MPU_RASR_SIZE_Pos)
#define MPU_DEFS_RASR_SIZE_64B (0x05 << MPU_RASR_SIZE_Pos)
#define MPU_DEFS_RASR_SIZE_128B (0x06 << MPU_RASR_SIZE_Pos)
#define MPU_DEFS_RASR_SIZE_256B (0x07 << MPU_RASR_SIZE_Pos)
#define MPU_DEFS_RASR_SIZE_512B (0x08 << MPU_RASR_SIZE_Pos)
#define MPU_DEFS_RASR_SIZE_1KB (0x09 << MPU_RASR_SIZE_Pos)
#define MPU_DEFS_RASR_SIZE_2KB (0x0A << MPU_RASR_SIZE_Pos)
#define MPU_DEFS_RASR_SIZE_4KB (0x0B << MPU_RASR_SIZE_Pos)
#define MPU_DEFS_RASR_SIZE_8KB (0x0C << MPU_RASR_SIZE_Pos)
#define MPU_DEFS_RASR_SIZE_16KB (0x0D << MPU_RASR_SIZE_Pos)
#define MPU_DEFS_RASR_SIZE_32KB (0x0E << MPU_RASR_SIZE_Pos)
#define MPU_DEFS_RASR_SIZE_64KB (0x0F << MPU_RASR_SIZE_Pos)
#define MPU_DEFS_RASR_SIZE_128KB (0x10 << MPU_RASR_SIZE_Pos)
#define MPU_DEFS_RASR_SIZE_256KB (0x11 << MPU_RASR_SIZE_Pos)
#define MPU_DEFS_RASR_SIZE_512KB (0x12 << MPU_RASR_SIZE_Pos)
#define MPU_DEFS_RASR_SIZE_1MB (0x13 << MPU_RASR_SIZE_Pos)
#define MPU_DEFS_RASR_SIZE_2MB (0x14 << MPU_RASR_SIZE_Pos)
#define MPU_DEFS_RASR_SIZE_4MB (0x15 << MPU_RASR_SIZE_Pos)
#define MPU_DEFS_RASR_SIZE_8MB (0x16 << MPU_RASR_SIZE_Pos)
#define MPU_DEFS_RASR_SIZE_16MB (0x17 << MPU_RASR_SIZE_Pos)
#define MPU_DEFS_RASR_SIZE_32MB (0x18 << MPU_RASR_SIZE_Pos)
#define MPU_DEFS_RASR_SIZE_64MB (0x19 << MPU_RASR_SIZE_Pos)
#define MPU_DEFS_RASR_SIZE_128MB (0x1A << MPU_RASR_SIZE_Pos)
#define MPU_DEFS_RASR_SIZE_256MB (0x1B << MPU_RASR_SIZE_Pos)
```

```
#define MPU_DEFS_RASR_SIZE_512MB (0x1C << MPU_RASR_SIZE_Pos)
#define MPU_DEFS_RASR_SIZE_1GB (0x1D << MPU_RASR_SIZE_Pos)
#define MPU_DEFS_RASR_SIZE_2GB (0x1E << MPU_RASR_SIZE_Pos)
#define MPU_DEFS_RASR_SIZE_4GB (0x1F << MPU_RASR_SIZE_Pos)
#define MPU_DEFS_RASE_AP_NO_ACCESS  (0x0 << MPU_RASR_AP_Pos)
#define MPU_DEFS_RASE_AP_PRIV_RW   (0x1 << MPU_RASR_AP_Pos)
#define MPU_DEFS_RASE_AP_PRIV_RW_USER_RO (0x2 << MPU_RASR_AP_Pos)
#define MPU_DEFS_RASE_AP_FULL_ACCESS  (0x3 << MPU_RASR_AP_Pos)
#define MPU_DEFS_RASE_AP_PRIV_RO   (0x5 << MPU_RASR_AP_Pos)
#define MPU_DEFS_RASE_AP_RO     (0x6 << MPU_RASR_AP_Pos)
#define MPU_DEFS_NORMAL_MEMORY_WT  (MPU_RASR_C_Msk)
#define MPU_DEFS_NORMAL_MEMORY_WB  (MPU_RASR_C_Msk | MPU_RASR_B_Msk)
#define MPU_DEFS_NORMAL_SHARED_MEMORY_WT (MPU_RASR_C_Msk |
MPU_RASR_S_Msk)
#define MPU_DEFS_NORMAL_SHARED_MEMORY_WB (MPU_DEFS_NORMAL_MEMORY_WB |
MPU_RASR_S_Msk)
#define MPU_DEFS_SHARED_DEVICE   (MPU_RASR_B_Msk)
#define MPU_DEFS_STRONGLY_ORDERED_DEVICE (0x0)
```

For a simple case of only four required regions, the MPU setup code can be written as a simple loop, with the configuration for the MPU->RBAR and MPU->RASR coded as a constant table:

```
// ------------------------------------------------------------
int mpu_setup(void)
{
 uint32_t i;
 uint32_t const mpu_cfg_rbar[4] = {
 0x08000000, // Flash address for STM32F4
 0x20000000, // SRAM
 GPIOD_BASE, // GPIO D base address
 RCC_BASE    // Reset Clock CTRL base address
 };
 uint32_t const mpu_cfg_rasr[4] = {
   (MPU_DEFS_RASR_SIZE_1MB      | MPU_DEFS_NORMAL_MEMORY_WT |
    MPU_DEFS_RASE_AP_FULL_ACCESS | MPU_RASR_ENABLE_Msk), // Flash
   (MPU_DEFS_RASR_SIZE_128KB     | MPU_DEFS_NORMAL_MEMORY_WT |
    MPU_DEFS_RASE_AP_FULL_ACCESS | MPU_RASR_ENABLE_Msk), // SRAM
   (MPU_DEFS_RASR_SIZE_1KB      | MPU_DEFS_SHARED_DEVICE |
    MPU_DEFS_RASE_AP_FULL_ACCESS | MPU_RASR_ENABLE_Msk), // GPIO D
   (MPU_DEFS_RASR_SIZE_1KB      | MPU_DEFS_SHARED_DEVICE |
    MPU_DEFS_RASE_AP_FULL_ACCESS | MPU_RASR_ENABLE_Msk) // RCC
   };
 if (MPU->TYPE==0) {return 1;} // Return 1 to indicate error
```

```
__DMB();                          // Make sure outstanding transfers are
                                  done
MPU->CTRL = 0;                    // Disable the MPU
for (i=0;i<4;i++) {               // Configure only 4 regions
MPU->RNR = i;                     // Select which MPU region to
                                  configure
MPU->RBAR = mpu_cfg_rbar[i];      // Configure region base address
                                  register
MPU->RASR = mpu_cfg_rasr[i];      // Configure region attribute and size
                                  register
}
for (i=4;i<8;i++) {// Disabled unused regions
  MPU->RNR = i;       // Select which MPU region to configure
  MPU->RBAR = 0;      // Configure region base address register
  MPU->RASR = 0;      // Configure region attribute and size register
  }
MPU->CTRL = MPU_CTRL_ENABLE_Msk; // Enable the MPU
__DSB();  // Memory barriers to ensure subsequence data &
              instruction
__ISB();  // transfers using updated MPU settings
return 0; // No error
}
// -------------------------------------------------------------
```

A simple check was added at the beginning of the function to detect if the MPU was present. If the MPU is not available, the function exits with a value of 1 to indicate the error. Otherwise, it returns 0 to indicate successful operation.

The example code also programs unused MPU regions to make sure that unused MPU regions are disabled. This is important for systems that configure MPU dynamically, because an unused region could have been programmed to be enabled previously.

The flow for this simple MPU setup function is illustrated in Figure 11.4.

To simplify the operation, the selection of the MPU region to be programmed can be merged into the programming of MPU->RBAR, as shown in the following code:

```
// -------------------------------------------------------------
int mpu_setup(void)
{
    uint32_t i;
    uint32_t const mpu_cfg_rbar[4] = {
      // Flash - region 0
      (0x08000000 | MPU_RBAR_VALID_Msk | (MPU_RBAR_REGION_Msk & 0)),
      // SRAM - region 1
      (0x20000000 | MPU_RBAR_VALID_Msk | (MPU_RBAR_REGION_Msk & 1)),
```

```
      // GPIO D base address - region 2
   (GPIOD_BASE | MPU_RBAR_VALID_Msk | (MPU_RBAR_REGION_Msk & 2)),
   // Reset Clock CTRL base address - region 3
   (RCC_BASE | MPU_RBAR_VALID_Msk | (MPU_RBAR_REGION_Msk & 3))
   };
  uint32_t const mpu_cfg_rasr[4] = {
   (MPU_DEFS_RASR_SIZE_1MB        | MPU_DEFS_NORMAL_MEMORY_WT |
    MPU_DEFS_RASE_AP_FULL_ACCESS | MPU_RASR_ENABLE_Msk), // Flash
   (MPU_DEFS_RASR_SIZE_128KB      | MPU_DEFS_NORMAL_MEMORY_WT |
    MPU_DEFS_RASE_AP_FULL_ACCESS | MPU_RASR_ENABLE_Msk), // SRAM
   (MPU_DEFS_RASR_SIZE_1KB        | MPU_DEFS_SHARED_DEVICE |
    MPU_DEFS_RASE_AP_FULL_ACCESS | MPU_RASR_ENABLE_Msk), // GPIO D
   (MPU_DEFS_RASR_SIZE_1KB        | MPU_DEFS_SHARED_DEVICE |
    MPU_DEFS_RASE_AP_FULL_ACCESS | MPU_RASR_ENABLE_Msk), // RCC
   };
  if (MPU->TYPE==0) {return 1;}    // Return 1 to indicate error
  __DMB();                         // Make sure outstanding transfers
                                   //   are done
  MPU->CTRL = 0;                   // Disable the MPU
  for (i=0;i<4;i++) {              // Configure only 4 regions
    MPU->RBAR = mpu_cfg_rbar[i];   // Configure region base address
                                   //   register
    MPU->RASR = mpu_cfg_rasr[i];   // Configure region attribute and
                                   //   size register

    }
  for (i=4;i<8;i++) {// Disabled unused regions
    MPU->RNR = i; // Select which MPU region to configure
    MPU->RBAR = 0; // Configure region base address register
    MPU->RASR = 0; // Configure region attribute and size register
    }
  MPU->CTRL = MPU_CTRL_ENABLE_Msk; // Enable the MPU
  __DSB(); // Memory barriers to ensure subsequence data & instruction
  __ISB(); // transfers using updated MPU settings
  return 0; // No error
  }
// --------------------------------------------------------------
```

To speed up the MPU programming process, we can unroll the loop. However, we can reduce the execution time even further by using the MPU alias registers. To do this, we define the MPU configuration as a data table:

```
// MPU configuration table
uint32_t const mpu_cfg_rbar_rasr[16] = {
   // Flash - region 0
   (0x08000000 | MPU_RBAR_VALID_Msk | (MPU_RBAR_REGION_Msk & 0)),
   // RBAR
```

```
  (MPU_DEFS_RASR_SIZE_1MB  | MPU_DEFS_NORMAL_MEMORY_WT | // RASR
  MPU_DEFS_RASE_AP_FULL_ACCESS | MPU_RASR_ENABLE_Msk)   ,
  // SRAM - region 1
  (0x20000000 | MPU_RBAR_VALID_Msk | (MPU_RBAR_REGION_Msk & 1)),
  // RBAR_A1
  (MPU_DEFS_RASR_SIZE_128KB  | MPU_DEFS_NORMAL_MEMORY_WT | // RASR_A1
  MPU_DEFS_RASE_AP_FULL_ACCESS | MPU_RASR_ENABLE_Msk)   ,
  // GPIO D base address - region 2
  (GPIOD_BASE | MPU_RBAR_VALID_Msk | (MPU_RBAR_REGION_Msk & 2)),
  // RBAR_A2
  (MPU_DEFS_RASR_SIZE_1KB  | MPU_DEFS_SHARED_DEVICE | // RASR_A2
  MPU_DEFS_RASE_AP_FULL_ACCESS | MPU_RASR_ENABLE_Msk)   ,
  // Reset Clock CTRL base address - region 3
  (RCC_BASE | MPU_RBAR_VALID_Msk | (MPU_RBAR_REGION_Msk & 3)),
  // RBAR_A3
  (MPU_DEFS_RASR_SIZE_1KB  | MPU_DEFS_SHARED_DEVICE | // RASR_A3
   MPU_DEFS_RASE_AP_FULL_ACCESS | MPU_RASR_ENABLE_Msk),
  (MPU_RBAR_VALID_Msk | (MPU_RBAR_REGION_Msk & 4)), 0, // Region 4 -
  unused
  (MPU_RBAR_VALID_Msk | (MPU_RBAR_REGION_Msk & 5)), 0, // Region 5 -
  unused
  (MPU_RBAR_VALID_Msk | (MPU_RBAR_REGION_Msk & 6)), 0, // Region 6 -
  unused
  (MPU_RBAR_VALID_Msk | (MPU_RBAR_REGION_Msk & 7)), 0, // Region 7 -
  unused
  };
```

Then we can create an assembly function (or embedded assembly function for ARM® Compilation toolchains including Keil™ MDK-ARM):

```
// -------------------------------------------------------------
__asm void mpu_cfg_copy(unsigned int src)
{
PUSH {R4-R9}
LDR R1, =0xE000ED9C // MPU->RBAR address
LDR R2, [R1,#-12]   // Get MPU->TYPE
CMP R1, #0          // If zero
ITT EQ              // If-Then
MOVSEQ R0, #1       // return 1
BEQ mpu_cfg_copy_end
DMB 0xF // Make sure outstanding transfers are done
MOVS R2, #0
STR R2, [R1, #-8]   // MPU->CTRL = 0
LDMIA R0!, {R2-R9} // Read 8 words from table (base update)
STMIA R1, {R2-R9}  // Write 8 words to MPU (no base update)
```

```
  LDMIA R0!, {R2-R9} // Read 8 words from table (base update)
  STMIA R1, {R2-R9}  // Write 8 words to MPU (no base update)
  DSB 0xF // Memory barriers to ensure subsequence data & instruction
  ISB 0xF // transfers using updated MPU settings
  MOVS R0, #0 // No error
mpu_cfg_copy_end
  POP {R4-R9}
  BX   LR
  ALIGN 4
  }
  // -------------------------------------------------------------
```

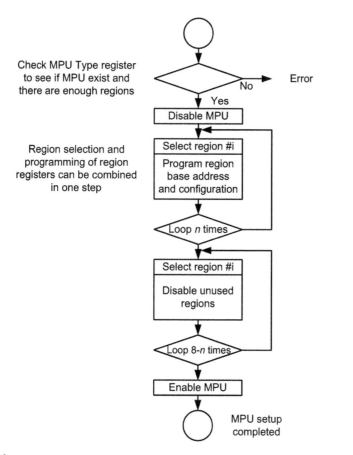

Check MPU Type register to see if MPU exist and there are enough regions

Error

No

Yes

Disable MPU

Region selection and programming of region registers can be combined in one step

Select region #i

Program region base address and configuration

Loop *n* times

Select region #i

Disable unused regions

Loop 8-*n* times

Enable MPU

MPU setup completed

**FIGURE 11.4**

Example step to set up the MPU

In the C code, we can call this MPU setup code as follows, passing the starting address of the table to the assembly function:

```
// Copy MPU setup table to MPU
mpu_cfg_copy((unsigned int) &mpu_cfg_rbar_rasr[0]);
```

The configuration methods shown so far assume that we know the required settings in advance. If not, we might need to create some generic functions to make the MPU configuration easier. For example, we can create the following C functions:

```
// ---------------------------------------------------------------
// Enable MPU with input options
// Options can be MPU_CTRL_HFNMIENA_Msk or MPU_CTRL_PRIVDEFENA_Msk
void mpu_enable(uint32_t options)
{
 MPU->CTRL = MPU_CTRL_ENABLE_Msk | options; // Disable the MPU
 __DSB(); // Ensure MPU settings take effects
 __ISB(); // Sequence instruction fetches using update settings
 return;
}
// Disable the MPU
void mpu_disable(void)
{
 __DMB();  // Make sure outstanding transfers are done
 MPU->CTRL = 0; // Disable the MPU
 return;
}
// Function to disable a region (0 to 7)
void mpu_region_disable(uint32_t region_num)
{
 MPU->RNR = region_num;
 MPU->RBAR = 0;
 MPU->RASR = 0;
 return;
}
// Function to enable a region
void mpu_region_config(uint32_t region_num, uint32_t addr, uint32_t
size, uint32_t attributes)
{
 MPU->RNR = region_num;
 MPU->RBAR = addr;
 MPU->RASR = size | attributes;
 return;
}
```

After these functions are created, we can configure the MPU using these functions:

```
int mpu_setup(void)
{
  if (MPU->TYPE==0) {return 1;} // Return 1 to indicate error
  mpu_disable();
  mpu_region_config(0, 0x08000000, MPU_DEFS_RASR_SIZE_1MB,
      MPU_DEFS_NORMAL_MEMORY_WT | MPU_DEFS_RASE_AP_FULL_ACCESS |
      MPU_RASR_ENABLE_Msk), // Region 0 - Flash
  mpu_region_config(1, 0x20000000, MPU_DEFS_RASR_SIZE_128KB,
      MPU_DEFS_NORMAL_MEMORY_WT | MPU_DEFS_RASE_AP_FULL_ACCESS |
      MPU_RASR_ENABLE_Msk), // Region 1 - SRAM
  mpu_region_config(2, GPIOD_BASE, MPU_DEFS_RASR_SIZE_1KB,
      MPU_DEFS_SHARED_DEVICE | MPU_DEFS_RASE_AP_FULL_ACCESS |
      MPU_RASR_ENABLE_Msk), // Region 2 - GPIO D
  mpu_region_config(3, RCC_BASE, MPU_DEFS_RASR_SIZE_1KB,
      MPU_DEFS_SHARED_DEVICE | MPU_DEFS_RASE_AP_FULL_ACCESS |
      MPU_RASR_ENABLE_Msk), // Region 3 - Reset Clock CTRL
  mpu_region_disable(4);// Disabled unused regions
  mpu_region_disable(5);
  mpu_region_disable(6);
  mpu_region_disable(7);
  mpu_enable(0); // Enable the MPU with no additional option
  return 0; // No error
}
```

## 11.4 Memory barrier and MPU configuration

In the examples shown, we have added a number of memory barrier instructions in the MPU configuration code:

- DMB (Data Memory Barrier). This is used before disabling the MPU to ensure that there is no reordering of data transfers and if there is any outstanding transfer, we wait until the transfer is completed before writing to the MPU Control Register (MPU->CTRL) to disable the MPU.
- DSB (Data Synchronization Barrier). This is used after enabling the MPU to ensure that the subsequent ISB instruction is executed only after the write to the MPU Control Register is completed. This also ensures all subsequence data transfers use the new MPU settings.
- ISB (Instruction Synchronization Barrier). This is used after the DSB to ensure the processor pipeline is flushed and subsequent instructions are re-fetched with updated MPU settings.

The use of these memory barriers is based on architecture recommendations. Omitting these memory barriers on the Cortex®-M3 and Cortex-M4 processors does not cause any failure due to the simple nature of the processor pipeline: the processor can only handle one data transfer at any time. The only case where an ISB is really needed is when the MPU settings are updated and the subsequent instructions access can only be carried out using the new MPU settings.

However, from a software portability point of view, these memory barriers are important because their use allows the software to be reused on all Cortex-M processors.

If the MPU is used by an embedded OS, and the MPU configuration is done inside the context-switching operation, which is typically within the PendSV exception handler, the ISB instruction is not required from an architecture point of view because the exception entrance and exit sequence also have the ISB effect.

Additional information about the use of memory barriers on the Cortex-M processors can be found in ARM® application note 321, "A Programing Guide to the Memory Barrier Instruction for ARM Cortex-M Family Processors" (reference 9).

## 11.5 Using sub-region disable

The Sub-Region Disable (SRD) feature is used to divide an MPU region into eight equal parts and set each of them enabled or disabled individually. This feature can be used in a number of ways, as discussed in the following sections.

### 11.5.1 Allow efficient memory separation

The SRD enables more efficient memory usage while allowing protection to be implemented. For example, assuming that task A needs 5KB of stack and task B needs 3KB of stack, and the MPU is used to separate the stack space, the memory arrangement without the SRD feature will need 8KB for task A's stack and 4KB for task B's stack, as shown in Figure 11.5.

With the SRD, we can reduce the memory usage by overlapping the two memory regions, and use SRD to prevent the application task from accessing the other task's stack space, as shown in Figure 11.6.

### 11.5.2 Reduce the total number of regions needed

When defining peripheral access permissions, very often you might find that some peripherals need to be accessed by unprivileged tasks and some must be protected and have to be privileged access only. To implement the protection without SRD, we might need to use a large number of regions.

Since the peripherals usually have the same address size, we can easily apply SRD to define the access permissions. For example, we can define a region (or use the background region feature) to enable privileged accesses to all peripherals.

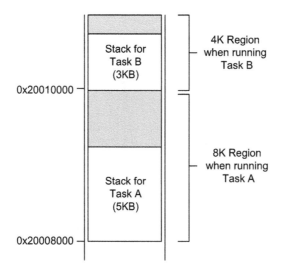

**FIGURE 11.5**

Without SRD, more memory space could be wasted because of region size and alignment requirements

Then define a higher numbered region that overlaps the peripheral address space as FULL ACCESS (accessible by unprivileged task), and use SRD to mask out the peripherals that have privileged access only. A simple illustration is shown in Figure 11.7.

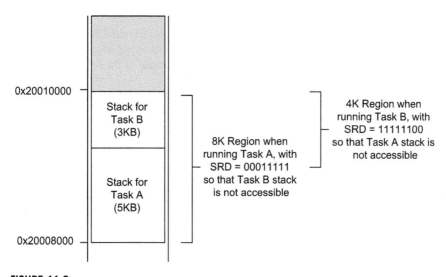

**FIGURE 11.6**

With SRD, regions can be overlapped but still separated for better memory usage efficiency

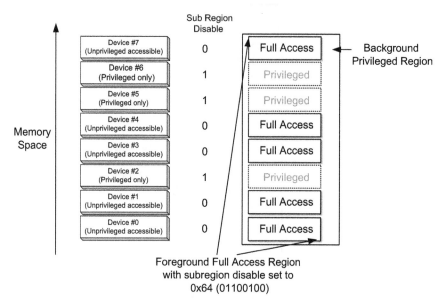

**FIGURE 11.7**

Using SRD to control access rights to separate peripherals

## 11.6 Considerations when using MPU

A number of aspects need to be considered when using the MPU. In many cases, when the MPU is used with an embedded OS, the OS being used needs to have MPU support built in. For example, a special version of FreeRTOS (called FreeRTOS-MPU, http://www.freertos.org), and the OpenRTOS from Wittenstein High Integrity Systems (http://www.highintegritysystems.com/) can make use of the MPU features.

### 11.6.1 Program code

In most cases, it can be impractical to try to partition the program memory into different MPU regions for different tasks because the tasks can share various functions, including run-time library functions and device-driver library functions. Also, if the application tasks and the OS are compiled together, it can be difficult to have clear and well-aligned address boundaries between each of the application tasks and the OS kernel, which is needed for setting up the MPU regions. Typically the program memory (e.g., flash) will be defined as just one region, and might be configured with read-only access permission.

### 11.6.2 Data memory

If an SRAM location is to be accessed by normal addressing as well as bit-band alias addresses, the MPU configuration needs to cover both address ranges.

If the application tasks and OS are compiled together in one go, it is likely that some of the data used by the application tasks and the OS will be mixed together. It is then impossible to isolate the access permissions of individual data elements. You might need to compile the tasks separately and then use linker scripts or other methods to place the data sections in the RAM manually. However, heap memory space might needed to be shared and cannot be protected using MPU.

Isolation of stack memory is usually easier to handle. You can reserve memory space in the linking stage and force the application tasks to use the reserved space for stack operations. Different embedded OSs and toolchains have different ways to allocate stack spaces.

### 11.6.3 Peripherals

Similar to data memory: if a peripheral is to be accessed by normal addressing as well as bit-band alias addresses, the MPU configuration needs to cover both address ranges.

## 11.7 Other usages of the MPU

Besides memory protection, the MPU can also be used for the following purposes:

- Disabling the write buffer inside the processor — in some cases, the write buffer inside the processor could make it a bit more difficult to identify software bugs because when a bus fault happen with a buffered write, the processor could

**Table 11.12** Comparison of MPU Features in Cortex-M0+ to Cortex-M3/M4

|  | ARMv6-M (Cortex-M0+) | ARMv7-M (Cortex-M3/M4) |
| --- | --- | --- |
| Number of regions | 8 | 8 |
| Unified I & D regions | Y | Y |
| Region address | Y | Y |
| Region size | 256 bytes to 4GB (can use SRD to get to 32 bytes) | 32 bytes to 4GB |
| Region memory attributes | S, C, B, XN | TEX, S, C, B, XN |
| Region Access Permission (AP) | Y | Y |
| Sub-Region Disable (SRD) | 8 bits | 8 bits |
| Background region | Yes (programmable) | Yes (programmable) |
| MPU bypass for NM/HardFault | Yes (programmable) | Yes (programmable) |
| Alias of MPU registers | N | Y |
| MPU registers accesses | Word size only | Word/Half-word/Byte |
| Fault Exception | HardFault only | HardFault/ MemManage |

have executed a number of instructions before the fault is detected. The MPU can be used to disable the internal write buffer by setting the Bufferable attribute to 0. (Note: You can use the Auxiliary Control Register to achieve this; see section 9.9.)

- Making a RAM memory space non-executable — in some applications, an external interface (e.g., Ethernet) can inject data into the buffer space allocated in RAM. The data could potentially contain malicious code and if the design contains other vulnerabilities, the malicious code injected into the system could be executed. The MPU can be used to force the RAM space to be eXecute Never (XN), which prevents data injected to the buffer from getting executed.

## 11.8 Comparing with the MPU in the Cortex®-M0+ processor

The Cortex®-M0+ processor has an optional MPU that is almost the same as the MPU in the Cortex-M3 and Cortex-M4 processors. There are a few differences, so if the MPU configuration software has to be used on Cortex-M0+ as well as on Cortex-M3 and Cortex-M4, the areas shown in Table 11.12 need to be taken care of.

The MPU memory attributes in ARMv6-M only support one level of cache policy. Therefore the TEX field is not available in the Cortex-M0+ processor. Overall the MPUs support mostly the same level of memory protection features, and the software porting between the two MPU types should be straightforward.

# Fault Exceptions and Fault Handling

## CHAPTER OUTLINE

The Definitive Guide to ARM® Cortex®-M3 and Cortex-M4 Processors. http://dx.doi.org/10.1016/B978-0-12-408082-9.00012-9

## 12.1 Overview of fault exceptions

Electronic systems can go wrong from time to time. The problems could be bugs in the software, but in many cases they can be caused by external factors such as:

- Unstable power supply
- Electrical noise (e.g., noise from power lines)
- Electromagnetic interference (EMI)
- Electrostatic discharge
- Extreme operation environment (e.g., temperature, mechanical vibrations)
- Wearing out of components (e.g., Flash/EEPROMs devices, crystal oscillators, capacitors) caused by repetitive programming or high-low temperature cycles
- Radiation (e.g., cosmic rays)
- Usage issues (e.g., end users didn't read the manual ☺) or invalid external data input

All these issues could lead to failure in the programs running on the processors. In many simple microcontrollers, you can find features like a watchdog timer and Brown-Out Detector (BOD). The watchdog can be programmed to trigger if the counter is not cleared within a certain time, and can be used to generate a reset or Non-Maskable Interrupt (NMI). The BOD can be used to generate a reset if the supply voltage drops to a certain critical level.

You can find a watchdog timer and BOD in many ARM® microcontrollers as well. However, when a failure occurs and the processor stops responding, it might take a bit of time for the watchdog to kick in. For most applications this is not a problem, but for some safety critical applications, a 1msec delay can be a matter of life or death.

In order to allow problems to be detected as early as possible, the Cortex®-M processors have a fault exception mechanism included. If a fault is detected, a fault exception is triggered and one of the fault exception handlers is executed.

By default, all the faults trigger the HardFault exception (exception type number 3). This fault exception is available on all Cortex-M processors including the Cortex-M0 and Cortex-M0+ processors. Cortex-M3 and Cortex-M4 processors have three additional configurable fault exception handlers:

- MemManage (Memory Management) Fault (exception type 4)
- Bus Fault (exception type 5)
- Usage Fault (exception type 6)

These exceptions are triggered if they are enabled, and if their priority is higher than the current exception priority level, as shown in Figure 12.1. These exceptions are called *configurable fault* exceptions, and have programmable exception priority levels (see section 7.9.5).

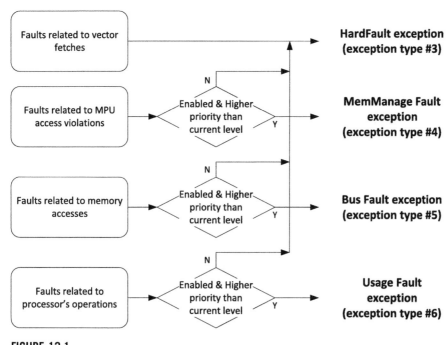

**FIGURE 12.1**

Fault exceptions available in ARMv7-M architecture

The fault handlers can be used in a number of ways:

- Shut down the system safely
- Inform users or other systems that it encountered a problem
- Carry out self-reset
- In the case of multi-tasking systems, the offending tasks could be terminated and restarted
- Other remedial actions can be carried out to try to fix the problem if possible (e.g., executing a floating point instruction with floating point unit turned off can cause an error, and can be easily solved by turning the floating point unit on)

Sometimes a system could carry out a number of different operations from the list above, depending on the type of fault detected.

To help detect what type of error was encountered in the fault handler, the Cortex-M3 and Cortex-M4 processors also have a number of Fault Status Registers (FSRs). The status bits inside these FSRs indicate the kind of fault detected. Although it might not pinpoint exactly when or where things went wrong, locating the source of the problem is made easier with these addition pieces of information. In addition, in some cases the faulting address is also captured by Fault Address Registers (FARs). More information about FSRs and FARs is given in section 12.4.

During software development, programming errors can also lead to fault exceptions. The information provided by the FARs can be very useful for software developers in identifying software issues in debugging.

The fault exception mechanism also allows applications to be debugged safely. For example, when developing a motor control system, you can shut down the motor by using the fault handlers before stopping the processor for debugging.

## 12.2 Causes of faults

### 12.2.1 Memory management (MemManage) faults

MemManage faults can be caused by violation of access rules defined by the MPU configurations. For example:

- Unprivileged tasks trying to access a memory region that is privileged access only
- Access to a memory location that is not defined by any defined MPU regions (except the Private Peripheral Bus (PPB), which is always accessible by privileged code)
- Writing to a memory location that is defined as read-only by the MPU

The accesses could be data accesses during program execution, program fetches, or stack operations during execution sequences. For instruction fetches that trigger a MemManage fault, the fault triggers only when the failed program location enters the execution stage.

For a MemManage fault triggered by stack operation during exception sequence:

- If the MemManage fault occurred during stack pushing in the exception entrance sequence, it is called a stacking error.
- If the MemManage fault occurrs during stack popping in the exception exit sequence, it is called an unstacking error.

The MemManage fault can also be triggered when trying to execute program code in eXecute Never (XN) regions such as the PERIPHERAL region, DEVICE region, or SYSTEM region (see section 6.9). This can happen even for Cortex®-M3 or Cortex-M4 processors without the optional MPU.

### 12.2.2 Bus faults

The bus faults can be triggered by error responses received from the processor bus interface during a memory access; for example:

- Instruction fetch (read), also called prefetch abort in traditional ARM® processors
- Data read or data write, also called data abort in in traditional ARM processors

In addition, the bus fault can also occur during stacking and unstacking of the exception handling sequence:

- If the bus error occurred during stack pushing in the exception entrance sequence, it is called a stacking error.
- If the bus error occurred during stack popping in the exception exit sequence, it is called an unstacking error.

If the bus error happened in the instruction fetch, the bus fault triggers only when the failed program location enters the execution stage. (A branch shadow access that triggers a bus error does not trigger the bus fault exception if the instruction does not enter execution stage.)

Please note that if a bus error is returned at vector fetch, the HardFault exception would be activated even when Bus Fault exception is enabled.

A memory system can return error responses if:

- The processor attempts to access an invalid memory location. In this case, the transfer is sent to a module in the bus system called default slave. The default slave returns an error response and triggers the bus fault exception in the processor.
- The device is not ready to accept a transfer (e.g., trying to access DRAM without initializing the DRAM controller might trigger the bus error. This behavior is device-specific.)
- The bus slave receiving the transfer request returns an error response. For example, it might happen if the transfer type/size is not supported by the bus slave, or if the peripherals determined that the operation carried out is not allowed.
- Unprivileged access to the Private Peripheral Bus (PPB) that violates the default memory access permission (see section 6.8).

Bus faults can be classified as:

- Precise bus faults — fault exceptions happened immediately when the memory access instruction is executed.
- Imprecise bus faults — fault exceptions happened sometime after the memory access instruction is executed.

The reason for a bus fault to become imprecise is due to the presence of write buffers in the processor bus interface (Figure 6.17). When the processor writes data to a bufferable address (see section 6.9 on memory access attributes, and section 11.2.5 on MPU Base Region attribute and Size register), the processor can proceed to execute the next instruction even if the transfer takes a number of clock cycles to complete.

The write buffer allows the processor to have higher performance, but this can make debugging a bit harder because by the time the bus fault exception is triggered, the processor could have executed a number of instructions, including branch instructions. If the branch target can be reached via several paths (Figure 9.16), it could be hard to tell where the faulting memory access took place unless you have an instruction trace (see Chapter 14, section 14.3.5). To help with debugging such situations, you can disable the write buffer using the DISDEFWBUF bit in the Auxiliary Control register (section 9.9).

Read operations and accesses to the Strongly Order region (e.g., Private Peripheral Bus, PPB) are always precise in the Cortex®-M3 and Cortex-M4 processors.

## 12.2.3 Usage faults

The Usage Fault exception can be caused by a wide range of factors:

- Execution of an undefined instruction (including trying to execute floating point instructions when the floating point unit is disabled).
- Execution of Co-processor instructions — the Cortex®-M3 and Cortex-M4 processors do not support Co-processor access instructions, but it is possible to use the usage fault mechanism to emulate co-processor instruction support.
- Trying to switch to ARM® state — classic ARM processors like ARM7TDMI™ support both ARM instruction and Thumb instruction sets, while Cortex-M processors only support Thumb ISA. Software ported from classic ARM processors might contain code that switches the processor to ARM state, and software could potentially use this feature to test whether the processor it is running on supports ARM code.
- Invalid EXC_RETURN code during exception-return sequence (see section 8.1.4 for details of EXC_RETURN code). For example, trying to return to Thread level with exceptions still active (apart from the current serving exception).
- Unaligned memory access with multiple load or multiple store instructions (including load double and store double; see section 6.6).
- Execution of SVC when the priority level of the SVC is the same or lower than current level.
- Exception return with Interrupt-Continuable Instruction (ICI) bits in the unstacked xPSR, but the instruction being executed after exception return is not a multiple-load/store instruction.

It is also possible, by setting up the Configuration Control Register (CCR; see sections 9.8.4 and 9.8.5) to generate usage faults for the following:

- Divide by zero
- All unaligned memory accesses

Please note that the floating point instructions supported by the Cortex-M4 are not co-processor instructions (e.g., MCR, MRC; see section 5.6.15). However, slightly confusingly, the register that enables the floating point unit is called the Coprocessor Access Control Register (CPACR; see section 9.10).

## 12.2.4 HardFaults

As illustrated in Figure 12.1, the HardFault exception can be triggered by escalation of configurable fault exceptions. In addition, the HardFault can be triggered by:

- Bus error received during a vector fetch
- Execution of breakpoint instruction (BKPT) with a debugger attached (halt debugging not enabled) and debug monitor exception (see section 14.3) not enabled

Note: In some development tool chains, breakpoints are used by the debugger to carry out semi-hosting. For example, when reaching a "printf" operation, the processor executes a BKPT instruction and halt, and the debugger can detect the halt and check the register status and the immediate value in the BKPT instruction. Then the debugger can display the message or character form the message in the printf statement. If the debugger is not attached, such operation results in HardFault and executes the HardFault exception handler.

## 12.3 Enabling fault handlers

By default the configurable fault exceptions are disabled. You can enable these exceptions by writing to System Handler Control and State Register (SCB->SHCSR). Be careful not to change the current status of system exception active status, since this can cause a fault exception.

### 12.3.1 MemManage fault

You can enable the MemManage Fault exception handler using:

```
SCB->SHCSR |= SCB_SHCSR_MEMFAULTENA_Msk; //Set bit 16
```

The default name for MemManage Fault exception handler (as defined in CMSIS-Core) is:

```
void MemManage_Handler(void);
```

You can set up the priority of the MemManage Fault using:

```
NVIC_SetPriority(MemoryManagement_IRQn, priority);
```

### 12.3.2 Bus fault

You can enable the Bus Fault exception handler using:

```
SCB->SHCSR |= SCB_SHCSR_BUSFAULTENA_Msk; //Set bit 17
```

The default name for the Bus Fault exception handler (as defined in CMSIS-Core) is:

```
void BusFault_Handler(void);
```

You can set up the priority of the Bus Fault using:

```
NVIC_SetPriority(BusFault_IRQn, priority);
```

### 12.3.3 Usage fault

You can enable the Usage Fault exception handler using:

```
SCB->SHCSR |= SCB_SHCSR_USGFAULTENA_Msk; //Set bit 18
```

The default name for the Usage Fault exception handler (as defined in CMSIS-Core) is:

```
void UsageFault_Handler(void);
```

You can set up the priority of the Usage Fault using:

```
NVIC_SetPriority(UsageFault_IRQn, priority);
```

### 12.3.4 HardFault

There is no need to enable the HardFault handler. This is always enabled and has a fixed exception priority of $-1$. The default name for the Hard Fault exception handler (as defined in CMSIS-Core) is:

```
void HardFault_Handler(void);
```

## 12.4 Fault status registers and fault address registers

### 12.4.1 Summary

The Cortex®-M3 and Cortex-M4 processors have a number of registers that are used for fault analysis. They can be used by the fault handler code, and in some cases, by the debugger software running on the debug host for displaying fault status. A summary of these registers is shown in Table 12.1. These registers can only be accessed in privileged state.

The Configurable Fault Status Register (CFSR) can be further divided into three parts, as show in Table 12.2. Besides accessing CFSR as a 32-bit word, each part of the CFSR can be accessed using byte and half-word transfers. There is no CMSIS-Core symbol for the divided MMSR, BFSR, and UFSR.

**Table 12.1** Registers for Fault Status and Address Information

| Address | Register | CMSIS-Core Symbol | Function |
|---------|----------|-------------------|----------|
| 0xE000ED28 | Configurable Fault Status Register | SCB->CFSR | Status information for Configurable faults |
| 0xE000ED2C | HardFault Status Register | SCB->HFSR | Status for HardFault |
| 0xE000ED30 | Debug Fault Status Register | SCB->DFSR | Status for Debug events |
| 0xE000ED34 | MemManage Fault Address Register | SCB->MMFAR | If available, showing accessed address that triggered the MemManage fault |
| 0xE000ED38 | BusFault Address Register | SCB->BFAR | If available, showing accessed address that triggered the bus fault |
| 0xE000ED3C | Auxiliary Fault Status Register | SCB->AFSR | Device-specific fault status |

**Table 12.2** Dividing Configurable Fault Status Register (SCB->CFSR) Into Three Parts

| Address | Register | Size | Function |
|---------|----------|------|----------|
| 0xE000ED28 | MemManage Fault Status Register (MMFSR) | Byte | Status information for MemManage Fault |
| 0xE000ED29 | Bus Fault Status Register (BFSR) | Byte | Status for Bus Fault |
| 0xE000ED2A | Usage Fault Status Register (UFSR) | Halfword | Status for Usage Fault |

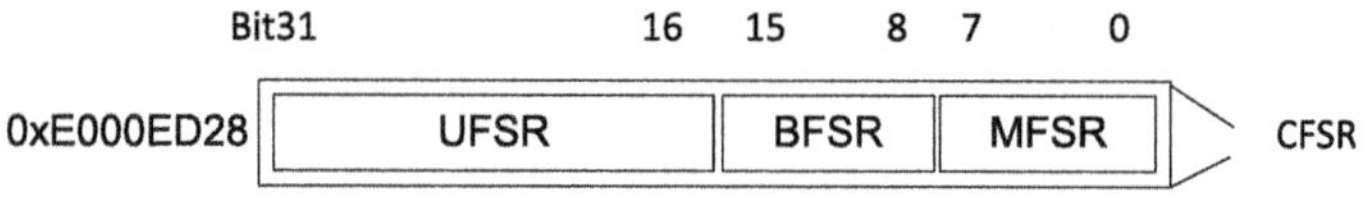

```
         Bit31              16  15    8  7      0
0xE000ED28 ┌──────────────┬┬───────┬┬───────┐
           │     UFSR     ││  BFSR ││  MFSR │ ▷  CFSR
           └──────────────┴┴───────┴┴───────┘
```

**FIGURE 12.2**

Configurable Fault Status Register partitioning

## 12.4.2 Information for MemManage fault

The programmer's model for the MemManage Fault Status Register is shown in Table 12.3.

Each fault indication status bit (not including MMARVALID) is set when the fault occurs, and stays high until a value of 1 is written to the register.

If the MMFSR indicates that the fault is a data access violation (DACCVIOL set to 1) or an instruction access violation (IACCVIOL set to 1), the faulting code can be located by the stacked program counter in the stack frame.

If the MMARVALID bit is set, it is also possible to determine the memory location that caused the fault by using the MemManage Fault Address Register (SCB->MMFAR).

**Table 12.3** MemManage Fault Status Register (lowest byte in SCB->CFSR)

| Bits | Name | Type | Reset Value | Description |
|------|------|------|-------------|-------------|
| 7 | MMARVALID | – | 0 | Indicates the MMFAR is valid |
| 6 | – | – | – (read as 0) | Reserved |
| 5 | MLSPERR | R/Wc | 0 | Floating point lazy stacking error (available on Cortex®-M4 with floating point unit only) |
| 4 | MSTKERR | R/Wc | 0 | Stacking error |
| 3 | MUNSTKERR | R/Wc | 0 | Unstacking error |
| 2 | – | – | – (read as 0) | Reserved |
| 1 | DACCVIOL | R/Wc | 0 | Data access violation |
| 0 | IACCVIOL | R/Wc | 0 | Instruction access violation |

MemManage faults which occur during stacking, unstacking, and lazy stacking (see sections 8.3.6 and 13.3) are indicated by MSTKERR, MUNSTKERR, and MLSPERR, respectively.

### 12.4.3 Information for bus fault

The programmer's model for the Bus Fault Status Register is shown in Table 12.4. Each fault indication status bit (not including BFARVALID) is set when the fault occurs, and stays high until a value of 1 is written to the register.

The IBUSERR indicates that the bus fault is caused by a bus error during an instruction fetch. Both PRECISERR and IMPRECISERR are for data accesses. PRECISERR indicates a precise bus error (see section 12.3.2), and the faulting instruction can be located from the stacked program counter value. The address of the faulting data access is also written to the Bus Fault Address Register (SCB->BFAR); however, the fault handler should still check if BFARVALID is still 1 after reading BFAR.

If the bus fault is imprecise (IMPRECISERR set to 1), the stacked program counter does not reflect the faulting instruction address, and the address of the faulting transfer will not show in the BFAR.

Bus faults occurring during stacking, unstacking, and lazy stacking (see sections 8.3.6 and 13.3) are indicated by STKERR, UNSTKERR, and LSPERR, respectively.

### 12.4.4 Information for usage fault

The programmer's model for the Usage Fault Status Register is shown in Table 12.5.

Each fault indication status bit is set when the fault occurs, and stays high until a value of 1 is written on the register.

Appendix I shows a breakdown of possible reasons for each type of usage fault.

**Table 12.4** Bus Fault Status Register (2nd byte in SCB->CFSR)

| CFSR Bits | Name | Type | Reset Value | Description |
|---|---|---|---|---|
| 15 | BFARVALID | – | 0 | Indicates BFAR is valid |
| 14 | – | – | – | – |
| 13 | LSPERR | R/Wc | 0 | Floating point lazy stacking error (available on Cortex®-M4 with floating point unit only) |
| 12 | STKERR | R/Wc | 0 | Stacking error |
| 11 | UNSTKERR | R/Wc | 0 | Unstacking error |
| 10 | IMPRECISERR | R/Wc | 0 | Imprecise data access error |
| 9 | PRECISERR | R/Wc | 0 | Precise data access error |
| 8 | IBUSERR | R/Wc | 0 | Instruction access error |

**Table 12.5** Usage Fault Status Register (Upper half-word in SCB->CFSR)

| CFSR Bits | Name | Type | Reset Value | Description |
|---|---|---|---|---|
| 25 | DIVBYZERO | R/Wc | 0 | Indicates a divide by zero has taken place (can be set only if DIV_0_TRP is set) |
| 24 | UNALIGNED | R/Wc | 0 | Indicates that an unaligned access fault has taken place |
| 23:20 | – | – | – | – |
| 19 | NOCP | R/Wc | 0 | Attempts to execute a coprocessor instruction |
| 18 | INVPC | R/Wc | 0 | Attempts to do an exception with a bad value in the EXC_RETURN number |
| 17 | INVSTATE | R/Wc | 0 | Attempts to switch to an invalid state (e.g., ARM) |
| 16 | UNDEFINSTR | R/Wc | 0 | Attempts to execute an undefined instruction |

### 12.4.5 HardFault status register

The programmer's model for the Usage Fault Status Register is shown in Table 12.6.

HardFault handler can use this register to determine whether a HardFault is caused by any of the configurable faults. If the FORCED bit is set, it indicates that the fault has been escalated from one of the configurable faults and it should check the value of CFSR to determine the cause of the fault.

Similar to other fault status registers, each fault indication status bit is set when the fault occurs, and stays high until a value of 1 is written to the register.

### 12.4.6 Debug fault status register (DFSR)

Unlike other fault status registers, the DFSR is intended to be used by debug tools such as a debugger software running on a debug host (e.g., a personal computer), or a

**Table 12.6** Hard Fault Status Register (0xE000ED2C ,SCB->HFSR)

| Bits | Name | Type | Reset Value | Description |
|---|---|---|---|---|
| 31 | DEBUGEVT | R/Wc | 0 | Indicates hard fault is triggered by debug event |
| 30 | FORCED | R/Wc | 0 | Indicates hard fault is taken because of bus fault, memory management fault, or usage fault |
| 29:2 | – | – | – | – |
| 1 | VECTBL | R/Wc | 0 | Indicates hard fault is caused by failed vector fetch |
| 0 | – | – | – | – |

**Table 12.7** Debug Fault Status Register (0xE000ED30 ,SCB->DFSR)

| Bits | Name | Type | Reset Value | Description |
|------|------|------|-------------|-------------|
| 31:5 | - | - | - | Reserved |
| 4 | EXTERNAL | R/Wc | 0 | Indicates the debug event is caused by an external signal (the EDBGRQ signal is a input on the processor, typically used in multi-processor design for synchronized debug). |
| 3 | VCATCH | R/Wc | 0 | Indicates the debug event is caused by a vector catch, a programmable feature that allows the processor to halt automatically when entering certain type of system exception including reset. |
| 2 | DWTTRAP | R/Wc | 0 | Indicates the debug event is caused by a watchpoint |
| 1 | BKPT | R/Wc | 0 | Indicates the debug event is caused by a breakpoint |
| 0 | HALTED | R/Wc | 0 | Indicates the processor is halted is by debugger request (including single step) |

debug agent software running on the microcontroller to determine what debug event has occurred.

The programmer's model for the Debug Fault Status Register is shown in Table 12.7.

Similar to other fault status registers, each fault indication status bit is set when the fault occurs, and stays high until a value of 1 is written to the register.

## 12.4.7 Fault address registers MMFAR and BFAR

When a MemManage fault or a bus fault occurs, you might be able to determine the address of the transfer that triggered the fault using MMFAR or BFAR registers.

The programmer's model for the MMFAR Register is shown in Table 12.8.

The programmer's model for the BFAR Register is shown in Table 12.9.

**Table 12.8** MemManage Fault Address Register (0xE000ED34, SCB->MMFAR)

| Bits | Name | Type | Reset Value | Description |
|------|------|------|-------------|-------------|
| 31:0 | ADDRESS | R/W | Unpredictable | When the value of MMARVALID is 1, this field holds the address of the address location that generates the MemManage fault. |

**Table 12.9** Bus Fault Address Register (0xE000ED38, SCB->BFAR)

| Bits | Name | Type | Reset Value | Description |
|------|------|------|-------------|-------------|
| 31:0 | ADDRESS | R/W | Unpredictable | When the value of BFARVALID is 1, this field holds the address of the address location that generates the Bus Fault. |

Inside the Cortex®-M3 and Cortex-M4 processors, the MMFAR and BFAR shared the same physical hardware. This reduces the silicon size of the processor. Therefore only one of the MMARVALID or BFARVALID can be 1 at a time. As a result, if one of the fault exceptions is pre-empted by another due to a new fault exception, the value in the MMFAR or BFAR could have become invalid. To ensure that the fault handlers are getting the accurate fault address information, it should:

1. First read the value of MMFAR (for MemManage fault), or BFAR (for bus fault), then
2. Read the value of MMFSR (for MemManage fault), or BFSR (for bus fault), to see if MMARVALID or BFARVALID is still 1. If they are still 1, then the fault address is valid.

Note that if an unaligned access faults, the address in the MMFAR is the actual address that faulted. The transfer is divided into a number of aligned transfers by the processor, and the MMFAR can be any value in the address range of these aligned transfers. For bus fault with BFARVALID set, the BFAR indicates the address requested by the instruction, but can be different from the actual faulting address. For example, in a system with a valid 64KB SRAM address 0x20000000 to 0x2000FFFF, a word-size access to 0x2000FFFE might fault in the second half-word at address 0x20010000. In this case, BFAR showing 0x2000FFFE is still in the valid address range.

## 12.4.8 Auxiliary fault status register

The AFSR was added from Cortex®-M3 r2p0 onwards. It allows silicon designers to add their own fault status information. The programmer's model for the AFSR Register is shown in Table 12.10.

Similar to other fault status registers, each fault indication status bit is set when the fault occurs, and stays high until a value of 1 is written to the register.

**Table 12.10** Auxiliary Fault Status Register (0xE000ED3C, SCB->AFSR)

| Bits | Name | Type | Reset Value | Description |
|------|------|------|-------------|-------------|
| 31:0 | Implementation Defined | R/W | 0 | Implementation defined fault status |

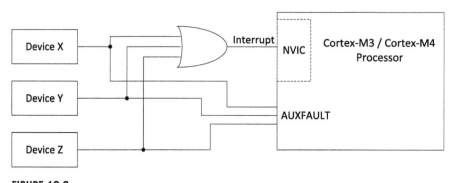

**FIGURE 12.3**

Additional fault generation sources connected to the processor via AUXFAULT and interrupt

On the processor interface, a 32-bit input (AUXFAULT) is available for silicon designers to connect to various devices that can be used to generate fault events, as shown in Figure 12.3.

When a fault event happens in one of these devices, it triggers an interrupt, and the interrupt handler can use the AFSR to determine which device generated the fault. Since this is not intended to be used for general interrupt processing, using the software to determine the cause of fault does not cause a latency issue.

## 12.5 Analyzing faults

It is not uncommon to encounter fault exceptions during software development. In some cases it can be a bit of challenge to find out what went wrong. In most cases, the information provided by the fault status registers and fault address registers is certainly very useful. In addition, more information can be obtained using various techniques and tools including:

- Stack Trace: After a fault exception is triggered, we can halt the processor and examine the processor status and the memory contents either by using breakpoint hardware, or manually inserting a breakpoint instruction. Besides the current register values, we can trace the stacked register values including the stacked Program Counter (PC) from the stack pointers. Combining the stacked PC and the fault status registers values can very often lead you to the right answers fairly quickly.
- Event Trace: The data trace feature in the Cortex®-M3 and Cortex-M4 processors allows you to collect exception history using low-cost debuggers. The exception trace can be output through the single pin Serial Wire Output pin (see Chapter 14). If a program failure is related to exception handling, the event trace feature allows you to see which exceptions occurred before the failure and hence make it easier to locate the issue.

- Instruction Trace: Use the Embedded Trace Macrocell (ETM) to collect information about instruction executed, and display it on a debugger to identify the processor operations before the failure. This requires a debugger with Trace Port capture function.

In typical stack trace operations, we can add a breakpoint to the beginning of the HardFault handler (or other configurable handlers if they are used). When a fault occurs, the processor enters the fault handler and halts.

First, we need to determine which stack pointer was being used when the fault occurred. In the majority of the applications without an OS, only the Main Stack Pointer (MSP) would be used. However, if the application uses PSP, we need to determine the SP used by checking bit 2 of the Link Register (LR), as shown in Figure 12.4.

From the stack pointer value, we can easily locate the stacked registers like stacked PC (return address) and stacked xPSR:

- In many cases the stacked PC provides the most important hint for debugging the fault. By generating a disassembled code listing of the program image in the toolchain, you can easily pinpoint the code fragment where the fault occurred, and understand the failure from the information provided in the fault status registers, and the current and stacked register values.
- Stacked xPSR can be useful for identifying if the processor was in handler mode when the fault occurred, and whether there has been an attempt to switch the processor into ARM® state (if the T-bit in the EPSR is cleared, there has been an attempt to switch the processor into ARM state).

Finally, the LR value when entering the fault handler might also provide hints about the cause of the fault. In the case of faults caused by invalid EXC_RETURN values, the value of LR when the fault handler is entered shows the previous LR value when the fault occurred. The fault handler can report the faulty LR value, and software programmers can then use this information to check why the LR ends up with an illegal return value.

In some debug tools, the debugger software contains features which allow you to access fault status information easily. For example, in Keil™ MDK-ARM, you can access to the fault status registers using the "Fault Report" window, as shown in Figure 12.5. This can be accessed from the pull-down menu "Peripherals"-> "Core Peripherals" -> "Fault Reports."

Various trace features in the debugger can also help to identify the source of the problem in your application code. More information on this is covered in Chapter 14. There is also an application note on the Keil website about debugging fault exceptions: Application Note 209 "Using Cortex-M3 and Cortex-M4 Fault Exceptions" (http://www.keil.com/appnotes/docs/apnt_209.asp).

Other debug tools also have the debug feature to assist fault analysis. For example, the debugger in the Atollic TrueStudio has a Fault Analyzer feature; it extracts information from the processor such as the fault status registers to identify the reasons that caused the fault.

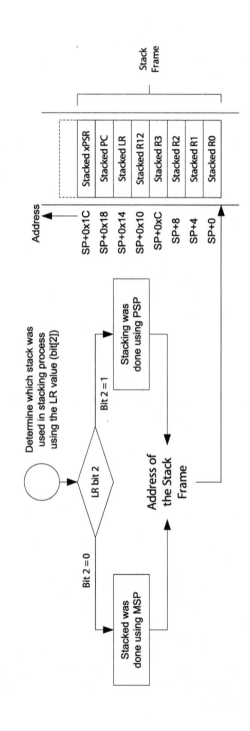

**FIGURE 12.4**

Stack trace flow to locate stack frame and stacked registers

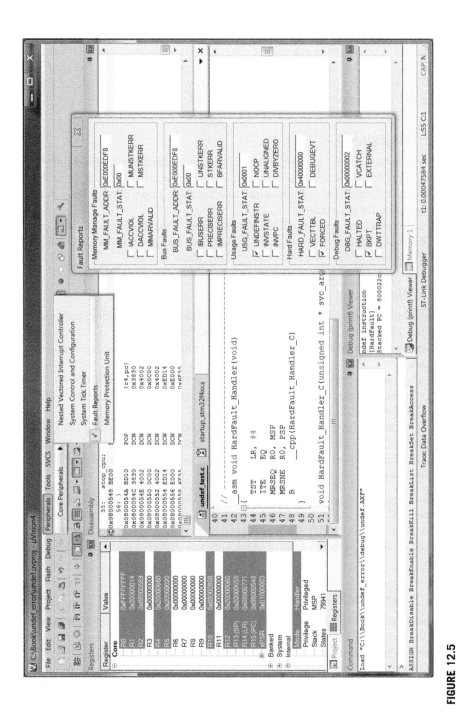

**FIGURE 12.5**

Fault Reports window in Keil MDK-ARM showing the fault status registers

## 12.6 Faults related to exception handling

In some cases, faults can be generated during exception handling. The most common case is incorrect stack setup; for example, the stack space reserved is too small and causes the stack space to run out. In this section we will look at what could have gone wrong and what fault exception could be triggered.

### 12.6.1 Stacking

During exception entry, a number of registers are pushed to the stack. Potentially this can trigger a Bus Fault if the memory system returns an error response, or a Mem-Manage fault if the MPU is programmed and the stack grows beyond the allocated memory space for the stack.

If a bus error is received, the STKERR bit (bit 4) in the Bus Fault Status Register (BFSR) is set. If an MPU violation is detected, the error is indicated by the MSTKERR bit (bit 4) in the MemManage fault status register.

### 12.6.2 Unstacking

During exception exits, the processor restores register values by reading back values from the stack frame. It is possible to trigger a Bus Fault if the memory system returned a bus error response, or a MemManage fault if the MPU detected an access violation.

If a bus error is received, the UNSTKERR bit (bit 3) in the BFSR is set. If an MPU violation is detected, the error is indicated by the MUNSTKERR bit (bit 3) in the MemManage fault status register.

It is uncommon to get an unstacking error without getting a stacking error. If the Stack Pointer (SP) value was incorrect, in most cases the fault would have happened during stacking. However, it is not impossible to get an unstacking fault without a stacking fault. For example, this can happen if:

- The value of SP was changed during the execution of the exception handler
- The MPU configuration was changed during the execution of the exception handler
- The value of EXC_RETURN was changed during the execution of the exception handler, so the SP being used in unstacking was different from the one used in stacking

### 12.6.3 Lazy stacking

For the Cortex®-M4 processor with floating point unit, Bus Fault and MemManage Fault could be triggered during lazy stacking. The lazy stacking feature allows the stacking of floating point registers to be deferred, and only push those registers to the allocated space if the exception handler uses the floating point unit. When this happens, the processor pipeline is stalled and carries out the stacking, and then executes the floating point instruction after the stacking is completed.

If a bus error is received during the lazy stacking operation, the Bus Fault exception is triggered and the error is indicated by LSPERR (bit 5) of the Bus Fault Status Register. If a MPU access violation occurs, the MemManage Fault exception is triggered and the error is indicated by MLSPERR (bit 5) of the MemManage Fault Status Register.

### 12.6.4 Vector fetches

If a bus error takes place during a vector fetch, the HardFault exception will be triggered, and the error will be indicated by the VECTTBL (bit 1) of the Hard Fault Status Register. The MPU always permits vector fetches and therefore there is no MPU access violation for vector fetches. If a vector fetch error occurs, one thing that needs checking is the value of the VTOR to see whether the vector table has been relocated to the correct address range.

If the LSB of the exception vector is 0, it indicates an attempt to switch the processor to the ARM® state (use ARM instructions instead of Thumb instructions), and this is not supported in Cortex®-M processors. When this happens, the processor will trigger a Usage Fault at the first instruction of the exception handler, with INVSTATE bit (bit 1) if the Usage Fault Status Register is set to 1 to indicate the error.

### 12.6.5 Invalid returns

If the EXC_RETURN value is invalid or does not match the state of the processor (as in using 0xFFFFFFF1 to return to Thread mode), it will trigger a Usage Fault. The bits INVPC (bit 2) or INVSTATE (bit 1) of the Usage Fault Status Register will be set, depending on the actual cause of the fault.

### 12.6.6 Priority levels and stacking or unstacking faults

Configurable fault handlers have programmable priority levels. If a fault happens and the current priority level is the same or higher than the associated configurable fault handler, the fault event is escalated to the HardFault exception, which has a fixed priority of −1.

If a stacking or unstacking error occurs during an exception sequence, the current priority level is based on the priority level of the interrupted process/task, as shown in Figure 12.6.

If the Bus Fault or MemManage Fault exception has the same or lower priority than the current priority level, the HardFault exception will be executed first.

If the Bus Fault or MemManage Fault exception is enabled and has higher priority than both the current level and the priority level of the exception to be serviced, then the Bus Fault or MemManage Fault exception would be executed first.

If the Bus Fault or MemManage Fault exception is enabled and has a priority level between the current level and the exception to be serviced, the handler for

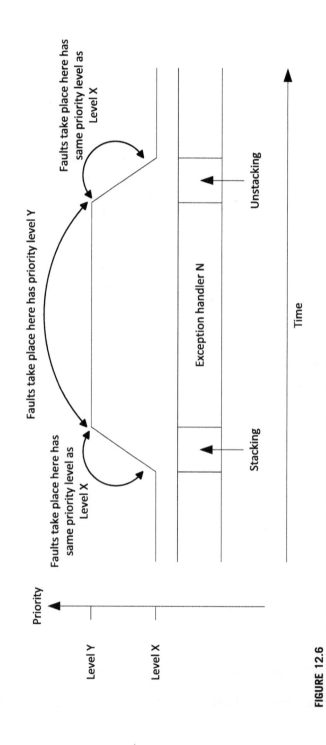

**FIGURE 12.6**

Priority level at stacking and unstacking

the exception to be serviced is executed first, and the Bus Fault or MemMange Fault handler is executed afterwards.

## 12.7 Lockup
### 12.7.1 What is lockup?

When an error condition occurs, one of the fault handlers will be triggered. If another fault happens inside a configurable fault handler, then either another configurable fault handler is triggered (if the fault is a different kind and the other fault handler and has higher priority than the current level), or the Hard Fault handler is triggered and executed. However, what happens if another fault happens during the execution of the HardFault handler? (This is a very unlucky situation, but it can happen.) In this case, a lockup will take place.

Lockup can happen if:

- A fault occurs during execution of the HardFault or NMI (Non-Maskable Interrupt) exception handler
- A bus error occurs during vector fetch for HardFault or NMI exceptions
- Trying to execute SVC instruction in the HardFault or NMI exception handler
- Vector fetch at startup sequence

During lockup, the processor stops program execution and asserts an output signal called LOCKUP. How this signal is used depends on the microcontroller design; in some cases it can be used to generate a system reset automatically. If the lockup is caused by an error response from the bus system, the processor might retry the access continuously, or if the fault is unrecoverable it could force the program counter to 0xFFFFFFFX and might keep fetching from there.

If the lockup is caused by a fault event inside the HardFault handler (double fault condition), the priority level of the processor is still at priority level −1, and it is still possible for the processor to respond to an NMI (priority level −2) and execute the NMI handler. But after the NMI handler finishes, it will return to the lockup state and the priority level will return to −1.

There are various ways to exit the lockup state:

- System reset or power on reset
- The debugger can halt the processor and clear the errors (e.g., using reset or clearing current exception handling status, update program counter value to a new starting point, etc.)

Typically a system reset is the best method as it ensures that the peripherals and all interrupt handling logic returns to the reset state.

You might wonder why we do not simply reset the processor when a lockup takes place. It might be good for a live system, but during software development, we should first try to find the cause of the problem. If we reset the system automatically, it will be impossible to analyze what went wrong because the hardware status will change.

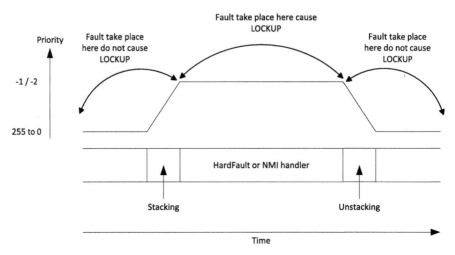

**FIGURE 12.7**

Only a Fault occurring during a HardFault or NMI handler will cause Lockup

The Cortex®-M processor designs export the lockup status to its interface and chip designers can implement a programmable auto reset feature, so that when this auto reset feature is enabled, the system can reset itself automatically.

Note that a bus error or MPU access violation occurs during stacking or unstacking (except vector fetch) when entering a HardFault handler or NMI handler does not cause the system to enter lockup state (see Figure 12.7). However, the Bus Fault exception could end up in a pending state and execute after the HardFault handler.

## 12.7.2 Avoiding lockup

In some applications, it is important to avoid lockup, and extra care is needed when developing the HardFault handler and NMI handler. For example, we might need to avoid stack memory access unless we know that the stack pointer is still in a valid memory range. For example, if the starting of the HardFault handler has a stack push operation and the MSP (Main Stack Pointer) was corrupted and pointed to invalid memory location, we could end up entering lockup state immediately at the start of the HardFault handler:

```
HardFault_Handler
     PUSH {R4-R7,LR}  ; Bad idea unless you are sure that the
                      ; stack is safe to use!

     . . .
```

Even if the stack pointer is in a valid memory range, we might still need to reduce the amount of stack used by the HardFault and NMI handler, if the available stack size is small.

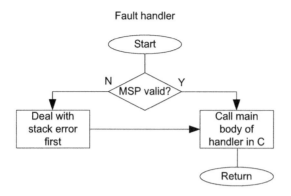

**FIGURE 12.8**

Adding a SP value check in fault handlers

In safety critical systems, we can add an assembly wrapper code for fault handlers (Figure 12.8) to check if the value of MSP is still in a valid range before calling the fault handlers in C code, which might have stack operations inserted by the C compilers.

One approach for developing HardFault and NMI handlers is to carry out only the essential tasks inside the handlers, and the rest of the tasks, such as error reporting, can be pended using a separate exception such as PendSV. This helps to ensure that the HardFault handler or NMI is small and robust.

Furthermore, we should ensure that the NMI and HardFault handler code will not try to use the SVC instruction. Since SVC always has lower priority than HardFault and NMI, using SVC in these handlers will cause lockup. This might look simple, but when your application is complex and you call functions from different files in your NMI and HardFault handlers, you might accidentally call a function that contains an SVC instruction. Therefore before you develop your software with SVC, you need to plan the SVC service implementation carefully.

## 12.8 Fault handlers

### 12.8.1 HardFault handler for debug purposes

A number of activities can be carried out in fault handlers, for example:

- Shutdown the system safely
- Report errors
- Self-reset
- Carry out remedial actions (if possible)
- For an OS environment, the task that triggered the fault can be terminated and restarted.

- Optionally clear the fault status in the fault status registers. This should be included in the fault handler if it carries out remedial action and resumes normal operations.

The implementations for most of these tasks are dependent on the application. The self-reset operation is discussed in section 9.6. Here, we will have a look at reporting information about the fault. One of the common ways to do this is to create a HardFault handler that:

- Reports that the HardFault happened
- Reports the Fault Status Register and Fault Address Register values
- Reports additional information from the stack frame

The following example HardFault handler assumes that you have some way to display messages generated from "printf" in C (see Chapter 18 for details). As mentioned in section 12.5, we will need to check the value of EXC_RETURN code to determine if MSP or PSP was used for the stacking.

In order to extract the value of EXC_RETURN from LR and to locate the starting of stack frame from MSP/PSP, we need a small assembly code wrapper. This extracts the starting address of the stack frame and passes it to the second part of the HardFault handler, which is programmed in C. This wrapper also passes the EXC_RETURN value as the second parameter:

```
/* Assembly wrapper for Keil™ MDK, ARM® Compilation tool chain
(including DS-5™ Professional and RealView® Development Suite) */
// -----------------------------------------------------------------
   ---------
// Hard Fault handler wrapper in assembly
// It extracts the location of stack frame and passes it to handler
// in C as a pointer. We also extract the LR value as second
// parameter.
__asm void HardFault_Handler(void)
{
    TST     LR, #4
    ITE     EQ
    MRSEQ   R0, MSP
    MRSNE   R0, PSP
    MOV     R1, LR
    B       __cpp(HardFault_Handler_C)
}
```

For users of gcc, you can create a separate assembly file to do the same thing:

```
/* Assembly file for gcc */
 .text
 .syntax unified
 .thumb
 .type   HardFault_Handler, %function
```

```
  .global HardFault_Handler
  .global HardFault_Handler_c

HardFault_Handler:
  tst    lr, #4
  ite    eq
  mrseq  r0, msp /* stacking was using MSP */
  mrseq  r0, psp /* stacking was using PSP */
  mov    r1, lr /* second parameter */
  ldr    r2,=HardFault_Handler_c
  bx     r2
  .end
```

And for users of IAR Embedded Workbench (many thanks for various Cortex[®]-M users for porting the example to IAR and posting it to the Internet[1]):

```
// Assembly wrapper for IAR Embedded Workbench
// ----------------------------------------------------------------
    ---------
// Hard Fault handler wrapper in assembly
// It extracts the location of stack frame and passes it to handler
// in C as a pointer. We also extract the LR value as second
// parameter.
void HardFault_Handler(void)
{
__asm("TST LR, #4");
__ASM("ITE EQ");
__ASM("MRSEQ R0, MSP");
__ASM("MRSNE R0, PSP");
__ASM("MOV R1, LR");
__ASM("B HardFault_Handler_C");
}
```

The second part of the HardFault handler is coded in C; it displays the fault status registers, fault address register, and the contents in the stack frame:

```
// Second part of the HardFault handler in C
void HardFault_Handler_C(unsigned long * hardfault_args, unsigned
int lr_value)
 {
  unsigned long stacked_r0;
  unsigned long stacked_r1;
  unsigned long stacked_r2;
  unsigned long stacked_r3;
```

---

[1]See http://blog.frankvh.com/2011/12/07/cortex-m3-m4-hard-fault-handler/

```
        unsigned long stacked_r12;
        unsigned long stacked_lr;
        unsigned long stacked_pc;
        unsigned long stacked_psr;
        unsigned long cfsr;
        unsigned long bus_fault_address;
        unsigned long memmanage_fault_address;

        bus_fault_address       = SCB->BFAR;
        memmanage_fault_address = SCB->MMFAR;
        cfsr                    = SCB->CFSR;

        stacked_r0 = ((unsigned long) hardfault_args[0]);
        stacked_r1 = ((unsigned long) hardfault_args[1]);
        stacked_r2 = ((unsigned long) hardfault_args[2]);
        stacked_r3 = ((unsigned long) hardfault_args[3]);
        stacked_r12 = ((unsigned long) hardfault_args[4]);
        stacked_lr = ((unsigned long) hardfault_args[5]);
        stacked_pc = ((unsigned long) hardfault_args[6]);
        stacked_psr = ((unsigned long) hardfault_args[7]);

        printf ("[HardFault]\n");
        printf ("- Stack frame:\n");
        printf (" R0 = %x\n", stacked_r0);
        printf (" R1 = %x\n", stacked_r1);
        printf (" R2 = %x\n", stacked_r2);
        printf (" R3 = %x\n", stacked_r3);
        printf (" R12 = %x\n", stacked_r12);
        printf (" LR = %x\n", stacked_lr);
        printf (" PC = %x\n", stacked_pc);
        printf (" PSR = %x\n", stacked_psr);
        printf ("- FSR/FAR:\n");
        printf (" CFSR = %x\n", cfsr);
        printf (" HFSR = %x\n", SCB->HFSR);
        printf (" DFSR = %x\n", SCB->DFSR);
        printf (" AFSR = %x\n", SCB->AFSR);
        if (cfsr & 0x0080) printf (" MMFAR = %x\n",
        memmanage_fault_address);
        if (cfsr & 0x8000) printf (" BFAR = %x\n", bus_fault_address);
        printf ("- Misc\n");
        printf (" LR/EXC_RETURN= %x\n", lr_value);

        while(1); // endless loop
}
```

Please note that this handler will not work correctly if the stack pointer is pointing to an invalid memory region (e.g., because of stack overflow). This affects all C code, as most C functions need stack memory. To help debug the issue, we can also generate a disassembled code list file so that we can locate the problem used to report the stacked program counter value.

The values of BFAR and MMFAR stay unchanged if the BFARVALID or MMAR-VALID is set. However, if a new fault occurs during the execution of this fault handler, the value of the BFAR and MMFAR could potentially be erased. In order to ensure the fault addresses accessed are valid, the following procedure should be used:

1. Read BFAR/MMFAR.
2. Read CFSR to get BFARVALID or MMARVALID. If the value is 0, the value of BFAR or MMFAR accessed can be invalid and can be discarded.
3. Optionally clear BFARVALID or MMARVALID.

Otherwise, it is possible to get an incorrect fault address if the following sequence occurs:

1. Read BFARVALID/MMARVALID,
2. Detected that valid bit is set, then going to read BFAR or MMFAR,
3. Just before reading BFAR or MMFAR, a higher-priority handler arrives at the current fault handler, and the higher-priority exception handler generates another fault,
4. Another fault handler is triggered, and this clears the BFARVALID or MMARVALID. This means that the value in BFAR and MMFAR will not be held constant, and will be lost.
5. After returning to the original fault handler, the value of BFAR or MMFAR is read, but now the value is invalid and leads to incorrect information in the fault report.

Therefore, it is important to read the BFAR or MMFAR first, and then read BFARVALID and MMARVALID in CFSR.

### 12.8.2 Fault mask

In a configurable fault handler, if needed we can set the FAULTMASK to:

- Disable all interrupts, thus allowing the processor to carry out remedial actions without getting interrupted (please note that the processor can still be interrupted by a NMI exception).
- Disable/Enable the configurable fault handler to bypass MPU, and to ignore bus faults (see BFHFNMIGN bit of Configuration Control Register in section 9.8.3)

These characteristics allow a configurable fault handler to try to access certain memory locations that may or may not be valid.

Potentially the FAULTMASK can also be used outside fault handlers. For example, if you have a piece of software that needs to run on a number of microcontrollers with various SRAM sizes, you can use the FAULTMASK to disable bus faults, and then carry out a RAM read-write test to detect the available RAM size during run-time.

## 12.9 Additional information
### 12.9.1 Running a system with two stacks

In Chapter 10 we covered the shadow stack point feature, which is useful for an OS. For systems without an embedded OS, the two-stack arrangement can have another usage: the separation of stacks used by Thread mode and Handler mode can help debugging stack issues in some cases, and allows exception handlers (including the fault handlers) to run normally even if the stack pointer for the Thread mode is corrupted and points to invalid memory locations. In safety critical systems this can be important.

To do this, we need to get the Thread mode code to switch from using the MSP (Main Stack Pointer) to using the PSP (Process Stack Pointer). It is relatively straightforward to do this in the reset handler. For example, if you are using Keil™ MDK-ARM, you can add code in the startup code to reserve an extra handler mode stack memory, and set the MSP, PSP, and CONTROL registers accordingly in the reset handler. You might also need to update the __user_initial_stackheap function at the end of the startup code.

```
; Modification to a startup code file (for Keil MDK-ARM) to switch
Thread mode to use PSP
 Handler_Stack_Size  EQU  0x00000200
 Thread_Stack_Size   EQU  0x00000400

               AREA STACK, NOINIT, READWRITE, ALIGN=3
Handler_Stack_Mem    SPACE Handler_Stack_Size
__initial_handler_sp
Thread_Stack_Mem     SPACE Thread_Stack_Size
__initial_sp

 ...
; Reset handler
Reset_Handler PROC
             EXPORT Reset_Handler    [WEAK]
             IMPORT SystemInit
             IMPORT __main
             LDR    R0, =__initial_sp
             MSR    PSP, R0
             LDR    R0, =__initial_handler_sp
             MSR    MSP, R0
             MOVS   R0, #2      ; Set SPSEL bit
             MSR    CONTROL, R0 ; Now Thread mode use PSP
             ISB
```

```
                LDR    R0, =SystemInit
                BLX    R0
                LDR    R0, =__main
                BX     R0
                ENDP
    ...
__user_initial_stackheap

        LDR  R0, = Heap_Mem
        LDR  R1, =(Thread_Stack_Mem + Thread_Stack_Size)
        LDR  R2, = (Heap_Mem + Heap_Size)
        LDR  R3, = Thread_Stack_Mem
        BX   LR
```

It is also possible to do this in C, but switching the stack pointer would be slightly more complex because it can be a bad idea to change the current stack point value after the C program started, as the stack might hold local variables that are already initialized and will be used later. To solve this problem, we need to change the PSP to where the current stack is (i.e., MSP current value), switch the SPSEL bit in the CONTROL register, then move the MSP to a memory space reserved for handler stack. For example, you can declare a memory space for the handler stack as static memory array.

```
Example C code to enable Thread mode to use the Process Stack with PSP
   uint64_t Handler_Stack[128]; // Handler Stack = 128x8 = 1024 bytes
   int main(void) {
   uint32_t tmp;
   ...
   tmp=(uint32_t)(Handler_Stack)+(sizeof Handler_Stack); // Get top of stack
   __set_PSP(__get_MSP());              // Set PSP to be the same as MSP
   __set_CONTROL(__get_CONTROL()|0x2); // Set SPSEL, Do not change
                                          other bits
   __ISB();                             // ISB after CONTROL change
                                        // (architectural recommendation)
   __set_MSP(tmp);                      // Move MSP to point to Handler
                                           stack
   ...
```

For Cortex®-M4 with floating point unit, since the floating point unit might have been activated and used, the bit 2 of the CONTROL register could already have been set. Therefore when setting SPSEL bit in the CONTROL register we need to perform a read-modify-write sequence to prevent clearing the FPCA bit accidentally.

## 12.9.2 Detect stack overflow

One of the common causes for software failure is stack overflow. To prevent this, traditionally, it is common for software developers to fill the SRAM with a predefined pattern (e.g., 0xDEADBEEF), then execute the program for a while, stop the target, and see how much stack has been used. This method works to an extent, but might not be accurate because the conditions for maximum stack usage might not have been triggered.

In some tool chains, you can get an estimation of the required stack size from report files after project compilation. For example, if you are using:

- Keil™ MDK-ARM®, after compilation you can find an HTML file in the project directory. One of the pieces of information provided in this file is the maximum stack size which the functions use.
- IAR Embedded Workbench, you need to enable two project options: the "Generate linker map file" option in the "List" tab for Linker, and the "Enable stack usage analysis" option in "Advanced" tab for Linker. After the compilation process, you can then see a "Stack Usage" section in the linker report (.map) in the "Debug\List" subdirectory.

Some software analysis tools can also give you a report on stack usages and a lot more information to help you improve the quality of the program code. However, if there is a stack issue such as stack leak in the software, the compilation report file cannot help you. So we need some ways to detect stack usage.

One method is to locate the stack near to the bottom of the SRAM space. When the stack is fully used, the processor gets a bus error in the next stack push because the transfer is no longer in a valid memory region, so the fault handler is executed. If the fault handler is not using the two-stacks arrangement, we need to reset the stack point to a valid memory location in the beginning of the fault handler, so that the remaining parts of the fault handler can run correctly.

Another method is to use the MPU to define a small, inaccessible or read-only memory region at the end of the stack space. If the stack overflows, the MemManage fault exception is triggered and the MPU can be turned off temporarily to allow additional stack space for the fault handler to execute.

If the system is connected to a debugger, you could set a data watch point (a debug feature) at the end of the stack memory so that the processor halts when all the stack space is used. For a standalone test environment, the data watchpoint feature can also potentially be used to trigger a debug monitor exception if no debugger is connected (if a debugger is connected, the debugger might overwrite the data watchpoint setting programmed by the application code).

For applications with an OS, the OS kernel can also carry out checking of the PSP value during each context switching to ensure that the application tasks only used the allocated stack space. While this is not as reliable as using the MPU, it is still a useful method and is easy to implement in many RTOS designs.

**CHAPTER OUTLINE**

## 13.1 About floating point data

### 13.1.1 Introduction

In C programming, you can define data as floating point values. For example, a value can be declared as single precision:

```
float pi = 3.141592F;
```

or double precision:

```
double pi = 3.14159265358979323846264338332795;
```

Floating point data allows the processor to handle a much wider data range (compared to integers or fixed point data) as well as very small values. There is also a half-precision floating point data format, which is 16-bit. The half-precision floating point format is not supported by some C compilers. In gcc or the ARM® C compiler, you can declare data as half precision using the __fp16 data type (additional command option needed; see section 13.4.5).

### 13.1.2 Single-precision floating point numbers

The single-precision data format is shown in Figure 13.1. In normal cases, the exponent is in the range of 1 to 254, and the value of the single-precision value would be represented by the equation in Figure 13.2.

**FIGURE 13.1**

Single-precision data format

$$\text{Value} = (-1)^{\text{Sign}} \times 2^{(\text{exponent}-127)} \times (1 + (\frac{1}{2} * \text{Fracion}[22]) + (\frac{1}{4} * \text{Fraction}[21]) + (1/8 * \text{Fraction}[20]) \ldots (1/(2^{23}) * \text{Fraction}[0]))$$

**FIGURE 13.2**

Normalized number format in single-precision format

**Table 13.1** Examples of Floating Point Values

| Floating Point Value | Sign | Exponent | Fraction | Hex Value |
|---|---|---|---|---|
| 1.0 | 0 | 127 (0x7F) | 000_0000_0000_0000_0000_0000 | 0x3F800000 |
| 1.5 | 0 | 127 (0x7F) | 100_0000_0000_0000_0000_0000 | 0x3FC00000 |
| 1.75 | 0 | 127 (0x7F) | 110_0000_0000_0000_0000_0000 | 0x3FE00000 |
| 0.04 → 1.28 * 2^(-5) | 0 | 127 − 5 = 122 (0x7A) | 010_0011_1101_0111_0000_1010 | 0x3D23D70A |
| -4.75 → -1.1875 *2^2 | 1 | 127 + 2 = 129 (0x81) | 001_1000_0000_0000_0000_0000 | 0xC0980000 |

$$\text{Value} = (-1)^{\text{Sign}} \times 2^{(-126)} \times ((\tfrac{1}{2} * \text{Fracion}[22]) + (\tfrac{1}{4} * \text{Fraction}[21]) + (1/8 * \text{Fraction}[20]) \dots (1/(2^{23}) * \text{Fraction}[0]))$$

**FIGURE 13.3**

Denormalized number in single-precision format

To convert a value to single-precision floating point, we need to normalize it to the range between 1.0 and 2.0, as shown in Table 13.1.

If the exponent value is 0, then there are several possibilities:

- If Fraction is equal to 0 and Sign bit is also 0, then it is a zero (+0) value.
- If Fraction is equal to 0 and Sign bit is 1, then it is a zero (−0) value. Usually +0 and −0 have the same behavior in operations. In a few cases, for example, when a divide-by-zero happened, the sign of the infinity result would depend on whether the divider is +0 or −0.
- If Fraction is not 0, then it is a denormalized value. It is a very small value between −(2^(−126)) and (2^(−126)).

A denormalized value can be represented by the equation in Figure 13.3.

If the exponent value is 0xFF, there are also several possibilities:

- If Fraction is equal to 0 and Sign bit is also 0, then it is an infinity (+∞) value.
- If Fraction is equal to 0 and Sign bit is 1, then it is a minus infinity (−∞) value.
- If Fraction is not 0, then it is a special code to indicate that the floating point value is invalid, more commonly known as NaN (Not a Number).

There are two types of NaN:

- If bit 22 of the Fraction is 0, it is a signaling NaN. The rest of the bit in the Fraction can be any value apart from all 0.
- If bit 22 of the Fraction is 1, it is a quiet NaN. The rest of the bit in the Fraction can be any value.

The two types of NaN can result in different floating exception behaviors in couple of floating point instructions (e.g., VCMP, VCMPE).

In some floating point operations, if the result is invalid it will return a "Default NaN" value. This has the value of 0x7FC00000 (Sign = 0, Exponent = 0xFF, bit 22 of Fraction is 1 and the rest of the Fraction bits are 0).

### 13.1.3 Half-precision floating point numbers

In many ways the half-precision floating point format is similar to single precision, but using fewer bits for the Exponent and Fraction field, as shown in Figure 13.4.

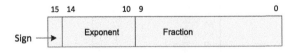

**FIGURE 13.4**

Half-precision format

Value = (-1) $^{Sign}$ x 2 $^{(exponent - 15)}$ x (1 + (½ * Fracion[9]) + (¼ * Fraction[8]) + (1/8 * Fraction[7]) ... (1/(2$^{10}$) * Fraction[0]))

**FIGURE 13.5**

Normalized number format in half-precision format

When 0 < Exponent < 0x1F, the value is a normalized value, and the value of the half-precision value would be represented by the equation in Figure 13.5:

If the exponent value is 0, then there are several possibilities:

- If Fraction is equal to 0 and Sign bit is also 0, then it is a zero (+0) value.
- If Fraction is equal to 0 and Sign bit is 1, then it is a zero (−0) value. Usually +0 and −0 behave in the same way in operations. In a few cases, for example, when a divide-by-zero happens, the sign of the infinity result would be dependent on whether the divider is +0 or −0.
- If Fraction is not 0, then it is a denormalized value. It is a very small value between −(2^(−14)) and (2^(−14)).

A denormalized value can be represented by the equation in Figure 13.6.

If the exponent value is 0x1F, the situation is a bit more complex. The floating point feature in ARMv7-M supports two operation modes for half-precision data:

- IEEE half precision
  - Alternate half precision. This does not support Infinity or NaN, but can have a larger number range and can have higher performance in some cases. However, if the application must be IEEE 754 compliant, this operation mode should not be used.

In IEEE half-precision mode, when the exponent value equals 0x1F:

- If Fraction is equal to 0 and Sign bit is also 0, then it is an infinity (+∞) value.
- If Fraction is equal to 0 and Sign bit is 1, then it is a minus infinity (− ∞) value.
- If Fraction is not 0, then it is a special code to indicate that the floating point value is invalid, more commonly known as NaN (Not a Number).

Similar to single precision, a NaN can be signaling or quiet:

- If bit 9 of the Fraction is 0, it is a signaling NaN. The rest of the bit in the Fraction can be any value apart from all 0.
- If bit 9 of the Fraction is 1, it is a quiet NaN. The rest of the bit in the Fraction can be any value.

In some floating point operations, if the result is invalid it will return a "Default NaN" value. This has the value of 0x7E00 (Sign = 0, Exponent = 0x1F, bit 9 of Fraction is 1, and the rest of the Fraction bits are 0).

Value = (-1) $^{Sign}$ x 2 $^{(- 14)}$ x((½ * Fracion[9]) + (¼ * Fraction[8]) + (1/8 * Fraction[7]) ... (1/(2$^{10}$) * Fraction[0]))

**FIGURE 13.6**

Denormalized number in half-precision format

Value = (-1)$^{Sign}$ x 2$^{16}$ x (1+ (½ * Fracion[9]) + (¼ * Fraction[8]) + (1/8 * Fraction[7]) ... (1/(2$^{10}$) * Fraction[0]))

**FIGURE 13.7**

Alternate normalized number in half precision format

In Alternate Half precision mode, when the exponent value equals 0x1F, the value is a normalized number and can be represented by the equation shown in Figure 13.7.

## 13.1.4 Double-precision floating point numbers

Although the floating point unit in Cortex®-M4 does not support double-precision floating point operations, you can still have double-precision data in your applications. In such cases the C compiler and linker will insert the appropriate run-time library functions to handle these calculations.

The double-precision data format is shown in Figure 13.8.

In a little endian memory system, the least significant word is stored in the lower address of a 64-bit address location, and the most significant word is stored in the upper address. In a big endian memory system it is the other way round.

When 0 < Exponent < 0x7FF, the value is a normalized value, and the value of the double-precision value is represented by the equation in Figure 13.9.

If the exponent value is 0, then there are several possibilities:

- If Fraction is equal to 0 and Sign bit is also 0, then it is a zero (+0) value.
- If Fraction is equal to 0 and Sign bit is 1, then it is a zero (−0) value. Usually +0 and −0 behave in the same way in operations. In a few cases, for example, when a divide-by-zero happens, the sign of the infinity result would be dependent on whether the divider is +0 or −0.
- If Fraction is not 0, then it is a denormalized value. It is a very small value between −(2ˆ(−1022)) and (2ˆ(−1022)).

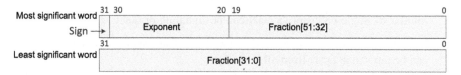

**FIGURE 13.8**

Double-precision data format

Value = (-1)$^{Sign}$ x 2$^{(exponent-1023)}$ x (1 + (½ * Fracion[51]) + (¼ * Fraction[50]) + (1/8 * Fraction[49]) ... (1/(2$^{52}$) * Fraction[0]))

**FIGURE 13.9**

Normalized number format in double-precision format

Value = (-1)$^{Sign}$ x 2$^{(-1022)}$ x ((½ * Fracion[51]) + (¼ * Fraction[50]) + (1/8 * Fraction[49]) ... (1/(2$^{52}$) * Fraction[0]))

**FIGURE 13.10**

Denormalized number in double-precision format

The value of a denormalized value can be represented by the equation in Figure 13.10.

If the exponent value is 0x7FF, there are also several possibilities:

- If Fraction is equal to 0 and Sign bit is also 0, then it is an infinity ($+\infty$) value.
- If Fraction is equal to 0 and Sign bit is 1, then it is a minus infinity ($-\infty$) value.
- If Fraction is not 0, then it is a NaN (Not a Number).

There are two types of NaN:

- If bit 51 of the Fraction is 0, it is a signaling NaN. The rest of the bit in the Fraction can be any value apart from all 0.
- If bit 51 of the Fraction is 1, it is a quiet NaN. The rest of the bit in the Fraction can be any value.

## 13.1.5 Floating point support in Cortex®-M processors

The Cortex®-M4 processor has the option of including a single-precision floating point unit. If the floating pointing unit is available, you can use it to accelerate single-precision floating point operations. Double-precision calculation still needs to be handled by C run-time library functions.

Even if the floating point unit is available and the operation is single precision, you might still need run-time library functions, for example, when dealing with functions like sinf(), cosf(), etc. These functions require a sequence of calculations and cannot be done by single instruction or a few instructions.

In all well-established toolchains you can compile an application and select NOT to use the floating point unit. This allows the compiled code to be used on another Cortex-M4 microcontroller product that does not have floating point unit support.

For Cortex-M4 microcontrollers without a floating point unit, or microcontrollers with Cortex-M3, Cortex-M0, Cortex-M0+ processors, or FPGA with Cortex-M1 processors, there is no floating point unit support and all floating point calculations have to be carried out using run-time library functions.

Alternatively, a software developer can use fixed point data. Fundamentally fixed point operations are just like integer operations but with additional shift adjustment operations. While this is faster than using floating point run-time library functions, it can only handle limited data ranges because the exponent is fixed. ARM® has an application note (reference 33) on creating fixed point arithmetic operations in ARM architecture.

## 13.2 Cortex®-M4 floating point unit (FPU)

### 13.2.1 Floating point unit overview

In ARMv7-M architecture, the floating point data and operations are based on IEEE Std 754-2008, IEEE Standard for Binary Floating Point Arithmetic. The Floating Point Unit (FPU) in the Cortex®-M4 processor is optional and supports single precision floating point calculations, as well as some conversion and memory access functions. The FPU design is compliant with the IEEE 754 standard, but is not a complete implementation. For example, the following operations are not implemented by hardware:

- Double-precision data calculations
- Floating point remainder (e.g., $z = \text{fmod}(x, y)$)
- Round floating point number to integer values floating point number
- Binary-to-decimal and decimal-to-binary conversions
- Direct comparison of single-precision and double-precision values

These operations need to be handled by software. The FPU in the Cortex-M4 processor is based on an extension of the ARMv7-M architecture called FPv4-SP (Floating Point version 4 − Single Precision). This is a subset of the VFPv4-D16 extension for ARMv-7A and ARMv7-R architecture (VFP stands for Vector Floating Point). As many floating point instructions are common to both versions, it is common to refer to floating point operations as VFP, and the floating point instruction mnemonics start with the letter "V."

The floating point design supports:

- A floating point register bank that contains thirty-two 32-bit registers, which can be used 32 registers, or be used in pairs as 16 double-word registers.
- Single-precision floating point calculations
- Conversion instructions for:
  - "integer ↔ single-precision floating point"
  - "fixed point ↔ single-precision floating point"
  - "half precision ↔ single-precision floating point"
- Data transfers of single-precision and double-word data between floating point register bank and memory
- Data transfer of single-precision between floating point register bank and integer register bank

In the architecture, the FPU is viewed as a co-processor. To be consistent with other ARM® architectures, the floating point unit is defined as Co-Processor #10 and #11 in the CPACR programmer's model (see section 13.2.3). The normal processing pipeline and the floating point unit share the same instruction fetch stage, but the instruction decode and execution stages are separate, as shown in Figure 13.11.

Some of the other classic ARM processors allow multiple co-processors in the programmer's model and support co-processor register access instructions.

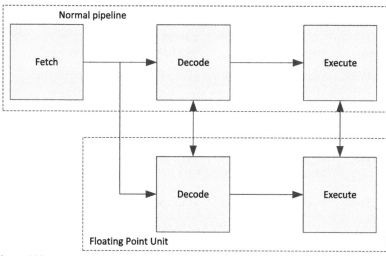

Cortex-M4

**FIGURE 13.11**

Co-processor pipeline concept

However, for floating point operations and floating point data transfers in the Cortex-M4 processor, a set of floating point instructions are used instead of co-processor access instructions.

### 13.2.2 Floating point registers overview

The FPU adds a number of registers to the processor system:

- CPACR (Co-processor Access Control Register) in SCB (System Control Block)
- Floating point register bank
- Floating point Status and Control Register (FPSCR)
- Additional registers in the FPU for floating point operations and control, as shown in Table 13.2.

### 13.2.3 CPACR register

The CPACR register allows you to enable or disable the FPU. It is located in address 0xE000ED88 and can be accessed as "SCB->CPACR" in CMSIS-Core. Bit 0 to bit 19 and bit 24 to bit 31 are not implemented and are reserved (Figure 13.12).

This programmer's model of this register provides enable control for up to 16 co-processors. On the Cortex®-M4, the FPU is defined as co-processor 10 and 11. Since there is no other co-processor, only CP10 and CP11 are available and both are for the FPU. When programming this register, the settings for CP10 and CP11 must be identical. The encoding for CP10 and CP11 is shown in Table 13.3.

**Table 13.2** Additional FPU Registers

| Address | Register | CMSIS-Core Symbol | Function |
|---|---|---|---|
| 0xE000EF34 | Floating Point Context Control Register | FPU->FPCCR | FPU control data |
| 0xE000EF38 | Floating Point Context Address Register | FPU->FPCAR | Hold the address of the unpopulated floating point register space in stack frame |
| 0xE000EF3C | Floating Point Default Status Control Register | FPU->FPDSCR | Default values for the floating point status control data (FPCCR) |
| 0xE000EF40 | Media and FP Feature Register 0 | FPU->MVFR0 | Read-only information about what VFP instruction features are implemented |
| 0xE000EF44 | Media and FP Feature Register 1 | FPU->MVFR1 | Read-only information about what VFP instruction features are implemented |

SCB->CPACR, 0xE000ED88

| 31 | | | 23 | 22 | 21 | 20 | | | | | | | | | 0 |
|---|---|---|---|---|---|---|---|---|---|---|---|---|---|---|---|
| | | | | CP11 | CP10 | | | | | | | | | | |

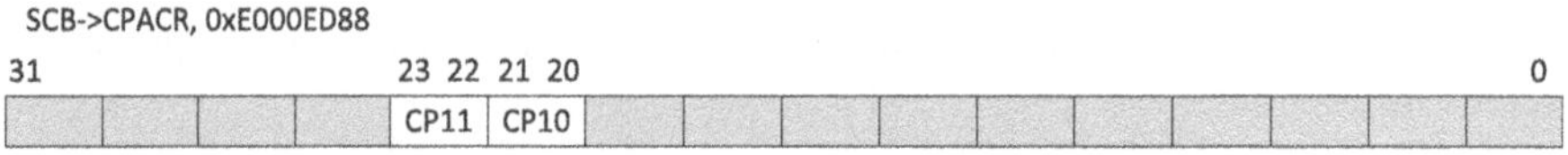

**FIGURE 13.12**

Co-processor access control register (SCB->CPACR, 0xE000ED88)

**Table 13.3** CP10 and CP11 Settings

| Bits | CP10 and CP11 Setting |
|---|---|
| 00 | Access denied. Any attempted access generate a Usage fault (type NOCP – No Co-processor) |
| 01 | Privileged Access only. Unprivileged access generate a Usage fault |
| 10 | Reserved – result unpredictable |
| 11 | Full access |

By default CP10 and CP11 are zero after reset. This setting disables the FPU and allows lower power consumption. Before using FPU, you need to program the CPACR to enable the FPU first. For example:

```
SCB->CPACR|= 0x00F00000; // Enable the floating point unit for full
access
```

This step is typically carried out inside the SystemInit() function provided in the device-specific software package file. SystemInit() is executed by the reset handler.

## 13.2.4 Floating point register bank

The floating point register bank contains thirty-two 32-bit registers, which can be organized as sixteen 64-bit double-word registers, as shown in Figure 13.13.

S0 to S15 are caller saved registers. So if a function A calls a function B, function A must save the contents of these registers (e.g., on the stack) before calling function B because these registers can be changed by the function call (e.g., return result).

S16 to S31 are callee saved registers. So if a function A calls a function B, and function B needs to use more than 16 registers for its calculations, it must save the contents of these registers (e.g., on the stack) first, and must restore these registers from the stack before returning to function A.

The initial values of these registers are undefined.

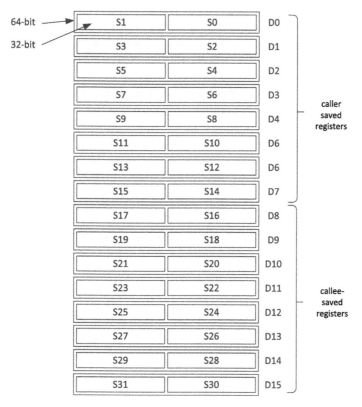

**FIGURE 13.13**

Floating point register bank

### 13.2.5 **Floating point status and control register (FPSCR)**

The FPSCR holds the arithmetic result flags and sticky status flags, as well as bit fields to control the behavior of the floating point unit (Figure 13.14 and Table 13.4). The N, Z, C, and V flags are updated by floating point comparison operations, as shown in Table 13.5.

You can use the results of a floating point compare for conditional branch/ conditional execution by copying the flags to APSR first:

```
VMRS APSR_nzcv, FPSCR ; Copy flags from FPSCR to flags in APSR
```

The bit fields AHP, DN, and FZ are control register bits for special operation modes. By default, all these bits default to 0 and their behavior is compliant with IEEE 754 single precision operation. In normal applications there is no need to modify the settings of the floating point operation control. Do not change these bits if your application requires IEEE 754 compliance.

The RMode bit field is for controlling the rounding mode for calculation results. The IEEE 754 standard defines several rounding modes, as shown in Table 13.6.

The bits IDC, IXC, UFC, OFC, DZC, and IOC are sticky status flags that show any abnormalities (floating point exceptions) during floating point operations. Software can check these flags after the floating point operations, and can clear them by writing zero to them. Section 13.5 has more information on floating point exceptions.

### 13.2.6 **Floating point context control register (FPU->FPCCR)**

The Floating Point Context Control Register (FPCCR) allows you to control the behavior of exception handling such as the lazy stacking feature (Table 13.7). It also allows you to access some of the control information.

In most applications there is no need to change the setting in this register. By default Automatic FPU context-saving, restoration and lazy stacking are enabled to reduce interrupt latency. You can set ASPEN and LSPEN in the configurations shown in Table 13.8.

### 13.2.7 **Floating point context address register (FPU->FPCAR)**

In Chapter 8 we covered the lazy stacking feature briefly (section 8.3.6). When an exception takes place and if the current context is active floating point context (the FPU has been used), then the exception stack frame contains registers from the

| | 31 | 30 | 29 | 28 | 27 | 26 | 25 | 24 | 23:22 | 21:8 | 7 | 6:5 | 4 | 3 | 2 | 1 | 0 |
|---|---|---|---|---|---|---|---|---|---|---|---|---|---|---|---|---|---|
| FPSCR | N | Z | C | V | | AHP | DN | FZ | RMode | Reserved | IDC | Reserved | IXC | UFC | OFC | DZC | IOC |

Reserved ⅃

**FIGURE 13.14**

Bit field in FPSCR

**Table 13.4** Bit Fields in FPSCR

| Bit | Description |
|---|---|
| N | Negative flag (update by floating point comparison operations) |
| Z | Zero flag (update by floating point comparison operations) |
| C | Carry/borrow flag (update by floating point comparison operations) |
| V | Overflow flag (update by floating point comparison operations) |
| AHP | Alternate half precision control bit:<br>0 – IEEE half-precision format (default)<br>1 – Alternative half-precision format, see section 13.1.3 |
| DN | Default NaN (Not a Number) mode control bit:<br>0 – NaN operands propagate through to the output of a floating point operation (default)<br>1 – Any operation involving one or more NaN(s) returns the default NaN |
| FZ | Flush-to-zero model control bit:<br>0 – Flush-to-zero mode disabled (default) (IEEE 754 standard compliant)<br>1 – Flush-to-zero mode enabled; denormalized values (tiny values with exponent equal 0) are flushed 0 |
| RMode | Rounding Mode Control field; the specified rounding mode is used by almost all floating point instructions:<br>00 – Round to Nearest (RN) mode (default)<br>01 – Round towards Plus Infinity (RP) mode<br>10 – Round towards Minus Infinity (RM) mode<br>11 – Round towards Zero (RZ) mode |
| IDC | Input Denormal cumulative exception bit; set to 1 when floating point exception occurred, clear by writing 0 to this bit |
| IXC | Inexact cumulative exception bit; set to 1 when floating point exception occurred, clear by writing 0 to this bit |
| UFC | Underflow cumulative exception bit; set to 1 when floating point exception occurred, clear by writing 0 to this bit |
| OFC | Overflow cumulative exception bit; set to 1 when floating point exception occurred, clear by writing 0 to this bit |
| DZC | Division by Zero cumulative exception bit; set to 1 when floating point exception occurred, clear by writing 0 to this bit |
| IOC | Invalid Operation cumulative exception bit; set to 1 when floating point exception occurred, clear by writing 0 to this bit |

**Table 13.5** Operation of N, Z, C, and V Flags in FPSCR

| Comparison Result | N | Z | C | V |
|---|---|---|---|---|
| Equal | 0 | 1 | 1 | 0 |
| Less than | 1 | 0 | 0 | 0 |
| Greater than | 0 | 0 | 1 | 0 |
| Unordered | 0 | 0 | 1 | 1 |

**Table 13.6** Rounding Modes Available on the Cortex®-M4 FPU

| Rounding Mode | Descriptions |
|---|---|
| Round to nearest | Rounds to the nearest value. This is the default configuration. IEEE 754 subdivides this mode to:<br>Round to nearest, ties to even: round to the nearest value with an even (zero) LSB. This is the default for binary floating point and recommended default for decimal floating point.<br>Round to nearest, ties away from zero: round to the nearest value above (for +ve numbers) or below (for –ve numbers). This is intended as an option for decimal floating point.<br>Since the floating point unit is using binary floating point only, the "Round to nearest, ties away from zero" mode is not available. |
| Round toward $+\infty$ | Also known as rounding up or ceiling. |
| Round toward $-\infty$ | Also known as rounding down or flooring. |
| Round toward 0 | Also known as truncation. |

**Table 13.7** Floating Point Context Control Register (FPU->FPCCR, 0xE000EF34)

| Bits | Name | Type | Reset Value | Descriptions |
|---|---|---|---|---|
| 31 | ASPEN | R/W | 1 | Enable/disable automatic setting of FPCA (bit 2 of the CONTROL register). When this is set (default), it enables the automatic state preservation and restoration of S0-S15 & FPSCR on exception entry and exception exit.<br>When it is clear to 0, automatic saving of FPU registers are disabled. Software using FPU might need to manage context saving manually. |
| 30 | LSPEN | R/W | 1 | Enable/disable lazy stacking (state preservation) for S0-S15 & FPSCR. When this is set (default), the exception sequence use lazy stacking feature to ensure low interrupt latency. |
| 29:9 | – | – | – | Reserved |
| 8 | MONRDY | | 0 | 0 = DebugMonitor is disabled or priority did not permit setting MON_PEND when the floatingpoint stack frame was allocated.<br>1 = DebugMonitor is enabled and priority permits setting MON_PEND when the floating point stack frame was allocated. |

**Table 13.7** Floating Point Context Control Register (FPU->FPCCR, 0xE000EF34)—
*Cont'd*

| Bits | Name | Type | Reset Value | Descriptions |
|------|------|------|-------------|--------------|
| 7 | – | – | – | Reserved |
| 6 | BFRDY | R | 0 | 0 = BusFault is disabled or priority did not permit setting the BusFault handler to the pending state when the floating point stack frame was allocated. 1 = BusFault is enabled and priority permitted setting the BusFault handler to the pending state when the floating point stack frame was allocated. |
| 5 | MMRDY | R | 0 | 0 = MemManage is disabled or priority did not permit setting the MemManage handler to the pending state when the floating point stack frame was allocated. 1 = MemManage is enabled and priority permitted setting the MemManage handler to the pending state when the floating point stack frame was allocated. |
| 4 | HFRDY | R | 0 | 0 = Priority did not permit setting the HardFault handler to the pending state when the floating point stack frame was allocated. 1 = Priority permitted setting the HardFault handler to the pending state when the floating point stack frame was allocated. |
| 3 | THREAD | R | 0 | 0 = Mode was not Thread Mode when the floating point stack frame was allocated. 1 = Mode was Thread Mode when the floating point stack frame was allocated. |
| 2 | – | – | – | Reserved |
| 1 | USER | R | 00 | 0 = Mode was not Thread Mode when the floating point stack frame was allocated. 1 = Mode was Thread Mode when the floating point stack frame was allocated. |
| 0 | LSPACT | R | 0 | 0 = Lazy state preservation is not active. 1 = Lazy state preservation is active. Floating point stack frame has been allocated but saving state to it has been deferred. |

**Table 13.8** Available Context Saving Configurations

| ASPEN | LSPEN | Configuration |
|-------|-------|---------------|
| 1 | 1 | Automatic state saving enabled, lazy stacking enabled (default) CONTROL.FPCA is automatically set to 1 when floating-point is used. If CONTROL.FPCA is 1 at the exception entry, the processor reserves space in the stack frame and sets LSPACT to 1. But the actual stacking does not happen unless the interrupt handler uses the FPU. |
| 1 | 0 | Lazy stacking disabled, automatic state saving enabled CONTROL.FPCA is automatically set to 1 when floating-point is used. At the exception entry, the floating-point registers S0-S15, and FPSCR are pushed to the stack if CONTROL.FPCA is 1. |
| 0 | 0 | No automatic state preservation. You can use this setting: 1. In applications without an embedded OS or multi-task scheduler, if none of the interrupt or exception handlers use the FPU. 2. In application code where only one exception handler uses the FPU and not used by Thread. If multiple interrupt handlers use the FPU, they must not be permitted to be nested. This can be done by setting them to the same priority level. 3. Alternatively, context saving can be manage in software manually in exception handlers. |
| 0 | 1 | Invalid configuration |

integer register bank (R0–R3, R12, LR, Return Address xPSR) as well as FPU registers (S0–S15, FPSCR). In order to reduce interrupt latency, by default lazy stacking is enabled and it means that the stacking mechanism will reserve the stack space for the FPU registers, but will not actually push these registers onto the stack until it really needs to.

The FPCAR register is part of this lazy stacking mechanism. It holds the address of the FPU registers in the stack frame so that the lazy stacking mechanism knows where to push the FPU registers to later. Bits 2 to 0 are not used because the stack frame is double-word aligned (Figure 13.15).

When an exception occurs in the lazy stacking scenario, the FPCAR is updated to the address of the FPU S0 register space in the stack frame, as shown in Figure 13.16.

Floating Point Context Address Register (FPU->FPCAR), address 0xE000EF38

**FIGURE 13.15**

Floating point context address register (FPCAR) bit assignment

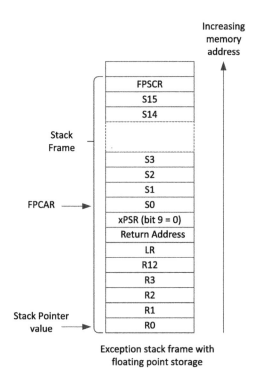

Increasing
memory
address

Stack
Frame

FPCAR

Stack Pointer
value

| |
| FPSCR |
| S15 |
| S14 |
| |
| S3 |
| S2 |
| S1 |
| S0 |
| xPSR (bit 9 = 0) |
| Return Address |
| LR |
| R12 |
| R3 |
| R2 |
| R1 |
| R0 |

Exception stack frame with
floating point storage

**FIGURE 13.16**

FPCAR points to the reserved FPU register memory space in the stack frame

## 13.2.8 Floating point default status control register (FPU-> FPDSCR)

The FPDSCR register holds the default configuration information (operation modes) for the floating point status control data. The values are copied to the FPSCR at exception entry (Figure 13.17).

In a complex system there can be different types of applications running in parallel, each with different FPU configurations (such as rounding mode). In order to allow this, the FPU configuration needs to be switched automatically at exception entry and exception return. The FPFSCR defines the FPU configuration when the exception handlers start, including the OS kernel as most parts of the OS are executing in the Handler mode.

| | 31 | 30 | 29 | 28 | 27 | 26 | 25 | 24 | 23:22 | 21:8 | 7 | 6:5 | 4 | 3 | 2 | 1 | 0 |
|---|---|---|---|---|---|---|---|---|---|---|---|---|---|---|---|---|---|
| FPDSCR | Reserved | | | | | AHP | DN | FZ | RMode | Reserved | | | | | | | |

**FIGURE 13.17**

Floating point default status control register (FPDSCR) bit assignment

**Table 13.9** Media and FP Feature Registers

| Address | Name | CMSIS-Core Symbol | Value |
|---------|------|-------------------|-------|
| 0xE000EF40 | Media and FP Feature Register 0 | FPU->MVFR0 | 0x10110021 |
| 0xE000EF44 | Media and FP Feature Register 1 | FPU->MVFR1 | 0x11000011 |

For application tasks, each of them have to set up the FPSCR when the tasks start. After that, the configuration will be saved and restored with the FPSCR during context switching.

### 13.2.9 Media and floating point feature registers (FPU->MVFR0, FPU->MVFR1)

The FPU has two read-only registers to allow software to determine what instruction features are supported. The values of MVFR0 and MVFR1 are hard coded (Table 13.9). Software can use these registers to determine which floating point features (Figure 13.18) are available.

If the bit field is 0, it means the feature is not available. If it is 1 or 2 it is supported. The single-precision field is set to 2 to indicate that apart from normal single-precision calculations, it can also handle floating divide and square root functions.

## 13.3 Lazy stacking in detail
### 13.3.1 Key elements of the lazy stacking feature

Lazy stacking is an important feature of Cortex®-M4. Without this feature, the time required for every exception increases to 29 cycles if the FPU is available and used: instead of pushing 8 registers to the stack, it has to push 25 registers to the stack. By having the lazy stacking feature, the exception latency is still just 12 clock cycles (for a zero wait state memory system), the same as the Cortex-M3 processor.

By default the lazy stacking feature is enabled, and software developers do not have to change anything to take full advantage of it. Also, there is no need to set up any registers during exception handling, as all the required operations are managed by hardware automatically.

There are several key elements in the lazy stacking mechanism.

**FPCA bit in the CONTROL register:** CONTROL.FPCA indicates if the current context (e.g., task) has a floating point operation. It is:

- Set to 1 when the processor executes a floating point instruction
- Cleared to zero at the beginning of an exception handler
- Set to the inverse of bit 4 in EXC_RETURN at exception return
- Cleared to zero after a reset

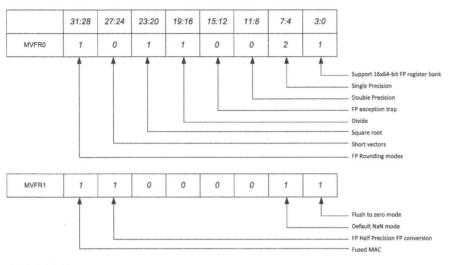

**FIGURE 13.18**

Media and floating point feature registers

**EXC_RETURN:** Bit 4 of the EXC_RETURN is set to 0 at exception entry if the interrupted task has floating point context (FPCA was 1). It indicates that the longer stack frame (contains R0–R3, R12, LR, Return Address, xPSR, S0–S15, FPSCR) was used at stacking. Otherwise, this bit is set to 1, which indicates that the stack frame was the shorter version (contains R0–R3, R12, LR, Return Address, xPSR).

**LSPACT bit in FPCCR:** When the processor enters an exception handler with lazy stacking enabled, and the interrupted task has a floating point context (FPCA was 1), the longer stack frame is used in stacking and LSPACT is set to 1. This indicates that the stacking of floating point registers is deferred and space is allocated in the stack frame, indicated by FPCAR. If the processor executes a floating point instruction while LSPACT is 1, the processor will stall the pipeline and start the stacking of floating point registers, and resume operation when this is done with the LSPACT also cleared to 0, to indicate that there is no more outstanding deferred floating point register stacking. This bit is also cleared to 0 in an exception return if bit 4 of EXC_RETURN value was 0.

**FPCAR register:** The FPCAR register holds the address to be used for stacking of floating point registers S0–S15 and FPSCR.

## 13.3.2 Scenario #1: No floating point context in interrupted task

If there was no floating point context before an interrupt, FPCA was zero and the short version of stack frame is used (Figure 13.19). This is the same for both Cortex®-M3 and Cortex-M4. If the exception handler or ISR used the FPU, the FPCA bit would be set to 1, but it will get cleared at the end of the ISR during the exception return.

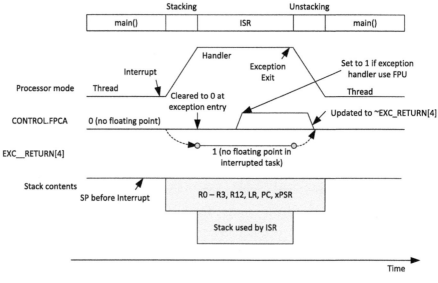

**FIGURE 13.19**

Exception handling with no floating point context in the interrupted task

### 13.3.3 **Scenario #2: Floating point context in interrupted task but not in ISR**

If there was a floating point context before an interrupt, FPCA was 1 and the long version of the stack frame is used (Figure 13.20). However, the stack frame contains the space for S0–S15 and FPSCR, and the values of these registers are not pushed to the stack. LSPACT is set to 1 to indicate the stacking of the floating point register is deferred.

At exception return, the processor sees that although EXC_RETURN[4] is 0 (long stack frame), LSPACT is 1, which indicates that the floating point registers were not pushed to the stack. So the unstacking of S0–S15 and FPSCR are not carried out and remain unchanged.

### 13.3.4 **Scenario #3: Floating point context in interrupted task and in ISR**

If there is a floating point operation inside the ISR, when the floating point instruction reaches the decode stage, the processor detects the presence of the floating point operation and stalls the processor, then pushes the floating point registers S0–S15 and FPSCR to the stack, as shown in Figure 13.21. After that the stacking is completed, and the ISR can then execute the floating point instruction.

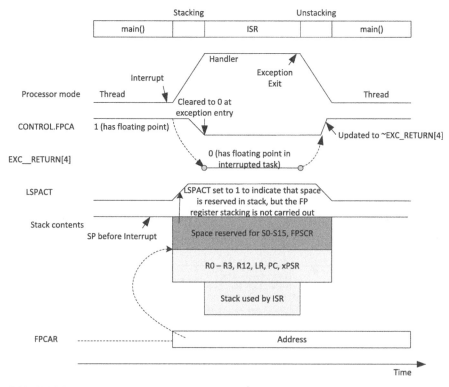

**FIGURE 13.20**

Exception handling with floating point context in the interrupted task, and no floating point operations in ISR

In Figure 13.21, a floating point instruction is executed during the ISR execution and triggers the deferred stacking. The address stored in FPCAR is used for storing the registers during this lazy stacking.

At the end of the ISR, the exception return takes place and the processor sees that EXC_RETURN[4] is 0 (long stack frame) and LSPACT is 0, which indicates that the processor has to unstack the floating point registers from the stack frame.

### 13.3.5 Scenario #4: Nested interrupt with floating point context in the second handler

Lazy stacking works across multiple levels of nested interrupts. For example, if the thread has floating point context, the low priority ISR does not have any floating point context, and the higher priority ISR has the floating point context, we can see that the deterred lazy stacking pushes the floating point registers to the first level of stack frame pointed by FPCAR, as shown in Figure 13.22.

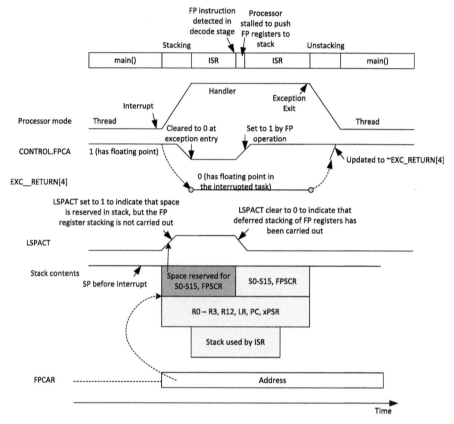

**FIGURE 13.21**

Exception handling with floating point context in the interrupted task as well as in the ISR

### 13.3.6 Scenario #5: Nested interrupt with floating point context in the both handlers

The lazy stacking mechanism also works for nested ISRs with FP context in both low- and high-priority ISRs. In that case the processor will reserve stack space for floating point registers several times, as shown Figure 13.23.

In each of the exception returns, bit 4 of the EXC_RETURN values are 0, and LSPACT is also zero, indicating that the long stack frames were used and the processor will have to unstack the floating point registers from each stack frame.

### 13.3.7 Interrupt of lazy stacking

If during lazy stacking, a new higher priority interrupt arrives, the lazy stacking is interrupted. The floating point instruction that triggered the lazy stacking was still in the decode stage and was not executed, so the return PC in the stack points to the address

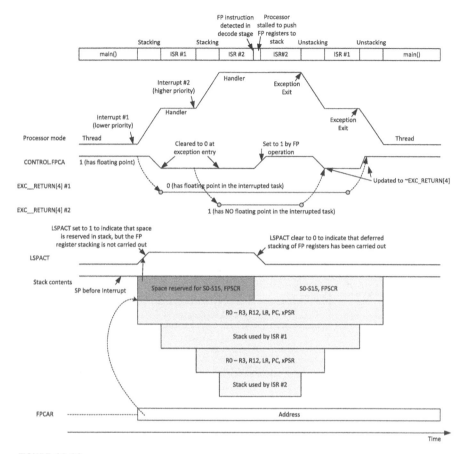

**FIGURE 13.22**

Nested Exception handling with floating point context in the higher priority interrupt as well as the Thread

of this instruction. If the higher priority interrupt does not use any floating point operation, after it is returned, the floating point instruction that triggered the lazy stacking the first time enters the processor pipeline again and triggers the lazy stacking the second time.

### 13.3.8 Interrupt of floating point instructions

Many floating point instructions take multiple clock cycles. If the interrupt takes place during VPUSH, VPOP, VLDM, or VSTM instructions (multiple memory transfers), the processor suspends the current instruction and uses the ICI bits in EPSR to store the status of these instructions, then executes the exception handler,

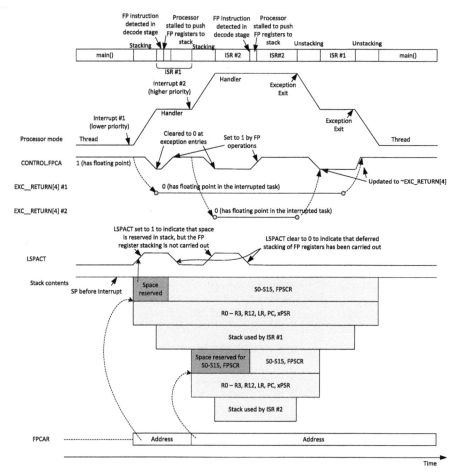

**FIGURE 13.23**

Nested Exception handling with floating point context in the Thread code as well as both levels of ISRs

and finally resumes these instructions from where it was suspended based on the restored ICI bits.

If the interrupt takes place during VSQRT and VDIV, the processor continues the execution of these instructions with the stacking operation going on in parallel.

## 13.4 Using the floating point unit
### 13.4.1 Floating point support in CMSIS-Core

To use the floating point unit, you first need to enable it. The CMSIS-Core has two pre-processing directives/macros related to FPU configuration (Table 13.10).

**Table 13.10** Pre-processing Directive in CMSIS-Core Related to FPU

| Pre-processing Directive | Descriptions |
| --- | --- |
| \_\_FPU_PRESENT | Indicate if the Cortex®-M processor in the microcontroller has FPU or not. If yes, this is set to 1 by the device specific header. |
| \_\_FPU_USED | Indicate whether an FPU is used or not. Must be set to 0 if \_\_FPU_PRESENT is 0. Can be either 0 or 1 if \_\_FPU_PRESENT is 1. This is set by the compilation tools (e.g., project setting). |

The data structure for FPU is available only if \_\_FPU_PRESENT is set to 1. If \_\_FPU_USED is set to 1, then the SystemInit() function enables the FPU by writing to CPACR when the reset handler is executed.

## 13.4.2 Floating point programming in C

In most applications, the accuracy of a single-precision floating point is sufficient. Using double-precision floating point data can increase code size and can take much longer, because the FPU in the Cortex®-M4 only supports single-precision calculation, so double-precision operations have to be done by software (using run-time library functions inserted by the development toolchain).

However, it is common for software developers to use double-precision floating point accidentally. For example, the following code from the Whetstone benchmark will be compiled with double precision, even if X, Y, T, and T2 variables are defined as "float" (single precision):

```
X=T*atan(T2*sin(X)*cos(X)/(cos(X+Y)+cos(X-Y)-1.0));
Y=T*atan(T2*sin(Y)*cos(Y)/(cos(X+Y)+cos(X-Y)-1.0));
```

This is because the math functions used here are double precision by default, and the constant 1.0 is also treated as double precision by default. To generate a pure single precision calculation, we need to modify this code to:

```
X=T*atanf(T2*sinf(X)*cosf(X)/(cosf(X+Y)+cosf(X-Y)-1.0F));
Y=T*atanf(T2*sinf(Y)*cosf(Y)/(cosf(X+Y)+cosf(X-Y)-1.0F));
```

Depending on the development tool being used, it might be possible to report if any double-precision calculation have been used, or force all calculation to single precision only. Alternatively, you can also generate disassembled code or linker report files to check whether the compiled image contains any double-precision run-time functions.

## 13.4.3 Compiler command line options

In most toolchains the command line options are set up automatically to enable you to use the FPU easily. For example, in Keil™ MDK-ARM the μVision IDE

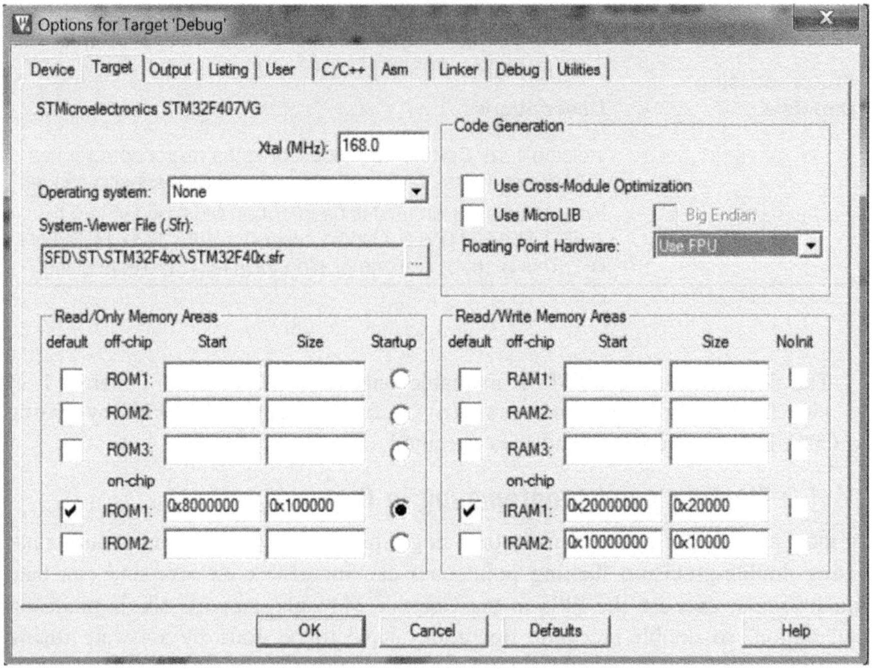

**FIGURE 13.24**

FPU option in Keil MDK-ARM

automatically sets the compiler option "–cpu Cortex®-M4.fp" to enable the use of floating point instructions when the "Use FPU" option is selected (Figure 13.24).

For users of ARM® DS-5™ or the older RealView™ Development Suite (RVDS), you can use the following command line options to use the FPU:

```
"--cpu Cortex-M4F", or
"--cpu=7E-M --fpu=fpv4-sp"
```

For gcc users, you can use the following command line options to use the FPU:

```
"-mcpu=cortex-m4 -mfpu=fpv4-sp-d16 -mfloat-abi=hard", or
"-mcpu=cortex-m4 -mfpu=fpv4-sp-d16 -mfloat-abi=softfp"
```

Note: Some free versions of gcc distributions might not provide the math run-time library for handling floating point calculations.

### 13.4.4 ABI Options: Hard-vfp and Soft-vfp

In most C compilers, you can specify how floating point calculations are handled using different ABI (Application Binary Interface) arrangements. In many cases,

**Table 13.11** Command Line Options for Various Floating Point ABI Arrangements

| ARM® C Compiler Floating Point ABI Option | Gcc Floating Point ABI Option | Descriptions |
|---|---|---|
| –fpu=softvfp | -mfloat-abi=soft | Soft ABI without FPU hardware – All floating point operations are handled by run-time library functions. Values are passed via integer register bank. |
| –fpu=softvfp+fpv4-sp | -mfloat-abi=softfp | Soft ABI with FPU hardware – This allows the compiled code to generate codes that access the FPU directly. But if a calculation needs to use a run-time library function, soft-float calling convention is used (using integer register bank). |
| –fpu=fpv4-sp | -mfloat-abi=hard | Hard ABI – This allows the compiled code to generate codes that access the FPU directly, and use FPU-specific calling conventions when calling run-time library functions. |

even if you have a FPU in the processor, you still need to use a number of run-time library functions, because many mathematical functions require a sequence of calculations. The ABI options affect:

- If the floating point unit is used
- How parameters are passed between caller functions and callee functions

With most development toolchains you can have three different options (Table 13.11). The differences in the operations of these options are shown in Figure 13.25.

For example, if a software library is to be compiled for various different Cortex®-M4 products including some with FPU and some without, we can use the soft ABI option. During the linking stage, if the target processor supports FPU, the run-time library functions inserted by the linker can then utilize the FPU for best performance.

The hard floating point ABI allows you to get the best performance if all the floating point calculations are single precision only. However, for applications using mostly double-precision calculations, the performance of hard ABI can be worse than using soft ABI. This is because when using hard ABI, the values to be processed are often transferred using the floating point register bank. Since the Cortex-M4 FPU does not support double-precision floating point calculations, the values have to be copied back to the integer register bank to be processed by software. This can result in additional overhead, and you might be better off using soft ABI with FPU hardware.

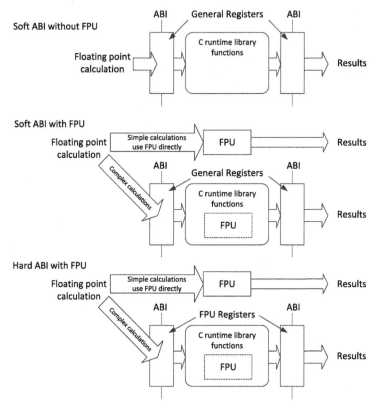

**FIGURE 13.25**

Various common floating point ABI options

In most applications, if there are few floating point run-time function accesses, the performance of hard ABI and soft ABI would be about the same.

### 13.4.5 Special FPU modes

By default, the FPU in the Cortex®-M4 is already compliant with the IEEE 754 requirements. In most cases there is no need to change the mode settings. If you need to use any of these modes in your application, typically you need to program the FPSCR and the FPDSCR. Otherwise the exception handlers could be using the default IEEE 754 behavior while the rest of the applications use other modes, resulting in inconsistency.

In the following sections we briefly cover these operation modes.

### *Flush-to-Zero mode*

Flush-to-Zero mode allows some floating point calculations to be faster by avoiding the need to calculate results in the denormalized value range (exponent = 0). When the value is too small to be represented by the normalized value range ($0 <$ exponent $<$ 0xFF), the value is replaced with zero. Flush-to-Zero mode is enabled by setting the FZ bit in FPSCR and FPDSCR.

### Default NaN mode

In default NaN (Not a Number) mode, if any of the inputs of a calculation is a NaN, or if the operation results in an invalid result, the calculation returns the default NaN (a non-signaling NaN, or quiet NaN). This is slightly different from the default configuration. By default, the default NaN mode is disabled and the behavior follows the IEEE 754 standard:

- An operation that produces an Invalid Operation floating point exception generates a quiet NaN as its result
- An operation involving a quiet NaN operand, but not a signaling NaN operand, returns an input NaN as its result

The default NaN mode is enabled by setting DN bit in FPSCR and FPDSCR. In some cases it enables quicker checking of NaN values in calculation results.

### Alternate half-precision mode

This mode only affects applications with half-precision data __fp16 (see section 13.1.3). By default, the FPU follows the IEEE 754 standard. If the exponent of the half-precision floating point data is 0x1F, the value is infinity or NaN. In the alternate half-precision mode, the value is a normalized value. The alternate half precision mode allows a wider value range, but does not support infinity and NaN.

The alternate half-precision mode can be enabled by setting the AHP bit in FPSCR and FPDSCR. To use half-precision data, you also need to set up your compiler command line options accordingly, as shown in Table 13.12.

### Rounding modes

The FPU supports four rounding modes as defined in the IEEE 754 standard. You can change the rounding mode at run-time. In C99, the fenv.h defines the four available modes, as listed in Table 13.13.

You can use these definitions with C99 functions defined in fenv.h:

- `int fegetround(void)` — return the currently selected rounding mode, represented by one of the values of the defined rounding mode macros

**Table 13.12** Command Line Options for Using Half Precision Data (__fp16)

| | IEEE Half Precision | Alternate Half Precision | None (Default) |
|---|---|---|---|
| ARM C compiler | –fp16_format=ieee | –fp16_format=alternative | –fp16_format=none |
| gcc | -mfp16-format=ieee | -mfp16-format=alternative | -mfp16-format=none |

**Table 13.13** C99 Definitions for Floating Point Rounding Modes

| fenv.h macros | Descriptions |
|---|---|
| FE_TONEAREST | Round to Nearest (RN) mode (default) |
| FE_UPWARD | Round towards Plus Infinity (RP) mode. |
| FE_DOWNWARD | Round towards Minus Infinity (RM) mode. |
| FE_TOWARDZERO | Round towards Zero (RZ) mode. |

- `int fesetround(int round)` — change the currently selected rounding mode. fesetround() returns zero if the change is successful, and nonzero if the change is not successful

You should use these C library functions when adjusting the rounding mode to ensure that the C run-time library functions are adjusted in the same way as the FPU.

## 13.5 Floating point exceptions

In section 13.2.5 and Table 13.4 we came across some floating point exception status bits. Here the term exception is not the same as the exceptions or interrupts in the NVIC. The floating point exceptions refer to issues in floating point processing. IEEE 754 defines the exceptions given in Table 13.14.

In addition to this, the FPU in the Cortex®-M4 also supports an additional exception for "Input Denormal" as given in Table 13.15.

The FPSCR provides six sticky bits so software code can check these values to see if the calculations were carried out successfully. In most cases these flags are ignored by software (compiler-generated code does not check these values).

**Table 13.14** Floating Point Exceptions Defined in the IEEE 754 Standard

| Exception | FPSCR Bit | Examples |
|---|---|---|
| Invalid operation | IOC | Square root of a negative number (return quiet NaN by default) |
| Division by zero | DZC | Divide by zero or log(0) (returns $\pm\infty$ by default) |
| Overflow | OFC | A result that is too large to be represented correctly (returns $\pm\infty$ by default) |
| Underflow | UFC | A result that is very small (returns denormalized value by default) |
| Inexact | IXC | The result has been rounded (return rounded results by default) |

**Table 13.15** Additional Floating Point Exceptions Provided in the Cortex-M4 FPU

| Exception | FPSCR Bit | Examples |
|---|---|---|
| Input Denormal | IDC | A denormalized input value is replaced with a zero in the calculation due to Flush-to-Zero mode. |

If you are designing software with high safety requirements, you could add checking to FPSCR. However, in some cases not all floating point calculations are carried out by the FPU. Some could be carried out by the C run-time library function. C99 has defined a number of functions for checking the floating exception status:

```
#include <fenv.h>
// check floating point exception flags
int fegetexceptflag(fexcept_t *flagp, int excepts);
// clear floating point exception flags
int feclearexcept(int excepts);
```

In addition, you can examine and change the configuration of the floating point run-time library using:

```
int fegetenv(envp);
int fesetenv(envp);
```

For detailed information on these functions, please refer to C99 documentation or manuals from toolchain vendors.

Alternatively, some development suites also provide additional functions for FPU control. For example, in ARM® C compilers (including Keil™ MDK) the __ieee_status() function allows you to configure the FPSCR easily:

```
// Modify the FPSCR (older version of __ieee_status() was
__fp_status())
unsigned int __ieee_status(unsigned int mask, unsigned int flags);
```

When using __ieee_status(), the mask parameter defines the bits that you want to modify, and the flags parameter specifies the new values of the bit covered by the mask. To make the use of these functions easier, fenv.h defines the following macros:

```
#define FE_IEEE_FLUSHZERO    (0x01000000)
#define FE_IEEE_ROUND_TONEAREST (0x00000000)
#define FE_IEEE_ROUND_UPWARD  (0x00400000)
#define FE_IEEE_ROUND_DOWNWARD  (0x00800000)
#define FE_IEEE_ROUND_TOWARDZERO (0x00C00000)
#define FE_IEEE_ROUND_MASK    (0x00C00000)
#define FE_IEEE_MASK_INVALID  (0x00000100)
```

```
#define FE_IEEE_MASK_DIVBYZERO  (0x00000200)
#define FE_IEEE_MASK_OVERFLOW  (0x00000400)
#define FE_IEEE_MASK_UNDERFLOW  (0x00000800)
#define FE_IEEE_MASK_INEXACT  (0x00001000)
#define FE_IEEE_MASK_ALL_EXCEPT (0x00001F00)
#define FE_IEEE_INVALID   (0x00000001)
#define FE_IEEE_DIVBYZERO   (0x00000002)
#define FE_IEEE_OVERFLOW   (0x00000004)
#define FE_IEEE_UNDERFLOW   (0x00000008)
#define FE_IEEE_INEXACT  (0x00000010)
#define FE_IEEE_ALL_EXCEPT   (0x0000001F)
```

For example, to clear the Underflow sticky flag, you can use:

```
__ieee_status(FE_IEEE_UNDERFLOW, 0);
```

In the Cortex-M4 hardware design, these exception status bits are also exported to the top level of the processor. Potentially these signals can be used to trigger an exception at the NVIC, as shown in Figure 13.26.

However, since the interrupt events are imprecise, the generated exception could be delayed by a few cycles even if the exception is not blocked. As a result, you cannot determine which floating point instruction triggered the exception. If the processor was executing a higher priority interrupt handler, the interrupt handler for the floating point exception cannot start until the other

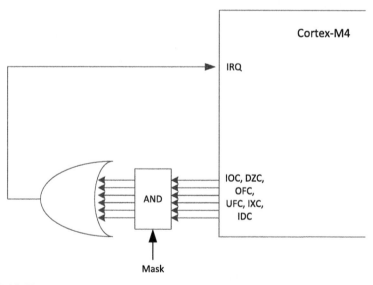

**FIGURE 13.26**

Use of floating point exception status bit for hardware exception generation interrupt handler is completed. However, this mechanism could be used in some systems that need to detect error conditions like divide-by-zero or overflow instantaneously.

When the FPU exception status is used to trigger exceptions at NVIC, please note that the exception handler needs to clear the exception status bits in FPSCR, as well as the stacked FPSCR before exception return. Otherwise the exception could be triggered again accidentally.

## 13.6 Hints and tips
### 13.6.1 Run-time libraries for microcontrollers

Some development suites provide special run-time libraries that are optimized for microcontrollers with small memory footprints. For example, with Keil™ MDK-ARM, ARM® DS-5™, or ARM RVDS you can select to use the MicroLIB (see Figure 13.24, just above the Use FPU option). In most cases, these libraries can provide the same floating point functionalities as the standard C libraries. However, there can be limitations on full IEEE 754 support.

For example, the MicroLIB has the following limitations in terms of IEEE 754 floating point support:

- Operations involving NaNs, infinities, or input denormals produce indeterminate results. Operations that produce a result that is nonzero but very small in value (normally represent with de-normalized value) return zero.
- IEEE exceptions cannot be flagged by Microlib, and there is no __ieee_status()/__fp_status() register in Microlib.
- The sign of zero is not treated as significant by Microlib, and zero that are output from Microlib floating point arithmetic have an unknown sign bit.
- Only the default rounding mode is supported.

In the majority of embedded applications such limitations do not cause any problem, and the MicroLIB allows applications to be compiled to a smaller size by reducing the size of the library.

### 13.6.2 Debug operation

Lazy stacking might add a bit of complexity to debugging. When the processor is halted in an exception handler, the stack frame might not contain the contents of the floating point registers. And when you single-step your code and the processor executes a floating point instruction, the deferred stacking will then take place.

# Introduction to the Debug and Trace Features

# 14

**443**

## 14.1 Debug and trace features overview

### 14.1.1 What are debug features?

The Cortex®-M3 and Cortex-M4 processors support a wide range of debug features. For readers who are new to microcontrollers, we first introduce some basic information about debug.

In Chapter 2 we mentioned that you might need some sort of debug adaptor to connect between the microcontroller and the debug host (personal computer). An example is shown in Figure 14.1. In some cases a debug adaptor is included in your development board. This debug adaptor converts the debug communication from USB to JTAG or Serial Wire debug protocol. The debug communication protocol is then converted again by hardware on-chip, so that various debug components and debug features can be accessed.

Some readers might wonder: since USB is so popular nowadays and many microcontrollers have a USB interface, why can't we use USB directly for debug operations?

Although USB is popular and widely available, USB communications require relatively complex hardware on the chip, and have various system requirements (e.g., clocks, voltage, power, etc.). As a result, USB is less suitable for direct use as a debug communication protocol, as it can increase the power consumption and the cost of the design. On the other hand, Serial Wire and JTAG interfaces have much smaller silicon areas and can be used with an extremely wide range of clock frequencies. The JTAG debug protocol also allows JTAG debug interfacing, in which multiple chips can be connected in a simple daisy chain arrangement, which is useful for some complex system designs.

**FIGURE 14.1**

An LPC4330-Xplorer development board connected to a Keil™ ULINK 2 debug adaptor

Once there is a communication connection between the debug host and the microcontroller, we can then:

- Download a compiled program image to the microcontroller.
- Control the microcontroller to perform a reset.
- Start the program execution, or if you like, execute instructions line by line (single stepping).
- Stop the processor (halting).
- Insert/remove breakpoints: the breakpoint feature allows you to define instruction addresses so that when the processor gets to that instruction, it stops there (enters halt, or sometimes referred to as debug state in ARM® documentation).
- Insert/remove watchpoints: the watchpoint feature allows you to define data addresses so that when the processor accesses this address location, it triggers a debug event that can be used to halt the processor.
- You can examine or change the contents of the memory or peripherals at any time, even when the processor is running. This feature is often called on-the-fly memory access.
- You can also examine or change the values inside the processor. This can only be done when the processor is stopped (halted).

All these debug features are available on all the Cortex-M processors, including the low-cost Cortex-M0 and Cortex-M0+ microcontroller products.

## 14.1.2 What are trace features?

Trace features on the Cortex®-M3 and Cortex-M4 processors use a separate interface to export information during program execution. It is real-time (with a small timing delay), and can provide a lot of useful information without stopping the processor.

Two types of trace interface operation modes are supported: either a single-pin model called Serial Wire Viewer (SWV) (using a signal called Serial Wire Output (SWO)), or a multi-pin Trace Port interface (typically four data pins + one clock pin). Although debug and trace interfaces operate independently, they are normally connected using a single connector. Figure 14.2 shows typical debug and trace connection configurations. Please refer to Appendix H for common debug and trace connector arrangements.

SWV provides a low pin count trace solution, and the SWO output signal can be shared with the TDO (Test Data Out, see Figure 14.2) pin when the Serial Wire debug protocol is used, and allows trace data to be collected with the same debug adaptor for debug operations. The baud rate of the SWO is typically limited to under 2Mbit/sec (limited by the debug adaptor hardware; the actual bandwidth could be even lower). However, it is very useful for exporting lots of information as discussed in the following:

- Exception events
  - Information associated with a data watchpoint event (e.g., data value, program counter value, address values, or combination of two pieces of such information).

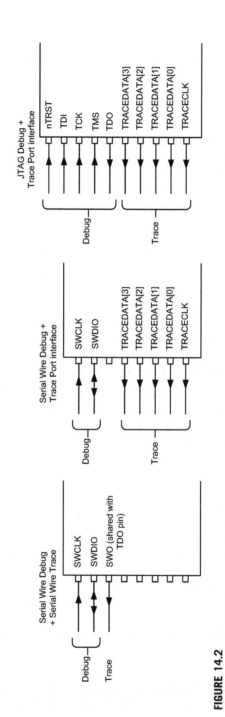

**FIGURE 14.2**

Debug and trace interface

- Events from profiling counters
  - Software-generated trace data (instrumentation trace; for example, a message output from a "printf" statement in the program code).
  - Timestamp information: for each trace data, you can enable a timestamp packet to go with it so that the timing of events can be reconstructed by the debug host.

The Trace Port interface provides a much higher trace data bandwidth. However, you need a more complex (and usually more expensive) trace adaptor to capture the trace information using the trace port interface. With the high bandwidth offered by the Trace Port output, you can collect additional trace information in addition to the information that you can get from SWO:

- Instruction trace. The instruction trace feature needs an optional on-chip hardware component called the Embedded Trace Macrocell (ETM). This can provide information about the program execution so the whole program execution history can be reconstructed by the debug host. This feature is very useful for debugging complex software bugs, code coverage measurement, and performance analysis. The ETM instruction trace also supports timestamp information.

Instruction trace is also available on the Cortex-M0+ processor as an optional component. However, the design is different. The Micro Trace Buffer (MTB) in the Cortex-M0+ processor uses a small part of the SRAM in the microcontroller as trace buffer, typically operating in circular buffer mode to store information about program flow changes. After the program stops, the debugger can then extract the trace data from SRAM using the debug connection. Table 14.1 compares the advantages of MTB and ETM instruction trace.

Today, debug adaptors like ULink Pro support streaming trace — so the length of trace history is only limited by the storage space on the debug host computer.

### 14.1.3 Debug and trace features summaries

There are many different debug and trace features in the Cortex®-M3 and Cortex-M4 processors. Many of these features are also available on the Cortex-M0 and Cortex-M0+ processors. Table 14.2 summarized these features on the current Cortex-M processors.

| **Table 14.1** Advantages of ETM Trace Compared to MTB Trace | |
|---|---|
| **Advantages of ETM (Available on Cortex-M3, Cortex-M4)** | **Advantages of MTB (Available on Cortex-M0+)** |
| Unlimited trace history. | Only need a low-cost debug adaptor, and use the same JTAG or Serial Wire debug connection. |
| Provide timing information via timestamp packets. | |
| Real time – information are collected while the processor is still running. | Small silicon footprint (just 1.5K gates for typical configuration) and therefore lower power. |

**Table 14.2** Debug and Trace Features on Various Cortex-M Processors

| Feature | Cortex-M0 | Cortex-M0+ | Cortex-M3 / Cortex-M4 |
|---|---|---|---|
| JTAG or Serial Wire debug protocol | Typically just one of the debug protocol | Typically just one of the debug protocol | Typically support both, and allow dynamic switching |
| Core debug – halting, single stepping, resume, reset, register accesses | Yes | Yes | Yes |
| One the fly memory and peripheral accesses | Yes | Yes | Yes |
| Hardware breakpoint comparators | Up to 4 | Up to 4 | Up to 8 (6 instruction addresses and 2 literal data addresses) |
| Software breakpoints (Breakpoint instruction) | Unlimited | Unlimited | Unlimited |
| Hardware watch point comparators | Up to 2 | Up to 2 | Up to 4 |
| Instruction Trace | No | MTB (optional) | ETM (optional) |
| Data trace | No | No | Yes |
| Software trace (Instrumentation Trace) | No | No | Yes |
| Debug monitor exception | No | No | Yes |
| Profiling counters | No | No | Yes |
| PC Sampling | Via debugger read accesses only | Via debugger read accesses only | Via debugger read accesses or via trace |
| Flash patch logic | No | No | Yes |
| CoreSight architecture compliant | Yes | Yes | Yes |

Many of these debug features are configurable by chip designers. For example, an ultra-low power sensor device based on a Cortex-M processor can have a reduced number of breakpoint and watchpoint comparators to reduce the power consumption. Some components like the ETM are optional; some

Cortex-M3 or Cortex-M4 microcontrollers might not have the ETM instruction trace.

From a different point of view, the debug and trace features can be divided into two types:

**Invasive debugging** — features that need to stop the processor or change the program execution flow significantly.

- Core debug features — Program halting, single-stepping, reset, resume
- Breakpoints
- Data watchpoints configured to halt the processor
- Processor internal register access (both read or write; this can only be carried out when the processor is halted)
- Debug monitor exception
- ROM base debugging using flash patch logic

**Non-invasive debugging** — features that have no or very little effect on the program flow.

- On the fly memory/peripheral accesses
- Instruction trace (through the Cortex-M3/M4 ETM or Cortex-M0+ MTB)
- Data trace (it use the same comparators for data watchpoint, but is configured for data trace)
- Software-generated trace (or called Instrumentation Trace, it needs some software code execution but the timing impact is relatively small)
- Profiling (using profiling counters, or PC sampling features)

In most cases, invasive debugging is easier to use as you can control the operation of each step in the program. However, invasive debugging can be unsuitable for some applications. For example, in some motor control applications it can be dangerous to stop the motor controller in the middle of operation, and therefore information about the program execution has to be collected using non-invasive debugging techniques.

## 14.2 Debug architecture
### 14.2.1 CoreSight™ debug architecture

The debug and trace features of the Cortex®-M processors are designed based on the CoreSight™ Debug Architecture. This architecture covers a wide area, including debug interface protocols, on-chip bus for debug accesses, control of debug components, security features, trace data interface, and more.

For normal software development, it is not necessary to have an in-depth understanding of CoreSight technology. For readers who would like to have a brief verview of this subject, the CoreSight Technology System Design Guide (reference 16) is a useful document for getting an overview of the architecture. Full details of the CoreSight Debug Architecture and Cortex-M specific debug system designs are documented in:

- CoreSight Architecture Specification (ARM® IHI 0029).
- ARM Debug Interface v5.0/5.1 (e.g., ARM IHI 0031A) — details of the programmer's model for debug connection components (Debug Port and Access Port, to be introduced later in this chapter, and the Serial Wire and JTAG communication details).
- Embedded Trace Macrocell Architecture Specification (ARM IHI 0014Q) — for details of the ETM trace packet format, programmer's model.
- ARMv7-M Architecture Reference Manual (reference 1) also covers the debug support available on the Cortex-M3 and Cortex-M4 processors.

One important aspect of the debug support in the Cortex-M processors is the support for multiple processor designs. The CoreSight architecture allows the sharing of debug connections and trace connections. By default, the Cortex-M3 and Cortex-M4 processors have a pre-configured processor system for single core environments, and the system can be modified to support multi-processor designs by adding additional CoreSight debug and trace components from ARM.

This is different from many classic ARM processors, such as ARM7TDMI™ and some of the ARM9™ processors, where the debug systems were not designed based on CoreSight, and the debug and trace interface is an integrated part of the processor designs. For ARM processors designed in recent years with CoreSight architecture, the designs use a modular approach so that the debug interfaces are decoupled from the processors, and can be linked together using a scalable bus system (Figure 14.3).

The debug components used in Cortex-M processors are designed specifically for low-gate count designs and therefore some of them are different from the standard CoreSight debug components in a number of ways. However, they are designed to be CoreSight compliant and can work with other CoreSight debug and trace components seamlessly.

### 14.2.2 Processor debug interface

Now let's look at the debug interface in a bit more detail.

In order to allow the debugger to access various debug features, a debug interface is needed. Many microcontrollers support a serial protocol called JTAG — Joint Test Action Group. The JTAG protocol is an industry standard protocol (IEEE 1149.1) and can be used for various functions such as chip-level or PCB-level testing, as well as giving access to debug features inside microcontrollers.

While JTAG is sufficient for many debug usage scenarios, it needs at least four pins: TCK, TDI, TMS, and TDO; the reset signal nTRST is optional. For some microcontroller devices with low pin counts, such as 28-pin packages, using four pins for debugging is too many. As a result, ARM® developed the Serial Wire debug protocol, which only needs two pins: SWCLK and SWDIO. The Serial Wire debug protocol provides the same debug access features and also supports parity error

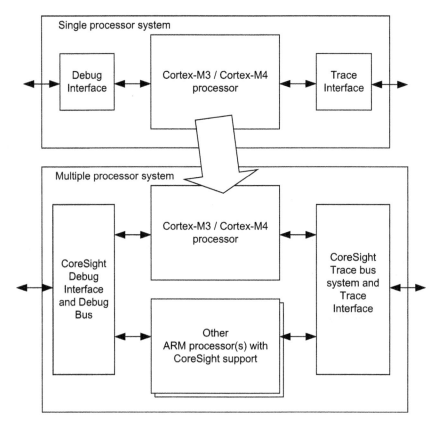

**FIGURE 14.3**

Decoupling of debug and trace interface from processor designs allows the processor to be used in both single-core and multi-core designs

detection, which enables better reliability in systems with higher electrical noise. Therefore the Serial Wire debug protocol is very attractive for many microcontroller vendors and users.

Typically, Serial Wire debug and JTAG debug protocols share the same connections: TCK and SWCLK use the same pin, and TMS and SWDIO use the same pin (Figure 14.4).

Many Cortex®-M3 and Cortex-M4 microcontrollers support both JTAG and Serial Wire debug protocols. You can dynamically switch between the two modes using special bit patterns on the TMS/SWDIO pin. In order to achieve lower power consumption, some microcontrollers omit the JTAG debug protocol capability and support only the Serial Wire debug protocol.

Typically Cortex-M0 and Cortex-M0+ microcontrollers only support one of these debug protocols to reduce power: mostly the Serial Wire debug protocol because fewer pins are needed.

JTAG and Serial Wire pin sharing

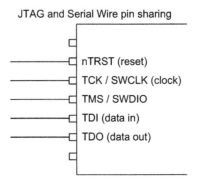

nTRST (reset)

TCK / SWCLK (clock)

TMS / SWDIO

TDI (data in)

TDO (data out)

**FIGURE 14.4**

Pin sharing between JTAG and Serial Wire debug protocol

When using the Serial Wire debug protocol, the TDO pin can be used for trace operation with SWO (Serial Wire Output), as outlined in section 14.1.2. Many low-cost debug adaptors support trace capture via SWO.

## 14.2.3 Debug Port (DP), Access Port (AP), and Debug Access Port (DAP)

Inside the chip, the Serial Wire/JTAG signals are connected to the debug system via a number of stages, as shown in Figure 14.5.

The first stage is a Debug Port (DP) component. This converts the debug interface protocol into a generic bus protocol for the internal debug bus. For Cortex®-M3 and Cortex-M4 designs, this is usually a module called SWJ-DP (Serial Wire JTAG DP), which supports both Serial Wire and JTAG protocols. In some cases, the design might use a SW-DP, which only supports Serial Wire protocol. Some older generations of ARM® microcontrollers might have a JTAG-DP module that only supports the JTAG protocol. Chip manufacturers can use the DP module that fits their needs.

The other side of the DP module is the internal debug bus, which is a 32-bit bus and has a bus protocol very similar to the Advanced Peripheral Bus (APB) in the AMBA® 3.0 specification. The internal debug bus supports up to 256 Access Ports (AP) devices, and the highest 8-bit of the address bus is used to determine which AP is accessed. Each Cortex-M3 or Cortex-M4 processor has a DAP interface that contains only one AP device, so theoretically you can have hundreds of processors sharing one debug connection. In typical single processor systems, the AP module for the Cortex-M3 or Cortex-M4 processor is the first AP device on the internal debug bus (i.e., the AP is selected when highest 8-bit of the address is zero).

The middle 16-bit of the address in the debug bus is not used, and each AP has only 64 words of address space. So in order to allow the whole 4GB address space to be accessed by the debugger, another level of address remapping is needed. The AHB-AP module is one type of memory Access Port module that converts commands from the debugger to memory accesses based on the Advanced

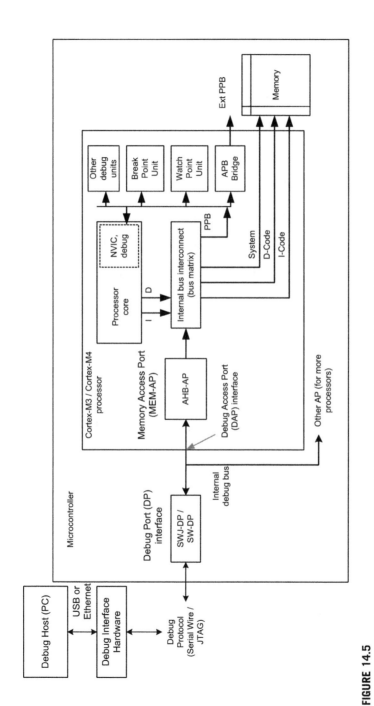

**FIGURE 14.5**

Connection from debug host to the Cortex®-M processor

High-performance Bus (AHB) protocol. The AHB-AP module is connected to the internal bus system of the Cortex-M3 or Cortex-M4 processor and has full visibility of the memory map. This allows the debugger to access all the memories, peripherals, debug components, and debug registers of the processor.

Besides AHB-AP, other forms of AP module are also available in the CoreSight product range. For example, an APB-AP module can be used to convert debugger accesses to APB accesses for CoreSight debug components, and a JTAG-AP module can be used to control traditional JTAG-based test interfaces such as the debug interface on ARM7TDMI™.

## 14.2.4 Trace interface

Another part of the CoreSight architecture concerns trace data handling. In the Cortex®-M3 and Cortex-M4 processors, there can be three types of trace source:

- Embedded Trace Macrocell (ETM) — this optional component generates instruction Trace.
- Data Watchpoint and Trace (DWT) unit — this unit can be used to generate data trace, event trace, and profiling trace information.
- Instrumentation Trace Macrocell (ITM) — this allows software-generated debug messages such as using printf, and is also used to generate timestamp information.

During tracing, the trace data in the form of packets are output from the trace sources using a set of on-chip trace data bus called Advanced Trace Bus (ATB). Based on the CoreSight architecture, trace data from multiple trace sources (e.g., multiple processors) can be merged using an ATB merger hardware called the CoreSight Trace Funnel. The merged data can then be converted and exported to the trace interface of the chip using another module called the Trace Port Interface Unit (TPIU), as shown in Figure 14.6. Once the converted data is exported, it can be captured using trace capturing devices and analyzed by the debug host (e.g., a personal computer). The data stream can then be converted back into multiple data streams.

In the Cortex-M3 and Cortex-M4 processor designs, in order to reduce the overall silicon size, the arrangement of the trace system is a bit different (Figure 14.7). The Cortex-M3/M4 TPIU module is designed with two ATB ports so that there is no need to use a separate trace funnel module. Also it supports both Trace Port mode and SWV mode (use SWO output signal), whereas in CoreSight systems the SWV operation requires a separate module.

## 14.2.5 CoreSight characteristics

The CoreSight™-based design has a number of advantages:

- The memory content and peripheral registers can be examined even when the processor is running.
- Multiple processor debug interfaces can be controlled with a single piece of debugger hardware. For example, if JTAG is used, only one Test Access Port (TAP) controller is required, even when there are multiple processors on the chip.

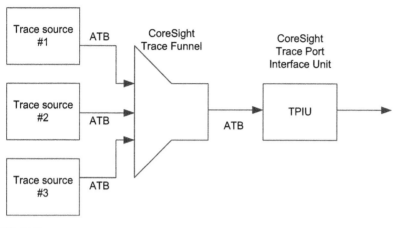

**FIGURE 14.6**

Typical trace stream merging in CoreSight systems

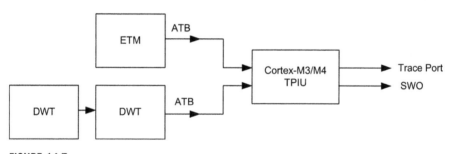

**FIGURE 14.7**

Trace stream merging in Cortex-M3 and Cortex-M4

- Internal debugging interfaces are based on a simple bus design, making it scalable and easy to develop additional test logic for other parts of the chip or SoC.
- It allows multiple trace data streams to be collected by one trace capture device and separated back into multiple streams on the debug host.
- New Debug Port, Access Port, and TPIU designs can be developed and used with existing processors and CoreSight components. So it is possible to have different types of debug interfaces when new technology is available.

The debugging system used in the Cortex®-M3 and Cortex-M4 processors is slightly different from the usual CoreSight implementation:

- Trace components are specially designed in the Cortex-M3/M4. Some of the ATB interface is 8 bits wide in the Cortex-M3/M4, whereas in CoreSight components the ATB width is 32 bits.
- The debug implementation in the Cortex-M3/M4 does not support TrustZone.[1]

[1]TrustZone is an ARM® Technology that provides security features to embedded products.

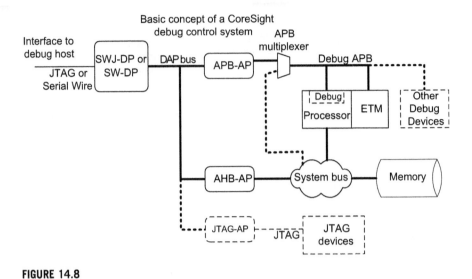

**FIGURE 14.8**

Design concept of a CoreSight system

- The debug components are part of the system memory map, whereas in standard CoreSight systems, a separate bus (potentially with a separate memory map) is used for controlling debug components. For example, the conceptual CoreSight debug system connection in a Cortex-R processor can be like the one shown in Figure 14.8, whereas the Cortex-M3/M4 debug components are part of the system memory space, as shown in Figure 14.9.

Although the debug components in Cortex-M3 and Cortex-M4 are built differently from usual CoreSight systems, the communication interface and protocols in the Cortex-M3 are compliant with CoreSight architecture and can be directly attached to other CoreSight systems. For example, the Cortex-M3/M4 trace interface can be connected to CoreSight components like CoreSight Trace Funnel and TPIU, and the DAP interface of the Cortex-M3/M4 can be connected to the CoreSight DAP. As a result, it is quite easy to integrate Cortex-M3 and Cortex-M4 processors into a multi-core system with other ARM® Cortex processors, and connect the debug and trace systems together.

## 14.3 Debug modes

There are two types of debug operation modes in Cortex®-M3 and Cortex-M4 processors:

- The first one is **halting**, whereby the processor stops program execution completely when a debug event occurs, or if the user requests the halting of program execution. This is the most commonly used debug method.

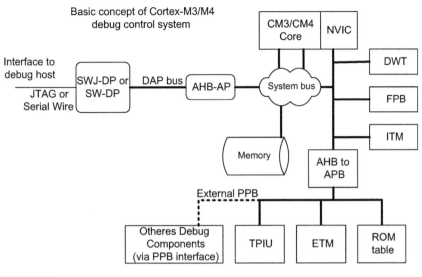

**FIGURE 14.9**

The Design System in the Cortex-M3 and Cortex-M4 processors

- The second one is the **debug monitor exception**, whereby the processor executes a special exception handler when a debug event occurs, or if the user requests the halting of application program execution (the application code apart from the debug monitor exception). The debug monitor exception handler then communicates with the debug host using a communication interface peripheral (e.g., UART). The debug monitor is exception type 12 and its priority is programmable, so it still allows higher-priority exceptions to take place while debugging tasks are being carried out.

Both debug mechanisms can be invoked by means of debug events, as well as by manually setting the control bit in debug control registers.

Halt-mode debugging using a Serial Wire or JTAG connection is very easy to use, powerful, and does not require additional program and data memories. You can:

- Stop program execution via the debugger, or when a debug event (e.g., breakpoint) occurs
- Single step each instruction, or each line of C code (depending on the debugger being used)
- Examine and modify the values of all registers in the processor when the processor is halted
- Examine and modify the values in memories or peripherals at any time, even when the processor is running

However, stopping the processor in the middle of an operation for debugging can be undesirable or even dangerous. For example, if the microcontroller is being used to control an engine or a motor, stopping the microcontroller for halt mode debugging might mean that we lose control of the engine or the motor, or in some case the sudden stopping of the controller could cause physical damage.

In order to solve this problem we use the debug monitor method. To stop a program from running, the processor executes the debug monitor exception handler. The debug monitor exception has programmable priority, and as a result, other interrupts and exceptions with priority levels higher than the debug monitor exception can still execute. In this way, critical operations like motor control can be handled by higher priority interrupt handlers and continue to operate even when the main program is stopped.

To use the debug monitor feature for debugging, we need to have a piece of debug agent code in the program image of the microcontroller (Figure 14.10). This code can be invoked by the debug monitor exception, as well as the interrupt handler of a communication interface. An initialization stage in the program execution is also needed to establish communication between the microcontroller and the debug host. Once the debug agent and the communication link are set up, the debug agent code can execute when there is a debug event, or when there are communication activities from the debug host.

The debug monitor exception solution has some limitations. For example, it cannot be used to debug operations in the Non-Maskable Interrupt (NMI) handler, HardFault handler, or any other exception handler with the same or higher priority than the Debug Monitor exception. Also, it can be blocked by exception masking registers such as PRIMASK. Finally, it needs additional resources including memory, execution time, and a communication peripheral.

The following is a summary of other differences between Halt mode debugging and Debug Monitor mode debugging.

**Halt mode:**

- Instruction execution is stopped
- The System Tick Timer (SYSTICK) counter is stopped
- Supports single-step operations

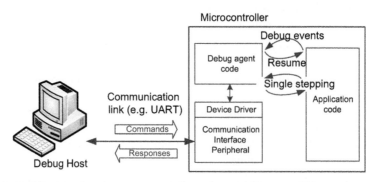

**FIGURE 14.10**

Debug monitor concept

- Interrupts can be pended and can be invoked during single stepping or be masked so that external interrupts are ignored during stepping

**Debug monitor mode:**

- Processor executes exception handler type 12 (debug monitor)
- SYSTICK counter continues to run
- New arrive interrupts may or may not preempt, depending on the priority of the debug monitor and the priority of the new interrupt
- If the debug event takes place when a higher-priority interrupt is running, the debug event will be missed
- Supports single-step operations
- Memory contents (e.g., stack memory) could be changed by the debug monitor handler during stacking and handler execution

## 14.4 Debug events

The Cortex®-M3/M4 processor will enter debug mode (either the halt or debug monitor exception) for a number of possible reasons. For halt mode debugging, the processor will enter halt mode if conditions resemble those shown in Figure 14.11. This behavior is controlled by a number of programmable bits in a number of debug control registers. C_DEBUGEN is one of the bit fields in the Debug Halting Control and Status Register (DHCSR, at address 0xE000EDF0); see Table 14.4. This bit is used to enable Halt mode debugging and can only be programmed by a debugger connected to the microcontroller. Another bit in this register is called C_HALT, and it can be set by the debugger to halt the processor manually.

One of the debug events is an external debug request signal, called EDBGREQ on the Cortex-M3/M4 processor. The actual connection of this signal depends on the microcontroller or SoC design. In some cases, this signal could be tied low and never occur. However, this can be connected to accept debug events from additional debug components (chip manufacturers can add extra debug components to the SoC) or, if the design is a multi-processor system, it could be linked to debug events from another processor.

Another debug event available is called Vector Catch. This is a programmable feature that, when enabled, can halt the processor right after a system reset, or when certain fault exceptions occur. It is controlled by a register called Debug Exception and Monitor Control Register (DEMCR, at address 0xE000EDFC); see Table 14.5. The vector catch mechanism is commonly used by the debugger to halt the processor at the beginning of a debug session, immediately after the program is downloaded and the processor is reset.

After debugging is completed, program execution can be returned to normal by clearing the C_HALT bit in DHCSR.

Similarly, for debugging with the debug monitor exceptions, a number of debug events can cause a debug monitor to take place (Figure 14.12).

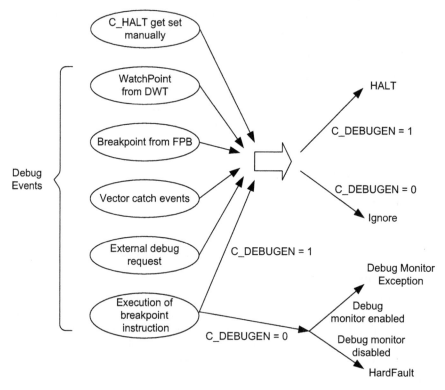

**FIGURE 14.11**

Debug events for halt mode debugging

For debug monitor, the behavior is a bit different from halt mode debugging. This is because the debug monitor exception is just one type of exception, and it can be affected by the current priority of the processor if it is running another exception handler.

After debugging is completed, the program execution can be returned to normal by carrying out an exception return.

## 14.5 Breakpoint feature

One of the most commonly used debug features in most microcontrollers is the breakpoint feature. In the Cortex®-M, the following two types of breakpoint mechanisms are supported:

- Breakpoint instruction BKPT, also known as software breakpoints. You can insert an unlimited number of BKPT instructions in your application for debugging (obviously subject to memory size).
- Breakpoint using address comparators in the Flash Patch and Breakpoint Unit (FPB). The number of hardware comparators is limited. For the Cortex-M3 and

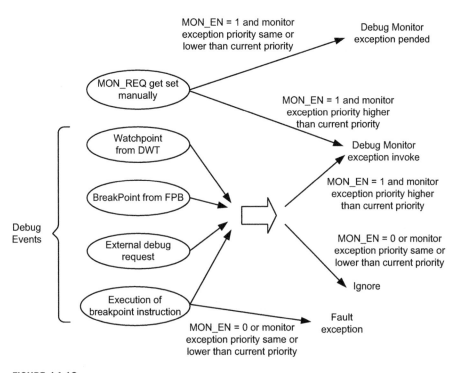

**FIGURE 14.12**

Debug events for Debug Monitor Exceptions

Cortex-M4, there can be up to eight comparators (only six of them can generate an instruction address breakpoint).

The breakpoint instruction (BKPT #immed8) is a 16-bit Thumb instruction with the encoding 0xBExx. The lower 8 bits depend on the immediate data given following the instruction. When this instruction is executed, it generates a debug event and can be used to halt the processor core if C_DBGEN is set, or if the debug monitor is enabled, it can be used to trigger the Debug Monitor exception. Because the debug monitor is one type of exception with programmable priority, it can only be used in thread or exception handlers with priority lower than itself. As a result, if the debug monitor is used for debugging, the BKPT instructions should not be used in exception handlers such as MonMaskable Interrupt (NMI) or HardFault.

When the processor is unhalted, or when the debug monitor exception returns, it is returned to the address of the BKPT instruction, not the address after the BKPT instruction. This is because in normal use of breakpoint instructions, the BKPT is used to replace a normal instruction, and when the breakpoint is hit and the debug action is carried out, the instruction memory is restored to the original instruction, and the rest of the instruction memory is unaffected.

If the BKPT instruction is executed with C_DEBUGEN = 0 and MON_EN = 0, it will cause the processor to enter a HardFault exception, with DEBUGEVT in the Hard Fault Status register (HFSR) set to 1, and BKPT in the Debug Fault Status register (DFSR) also set to 1.

The FPB unit can be programmed to generate breakpoint events even if the program memory cannot be altered. However, it is limited to six instruction addresses and two literal addresses. More information about the FPB is covered later, in section 14.6.2.

## 14.6 Debug components introduction

### 14.6.1 Processor debug support

The processor core contains a number of debug control registers (See Table 14.3). They provide controls for debug operations including halting, resume, single stepping, and configuration of Debug Monitor exception, Vector Catch feature, etc.

Details of DHCSR are shown in Table 14.4. Note: For DHCSR, bits 5,3,2,0 are reset by power on reset only. Bit 1 can be reset by power on reset (cold reset) and system reset (including VECTRESET using SCB->AIRCR).

To enter halt mode, the C_DEBUGEN bit in the Debug Halting Control and Status register (DHCSR) must be set. This bit can only be programmed through the debugger connection (via the Debug Access Port, DAP), so you cannot halt the Cortex®-M3/M4 processor without a debugger. After C_DEBUGEN is set, the core can be halted by setting the C_HALT bit in DHCSR. This bit can be set by either the debugger or by the software running on the processor itself.

The bit field definition of DHCSR differs between read operations and write operations. For write operations, a debug key value must be used on bit 31 to bit 16. For read operations, there is no debug key and the return value of the upper half-word contains the status bits (see Table 14.4).

**Table 14.3** Core Debug Registers in Cortex®-M3 and Cortex-M4

| Address | Name | Type | Reset Value |
|---------|------|------|-------------|
| 0xE000EDF0 | Debug Halting Control Status Register (DHCSR) | R/W | 0x00000000 |
| 0xE000EDF4 | Debug Core Register Selector Register (DCRSR) | W | - |
| 0xE000EDF8 | Debug Core Register Data Register (DCRDR) | R/W | - |
| 0xE000EDFC | Debug Exception and Monitor Control Register (DEMCR) | R/W | 0x00000000 |

**Table 14.4** Debug Halting Control and Status Register (CoreDebug->DHCSR, 0xE000EDF0)

| Bits | Name | Type | Reset Value | Description |
|------|------|------|-------------|-------------|
| 31:16 | KEY | W | — | Debug key; value of 0xA05F must be written to this field to write to this register, otherwise the write will be ignored |
| 25 | S_RESET_ST | R | — | Core has been reset or being reset; this bit is clear on read |
| 24 | S_RETIRE_ST | R | — | Instruction is completed since last read; this bit is clear on read |
| 19 | S_LOCKUP | R | — | When this bit is 1, the core is in a locked-up state |
| 18 | S_SLEEP | R | — | When this bit is 1, the core is in sleep mode |
| 17 | S_HALT | R | — | When this bit is 1, the core is halted |
| 16 | S_REGRDY | R | — | Register read/write operation is completed |
| 15:6 | Reserved | — | — | Reserved |
| 5 | C_SNAPSTALL | R/W | 0* | Use to break a stalled memory access |
| 4 | Reserved | — | — | Reserved |
| 3 | C_MASKINTS | R/W | 0* | Mask interrupts while stepping; can only be modified when the processor is halted |
| 2 | C_STEP | R/W | 0* | Single step the processor; valid only if C_DEBUGEN is set |
| 1 | C_HALT | R/W | 0 | Halt the processor core; valid only if C_DEBUGEN is set |
| 0 | C_DEBUGEN | R/W | 0* | Enable halt mode debug |

In normal situations, the DHCSR is used only by the debugger. Application code should not change DHCSR contents to avoid causing problems to debugger tools.

Details of the DEMCR are shown in Table 14.5. Note: For DEMCR, bits 16 to 19 are reset by system reset as well as power on reset. Other bits are reset by power on reset only.

The DEMCR register is used to control the Vector Catch feature, Debug Monitor exception, as well as enabling the trace subsystem. Before using any trace features

**Table 14.5** Debug Exception and Monitor Control Register (CoreDebug->DEMCR, 0xE000EDFC)

| Bits | Name | Type | Reset Value | Description |
|------|------|------|-------------|-------------|
| 24 | TRCENA | R/W | 0* | Trace system enable; to use DWT, ETM, ITM and TPIU, this bit must be set to 1 |
| 23:20 | Reserved | — | — | Reserved |
| 19 | MON_REQ | R/W | 0 | Indication that the debug monitor is caused by a manual pending request rather than hardware debug events |
| 18 | MON_STEP | R/W | 0 | Single step the processor; valid only if MON_EN is set |
| 17 | MON_PEND | R/W | 0 | Pend the monitor exception request; the core will enter monitor exceptions when priority allows |
| 16 | MON_EN | R/W | 0 | Enable the debug monitor exception |
| 15:11 | Reserved | — | — | Reserved |
| 10 | VC_HARDERR | R/W | 0* | Debug trap on hard faults |
| 9 | VC_INTERR | R/W | 0* | Debug trap on interrupt/ exception service errors |
| 8 | VC_BUSERR | R/W | 0* | Debug trap on bus faults |
| 7 | VC_STATERR | R/W | 0* | Debug trap on usage fault state errors |
| 6 | VC_CHKERR | R/W | 0* | Debug trap on usage fault-enabled checking errors (e.g., unaligned, divide by zero) |
| 5 | VC_NOCPERR | R/W | 0* | Debug trap on usage fault, no coprocessor errors |
| 4 | VC_MMERR | R/W | 0* | Debug trap on memory management fault |
| 3:1 | Reserved | – | – | Reserved |
| 0 | VC_CORERESET | R/W | 0* | Debug trap on core reset |

(e.g., instruction trace, data trace) or accessing any trace component (e.g., DWT, ITM, ETM, TPIU), the TRCENA bit must be set to 1.

Two more registers are included to provide access to registers inside the processor core. They are the Debug Core Register Selector register (DCRSR) and the Debug Core Register Data register (DCRDR) (see Tables 14.6 and 14.7). The register transfer feature can be used only when the processor is halted. For Debug Monitor mode debugging, the debug agent code can have access to all registers by software.

**Table 14.6** Debug Core Register Selector Register (CoreDebug->DCRSR, 0xE000EDF4)

| Bits | Name | Type | Reset Value | Description |
|------|------|------|-------------|-------------|
| 16 | REGWnR | W | — | Direction of data transfer: Write = 1, Read = 0 |
| 15:7 | Reserved | — | — | - |
| 6:0 | REGSEL | W | — | Register to be accessed: 0000000 = R0 0000001 = R1 … 0001111 = R15 0010000 = xPSR/flags 0010001 = MSP (Main Stack Pointer) 0010010 = PSP (Process Stack Pointer) 0010100 = Special registers: [31:24] Control [23:16] FAULTMASK [15:8] BASEPRI [7:0] PRIMASK 0100001 = Floating Point Status & Control Register (FPSCR) 1000000 = Floating point register S0 … 1011111 = Floating point register S31 Other values are reserved |

**Table 14.7** Debug Core Register Data Register (CoreDebug->DCRDR, 0xE000EDF8)

| Bits | Name | Type | Reset Value | Description |
|------|------|------|-------------|-------------|
| 31:0 | Data | R/W | — | Data register to hold register read result or to write data into selected register |

To use these registers to read register contents, the following procedure must be followed:

1. Make sure the processor is halted.
2. Write to the DCRSR with bit 16 set to 0, indicating it is a read operation.
3. Poll until the S_REGRDY bit in DHCSR (0xE000EDF0) is 1.
4. Read the DCRDR to get the register content.

Similar operations are needed for writing to a register:

1. Make sure the processor is halted.
2. Write the data value to the DCRDR.
3. Write to the DCRSR with bit 16 set to 1, indicating it is a write operation.
4. Poll until the S_REGRDY bit in DHCSR (0xE000EDF0) is 1.

The DCRSR and the DCRDR registers can only transfer register values during halt mode debug. For debugging using a debug monitor handler, the contents of some of the registers can be accessed from the stack memory; the others can be accessed directly within the monitor exception handler.

The DCRDR can also be used for semi-hosting if suitable function libraries and debugger support are available. For example, when an application executes a printf statement, the text output could be generated by a number of putc (put character) function calls. The putc function calls can be implemented as functions that store the output character and status to the DCRDR and then trigger the debug mode. The debugger can then detect the core halt and collect the output character for display. This operation, however, requires the core to halt, whereas the printf solution using ITM (section 18.2) does not have this limitation.

There are various additional features in the Cortex-M3 and Cortex-M4 processors to support debug operations:

- External debug request signal: The processor provides an external debug request signal that allows the Cortex-M3/M4 processor to enter debug mode through an external event such as debug status of other processors in a multi-processor system. This feature is very useful for debugging a multi-processor system. In simple microcontrollers, this signal is likely to be tied low.
- Debug restart interface: The processor provides a hardware handshaking signal interface to allow the processor to be unhalted using other hardware on the chip. This feature is commonly used for synchronized debug restart in a multi-processor system. In a single processor system the handshaking interface is normally unused.
- DFSR: Because of the various debug events available on the Cortex-M3/M4, a DFSR (Debug Fault Status Register) is available for the debugger to determine the debug event that has taken place (see section 12.4.6).
- Reset control: During debugging, the processor core can be restarted using the VECTRESET control bit or SYSRESETREQ control bit in the Application Interrupt and Reset Control register (0xE000ED0C). Using this reset control register, the processor can be reset without affecting the debug components in the system (see section 9.6).
- Interrupt masking: This feature is very useful during stepping. For example, if you need to debug an application but do not want the code to enter the interrupt service routine during the stepping, the interrupt request can be masked. This is done by setting the C_MASKINTS bit in the DHCSR (0xE000EDF0) (see Table 14.4).

- Stalled bus transfer termination: If a bus transfer is stalled for a very long time, it is possible for the debugger to terminate the stalled transfer by a debug control register. This is done by setting the C_SNAPSTALL bit in the DHCSR (0xE000EDF0). This feature can be used only by a debugger during halt (see Table 14.4).

### 14.6.2 Flash patch and breakpoint (FPB) unit

#### *Purpose of the FPB*

The Flash Patch and Breakpoint (FPB) unit has two functions:

- To provide a hardware breakpoint feature — generate a breakpoint event to the processor core to invoke debug modes such as halting debug or debug monitor exception
- To patch instruction or literal data by remapping a memory access in the CODE region (first 0.5GB of memory space) to the SRAM region (the next 0.5GB of memory space)

The breakpoint function is fairly easy to understand: During debugging, you can set one or several breakpoints to program addresses or literal constant addresses. If the program code at the breakpoint addresses gets executed, this triggers the breakpoint debug event and causes the program execution to halt (for halt mode debug) or triggers the debug monitor exception (if debug monitor is used). Then, you can examine the register's content, memories, and peripherals, and carry out debug using single stepping, etc.

The Flash Patch function allows the use of a small programmable memory in the system to apply patches to a program memory that cannot be modified. For products to be produced in high-volume, using mask ROM or one-time-programmable ROM can reduce the cost of the product. But, if a software bug is found after the device is programmed, it could be costly to replace the devices. By integrating a small reprogrammable memory, for example, a very small Flash or Electrically-Erasable Programmable Read-Only Memory (EEPROM), patches can be made to the original software programmed in the device. For microcontrollers that only use flash memory to store software, the Flash Patch is not required as the whole flash memory can be erased and reprogrammed easily.

The FPB can either be used for breakpoint generation, or used for Flash Patch, but not at the same time. When a device is configured to use FPB for Flash Patch, connecting a debugger to it during run-time will end up having the flash patch configuration overridden by the debugger.

#### *FPB comparators*

The FPB in Cortex®-M3 and Cortex-M4 processors contains up to eight comparators:

- Comparators #0 to #5 are instruction address comparators. Each of them can be programmed to generate breakpoint events, or remap instruction access to SRAM (but not at the same time).
- Comparators #6 to #7 are literal comparators. They are used to remap literal data accesses in CODE region to SRAM.

## WHAT ARE LITERAL LOADS?

When we program in assembler language, very often we need to set up immediate data values in a register. When the value of the immediate data is large, the operation cannot be fitted into one instruction space. For example:

```
LDR R0, =0xE000E400 ; External Interrupt Priority Register
                    ; starting address
```

Because no instruction has an immediate value space of 32, we need to put the immediate data in a different memory space, usually after the current fragment of program code, and then use a PC relative load instruction to read the immediate data into the register. So what we get in the compiled binary code will be something like the following:

```
LDR R0, [PC, #<immed_8>*4]
    ; immed_8 = (address of literal value -- PC)/4
...
; literal pool
...
DCD 0xE000E400
...
```

or if the literal data is further away, use a 32-bit PC related LDR instruction:

```
LDR.W R0, [PC, #+/- <offset_12>]
    ; offset_12 = address of literal value - PC
...
; literal pool
...
DCD 0xE000E400
...
```

Because we are likely to use more than one literal value in our code, the assembler or compiler will usually generate a block of literal data, which is commonly called a literal pool.

In Cortex-M3 or Cortex-M4 processors, the literal loads are data-read operations carried out on the data bus (either D-CODE bus or System bus depending on memory location).

The FPB has a Flash Patch control register that contains an enable bit to enable the FPB. In addition, each comparator comes with a separate enable bit in its comparator control register. Both of the enable bits must be set to 1 for a comparator to operate.

The comparators can be programmed to remap addresses from Code space to the SRAM memory region. When this function is used, the REMAP register needs to be programmed to provide the base address of the remapped contents. The upper three bits of the REMAP register (bit[31:29]) are hardwired to b001, limiting the remap base address location to within 0x20000000 to 0x3FFFFF80, which is always within the SRAM memory region.

The data holding the remapped values is arranged as a continuous eight words of data in the SRAM region, which must be aligned to an eight-word boundary, with the starting address pointed by the REMAP register. When the instruction address or the

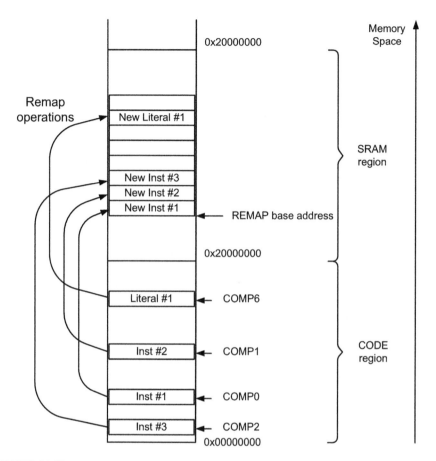

**FIGURE 14.13**

Flash Patch — Remap of Instructions and Literal Read

literal address hits the address defined by the comparator, and if the comparator is programmed for remap function, the read access is remapped to the table pointed to by the REMAP register (Figure 14.13).

The remapped address can be expressed as: {3'b001, REMAP, COMP[2:0], ADDR[1:0]}, where COMP[2:0] is the number of the matching comparator, and ADDR[1:0] is the lowest two bits of the original address value of the access.

Using the remap function, it is possible to change a function call instruction so that a modified function in a different memory (e.g., a small programmable patch memory with flash or EEPROM) is used, and even the program code is in ROM and cannot be changed.

You can also create some "what-if" test cases in which the original instruction or a literal value is replaced by a different one.

Another example use is to place test functions or subroutines in the SRAM region, and patch the program ROM in the CODE region so that a branch to the test functions or subroutine can take place. These usage possibilities make it possible to debug a ROM-based device. (More information on this topic can be found in section 23.10.)

Alternatively, the six instruction address comparators can be used to generate breakpoints as well as to invoke halt mode debug or debug monitor exceptions.

### 14.6.3 Data watchpoint and trace (DWT) unit

The DWT contains up to four hardware comparators for data watchpoint generation, data trace, debug event generation, and a number of profiling counters. The DWT has many different functions, as discussed in the following sections.

#### Debug event generation
- Generate Debug event (such as data watchpoint event, PC matching event) that can be used to halt the processor or invoke Debug Monitor exception.
- The Debug event generated can also be used by the ETM for trace start/stop control, or generate the ETM trigger (cause the ETM to emit a trigger packet in the instruction trace stream).

#### Data trace
- Generate data trace packets when the comparator matches. The packets can contain data values, data addresses, or current PC values.

#### PC sampling
- Generate periodic PC sampling to the trace stream for profiling usage.
- Allow the debugger to sample PC value by reading a register called PCSR (PC Sampling register) periodically. This allows some basic profiling to be carried out even if the hardware does not support trace capture.

#### Profiling
- A number of counters (mostly 8-bit) are available for counting various types of cycle information. When the counter underflows, a trace packet is generated so that by combining the trace packet collected and the counter values, the debugger can determine the activities of the processor over a period of time (e.g., how many cycles were spent in memory accesses, sleep, exception overhead, etc.).

#### Exception trace
- The DWT can generate a trace packet at exception entrance and exit, allowing you to determine which exceptions have occurred, which is very useful for debug. When combining timestamping packets in the ITM, it can even allow you to measure how much time the processor spent on each exception.

Before accessing any DWT registers or using any DWT features, the TRCENA bit (Trace Enable) in CoreDebug->DEMCR (see Table 14.5) must be set to 1. To use trace functions (including profiling trace, data trace, etc.), the DWTENA bit in the ITM Trace Control Register must also be set to 1 (see appendix G, Table G.27).

Each of the comparators has three registers, which are as follows:

- DWT_COMP (comparator registers)
- DWT_MASK register
- DWT_FUNCTION register

The comparators have a 32-bit compare value that can be used for comparing:

- Data address
- PC value
- DWT_CYCCNT value (comparator 0 only)
- Data value (comparator 1 only)

The DWT_MASK register determines the bit compare mask to be used during the comparison. This allows the comparator to generate a data trace for a data address range by masking off the lowest bit of the addresses during the comparison. The DWT_MASK register is 4 bits wide, and defaults to 0, which means comparing all bits. When set to 1, bit 0 of the value is ignored during the comparison. By setting DWT_MASK to 15, it is possible to trace data access in an address range of 32KB maximum size (Table 14.8). However, because of the limitation of the FIFIO size in the trace system, it is not practical to trace lots of data transfers, as this will cause trace overflow and result in the loss of some trace data.

The DWT_FUNCTION registers determine the function of the comparators. To avoid unexpected behavior, the MASK register and the COMP register should be programmed before this register is set. If the comparator's function is to be changed, you should disable the comparator by setting FUNCTION to 0 (disable), then program the MASK and COMP registers, and then enable the FUNCTION register in the last step.

In systems with multiple processors, the debug event generated by the comparators can be linked to other processor systems to trigger debug operations in other processors.

The rest of the DWT counters are typically used for profiling the application code. They can be programmed to emit events (in the form of trace packets)

**Table 14.8** Encoding of the DWT_MASK Registers

| MASK | Ignore Bit |
|------|------------|
| 0 | All bits are compared |
| 1 | Ignore bit [0] |
| 2 | Ignore bit [1:0] |
| 3 | Ignore bit [2:0] |
| ... | ... |
| 15 | Ignore bit [14:0] |

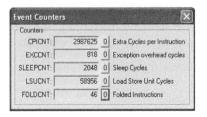

**FIGURE 14.14**

Program Execution statistics in Keil µVision debugger using DWT profiling counters

when the counter overflows. One typical application is to use the CYCCNT register to count the number of clock cycles required for a specific task, for benchmarking purposes. Although most of these profiling counters are only 8 bits wide, when combining the counter value and the trace packet information emitted, the counting range is unlimited.

For example, the Keil™ µVision development tool can use these profiling counters to generate statistical information (Figure 14.14). These counters trigger event packets to be generated and are collected by the debugger through the Serial Wire Output (SWO).

The counters can also be used for calculating CPI (Cycle-Per-Instruction). For example, the number of instructions executed over a period of time can be measured using:

Total instruction executed = Total cycle count − CPICNT − EXCCNT − SLEEPCNT - LSUCNT + FOLDCNT

Note: FOLDCNT is for folded instruction cycles, which indicate an instruction execution such as IT has overlapped with the execution cycle of another instruction.

The profiling counters are stopped when the processor is halted.

PC sampling is a simple method for basic profiling. The DWT supports PC sampling in two ways:

- Output PC samples via the trace interface periodically
- Debugger can read the PC Sample Register (DWT->PCSR) periodically

By sampling the PC value over a long period of execution time, it is possible to estimate:

- which functions are executed (it could miss some functions if they are short, but some of those can be determined by analysis of the execution path). In this way it helps to identify parts of the program that are never tested.
- how much time is spent on which functions. Over a long period of sampling, you can estimate statistically how much (in terms of percentage) execution time is spent on a given function.

### 14.6.4 **Instrumentation trace macrocell (ITM)**
#### *Overview*
The ITM has multiple functionalities:

- Software trace — software can directly write messages to ITM stimulus port registers and the ITM can encapsulate the data in trace packets and output them through the trace interface.
- The ITM works as a trace packet merging device inside the processor to merge trace packets from DWT, stimulus port, and the timestamp packet generator (Figure 14.15). There is also a small FIFO in the ITM to help reducing trace packet overflow.
- The ITM can generate timestamp packets that are inserted in to the trace stream to help the debugger to reconstruct the timing information of the events.

To use ITM for debugging, the microcontroller or SoC device must have a trace port interface. If the device you use does not have a trace interface, or the debug adaptor you use does not support trace capture, you can still output console text messages using other peripheral interfaces such as UART or LCD. However, other features such as DWT profiling will not work. Some debuggers also support printf (and other semi-hosting features) by using core debug registers (e.g., CoreDebug->DCRDR) as a communication channel.

Before accessing any ITM registers or use any ITM features, the TRCENA bit (Trace Enable) in CoreDebug->DEMCR (see Table 14.5) must be set to 1.

In addition, there is a lock register in the ITM. You need to write the access key 0xC5ACCE55 (CoreSight Access) to this register before programming the ITM Trace Control Register (this could be done automatically by the debugger). Otherwise the write operations will be ignored:

```
ITM->LAR = 0xC5ACCE55; // Enable access to ITM register
```

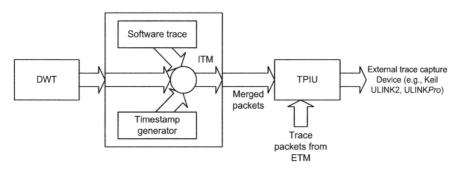

**FIGURE 14.15**

Merging of Trace Packets in the ITM

In a CoreSight™ trace system, each trace source must be assigned with a trace source ID value. This is a programmable value and is one of the bit fields (ATBID) in the ITM Trace Control register. Normally this trace ID value is automatically setup by a debugger. This ID value must be unique from the IDs for other trace sources, so that the debug host receiving the trace packet can separate the ITM's trace packets from other trace packets.

### Software trace with the ITM

One of the main uses of the ITM is to support debug message output (such as printf). The ITM contains 32 stimulus ports, allowing different software processes to output to different ports. The messages can be separated later at the debug host. Each port can be enabled or disabled by the Trace Enable register and can be programmed (in groups of eight ports) to allow or disallow user (unprivileged) processes to write to it.

Unlike UART-based text output, using the ITM to output does not cause much delay for the application. A FIFO buffer is used inside the ITM, so writing output messages can be buffered. However, it is still necessary to check whether the FIFO is full before you write to it.

The output messages can be collected at the trace port interface or the Serial Wire Viewer (SWV) interface on the TPIU. There is no need to remove code that generates the debug messages from the final code, because if the TRCENA control bit is low, the ITM will be inactive and debug messages will not be output. You can also switch on the output message in a "live" system and use the Trace Enable register in the ITM to limit which ports are enabled so that only some of the messages can be output.

For example, the Keil™ μVision development tool can collect and display the text output using the ITM viewer shown in Figure 14.16.

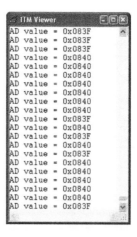

**FIGURE 14.16**

μVision ITM viewer display shows the software-generated ITM text output

CMSIS-Core provides a function for handling text messages using the ITM stimulus port:

```
unit32_t ITM_SendChar (uint32_t ch)
```

This function uses stimulus port #0. It returns the value of "ch" input. Normally the debugger can set up the trace port and ITM for you, so you only need to call this function to output each character you want to display. To use this function you must also set up the debugger to enable trace capture. For example, if the SWO signal is used, the debugger must capture the trace using the correct baud rate. Usually the debugger GUI allows you to configure the processor frequency and Serial Wire Viewer (part of the TPIU) baud rate divide ratio. Also, if the SWO output is shared with TDO pin, the Serial Wire debug communication protocol must be selected.

Although the ITM only allows data output, the CMSIS-Core also includes a function for the debugger to output a character to the application running on the microcontroller. The function is:

```
int32_t ITM_ReceiveChar (void)
```

The actual communication is handled by the debug interface (i.e., Serial Wire or JTAG). If there is no data to receive, the function returns −1. If data is available it returns the character received. Another function is available for checking if a character is received:

```
int32_t ITM_CheckChar (void)
```

The ITM_CheckChar() returns 1 if a character is available. Otherwise, it returns 0.

The ITM Stimulus Port #31 is typically used by RTOS and debug to allow OS-aware debugging. The OS outputs information about the OS status so the debugger can tell when context switching has taken place and which task the processor is running.

Note: The ITM FIFO is shared between multiple Stimulus Port channels. If the ITM stimulus ports are used by different tasks in a multi-tasking environment, you should use the semaphore feature in the OS to ensure that only one task has access to the ITM stimulus port at a time. The need for semaphore also applies when program code in Thread mode and exception handlers can use the ITM stimulus port simultaneously. Otherwise, the follow sequence could happen:

1. Task A polled the FIFO status and confirmed that the FIFO is not full.
2. A context switch takes place, Task B (or an exception handler) then polled a different ITM stimulus port and found that the FIFO is not full, so it output data to the stimulus port. Now the FIFO is full.
3. A context switch takes place and returns to task A. Since this task has already polled the Stimulus Port status, it thinks that the FIFO is not full and writes data to the Stimulus Port. This triggers a trace overflow.

In practice the risk of this happening is fairly low as it does take some time for context switching to be completed. However it is still a possible problem, especially if the trace bandwidth is low (e.g., using Serial Wire Viewer feature, i.e., SWO signal, for trace capture).

| Num | Name | Count | Total Time | Min Time In | Max Time In | Min Time Out | Max Time Out | First Time [s] | Last Time [s] |
|-----|------|-------|-----------|-------------|-------------|--------------|--------------|----------------|---------------|
| 2 | NMI | 0 | 0 s | | | | | | |
| 3 | HardFault | 0 | 0 s | | | | | | |
| 4 | MemManage | 0 | 0 s | | | | | | |
| 5 | BusFault | 0 | 0 s | | | | | | |
| 6 | UsageFault | 0 | 0 s | | | | | | |
| 11 | SVCall | 475 | 158.236 us | 77.500 us | 80.736 us | 135.861 us | 14.549 s | 0.00021660 | 25.44279225 |
| 12 | DbgMon | 0 | 0 s | | | | | | |
| 14 | PendSV | 0 | 0 s | | | | | | |
| 15 | SysTick | 2576 | 4.309 ms | 1.417 us | 93.694 us | 765.222 us | 10.066 ms | 0.00087276 | 25.47015878 |
| 16 | ExtIRQ 0 | 0 | 0 s | | | | | | |
| 17 | ExtIRQ 1 | 0 | 0 s | | | | | | |
| 18 | ExtIRQ 2 | 0 | 0 s | | | | | | |
| 19 | ExtIRQ 3 | 0 | 0 s | | | | | | |
| 20 | ExtIRQ 4 | 0 | 0 s | | | | | | |
| 21 | ExtIRQ 5 | 0 | 0 s | | | | | | |
| 22 | ExtIRQ 6 | 0 | 0 s | | | | | | |
| 23 | ExtIRQ 7 | 0 | 0 s | | | | | | |

**FIGURE 14.17**

Exception trace in μVision debugger

### Hardware trace with ITM and DWT

The ITM handles merging of packets from the DWT. To use the DWT trace, you need to set the DWTENA bit in the ITM Trace Control Register; the rest of the DWT trace settings are still required. All these are likely to be handled by the debugger automatically.

### ITM timestamp

The ITM has a timestamp feature that allows trace capture tools to obtain timing information by inserting delta timestamp packets into the traces when a new trace packet enters the FIFO inside the ITM. The timestamp packet is also generated when the timestamp counter overflows.

The timestamp packets provide the time difference (delta) with respect to previous events. Using the delta timestamp packets, the trace capture tools can then establish the time when each packet is generated, and hence reconstruct the timing of various debug events. Cortex®-M4 and Cortex-M3 r2p1 also have a global timestamp mechanism, which allows correlation of trace information between different trace sources (e.g., between ITM and ETM, or even between multiple processors).

Combining the trace functionality of DWT and ITM, we can collect a lot of useful information. For example, the exception trace windows in the Keil μVision development tool can tell you what exceptions have taken place and how much time was spent on the exceptions, as shown in Figure 14.17.

### 14.6.5 Embedded trace macrocell (ETM)

The ETM block is used for providing instruction traces. It is optional and might not be available on some Cortex®-M3 and Cortex-M4 products. When it is enabled and when the trace operation starts, it generates instruction trace packets. A FIFO buffer is provided in the ETM to allow enough time for the Trace Port Interface Unit (TPIU) to process and serialize the trace data while the trace stream is being captured. Figure 14.18 shows instruction trace display in Keil μVision debugger.

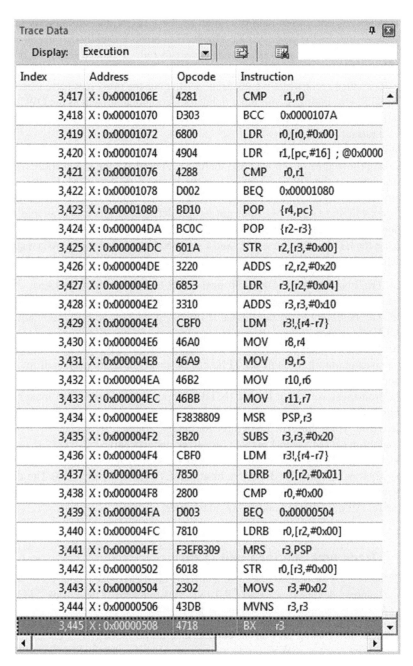

**FIGURE 14.18**

Instruction Trace from ETM

To reduce the amount of data generated by the ETM, it does not always output exactly what address the processor has reached/executed. It usually outputs information about program flow and outputs full addresses only if needed (e.g., if a branch has taken place). Because the debugging host should have a copy of the binary image, it can then reconstruct the instruction sequence the processor has carried out.

The ETM also interacts with other debugging components such as the DWT. The comparators in the DWT can be used to generate trigger events in the ETM or to control the trace start/stop.

Unlike the ETM in traditional ARM® processors, the Cortex-M3 and Cortex-M4 ETM are optimized for low power and therefore do not have their own address comparators, because the DWT can carry out the comparison for ETM. Furthermore, because the data trace functionality is carried out by the DWT, the ETM design in the Cortex-M3 and Cortex-M4 does not have a data trace, which is available from traditional ETM for some other ARM cores.

To use the ETM in the Cortex-M3 and Cortex-M4, the following setup is required (handled by debug tools):

- The TRCENA bit in the DEMCR must be set to 1 (refer to Table 14.5).
- The ETM needs to be unlocked so that its control registers can be programmed. This can be done by writing the value 0xC5ACCE55 to the ETM LOCK_ACCESS register.
- The ATB ID register (ATID) should be programmed to a unique value so that the trace packet output through the TPIU can be separated from packets from other trace sources.
- The Non-Invasive Debug Enable (NIDEN) input signal of the ETM must be set to high. The implementation of this signal is device-specific. Refer to the datasheet from your chip's manufacturer for details.
- Program the ETM control registers for trace generation.
- The ETM in Cortex-M3 r0p0 to r2p0 is based on ETM architecture v3.4. Cortex-M3 r2p1 and Cortex-M4 ETM are based on ETM architecture v3.5. The difference for Cortex-M3/M4 in ETM v3.5 is the inclusion of global timestamp feature. All other trace packets are compatible.

### 14.6.6 Trace port interface unit (TPIU)

The TPIU is used to output trace packets from the ITM, DWT, and ETM to the external capture device (e.g., Keil™ ULINK2/ULINKPro). The Cortex®-M3 and Cortex-M4 TPIU supports two output modes:

- Clocked mode, using up to 4-bit parallel data output ports
- Serial Wire Viewer (SWV) mode, using a single-bit output called Serial Wire Output (SWO)

In clocked mode, the actual number of bits being used on the data output port can be programmed to different sizes. This will depend on the chip package as well as the number of signal pins available for trace output in the application. The maximum trace port size supported by the chip can be determined from one of the registers in the TPIU. In addition, the speed of trace data output can also be programmed using a pre-scaler.

In SWV mode, a 1-bit serial protocol is used and this reduces the number of output signals to 1, but the maximum bandwidth for trace output will also be reduced. When combining SWV with Serial-Wire debug protocol, the Test Data Output (TDO) pin normally used for Joint Test Action Group (JTAG) protocol can be shared with SWO (Figure 14.19). For example, the trace output in SWV mode can be collected using a standard debug connector for JTAG using a Keil ULINK2 module.

Alternatively, the SWV output mode can also share a pin with the trace output pin in clocked mode. The trace data (either in clocked mode or SWV mode) can be collected by external Trace Port Analyzer (TPA) like the ARM® D-Stream or Keil ULINKPro.

When instruction trace (using ETM) is required, the clocked mode is more suitable than SWV mode as it provides higher trace bandwidth. For a simple data trace and event trace (e.g., tracing of exception events), the SWV mode is usually sufficient and can be used with fewer connection pins.

The TPIU supports asynchronous clocking on the trace interface (Figure 14.20). The Trace Port interface or the SWO output can operate at a different clock frequencies compared to the processor. This allows the trace port to operate at a higher clock frequency than the processor in order to provide higher trace bandwidth.

Another available option on the TPIU is the use of a formatter. The formatter encapsulates the trace data into a format that combines the data and the trace bus ID. This allows trace streams to be merged and separated by the debug host. When using SWV mode for trace without ETM, there is only one active trace bus and the formatter can be switched off to enable higher data throughput.

To use the TPIU, the TRCENA bit in the DEMCR must be set to 1, and the protocol (mode) selection register and trace port size control registers need to be programmed by the trace capture software.

### 14.6.7 ROM table

CoreSight™ Debug Architecture is very scalable and can be used in complex System-on-Chip designs with a large number of debug components. In order to support a wide range of system configuration, CoreSight Design Architecture provides a mechanism to allow the debugger to automatically locate debug components in the system, and the ROM table is part of this mechanism.

The Cortex®-M3 and Cortex-M4 processors have a pre-defined memory map and include a number of debug components. However, chip designers can choose to omit some of the debug components or add additional ones if preferred. To allow debug tools to detect the debug component configuration in the debug system, a ROM table is included; it provides information on the System Control Block (SCB) and debug block addresses.

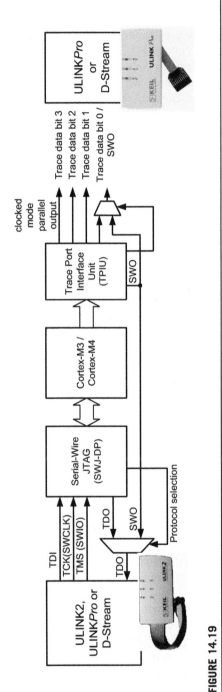

**FIGURE 14.19**

Trace connection and pin sharing of Serial Wire Output (SWO)

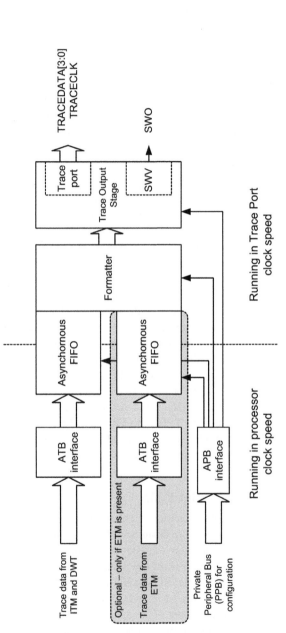

**FIGURE 14.20**

The ROM table is just a simple look-up table and contains a few entries of address information, and a few ID registers to allow the debugger to identify it as a ROM table device. In typical Cortex-M3 or Cortex-M4 systems, the ROM table is located in address 0xE00FF000. The debugger can identify this primary ROM table address value from a register in the AHB-AP module when it connects to the system.

Using the contents of the ROM table, the memory locations of system and debug components can be calculated. The debug tool can then check the ID registers of the discovered components and determine what is available on the system.

For the Cortex-M3 or Cortex-M4 processors, the first entry in the ROM table (0xE00FF000) should contain the offset to the SCS (System Control Space) memory location. (The default value in the ROM table's first entry is 0xFFF0F003; bit[1:0] means that the device exists and there is another entry in the ROM table following. The SCS offset can be calculated as 0xE00FF000 + 0xFFF0F000 = 0xE000E000.)

The default ROM table for the Cortex-M3 and Cortex-M4 is shown in Table 14.9. However, because chip manufacturers can add, remove, or replace some of the optional debug components with other CoreSight debug components, the value you will find on your Cortex-M3 or Cortex-M4 device could be different.

The lowest two bits of the value indicate whether the device exists. In normal cases, the SCS, DWT, and FPB should always be there, so the last two bits are always 1. However, the TPIU and the ETM could be taken out by the chip manufacturer and might be replaced with other debugging components from the CoreSight product family.

The upper part of the value indicates the address offset from the ROM table base address. For example:

```
SCS address = 0xE00FF000 + 0xFFF0F000 = 0xE000E000 (truncated to
32-bit)
```

For debug tool development using CoreSight technology, it is necessary to determine the address of debug components from the ROM table. Some Cortex-M3/M4 devices might have a different setup for the debug component connection that can result in additional base addresses. By calculating the correct device address from this ROM table, the debugger can determine the base address of the provided debug component, and then from the component ID of those components the debugger can determine the type of debug components that are available (Figure 14.21).

It is also possible to have multiple levels of ROM tables, where the address of a secondary ROM table can be located using one of the entries in the primary ROM table. This can often be found in complex SoC devices.

## 14.6.8 AHB access port (AHB-AP)

The Advanced High-Performance Bus Access Port (AHB-AP) is a bridge between the debug interface module (Serial-Wire JTAG Debug Port or Serial-Wire Debug Port) and the Cortex®-M3/M4 memory system (see Figure 14.5). Its programmer's

**Table 14.9** Default ROM Table Values in Cortex-M3/M4

| Address | Value | Name | Description |
|---------|-------|------|-------------|
| 0xE00FF000 | 0xFFF0F003 | SCS | Points to the SCS base address at 0xE000E000 |
| 0xE00FF004 | 0xFFF02003 | DWT | Points to the DWT base address at 0xE0001000 |
| 0xE00FF008 | 0xFFF03003 | FPB | Points to the FPB base address at 0xE0002000 |
| 0xE00FF00C | 0xFFF01003 | ITM | Points to the ITM base address at 0xE0000000 |
| 0xE00FF010 | 0xFFF41003/0xFFF41002 | TPIU | Points to the TPIU base address at 0xE0040000 |
| 0xE00FF014 | 0xFFF42003/0xFFF42002 | ETM | Points to the ETM base address at 0xE0041000 |
| 0xE00FF018 | 0 | End | End-of-table marker |
| 0xE00FFFCC | 0x1 | MEMTYPE | Indicates that system memory can be accessed on this memory map |
| 0xE00FFFD0 | 0/0x04 | PID4 | Peripheral ID space; reserved |
| 0xE00FFFD4 | 0/0x00 | PID5 | Peripheral ID space; reserved |
| 0xE00FFFD8 | 0/0x00 | PID6 | Peripheral ID space; reserved |
| 0xE00FFFDC | 0/0x00 | PID7 | Peripheral ID space; reserved |
| 0xE00FFFE0 | 0/0xC3 | PID0 | Peripheral ID space; reserved |
| 0xE00FFFE4 | 0/0xB4 | PID1 | Peripheral ID space; reserved |
| 0xE00FFFE8 | 0/0x0B | PID2 | Peripheral ID space; reserved |
| 0xE00FFFEC | 0/0x00 | PID3 | Peripheral ID space; reserved |
| 0xE00FFFF0 | 0/0x0D | CID0 | Component ID space; reserved |
| 0xE00FFFF4 | 0/0x10 | CID1 | Component ID space; reserved |
| 0xE00FFFF8 | 0/0x05 | CID2 | Component ID space; reserved |
| 0xE00FFFFC | 0/0xB1 | CID3 | Component ID space; reserved |

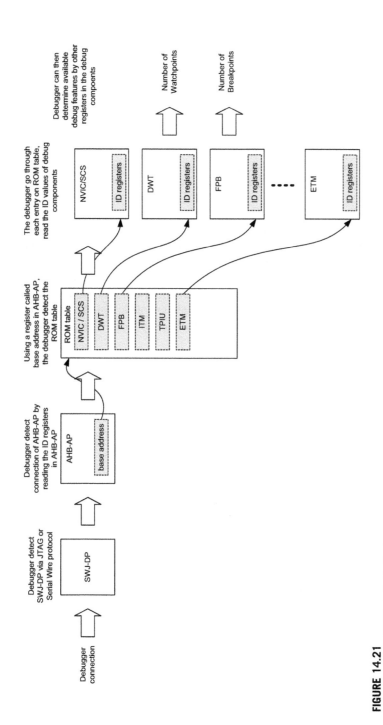

**FIGURE 14.21**

Automatic detection of components via CoreSight architecture model is invisible from the software running on the processor, and can only be accessed from the debugger on the internal debug APB.

For the most basic data transfers between the debug host and the Cortex-M3/M4 system, the following three registers in the AHB-AP are used (more details can be found in Appendix G, section G.6):

- Control and Status Word (CSW)
- Transfer Address register (TAR)
- Data Read/Write (DRW)

The CSW register can control the transfer direction (read/write), transfer size, transfer types, and so on. The TAR register is used to specify the transfer address, and the DRW register is used to carry out the data transfer operation (transfer starts when this register is accessed).

The data register DRW represents exactly what is shown on the bus. For half-word and byte transfers, the required data will have to be manually shifted to the correct byte lane by debugger software. For example, if you want to carry out a data transfer of half-word size to address 0x1002, you need to have the data on bit [31:16] of the DRW register. The AHB-AP can generate unaligned transfers, but it does not rotate the result data based on the address offset. So, the debugger software will have to either rotate the data manually or split an unaligned data access into several accesses if needed.

Other registers in the AHB-AP provide additional features. For example, the AHB-AP provides four banked registers and an automatic address increment function so that access to memory within close range or sequential transfers can be sped up. The AHB-AP also contains a register called base address to indicate the address of primary ROM table.

In the CSW register, there is one bit called MasterType. This is normally set to 1 so that hardware receiving the transfer from AHB-AP knows that it is from the debugger. However, the debugger can pretend to be the core by clearing this bit (silicon designers can choose to disable this feature). In this case, the transfer received by the device attached to the AHB system should behave as though it is accessed by the processor. This is useful for testing peripherals with FIFO that can behave differently when accessed by the debugger. This feature can be disabled by chip designer if needed.

## 14.7 Debug operations
### 14.7.1 Debug connection

When a debugger connects to a microcontroller device, it might carry out the following sequence:

1. First, attempt to detect the ID register value in the SWJ-DP.
2. Issue a debug power request in the SWJ-DP to ensure that the system is ready for debug connection.

3. Optionally scan through the debug APB to check what type of AP is connected.
4. Detect the ID register values in the AHB-AP and check that it can access the memory map.
5. Optionally obtain the address of the primary ROM table from AHB-AP and detect the debug components in the system via the ROM table.
6. Optionally download a program image to the device based on the project settings.
7. Optionally enable the reset Vector Catch feature, and reset the system using SCB->AIRCR (SYSRESETREQ).

Now the processor is halted at the beginning of program execution.

### 14.7.2 Flash programming

Typically the flash programming operation is carried out in blocks of data. The debugger downloads a flash programming algorithm code and a block of a program image to be programmed to the SRAM, and then executes the flash programming code so that the processor carries out the flash programming operation by itself. Since each time the SRAM can only hold a small part of the program image, the steps have to be repeated a number of times until the whole program image is programmed to the flash.

The debugger can then repeat the steps in section 14.7.1 to enable the start of the debug session.

### 14.7.3 Breakpoints

Since the processors have a limited number of breakpoint comparators, when a breakpoint is set, the debugger might first try to write a breakpoint instruction to the program location to see if the program is running from SRAM (which can be modified). If the write cannot be done, then the debugger will use a FPB hardware comparator to set the breakpoint to this memory location. If the breakpoint location is not in the CODE region, the debugger could set up a DWT comparator to use the PC match function. This is not the same as the breakpoint mechanism because the processor halts after the instruction is executed, but at least it can still be useful for simple debugging.

If the program is running from SRAM, and the breakpoint is hit, the debugger will then have to replace the breakpoint instruction with the original instruction before the user resumes program execution.

# Getting Started with Keil Microcontroller Development Kit for ARM®

# 15

## CHAPTER OUTLINE

## 15.1 Overview

Many commercial development platforms are available for the Cortex®-M processors. One of the popular choices is the Keil™ Microcontroller Development Kit for ARM® (MDK-ARM). The MDK-ARM contains various components including:

- μVision Integrated Development Environment (IDE)
- ARM Compilation Tools including
  - C/C++ Compiler
  - Assembler
  - Linker and utilities

The Definitive Guide to ARM® Cortex®-M3 and Cortex-M4 Processors. http://dx.doi.org/10.1016/B978-0-12-408082-9.00015-4

- Debugger
- Simulator
- RTX Real-Time OS Kernel
- Reference start-up code for 1000s of microcontrollers
- Flash programming algorithms
- Program examples

To learn about Cortex-M microcontroller programming with MDK-ARM, it is not necessary to have the actual hardware. The µVision environment contains an instruction set simulator that allows testing of simple programs that do not require a development board. However, there are also a number of low-cost development kits available for $30 or less, which are suitable for beginners and for evaluation.

The debugger in the Keil MDK-ARM can be used with a number of different debug adaptors. This includes commercial debug adaptors such as:

- Keil ULINK2, ULINK*Pro*, ULINK-ME
- Signum Systems JTAGjet
- J-Link, J-Trace from Segger

There are also a number of debug adaptors that come with development boards:

- CMSIS-DAP
- ST-LINK, ST-LINK V2
- Silicon Labs UDA Debugger
- Stellaris ICDI (Texas Instrument)
- NULink Debugger

It is also possible to use other debug adaptors if a third-party debugger plug-in is available. For example, CooCox (http://www.coocox.org) provides open debug probes called CoLink and CoLinkEx. The design information and schematics for these hardware probes are freely available, so that anyone can build their own debug adaptor in a "DIY" manner.

A Lite version of the Keil MDK-ARM can be downloaded from the Keil website (http://www.keil.com). The Lite version of the tool is limited to 32KB program code (compiled size), but has no time limitation, so you do not need to spend much money to get started. The Lite version of Keil MDK-ARM is also included in a number of Cortex-M evaluation kits from various microcontroller vendors.

## 15.2 Typical program compilation flow

The typical program compilation flow of a project in Keil™ MDK is illustrated in Figure 15.1. Once a project is created, the compilation flow can be handled by the IDE and therefore you can program your microcontroller and test it with just a few steps.

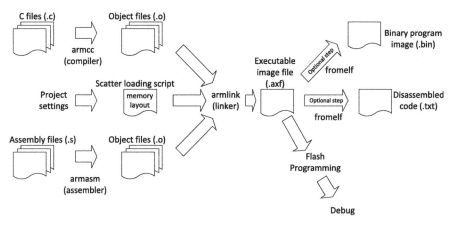

**FIGURE 15.1**

Example compilation flow with Keil MDK

Although you can program almost everything in C, the startup code (which is provided by microcontroller vendors, and usually included in the Keil MDK installation) is usually in assembly language. In addition, you will normally need a few more files from the microcontroller vendors. At a minimum, you can create a project with just one application file and a few files from the microcontroller vendor, as shown in Figure 15.2.

Behind the scenes, the device-specific header file pulls in further CMSIS-Core header files including some generic CMSIS-Core files from ARM®, as shown in Figure 15.3. Usually these files are part of the toolchain installation so it is not

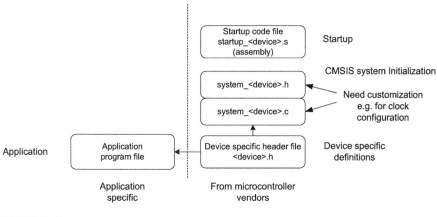

**FIGURE 15.2**

Example project with CMSIS-Core

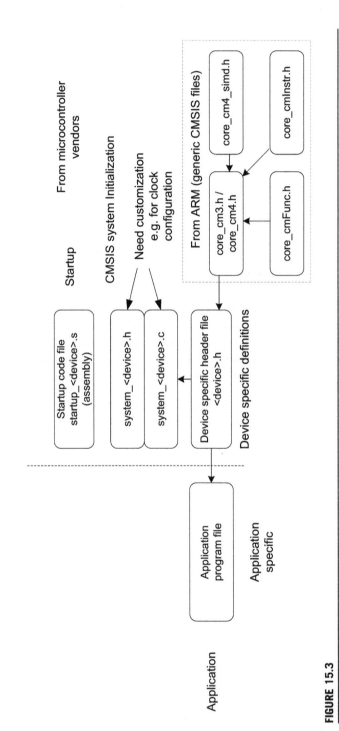

**FIGURE 15.3**

Example project view when including CMSIS-Core files from ARM

necessary to include them in the project explicitly. In some cases you might need to include additional files if you need to stick to a particular version of CMSIS-Core.

If you are using older versions of CMSIS-Core (version 2.0 or older), you might also find that you need to include a file called core_cm3.c or core_cm4.c in the CMSIS-Core package for some of the core functions like access to special registers and a couple of intrinsic functions. These files are no longer required in newer versions of CMSIS-Core as the functions have been incorporated directly into the header files.

## 15.3 Getting started with µVision

Many development kit vendors provide example projects in their software packages, and you can find a number of example projects for various popular microcontroller boards in the Keil™ MDK-ARM installation. You can open the project by double-clicking on the project file (a file with .uvproj extension).

But let's say you want to create a new project from scratch. Here is a step-by-step guide to what you need to do for a minimum project. In this example we will use the STM32F4 Discovery board, and create a simple project called blinky, which toggles some LEDs. The files needed are shown in Figure 15.4. This development includes a debug adaptor called ST-LINK V2.

After starting the µVision IDE, you might see a screen similar to the one shown in Figure 15.5.

Alternatively the IDE may start with the last opened project. You can first close the old project using the pull-down menu: select Project → Close project. Then open a new project.

For the first project, we are going to create it in a folder as follows: "C:\Book\-ch_15_blink_1\blinky.uvproj." This is shown in Figure 15.6.

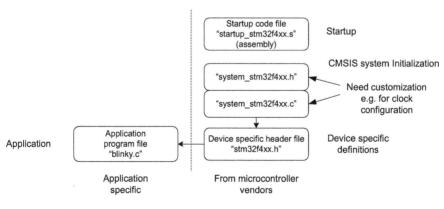

**FIGURE 15.4**

Example project for STM32F4 Discovery

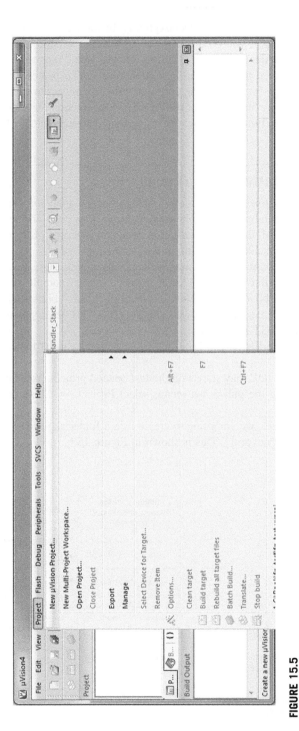

**FIGURE 15.5**

The µVision IDE starting screen, and creating a new project

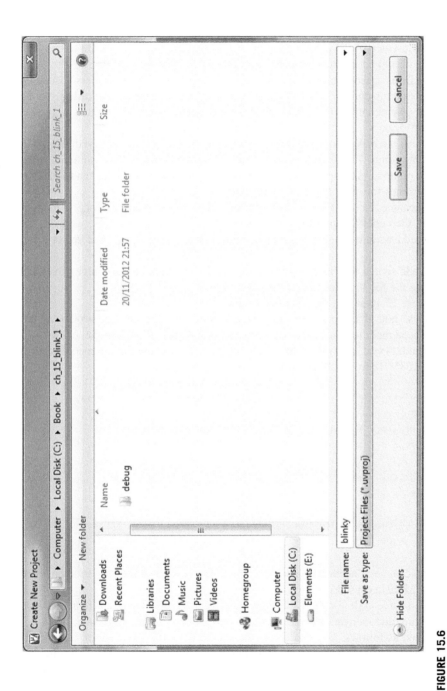

**FIGURE 15.6**

Choose the project directory and project name (blinky)

Then we need to select the microcontroller device. From the device list, we select STM32F417VG, which is used on the STM32F4Discovery board (Figure 15.7).

The last step of the project creation wizard will ask you if you want to copy the default startup code into your project (Figure 15.8). The Keil MDK-ARM installation includes a wide range of startup code for different microcontrollers. Usually a device's startup code in MDK-ARM should be the same as the startup code provided by the microcontroller vendors. One this is done, we have created a project (Figure 15.9).

In order to make project management easier, the project files can be divided into different groups. The default group is called "Source Group 1." We can rename the current group to "startup" to make it clear that it contains startup code, and add additional groups for other types of source files.

To rename "Source Group 1" to "Startup," click on the "Source Group 1" to highlight it, then click it again to edit the name.

To add additional source groups, right-click on "Target 1," and select "Add Group."

To add source files in a source group, just double-click on the source group and then use the file browser to add the source file.

Figure 15.10 shows the modified project.

Please note that the system initialization files from the microcontroller vendor might need modifications for clock frequency adjustment. For example, in the "system_stm32f4xx.c," we made the following changes to allow the microcontroller to run at 168MHz from a 8MHz crystal:

```
/***************** PLL Parameters *********************/
/* PLL_VCO = (HSE_VALUE or HSI_VALUE / PLL_M) * PLL_N */
#define PLL_M   8
#define PLL_N   336

/* SYSCLK = PLL_VCO / PLL_P */
#define PLL_P   2

/* USB OTG FS, SDIO and RNG Clock = PLL_VCO / PLLQ */
#define PLL_Q   7
```

The change is needed because the source file was developed for another board with a 25 MHz crystal.

The source code for the LED toggling is very simple. There are four LEDs on this board, and we want to turn them on one by one. The program code "blinky.c" is created as follows. You can create the source file with a text editor, then add it to the project. You can also create new source files directly from µVision by clicking "File" then "New;" just be sure that you remember to add the new file to your project after saving.

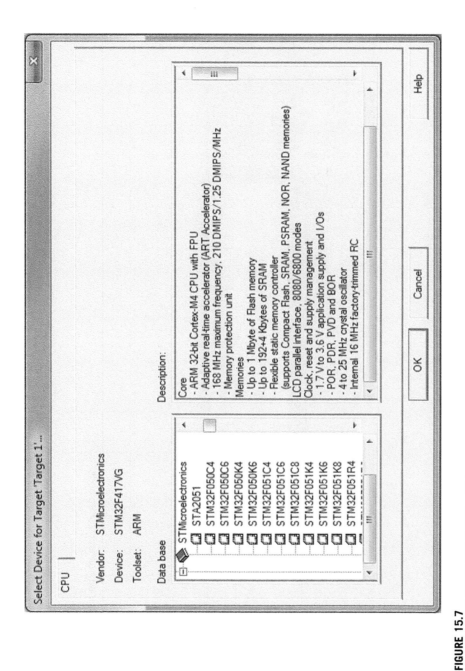

**FIGURE 15.7**

Device selection window

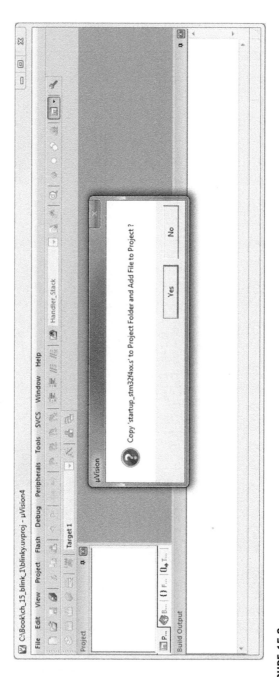

**FIGURE 15.8**

Project created with default startup code

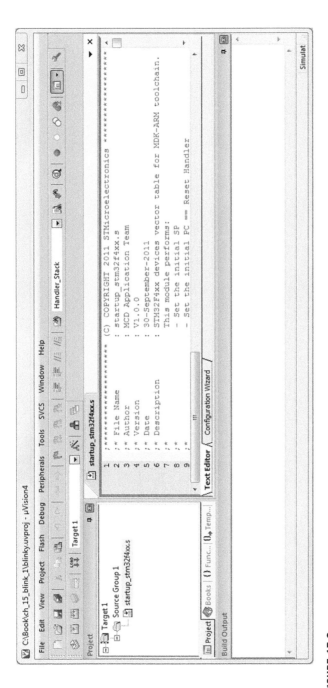

**FIGURE 15.9**

Blinky project created

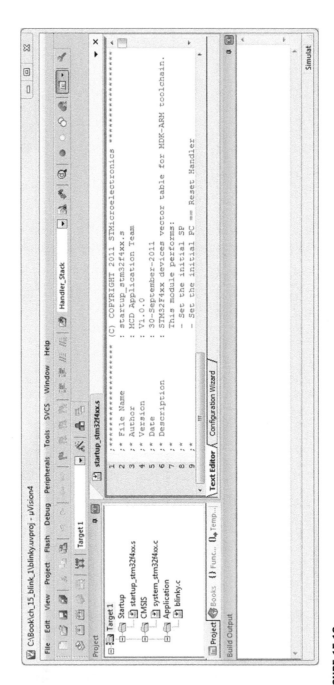

**FIGURE 15.10**

The blinky project

```
/*blinky.c for STM32F4 Discovery board*/
#include "stm32f4xx.h"

void Delay(uint32_t nCount);

int main(void)
{
 // Enable double word stack alignment
 //(recommended in Cortex-M3 r1p1, default in
 // Cortex-M3 r2px and Cortex-M4)

 SCB->CCR |= SCB_CCR_STKALIGN_Msk;

 // Configure LED outputs
 RCC->AHB1ENR |= RCC_AHB1ENR_GPIODEN; // Enable Port D clock
 // Set pin 12 to 15 as general purpose output mode (pull-push)
 GPIOD->MODER |= (GPIO_MODER_MODER12_0 |
                  GPIO_MODER_MODER13_0 |
                  GPIO_MODER_MODER14_0 |
                  GPIO_MODER_MODER15_0 ) ;
 // GPIOD->OTYPER |= 0; // No need to change - use pull-push
#define USE_2MHZ_OUTPUT
#ifdef USE_2MHZ_OUTPUT
 // GPIOD->OSPEEDR |= 0; // No need to change - slow speed
#endif
#ifdef USE_25MHZ_OUTPUT
 GPIOD->OSPEEDR |= (GPIO_OSPEEDER_OSPEEDR12_0 |
                    GPIO_OSPEEDER_OSPEEDR13_0 |
                    GPIO_OSPEEDER_OSPEEDR14_0 |
                    GPIO_OSPEEDER_OSPEEDR15_0 );
#endif
#ifdef USE_50MHZ_OUTPUT
 GPIOD->OSPEEDR |= (GPIO_OSPEEDER_OSPEEDR12_1 |
                    GPIO_OSPEEDER_OSPEEDR13_1 |
                    GPIO_OSPEEDER_OSPEEDR14_1 |
                    GPIO_OSPEEDER_OSPEEDR15_1 );
#endif
#ifdef USE_100MHZ_OUTPUT
 GPIOD->OSPEEDR |= (GPIO_OSPEEDER_OSPEEDR12 |
                    GPIO_OSPEEDER_OSPEEDR13 |
                    GPIO_OSPEEDER_OSPEEDR14 |
                    GPIO_OSPEEDER_OSPEEDR15 );
#endif
 GPIOD->PUPDR = 0; // No pull up , no pull down
```

```
#define LOOP_COUNT 0x3FFFFF
  while(1){
    GPIOD->BSRRL = (1<<12); // Set bit 12
    Delay(LOOP_COUNT);
    GPIOD->BSRRH = (1<<12); // Clear bit 12
    GPIOD->BSRRL = (1<<13); // Set bit 13
    Delay(LOOP_COUNT);
    GPIOD->BSRRH = (1<<13); // Clear bit 13
    GPIOD->BSRRL = (1<<14); // Set bit 14
    Delay(LOOP_COUNT);
    GPIOD->BSRRH = (1<<14); // Clear bit 14
    GPIOD->BSRRL = (1<<15); // Set bit 15
    Delay(LOOP_COUNT);
    GPIOD->BSRRH = (1<<15); // Clear bit 15
    };
}
/**
 * @brief Delay Function.
 * @param nCount:specifies the Delay time length.
 * @retval None
 */
void Delay(uint32_t nCount)
{
  while(nCount--)
  {
    __NOP();
  }
}
```

Since the GPIO can be configured for various speeds, we have put in some conditionally compiled code to allow the GPIO pin characteristics to be changed easily.

Once all the source files are ready, we can now compile the project. This can be done by:

- From the pull-down menu, select "Project → Build Target," or
- Right-click on Target 1 in the project browser, and select "Build Target," or
- Use the "F7" hot key.

Once the project is compiled, you can see the compilation output as shown in Figure 15.11.

After generating the program image "blinky.axf," we need to carry out a few more steps before we can download the program to the microcontroller board and test it:

1. Install the USB Driver for the ST-Link V2 debug adaptor on the board. This step is specific to the microcontroller development board you use. In the case of the

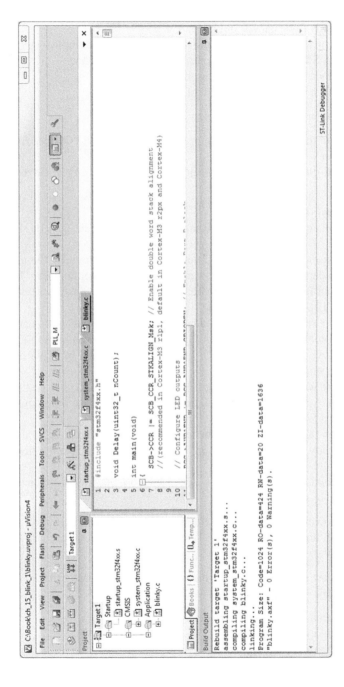

**FIGURE 15.11**

Compilation output

STM32F4 Discovery board, the debug interface requires a USB device driver and the installation program can be found in C:\Keil\ARM\STLink\USBDriver. This only needs to be carried out once.

2. Potentially you might need to execute the ST-Link Upgrade utility in C:\Keil\ARM\STLink. This only needs to be carried out once.
3. Select ST-LINK V2 for debugging in the project options.
4. Select ST-LINK V2 for flash download.

To change the project options, you can:

- Right-click on the "Target 1" on the project browser, select "Options for target 'Target 1'," or
- From the pull-down menu, select "Project → Options for target 'Target 1'," or
- Use Alt-F7.

Click on target option button

Then you will see the project option dialog with 10 different tabs, as shown in Figure 15.12.

Now click on the "Debug" tab, and you will see the options for the available types of debug adaptors. Potentially you might see multiple options for "ST-Link Debugger." Do not forget to click on the option for using hardware rather than simulator (Figure 15.13).

After you have select the ST-LINK adpator, click the "Settings" button next to it, and you should see the options shown in Figure 15.14. If not, you have selected the older version of ST-Link adaptor, which only offers the option of JTAG or Serial Wire debug protocol.

Now close this window and go back to the project options, and click on the "Utilities" tab. Again, we select the same ST-Link Debugger for flash programming, as shown in Figure 15.15.

If we click on the "Settings," we can see that the flash programming algorithm should have already been set up based on the choice of the microcontroller device (Figure 15.16).

Once these configurations are done, we can then close the project option window, and start the debug session by:

- Using pull-down menu "Debug → Start/Stop Debug Session," or
- Using Ctrl-F5 hot key, or
- Clicking on the 🔍 button.

You should now see the debugger window (Figure 15.17). You can start the program execution by:

- Pull-down menu "Debug → Run," or
- Using F5 hot key, or
- Clicking on the Run button 📲 .

If everything is set up correctly, the LEDs on the board should start flashing. Congratulations! You have just created your first Cortex®-M project!

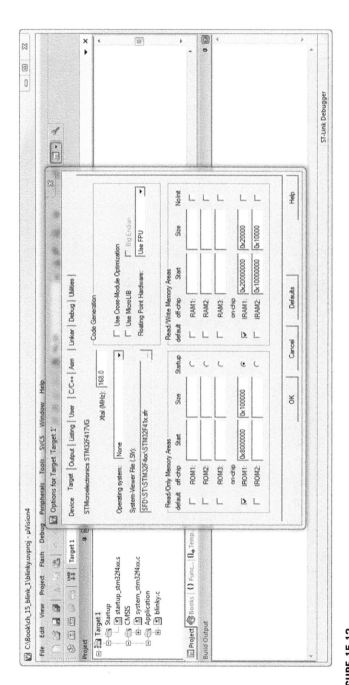

**FIGURE 15.12**

Project options dialog

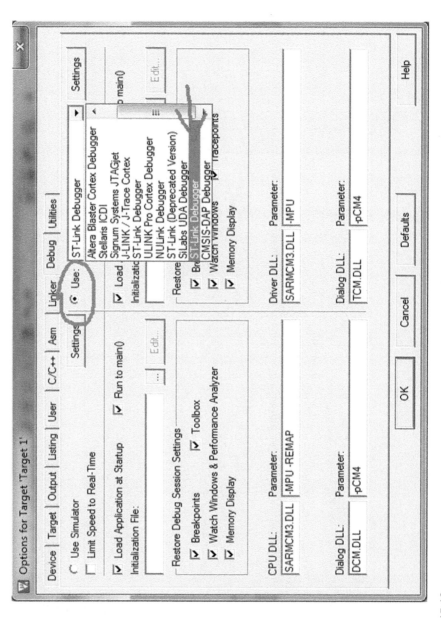

**FIGURE 15.13**

Change debug options to use ST-LINK debugger

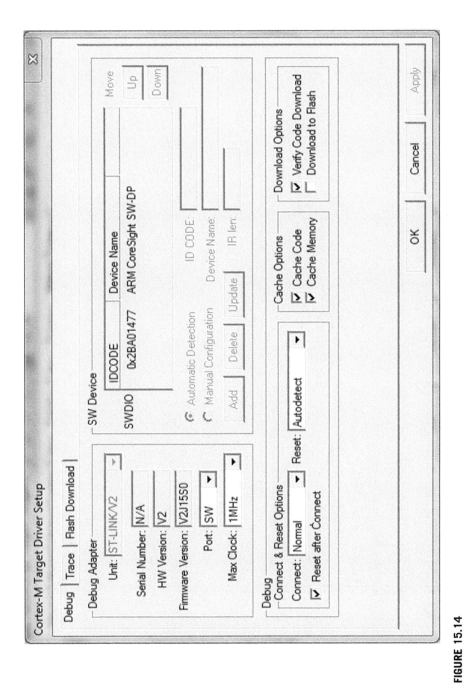

**FIGURE 15.14**

ST-LINK V2 options

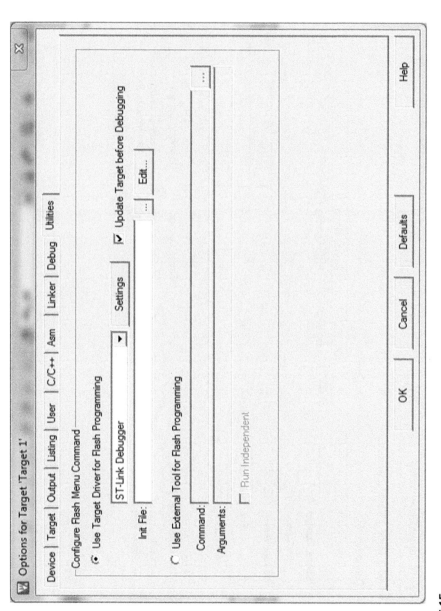

**FIGURE 15.15**

Flash programming options

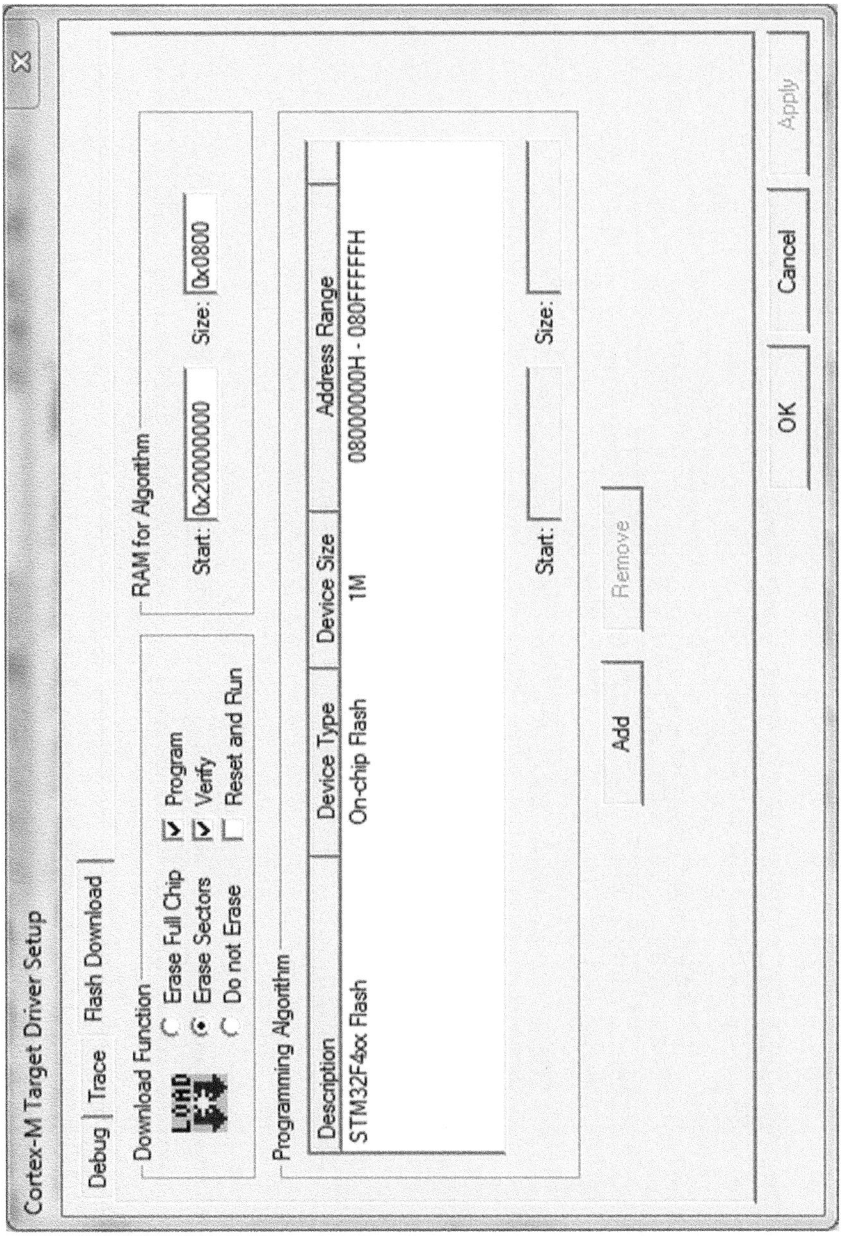

**FIGURE 15.16**

Flash programming algorithm options

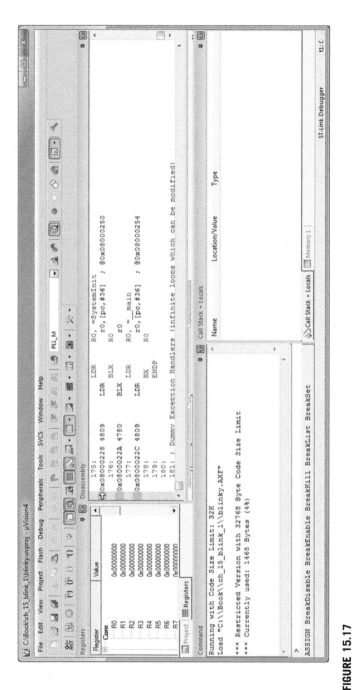

**FIGURE 15.17**

Debugger window

You can now stop the debug session using the following methods (same as starting the debug session):

- Using pull-down menu "Debug $\rightarrow$ Start/Stop Debug Session," or
- Using Ctrl-F5 hot key, or
- Clicking on 🔍 button.

## 15.4 Project options

In Figure 15.12 we can see that there are a range of project options. Figure 15.18 lists the categories of the options based on the tabs in the "Options" dialog.

Please note that μVision projects support multiple targets (in most simple projects there is only one target). Different targets can have different compiler and debug settings. By creating multiple targets you can switch between multiple sets of project configurations easily.

To add a target, right-click on the target name and select "Manage Components," and then click on the button for "New (Insert)" target.

An example of multiple targets is to allow the same application to be compiled either with debug symbols or without debug symbols (for release). When there are multiple targets in a project, you can switch between them using the target selection box on the toolbar. It is also recommended that you give your targets meaningful names, such as "Debug," "Release," etc.

### 15.4.1 Device option

This tab defines the microcontroller device you use for the project (Figure 15.19). When selecting a device from this dialog, settings like the compiler flags, memory map, and flash algorithms are pre-configured for that device. If the microcontroller device that you use is not listed, you can still select Cortex®-M3 or Cortex-M4 under ARM® section and manually set the configuration options.

### 15.4.2 Target options

The "Target Options" tab (Figure 15.20) allows you to define the memory map of the device, options to utilize the FPU (floating point unit) on the Cortex®-M4 if present on your device, and an option to use the RTX Kernel, a RealTime Operating System (RTOS) that comes with Keil™ MDK-ARM® installation.

The memory map setup is usually automatically generated when you select the microcontroller device.

### 15.4.3 Output options

The "Output options" tab (Figure 15.21) allows you to select whether the project should generate an executable image or a library. It also allows you to specify the

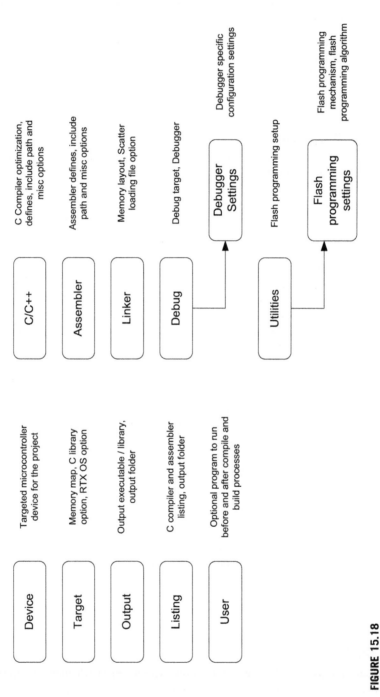

**FIGURE 15.18**

Project options tab

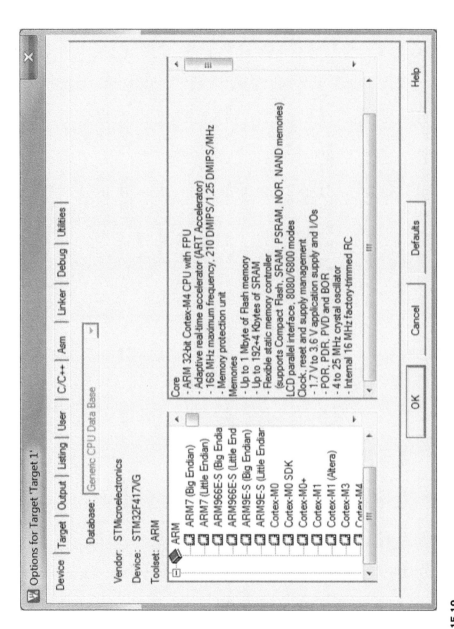

**FIGURE 15.19**

Device options tab

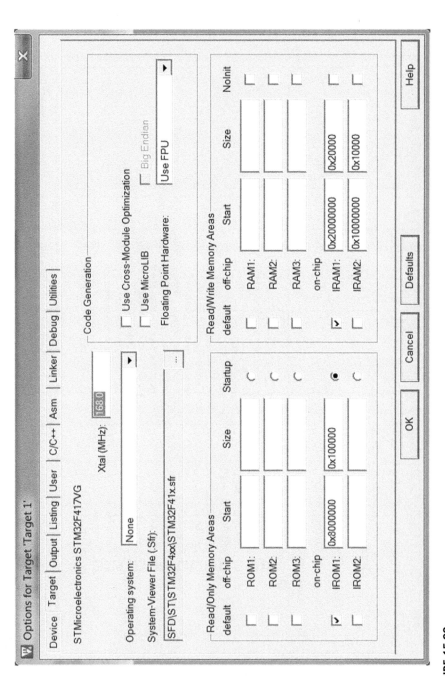

**FIGURE 15.20**

Target options tab

directory where the generated files are created. For example, you can create a subdirectory in your project area, and set the output directory to this location using the "Select Folder for Objects" dialog. This is very useful, as it can help you to keep your main project directory clean and tidy, instead of having lots of object files and generated files cluttering it.

The "Create Batch File" option generates a batch file (a script file for the Windows/DOS command prompt), which allows you to rerun the compilation without using the µVision IDE.

### 15.4.4 Listing options

The "Listing Options" tab (Figure 15.22) allows you to enable/disable assembly listing files. By default the C Compiler listing file is turned off. When debugging software bugs it can be useful to turn on this option so that you can see exactly what assembly instruction sequence is generated. Similar to the "Output" options, you can click on "Select Folder for Objects" to define where the output listings should be stored. You can also generate a disassembly listing after the linking stage using a different method, as discussed in section 14.4.5.

### 15.4.5 User options

The "User Options" tab allows you to specify additional commands to be executed. For example, in Figure 15.23, we added a command line to generate the disassembled listing of the complete program image. This gets executed after the compilation stage is completed and can be very useful for debugging (see section Appendix I.5). In the user option example shown below, "$K" is the root folder of the Keil™ development tool and "#L" is the linker output file. These key sequences can also be used to pass arguments to external user programs. You can find a list of key sequence code on the Keil website: http://www.keil.com/support/man/docs/uv4/uv4_ut_keysequence.htm.

### 15.4.6 C/C++ options

The "C/C++ Options" tab (Figure 15.24) allows you to define optimization options, C pre-processing directives (defines), search path for include files, and miscellaneous compile switches. Please note that by default a number of include file directories are automatically included in the project (see the Compile control string list at the bottom). For example, the CMSIS-Core includes files, and sometimes device-specific header files could be included automatically. If you want to use a specific version of the CMSIS-Core files, you might need to disable this automatic include path feature by clicking on the "No Auto Includes" box in this dialog.

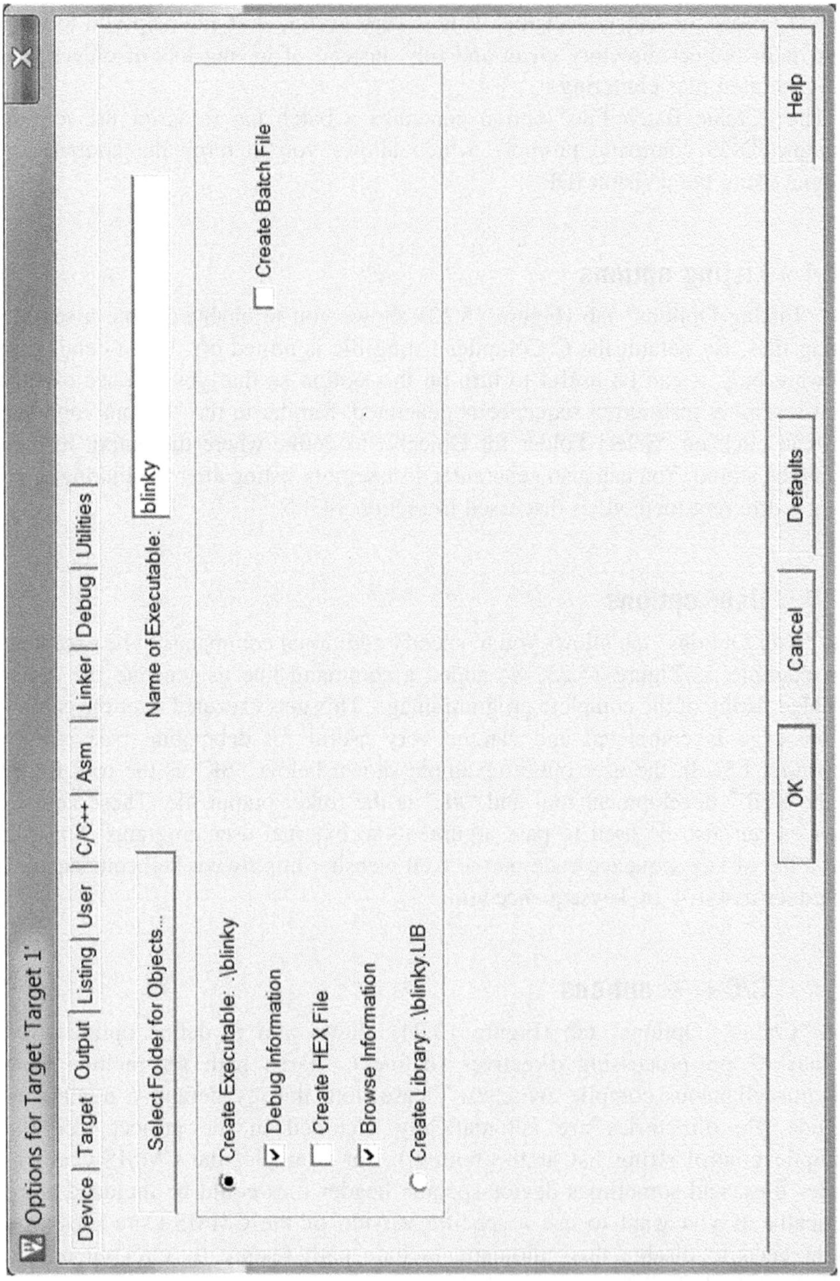

**FIGURE 15.21**

Output options tab

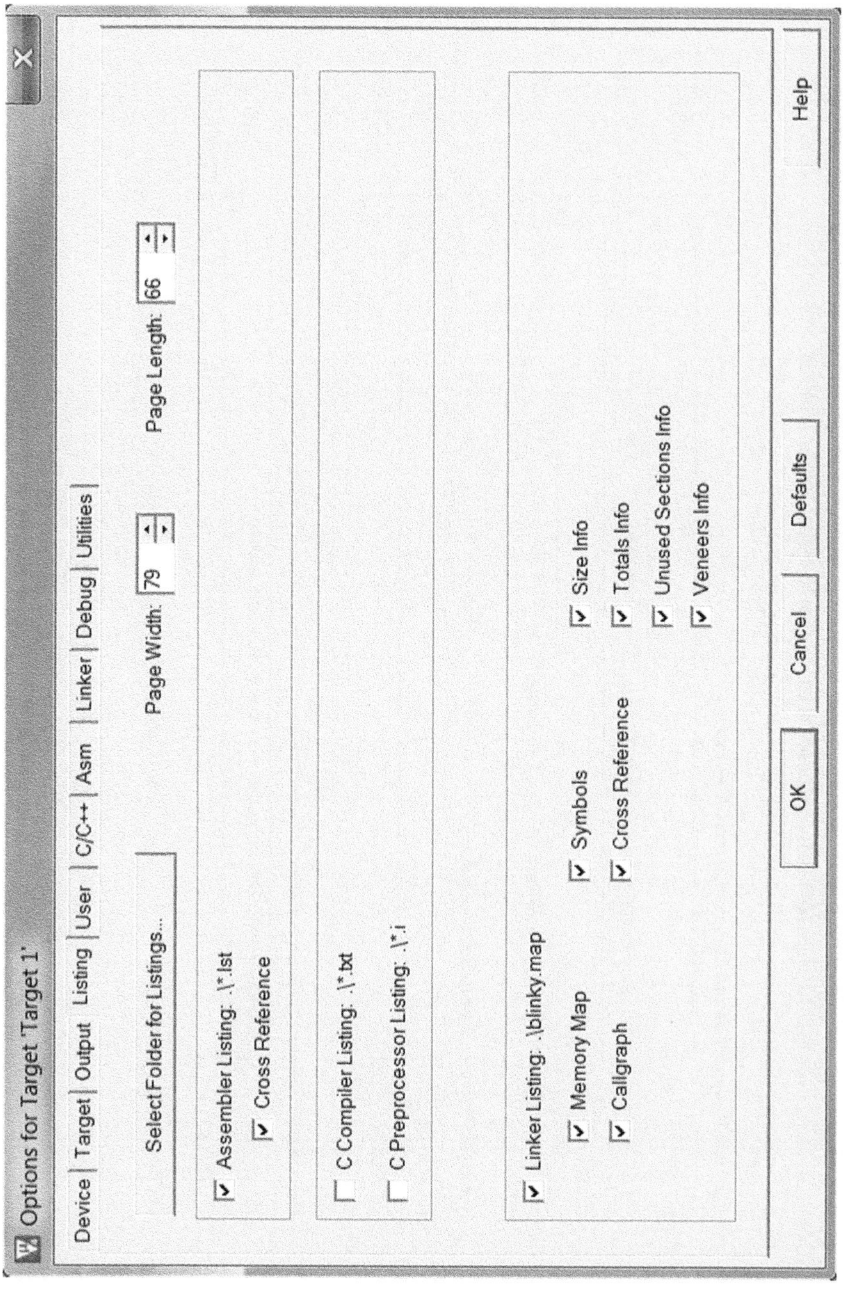

**FIGURE 15.22**

Listing options tab

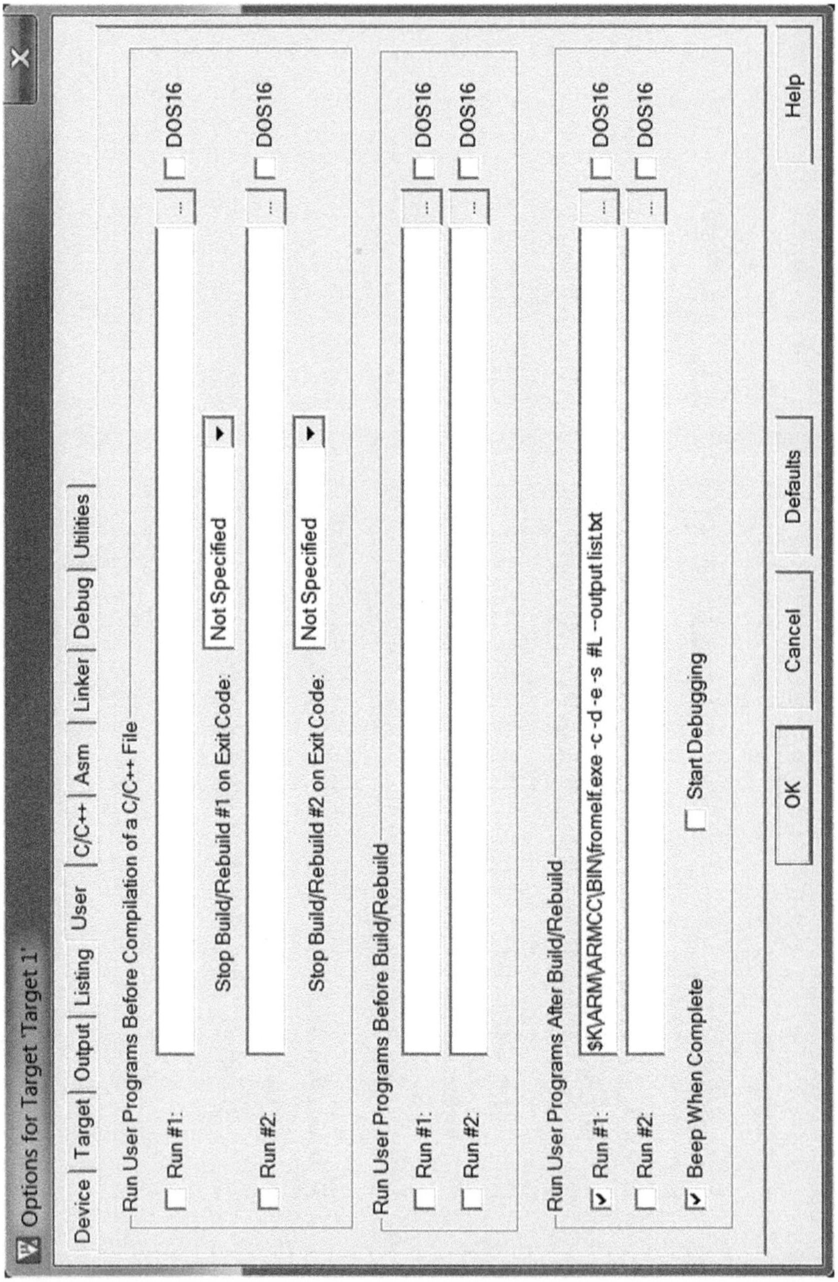

**FIGURE 15.23**

User options tab

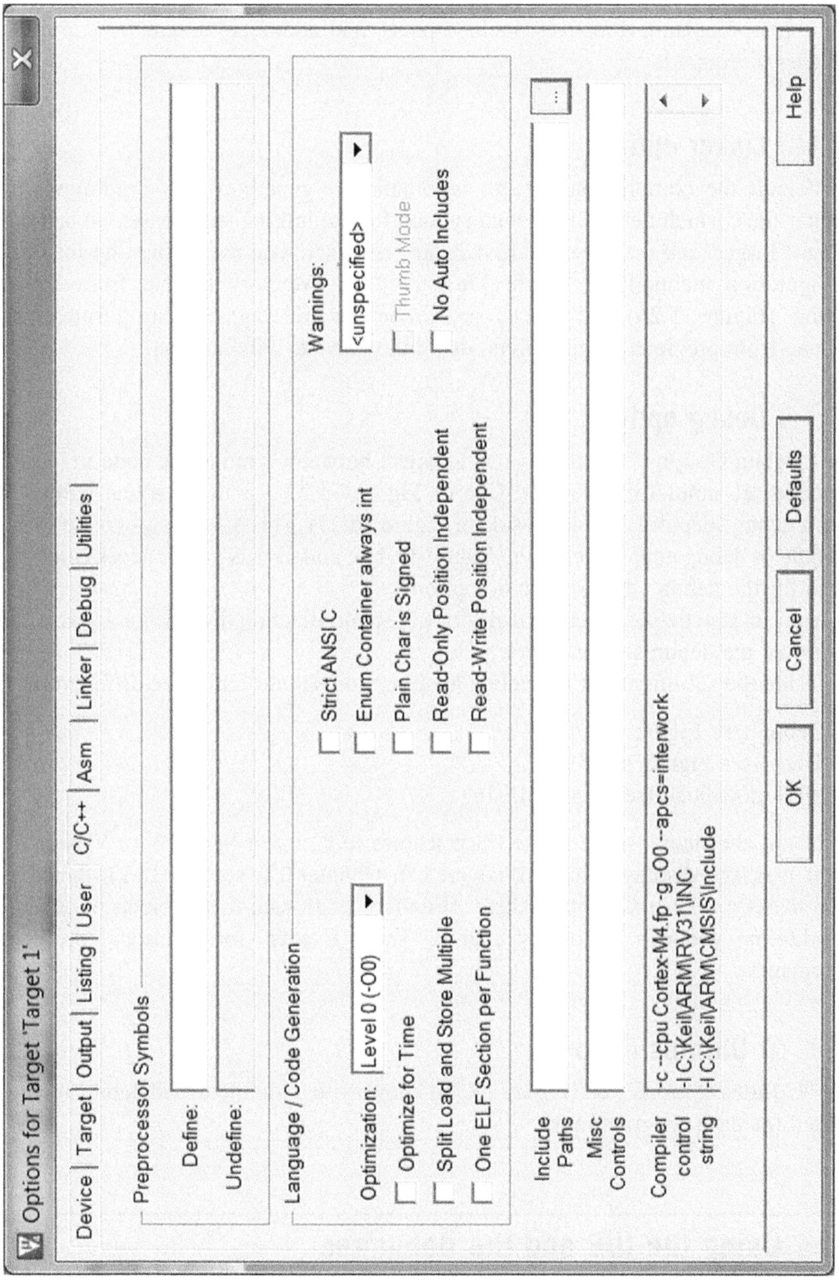

**FIGURE 15.24**

Options in the C/C++ option tab

### 15.4.7 **Assembler options**

Similarly there is also an option tab for the assembler (Figure 15.25). This allows you to define pre-processing directives, include paths, and additional assembler command switches if required.

### 15.4.8 **Linker options**

By default the compilation process automatically generates a configuration file (scatter file), which defines the memory map for the linking stage based on settings in the "Target" tab (see section 15.4.2). If necessary, you can define the memory configuration manually by unchecking the "Use Memory Layout from Target Dialog" (Figure 15.26), and select your own Scatter File (e.g., by editing from existing one from previous compilations) using the "Scatter File" option.

### 15.4.9 **Debug options**

The "Debug Options" tab allows you to select between running the code in the instruction set simulator (left-hand side of Figure 14.27), or using actual hardware with a debug adaptor (right-hand side of Figure 15.27). This also allows you to select the type of debug adaptor (e.g., see Figure 15.13), and allows you to access the sub-menu for the debug adaptor specific options.

You can also define an additional script file (Initialization file), which is executed each time the debug session starts.

Inside the sub-menu for the debug adaptor, you should find three different tabs:

- Debug (see Figure 15.14)
- Trace (see Figure 15.28)
- Flash download (see Figure 15.16)

If you are planning to use the trace feature (e.g., using Serial Wire Viewer for printf message display, which is covered in Chapter 18, section 18.2), then you need to set up the trace options here, (Figure 15.28) such as the clock frequency. In addition, you can optionally enable "Trace Events" for additional profiling information.

### 15.4.10 **Utilities options**

The "Utilities Options" tab (Figure 15.29) allows you to define which debug adaptor is used for flash programming.

## 15.5 **Using the IDE and the debugger**

We looked at the debugger briefly in section 15.3. Here we will look at it in a bit more detail. After the project is compiled successfully, in most cases the program

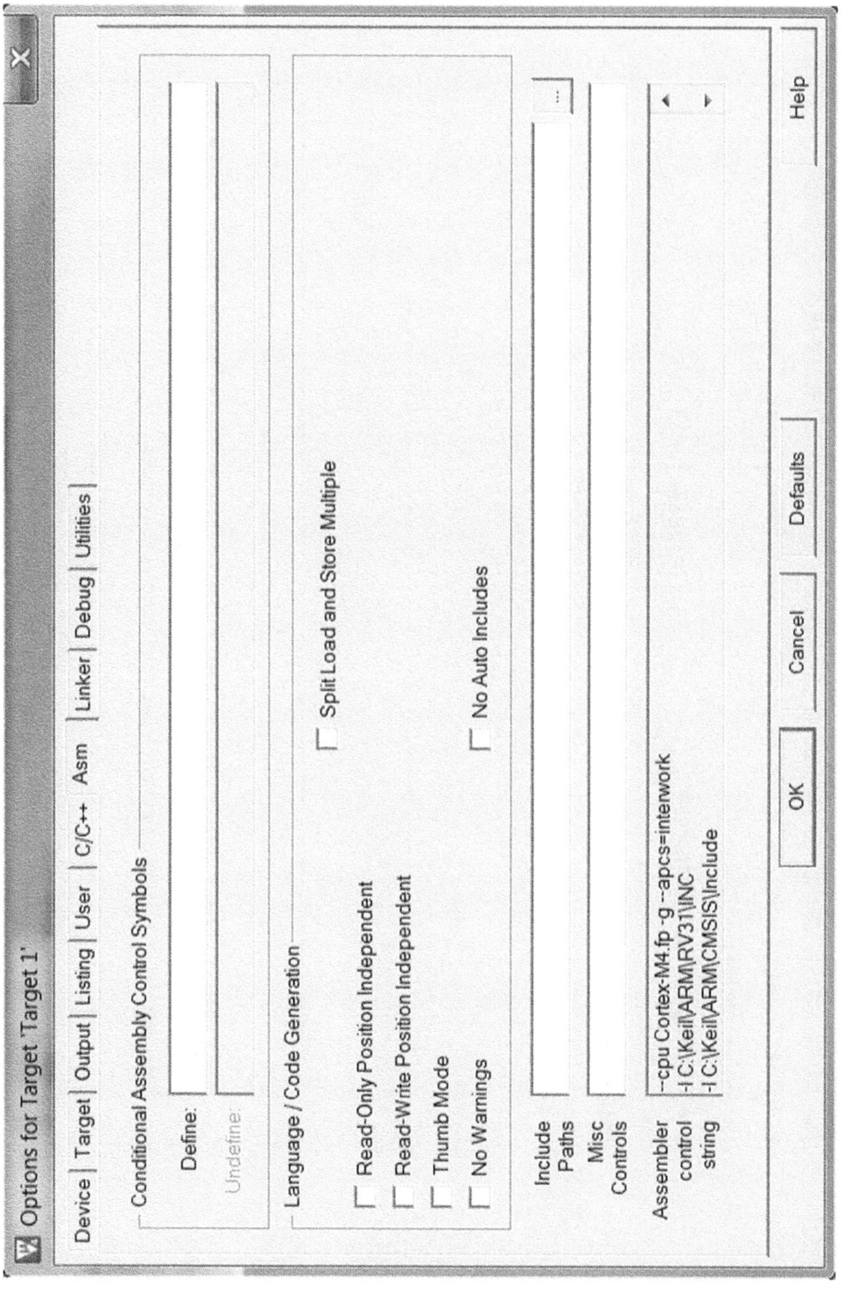

**FIGURE 15.25**

Assembler options

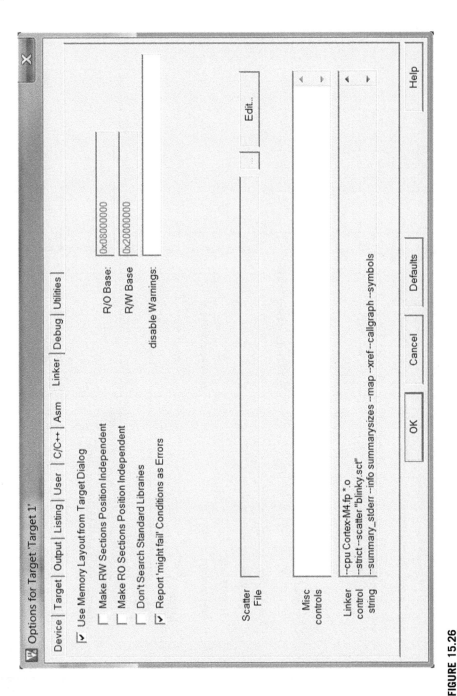

**FIGURE 15.26**

Linker options tab

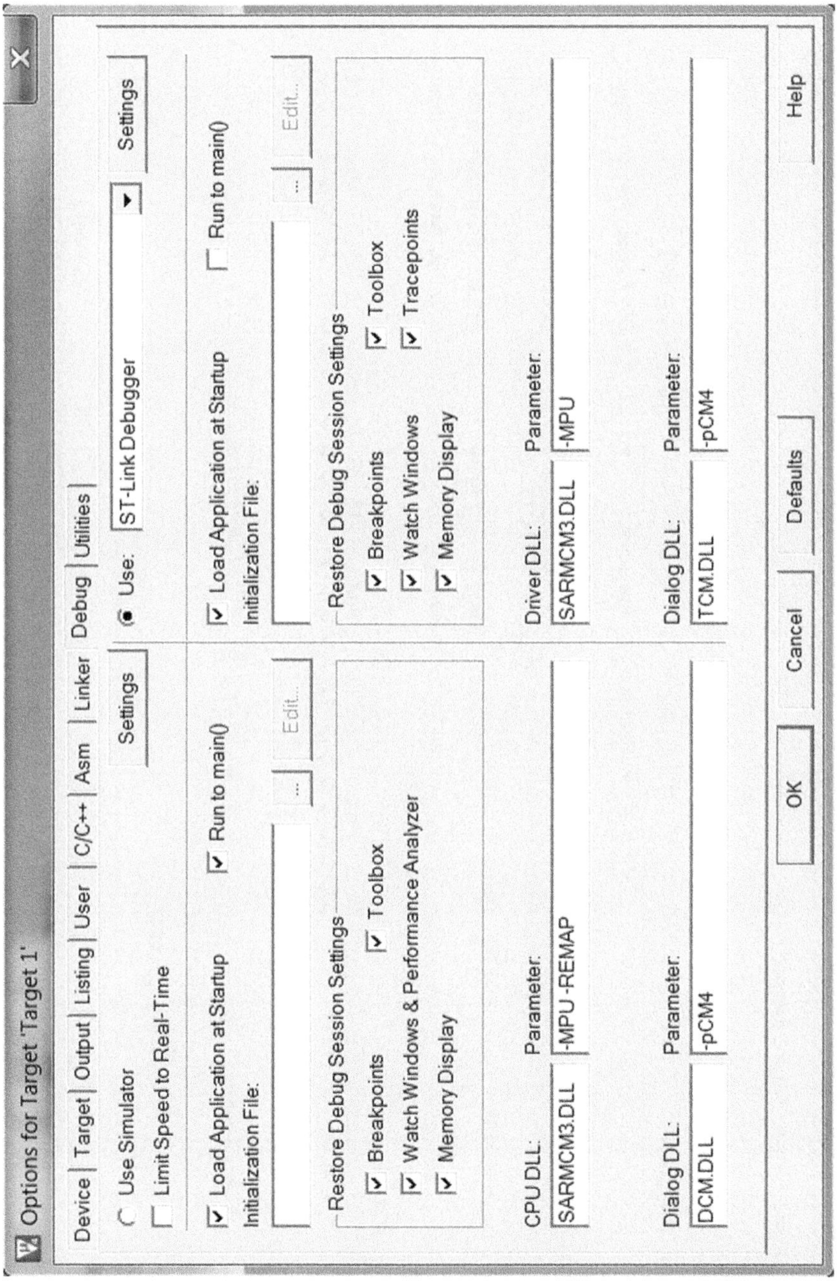

**FIGURE 15.27**

Debug options tab

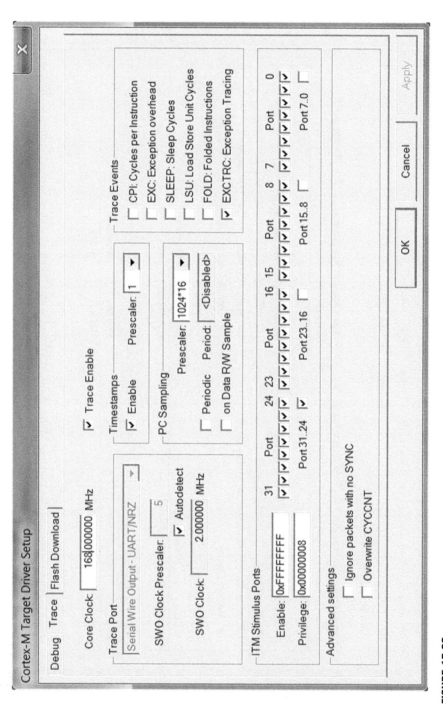

**FIGURE 15.28**

Trace options tab

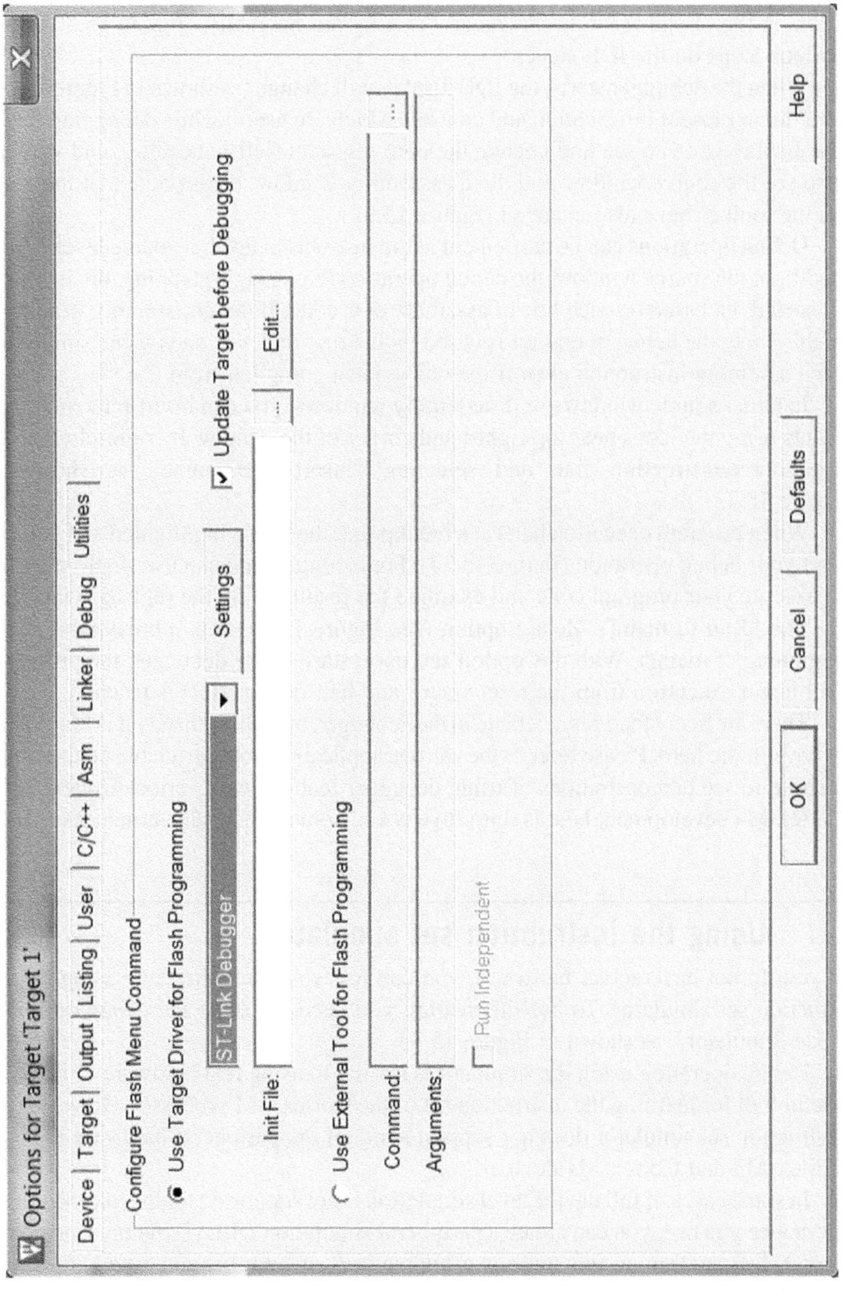


**FIGURE 15.29**
Utilities options tab

code will be loaded to the flash memory automatically when we start the debugger (see Figure 15.27, check "Load Application in Startup" option). Otherwise, we can click on the "Load" icon to download the program manually. (Figure 15.30 shows various icons on the IDE screen).

When the debugger starts, the IDE display will change (as shown in Figure 15.31) in order to present information and controls which are useful while debugging. From the display you can see and change the core registers (left-hand side), and you can also see the source window and the disassembly window. Please note that the icons on the toolbar have also changed (Figure 15.32).

Debug operations can be carried out at the instruction level or source level: If you highlight the source window, the debug operation (e.g., single stepping, breakpoints) is carried out based on each line of assembly or C code. If the disassembly window is highlighted, the debug operation is based on instruction level, so you can single step each assembly instruction even if they have been compiled from C code.

In either source windows or disassembly windows, you can insert/remove breakpoints using the icons near top right-hand corner of the window by right-clicking on the source/instruction line and selecting "insert breakpoint", as shown in Figure 15.33.

When program execution halts at a breakpoint, the line is highlighted and you can start your debug operation (Figure 15.34). For example, you can use single stepping to execute your program code and examine the results using the register window.

The "Run to main()" debug option (see Figure 15.27) sets a breakpoint at the beginning of main(). With this option set, upon starting the debugger, the processor will begin execution from the reset vector and halt once main() is reached.

There are many features available in the debugger, but unfortunately it is impossible to cover more here. Please refer to the various application notes available on the Keil™ website to see demonstrations of using debugger features with various Cortex®-M3/Cortex-M4 development boards (http://www.keil.com/appnotes/list/arm.htm).

## 15.6 Using the instruction set simulator

If you do not have actual hardware, you can test your program code using the instruction set simulator. To use this feature you need to set up the debug option to "Use Simulator," as shown in Figure 15.35.

Debug operation using the simulator is similar to using real hardware. It is a very useful tool for learning the instruction set of the Cortex®-M processor. However, the instruction set simulator does not support some of the peripheral hardware in some Cortex-M3 and Cortex-M4 devices.

In some cases, if full device level simulation is not supported for the microcontroller device you use, you can create a peripheral simulation DLL (Dynamic Linked Library). Information on this topic is available in Keil™ application note 196 (http://www.keil.com/appnotes/docs/apnt_196.asp). In addition, you might need to adjust the memory setting so that additional memory ranges are recognized by the simulator.

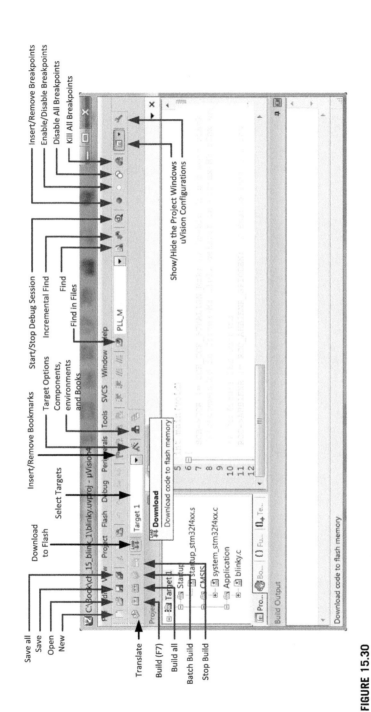

**FIGURE 15.30**

Various icons in the μVision IDE

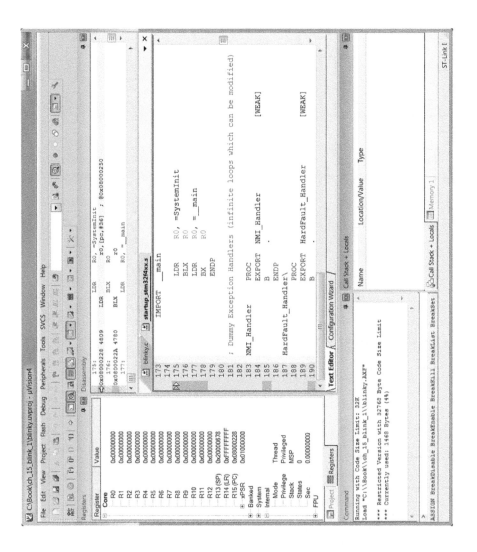

**FIGURE 15.31**

μVision debugger

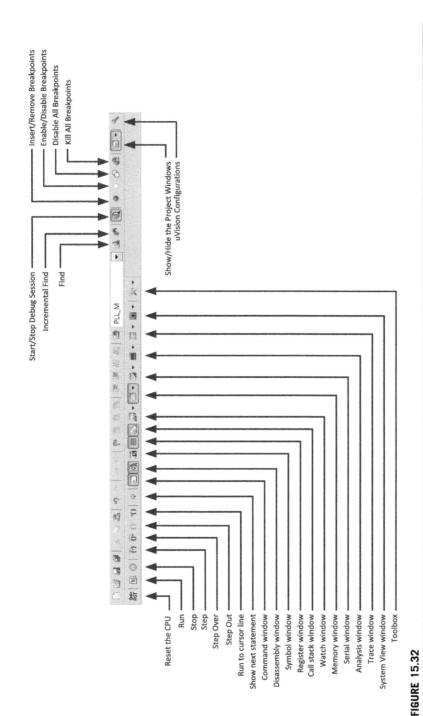

Start/Stop Debug Session
Incremental Find
Find

Insert/Remove Breakpoints
Enable/Disable Breakpoints
Disable All Breakpoints
Kill All Breakpoints

Show/Hide the Project Windows
uVision Configurations

Reset the CPU
Run
Stop
Step
Step Over
Step Out
Run to cursor line
Show next statement
Command window
Disassembly window
Symbol window
Register window
Call stack window
Watch window
Memory window
Serial window
Analysis window
Trace window
System View window
Toolbox

**FIGURE 15.32**

Icons in the toolbar during debugger mode

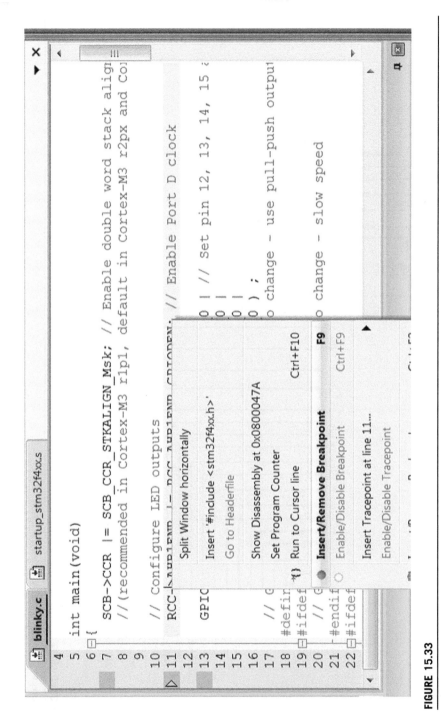

**FIGURE 15.33**

Insert breakpoint by right-clicking on the line and selecting insert breakpoint

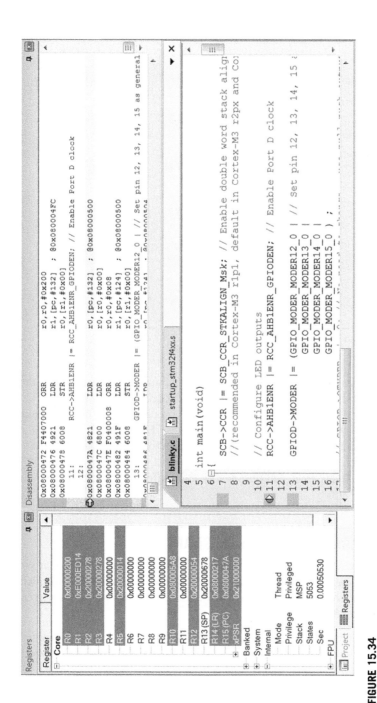

**FIGURE 15.34**

Processor halted after hitting a breakpoint

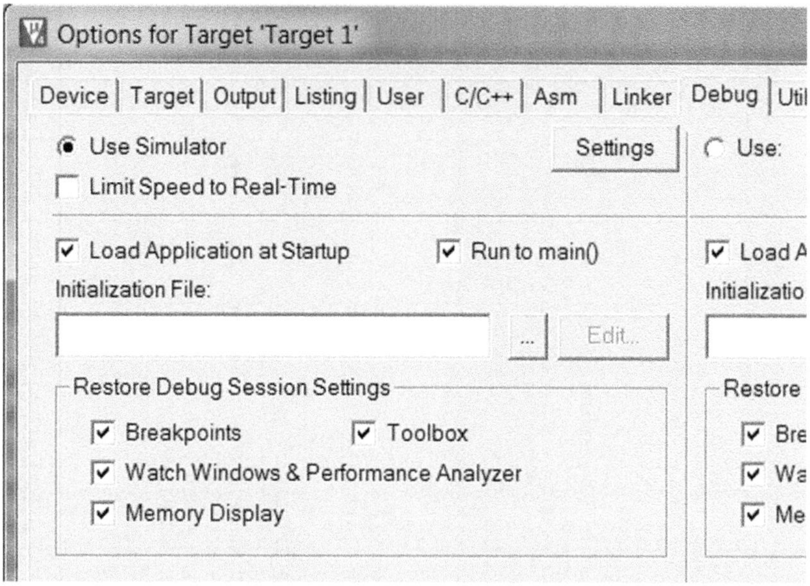

**FIGURE 15.35**

Select instruction set simulator for debug

This can be done by using the Memory Map configuration menu by accessing the pull-down menu "Debug → Memory Map" during the debug session (Figure 15.36).

When using the instruction set simulator, you can display the execution time display using the pull-down menu: "Debug → Execution Profiling → Show Time." This can be useful to get a rough estimation of how long a certain task takes. However, please note that the timing information of the instruction set simulator is not cycle accurate, and does not factor in effects of wait states in the memory system.

## 15.7 Running programs from SRAM

One of the interesting capabilities of Cortex®-M microcontrollers is that you can execute programs from RAM, which is not possible in a few non-ARM microcontroller architectures. There could be many reasons why you would want to execute a program from SRAM:

- The device that you are using could have an OTP (One Time Programmable) ROM so you don't want to program the chip unless the code is finalized.
- Some microcontrollers do not have internal flash memories and need to use external memory storages. During software development you might want to start off by using internal SRAM for testing.

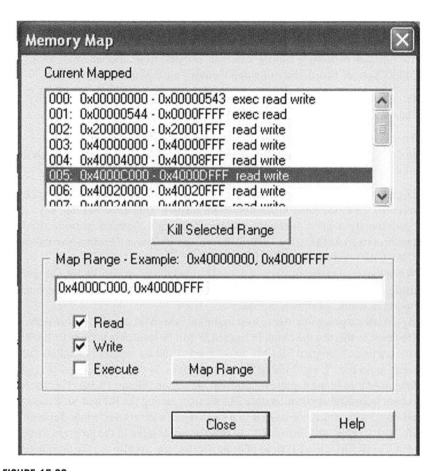

**FIGURE 15.36**

Adding new memory range to simulator memory setup

- For production testing or certain field testing, you might want to just run some test codes on the system without changing the program code already programmed on the device. In this case you can download and run the test code from SRAM.
- In systems with slow flash memories, you might want to improve the performance by copying the program image from flash to SRAM during the startup sequence, and executing the program from SRAM for highest performance.

For example, if you are using the LPCXplorer board (Figure 14.1), because the LPC4330 microcontroller does not have internal flash memory, you might want to test your program code by running it from the internal SRAM. The best starting point for using this board is to download the example in Keil™ application 233 (http://www.keil.com/appnotes/docs/apnt_233.asp). The example project in this application note provides two targets; one is running from the external SRAM and the second target is for running the code using the external flash memory.

To run the following blinky example project from the internal SRAM of LPC4300 Xplorer board, the following memory settings are used:

- Program code: starting from 0x10000000, size 0x20000
- Data RAM: starting from 0x20000000, size 0x10000

Therefore the memory setting becomes the one shown in Figure 15.37.

In order to get the program to load on to the board and run correctly, a few additional setting adjustments are needed:

- Disable CRP key section in startup code — In NXP LPC4330, address 0x2FC is reserved for Code Read Protection (CRP). The startup code has this key conditionally defined. Since the program is going to be running from SRAM (address 0x10000000), the CRP feature is not used and therefore we add the "NO_CRP" defines in the assembler options.
- When the debug session starts, we need to load the program into SRAM without using any flash programming algorithms. So we switch the flash programming option to blank, as shown in Figure 15.38.
- By default the processor tries to load the initial value of PC and MSP from address 0. However, in this test the initial PC and MSP will be located in address 0x10000000 (the starting of program SRAM), so we need to add a simple debug initialization script to set the PC and MSP to the correct values. In the debug option we add the Dbg_RAM.ini in the debug initialization file option (Figure 15.39). This file is copied from Keil application note 233. Besides setting the PC and MSP values, it also forces the processor's xPSR to a valid value, and sets the Vector Table Offset Register (VTOR; see section 7.5) so that the vector table in the program image stored in SRAM is used. Instead of using a script, you can also change these registers values inside the debugger manually. But using a script is much more convenient.

The code inside Dbg_RAM.ini forces the debugger to load the program image in to the RAM, then issues a reset and sets up the stack point value (SP), program counter value (PC), vector table offset (VTOR), and other device-specific registers. Afterward it starts the program execution using the go ("g") command.

```
/*Dbg_RAM.ini — a script for µVision debugger. This code is specific
to NXP LPC4300.*/
/*------------------------------------------------------------
* Name: Dbg_RAM.ini
* Purpose: RAM Debug Initialization File
* Note(s):
*------------------------------------------------------------
* This file is part of the uVision/ARM development tools.
* This software may only be used under the terms of a valid, current,
* end user licence from KEIL for a compatible version of KEIL software
```

```
    * development tools. Nothing else gives you the right to use this
    software.
    *
    * This software is supplied "AS IS" without warranties of any kind.
    *
    * Copyright (c) 2012 Keil - An ARM Company. All rights reserved.
    *-------------------------------------------------------------*/


    /*--------------------------------------------------
     Setup() configure PC & SP for RAM Debug
    *-----------------------------------------------------*/
FUNC void Setup (void) {
    // Reset peripherals: LCD, USB0, USB1, DMA, SDIO, ETHERNET, GPIO
    _WDWORD(0x40053100, 0x105F0000);  // Issue reset
    _sleep_(1);

    SP = _RDWORD(0x10000000);     // Setup Stack Pointer
    PC = _RDWORD(0x10000004);     // Setup Program Counter
    XPSR = 0x01000000;            // Set Thumb bit
    _WDWORD(0xE000ED08, 0x10000000);  // Setup Vector Table Offset
                                      //            Register
    _WDWORD(0x40043100, 0x10000000);  // Set the shadow pointer
    }

    LOAD %L INCREMENTAL       // Download

    Setup();                  // Setup for Running

    g, main
```

After these adjustments, we can start the debug session, which downloads the program to SRAM and executes the blinky program.

```
    /*Blinky program for LPC4330 on NGX LPCXplorer board*/
    #include <LPC43xx.h>

    void Delay(uint32_t nCount);

    const uint32_t led_mask[] = { 1UL << 11,    /* GPIO1.11  */
                                  1UL << 12 };   /* GPIO1.12  */

    /* Clock Control Unit register bits */
    #define CCU_CLK_CFG_RUN (1 << 0)
    #define CCU_CLK_CFG_AUTO (1 << 1)
```

```
#define CCU_CLK_STAT_RUN (1 << 0)
#define LOOP_COUNT 0x3FFFFF

int main (void) {
  SCB->CCR |= SCB_CCR_STKALIGN_Msk; // Enable double word stack
alignment
  //(recommended in Cortex-M3 r1p1, default in Cortex-M3 r2px and
Cortex-M4)

  // Configure LED outputs
    // Enable clock and init GPIO outputs */
    LPC_CCU1->CLK_M4_GPIO_CFG = CCU_CLK_CFG_AUTO | CCU_CLK_CFG_RUN;
    while (!(LPC_CCU1->CLK_M4_GPIO_STAT & CCU_CLK_STAT_RUN));

    LPC_SCU->SFSP2_11 = 0;          /* GPIO1[11]   */
    LPC_SCU->SFSP2_12 = 0;          /* GPIO1[12]   */

    LPC_GPIO_PORT->DIR[1] |= (led_mask[0] | led_mask[1]);
    LPC_GPIO_PORT->SET[1] = (led_mask[0] | led_mask[1]); /* switch
                                                      LEDs off
                                                      */
  while(1){
    LPC_GPIO_PORT->CLR[1] = led_mask[0];// Set LED 0
    Delay(LOOP_COUNT);
    LPC_GPIO_PORT->SET[1] = led_mask[0];// Clr LED 0
    LPC_GPIO_PORT->CLR[1] = led_mask[1];// Set LED 1
    Delay(LOOP_COUNT);
    LPC_GPIO_PORT->SET[1] = led_mask[1];// Clr LED 1
  };
}
void Delay(uint32_t nCount)
{
  while(nCount--)
  {
    __NOP();
  }
}
```

## 15.8 Optimization options

A number of compiler and code generation options are available to allow different optimizations. The first group of these options is the C compiler options, as shown in Figure 15.24. The C compiler options allow you to select

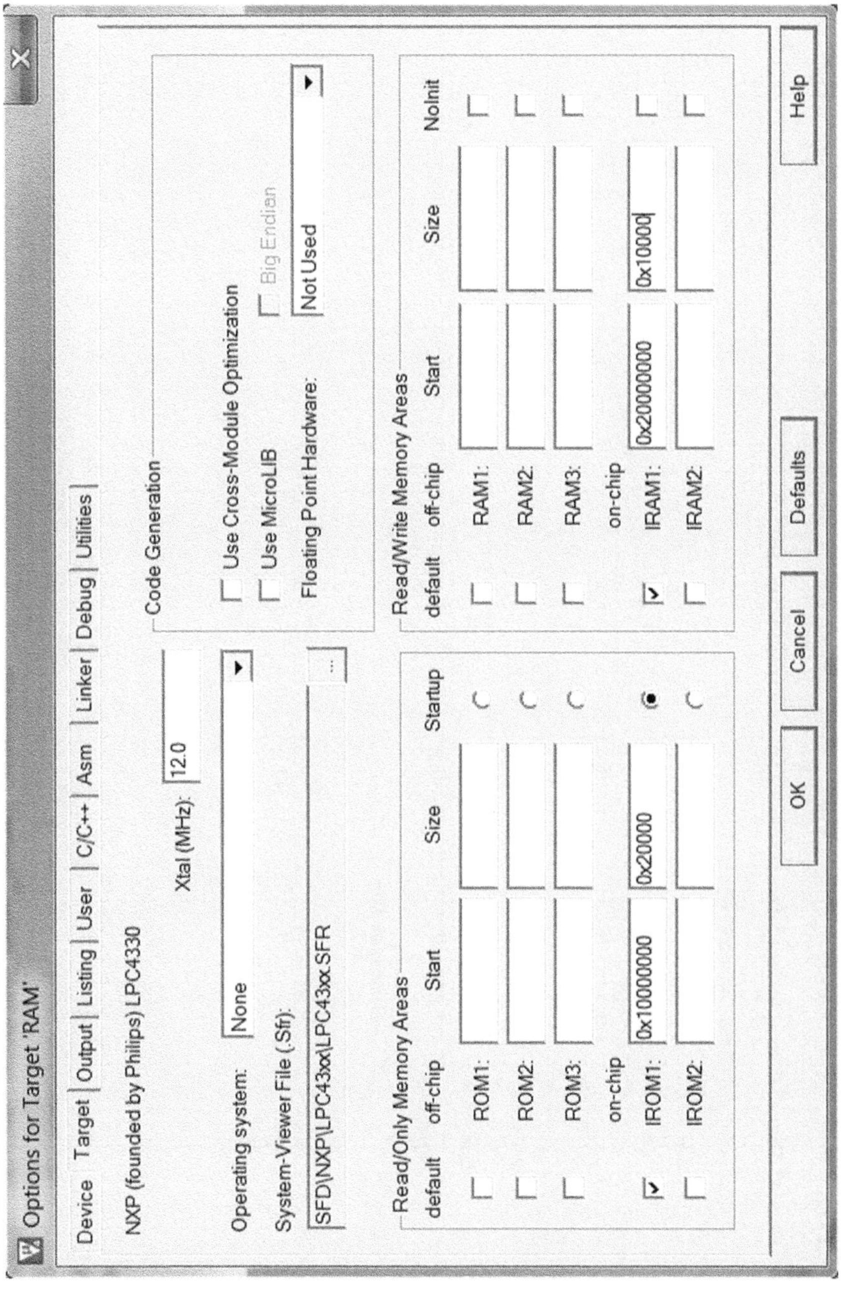

**FIGURE 15.37**

Memory configuration when running program from SRAM

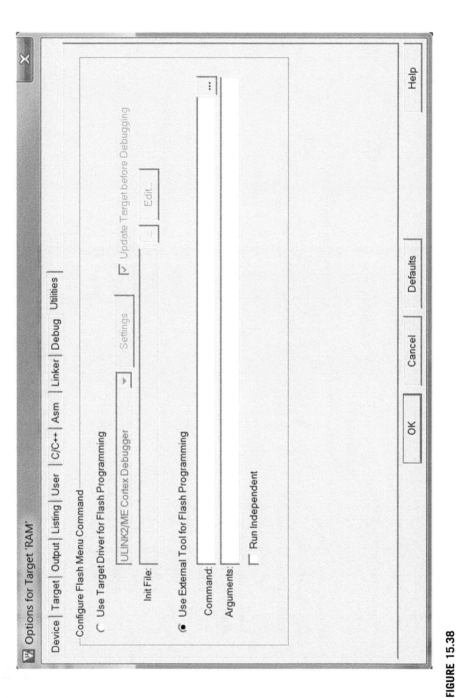

**FIGURE 15.38**

Blank flash programming option for LPC4330 when running program code from SRAM

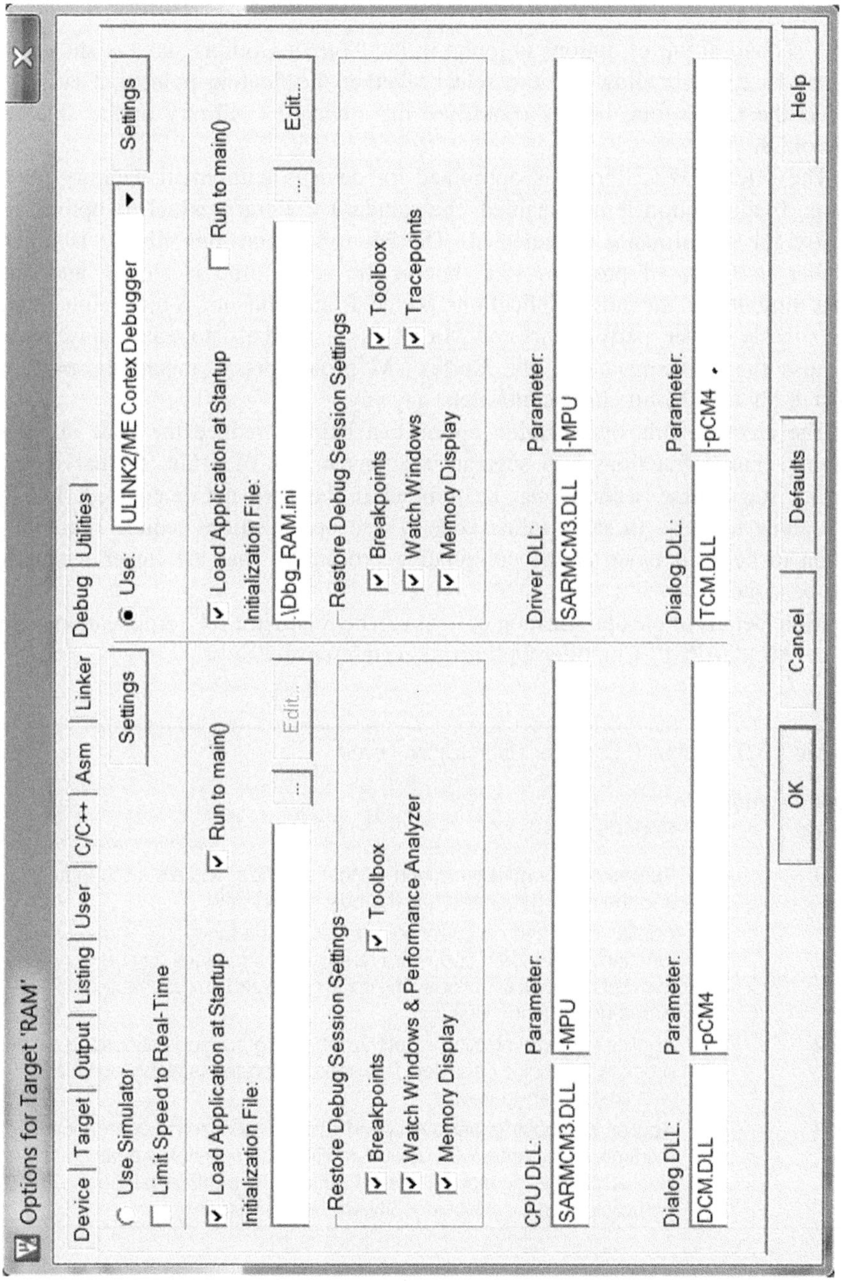

**FIGURE 15.39**

Debug script option

optimization levels (0 to 3, see Table 15.1) through a drop-down menu. Optimization is set to reduce code size by default unless the tick box "Optimize for Time" is set.

A second group of options is found in the "Target Options" tab, as shown in Figure 15.20. This allows you to select whether the floating point unit is used, and if the C run-time library should be the standard C library or the smaller MicroLIB.

The MicroLIB C library is optimized for devices with small memory footprints. If this option is not enabled, the standard C library, which is optimized mostly for performance, is selected. The MicroLIB run-time library is much smaller in terms of program size, but at the same time is slower and has some limitations. In most applications ported from 8-bit or 16-bit architecture, the slightly lower performance of MicroLIB is unlikely to cause any issue because the performance of the Cortex®-M processors is much higher than most 8-bit and 16-bit microcontrollers anyway.

The cross-module optimization option can help to reduce the code size by placing unused functions into separate sections in the ELF file, so that in the linking stage these sections can be omitted if they are not referenced. It can also allow modules to share inline code. These optimizations require the application to be built twice in the compilation process so that the linker feedback can be done.

More details of the optimization options can be found in Keil™ Application Note 202 — MDK-ARM® Compiler Optimization (reference 20).

**Table 15.1** Various C Compiler Optimization Levels

| Optimization Level | Descriptions |
| --- | --- |
| -O0 | Applies minimum optimization – Most optimizations are switched off, and the code generated has the best debug view. |
| -O1 | Applies restricted optimization – unused inline functions, unused static functions, redundant codes are removed. Instructions can be reordered to avoid interlock situations. The code generated is reasonably optimized with a good debug view. |
| -O2 | Applies high optimization – optimize the program code according to the processor specific behavior. The code generated is highly optimized, with limited debug view. |
| -O3 | Applies the most aggressive optimization – optimize in accordance with the time/space option. By default, multi-file compilation is enabled at this level. This give the highest level of optimization, but take longer compilation time and lower software debug visibility. |

## 15.9 **Other hints and tips**

### 15.9.1 **Stack and heap memory size configurations**

The stack memory size and heap memory size are defined inside the startup code. For example, you can find the follow code fragment in `startup_stmf4xx.s`:

```
; Amount of memory (in bytes) allocated for Stack
; Tailor this value to your application needs
;  <h> Stack Configuration
;    <o> Stack Size (in Bytes) <0x0-0xFFFFFFFF:8>
;  </h>

Stack_Size EQU   0x00000400

          AREA STACK, NOINIT, READWRITE, ALIGN=3
Stack_Mem  SPACE Stack_Size
__initial_sp


;  <h> Heap Configuration
;    <o> Heap Size (in Bytes) <0x0-0xFFFFFFFF:8>
;  </h>

Heap_Size   EQU   0x00000200

          AREA HEAP, NOINIT, READWRITE, ALIGN=3
__heap_base
Heap_Mem    SPACE Heap_Size
__heap_limit
```

You need to adjust the stack and heap memory size according to your application. You can find out the stack size requirements for various functions from an HTML file generated after the compilation.

The heap memory is typically used by memory allocation functions, and in some cases also by other C run-time functions, including "printf", when certain data formatting strings are specified.

### 15.9.2 **Other information**

There are many debug features available in the Keil™ MDK-ARM that work with various debug components on the Cortex®-M processors. On the ARM® application note section of the Keil website, you can find a number of application notes about using Keil MDK-ARM with a number of commonly used Cortex-M development boards. This can be a good starting point for using the boards as well as a way to learn how to use various debug features.

In addition, Chapter 18 covers some information about using ITM for printf message output. This can be a very useful feature for software development.

# Getting Started with the IAR Embedded Workbench for ARM®

# 16

## 16.1 Overview of the IAR embedded workbench for ARM®

The IAR Embedded Workbench for ARM® is a popular development suite for ARM-based microcontrollers. It contains:

- A C and C++ compiler for various ARM processors
- An Integration Development Environment (IDE) with project management and editor
- C-SPY® debugger with ARM simulator, JTAG support, and support for RTOS-aware debugging on hardware (a number of RTOS plug-ins are available). The debugger supports various debug adaptors including:
  - I-Jet (IAR)
  - Segger J-Link/J-Link Ultra/J-Trace
  - Signum JtagJet/JtagJet-Trace
  - GDB server
  - ST Link/ST Link v2
  - Stellaris FTDI (a debug adaptor on some TI Stellaris boards)
  - Stellaris ICDI (a debug adaptor on some TI Stellaris boards)
  - SAM-ICE (Atmel)
  - …etc.
- Additional components including ARM assembler, linker and librarian tools, flash programming support
- Examples for various development boards from multiple manufacturers
- Documentation

The full version of IAR Embedded Workbench also supports:

- Automatic checking of MISRA C rules (MISRA C:2004)
- Source code for runtime libraries

The IAR Embedded Workbench is a commercial tool. Various editions are available, including a free version called Kickstart, which is limited to 32KB code size and has some of the advance features disabled. You can also download a fully featured version for a 30-day evaluation.

The IAR Embedded Workbench is easy to use and supports many debug features available in the Cortex®-M processors. In this chapter we will demonstrate the use of IAR Embedded Workbench for ARM with the STM32F4 Discovery board.

## 16.2 Typical program compilation flow

Just like most commercial development suites, the compilation process is handled automatically by the IDE and can be invoked easily by the GUI. So in most cases you don't need to understand the details of the compilation flow. Once the project is created, the IDE automatically invokes various tools to compile the code and generate the executable image, as shown in Figure 16.1.

Most of the device configuration, such as configuration files for memory layout and flash programming details are pre-installed, so you only need to select the right microcontroller devices in the project settings to enable the correct compilation flow.

In order to simplify the application development and allow quicker software development, in most cases you will be using a number of files prepared by the microcontroller vendor so that you don't have to waste time in creating definition files

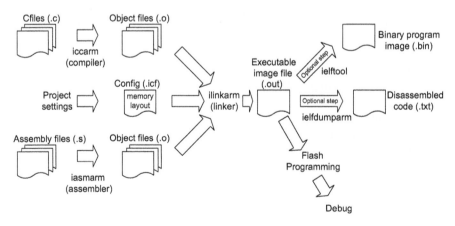

**FIGURE 16.1**

Example compilation flow with IAR Embedded Workbench

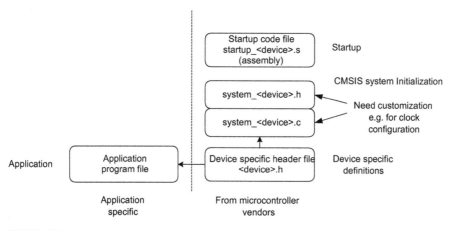

**FIGURE 16.2**

Example project with CMSIS-Core

for peripheral registers. These files are normally part of the CMSIS-compliant device driver library supplied by the microcontroller vendor. In many cases these are referred as software packages that might also include additional components such as examples, tutorials, and additional software libraries.

A simple example project using the CMSIS device library is illustrated in Figure 16.2.

While your application might only contain one file (left-hand side of Figure 16.2), the project also includes a number of files from the microcontroller vendor. While you can create your applications almost entirely in C language, the startup code that contains the vector table is often provided in the form of assembly code. The startup code is toolchain specific. However, the rest of the files in the project are toolchain independent. In fact, in the blinky project example that we will cover in section 16.3, apart from the assembly startup code, all the other program code files are identical to the example project for the Keil™ MDK-ARM discussed in Chapter 15. This is an important advantage of the CMSIS because it makes most of the software components independent of the toolchain and hence the software code is much more portable and reusable.

Additional CMSIS-Core files are referenced by some of these CMSIS-Core files. These are generic CMSIS-Core files from ARM® (bottom right-hand corner of Figure 16.3) and are integrated in the IAR Embedded Workbench installation. One project option allows these files to be automatically included during the compilation stage. If necessary, you can disable this project option and add the generic CMSIS-Core files into the project manually. This might be needed if you need to use a specific version of CMSIS-Core files.

If you are using older versions of CMSIS-Core (version 2.0 or older), you might also find that you need to include a file called core_cm3.c or core_cm4.c in the

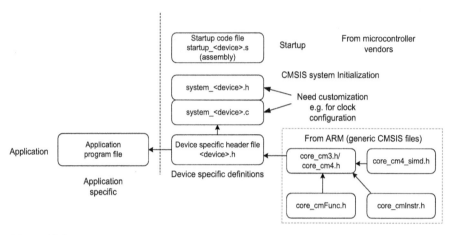

**FIGURE 16.3**

Example project view when including CMSIS-Core files from ARM

CMSIS-Core package for some of the core functions like access to special registers and a couple of intrinsic functions. These files are no longer required in newer versions of CMSIS-Core, and the CMSIS-Core functions are still 100% compatible with older versions.

## 16.3 Creating a simple blinky project

When the IAR Embedded Workbench is started, you will see the screen shown in Figure 16.4. You can open an existing example project by clicking on "EXAMPLE PROJECTS." There are many ready-made examples that can serve as a starting point for application development. In this section, we look at how to create a new project from scratch.

We can create a new project using the pull-down menu: "Project → Create New Project…"

A new window will appear to allow you to select the type of project to create, as shown in Figure 16.5.

We select to create an empty project, and will then be asked to define the location of the project file. Here in this stage we create a project called blinky (Figure 16.6).

Once this is done, we have an empty project, and we can start adding files to it. In order to make the project files more organized, we can add a number of file groups in the project and put different types of files into these groups. You can access the add group/file function by right clicking on the project target (Debug) and selecting "Add," as shown in Figure 16.7, or from the pull-down menu: "Project → Add Group/Add File."

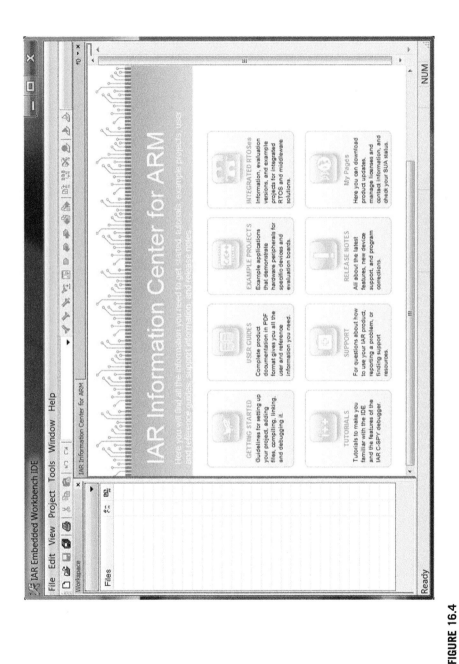

**FIGURE 16.4**

Start screen of the IAR Embedded Workbench for ARM

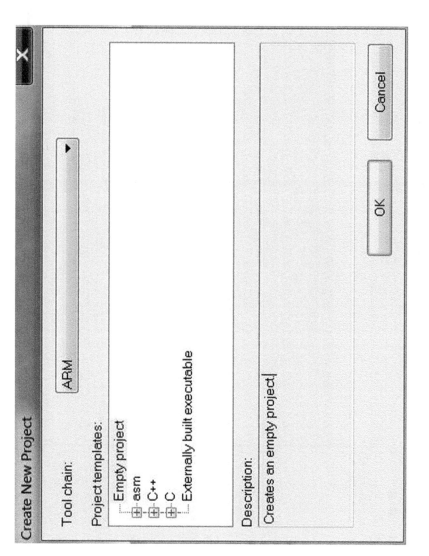

**FIGURE 16.5**

New Project window

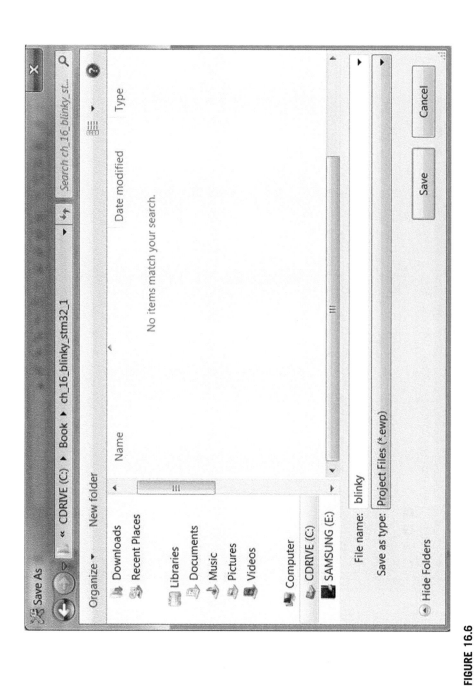

**FIGURE 16.6**

Creating an empty blinky project file

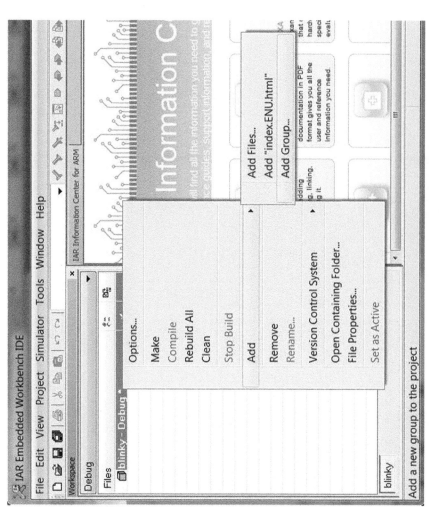

**FIGURE 16.7**

Adding groups and files to a project

**Table 16.1** Files in the Blinky Project

| File | Description |
|------|-------------|
| startup_stm32f4xx.s | Startup code (in Assembly) for STM32F4. This file is specific to IAR Embedded Workbench and is available in STM32F4 software package. |
| stm32f4xx.h | Device definition files including peripheral register definitions, exception type definitions. This file is available in the STM32F4 software package. |
| system_stm32f4xx.c | System initialization function (SystemInit()) for STM32F4 and related functions as specified in CMSIS-Core. This file is available in the STM32F4 software package. |
| system_stm32f4xx.h | Header file for defining function prototypes of functions in system_stm32f4xx.c. This file is available in the STM32F4 software package. |
| blinky.c | The Blinky application which toggles LEDs on the board |

**Table 16.2** Project Options to get the Blinky Project Working

| Category | Tab | Details |
|----------|-----|---------|
| General options | Target | Device → STM32F407VG |
| General options | Target | FPU: Either none or VFPv4 options can be used for the blinky project. |
| General options | Library Configuration | Use CMSIS. This automatically includes essential CMSIS-Core Header files in the project. |
| C/C++ Compiler | List | Output list file. This is optional, but can be useful for debugging. |
| Linker | Config | Optional settings: Override default if you need to change the memory map (e.g., different stack and heap size). |
| Debugger | Setup | ST-Link. Use the on-board debug adaptor for debug. |
| Debugger | Download | Use flash loader. This enable flash download to the STM32F4 microcontroller. |
| ST-Link | ST-Link | Select SWD debug protocol. |
| ST-Link | ST-Link | Optional settings: CPU clock → 168MHz Auto in SWO clock |

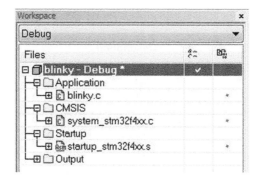

**FIGURE 16.8**

Blinky project

Inside the project folder, we have the files shown in Table 16.1.

We add the blinky program and the additional files to the project. We haven't explicitly included other CMSIS Header files as these can be automatically included using a project option ("General Options"; see Table 16.2).

In the next steps, we need to set up various project options. These can be accessed by right-clicking on the project target ("blinky − Debug" as shown in Figure 16.8), by the pull-down menu (Projects → Options), or by shortcut key ALT-F7. There are wide range of project options available; for many categories on the left, you can find multiple tabs. For example, in the "General options" category, you can find: Target, Output, Library Configuration, Library Options, MISRA C-2004, MISRA C-1999 (Figure 16.9).

In this blinky project, we need to set up a number of options as shown in Table 16.2.

After setting up the project options, we can now try to compile the project and test it. To start, we right-click on the project target (Blinky − Debug) and select Build, then the IDE will ask us to save the current workspace; we will save it in the same project directory as "blinky.eww," as shown in Figure 16.10.

If everything is set up correctly, you should see the report at the IDE as shown in Figure 16.11. Congratulations! You have successfully built your first ARM® project with the IAR Embedded Workbench.

Now you need to download the program to the microcontroller board and test it. This can be done in three ways: Using the pull-down menu to select "Project → Download and Debug," click on the ⬇ (Download and Debug icon) on the toolbar or use the keyboard shortcut Ctrl-D.

After the program image is downloaded to the board, the debugger screen should appear as shown in Figure 16.12. The program is currently halted just before the first line of C code in main(), as indicated by the green arrow.

You can start running the program by clicking on the "go" icon 🔁 on the toolbar. The LEDs on the board should start flashing. You can halt, resume, reset, or single step the program using various icons on the debugger screen, as shown in the left-hand side of Figure 16.13.

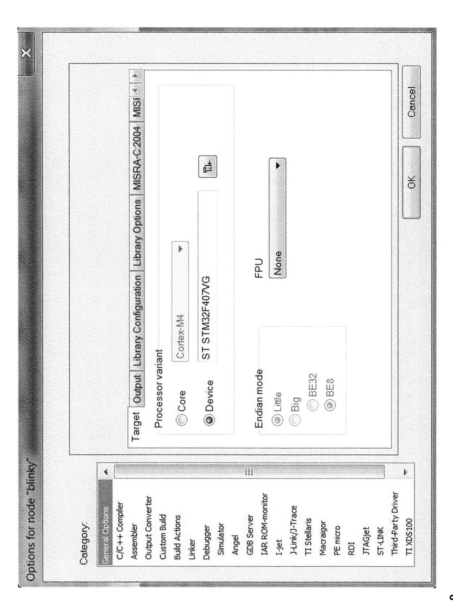

**FIGURE 16.9**

Project options

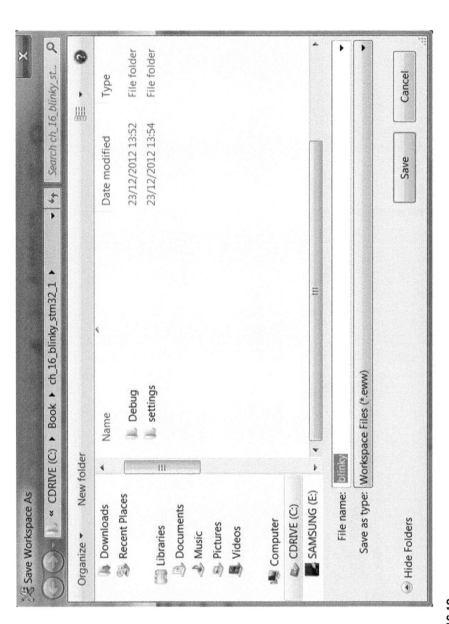

**FIGURE 16.10**

Saving the workspace before compiling starts

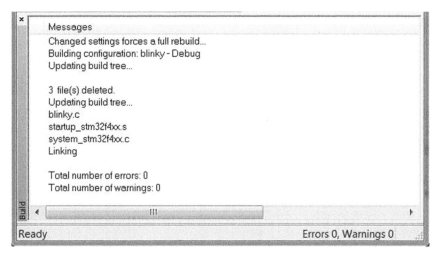

**FIGURE 16.11**

Compilation result

You can insert or remove breakpoints by right-clicking a line on the source window and selecting "toggle breakpoint." When the processor is halted, you can see the processor's registers using the register window, which can be accessed by the pull-down menu "View → Register."

## 16.4 Project options

The IDE in the IAR Embedded Workbench provides many options. Figure 16.14 shows the main option categories and the tabs available.

For example, the IAR C Compiler allows various levels of optimization efforts, and when the optimization level is set to high, you can select between size optimization, speed optimization, and balance optimization. You can also enable or disable some of the individual optimization techniques (Figure 16.15).

There are also a number of additional settings for each supported debug adaptor. Very often you need to set up the debug adaptor settings if you need to use any advanced debug features such as instrumentation trace (i.e., ITM).

## 16.5 Hints and tips

The stack size and heap size requirement of a project is defined in the "Linker" options. You need to select the option to override the default Linker Configuration File, set up the stack and heap memory size (Figure 16.16), and save the settings in a new configuration file in your project directory.

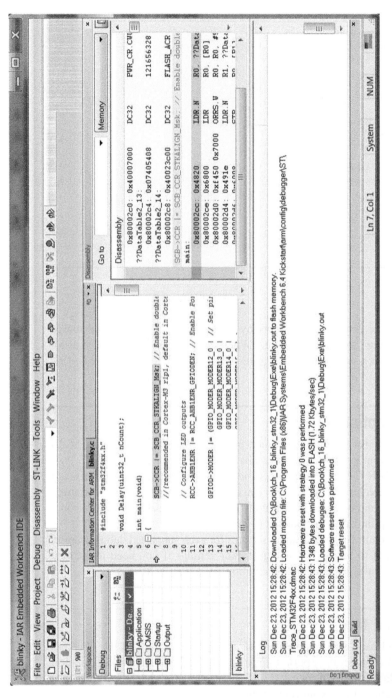

**FIGURE 16.12**

Debugger screen

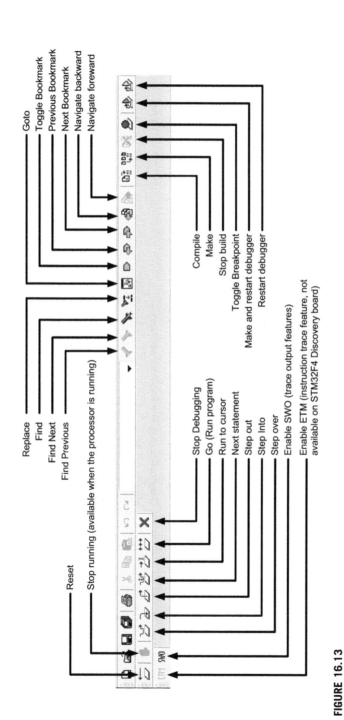

**FIGURE 16.13**

Icons on the toolbar of the debug screen

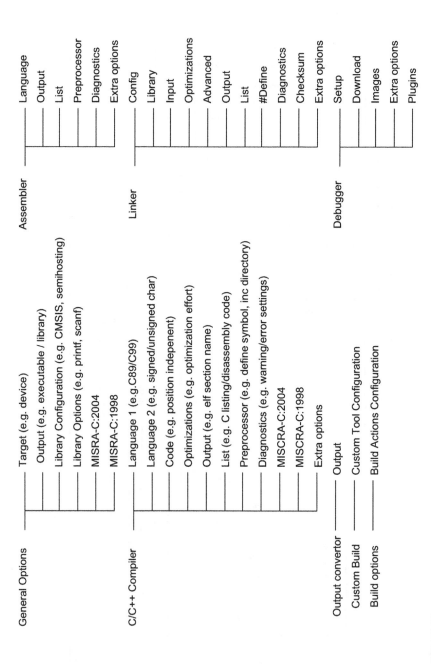

**FIGURE 16.14**

Project options, categories and tabs

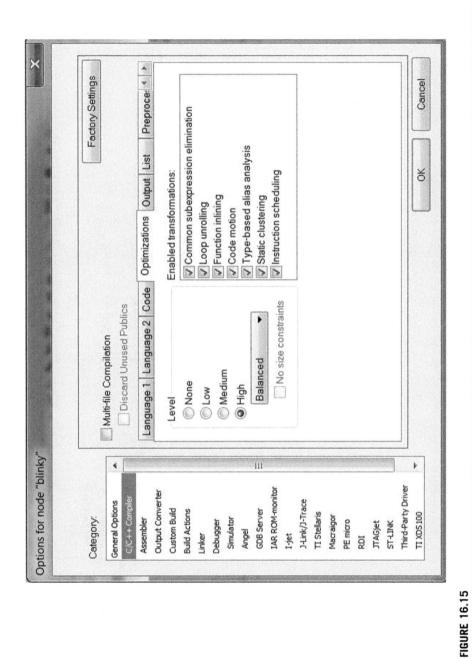

**FIGURE 16.15**

Optimization choice for C/C++ compiler

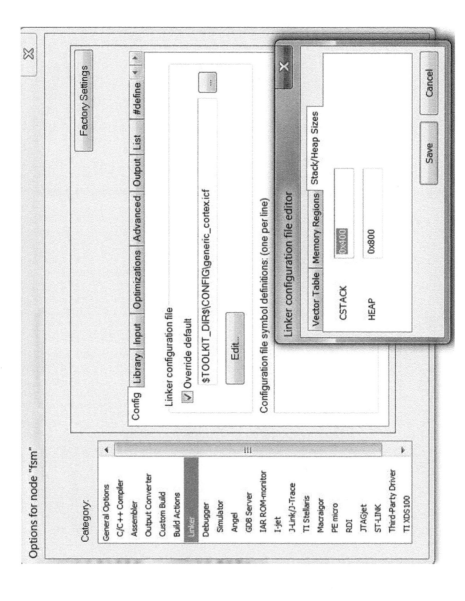

**FIGURE 16.16**

Stack and Heap memory size settings

Much of the useful information about the IAR Embedded Workbench is included in the documentation that is part of the installation. Typically you can find these files from the "Help" menu.

On the IAR website, under the "Resources" section, there are also a number of useful and informative technical articles. An example is "Mastering stack and heap for system reliability", reference 24.

# Getting Started with the GNU Compiler Collection (gcc)

# 17

## CHAPTER OUTLINE

## 17.1 The GNU Compiler Collection (gcc) toolchain

The GNU C Compiler is the de facto compiler choice for many open source projects. Since you can get gcc for free, it is very popular with hobbyists and academic users. Although you can built the gcc toolchain for Cortex®-M processors using the gcc source packages,[1] the build process requires an in-depth understanding of the tools. To make this easier, there are a number of pre-built packages available.

In this chapter we will cover the use of GNU Tools for ARM® Embedded Processors. You can download a pre-built package from the LaunchPad[2] website. The is package only contains the command line tools. However, you can use third-party IDE (Integrated Development Environment) tools with them. For example, you can use the GNU Tools for ARM Embedded Processors with Keil™ MDK, or Coo-Cox CoIDE, a free IDE.

---

[1]You can get the packages from http://gcc.gnu.org and http://www.gnu.org/software/binutils/.
[2]At the moment this is hosted at https://launchpad.net/gcc-arm-embedded. In the long term the URL might change.

**The Definitive Guide to ARM® Cortex®-M3 and Cortex-M4 Processors. http://dx.doi.org/10.1016/B978-0-12-408082-9.00017-8**

## 17.2 Typical development flow

The gcc toolchain contains a C compiler, assembler, linker, libraries, debugger, and additional utilities. You can develop applications using C language, assembly language, or a mixture of both. The typical command names are shown in Table 17.1.

The prefix of commands reflects the type of the pre-built toolchain. In this case, the command names shown in the third column of Table 17.1 are pre-built for ARM® EABI[3] without a specific target OS platform, hence the prefix "none." Some GNU toolchains could be created for developing applications for Linux platforms, and in those cases the prefix would be "arm-linux-."

A typical flow of software development using gcc is shown in Figure 17.1. Unlike using the ARM Compilation toolchain (i.e., armcc), it is common for have the compile and link operations combined in one gcc run. This is easier and less error prone, as the compiler can invoke the linker automatically, generate all the required link options, and pass on all required libraries.

To compile a typical project, you will need to have the files listed in Table 17.2.

In order to make software development easier, the microcontroller vendors normally provide a set of files which include some of the items listed in Table 17.2. Sometimes these are called CMSIS-compliant device driver libraries, or microcontroller software packages. These packages might also include example projects or additional driver libraries.

For example, in a simple project that toggles LEDs on a STM32F4 Discovery board (based on the Cortex®-M4 processor), you might have the following files in your project (as shown in Figure 17.2).

**Table 17.1** Command Names (Note: the command names for toolchains from other vendors can be different)

| Tools | Generic Command Name | Command Name in GNU Tools for ARM Embedded Processors |
|---|---|---|
| C compiler | gcc | arm-none-eabi-gcc |
| Assembler | as | arm-none-eabi-as |
| Linker | ld | arm-none-eabi-ld |
| Binary file generation tool | objcopy | arm-none-eabi-objcopy |
| Disassembler | objdump | arm-none-eabi-objdump |

---

[3]The Embedded-Application Binary Interface (EABI) specifies standard conventions for file formats, data types, register usage, stack frame organization, and function parameter passing of an embedded software program.

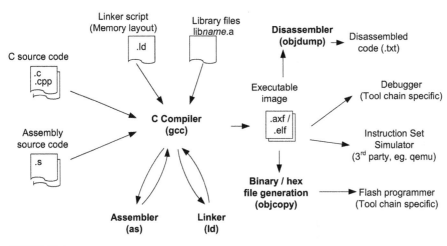

**FIGURE 17.1**

Typical program development flow

**Table 17.2** Typical Required Files for the Project

| File Type | Descriptions |
|---|---|
| Application code | Source code of your application. |
| Device-specific CMSIS Header files | The definition header files for the microcontroller you use. This is provided by the microcontroller vendor. |
| Device-specific startup code for gcc | The device specific startup code for the microcontroller you use. This is provided by the microcontroller vendor. |
| Device-specific system initialization files | This contains the SystemInit() function (system initialization) which is specified by CMSIS-Core, and additional functions for system clock updates. This is provided by the microcontroller vendor. |
| Generic CMSIS Header files | This is typically included in the device driver library package or included in tool installation. Or you can download it from ARM (www.arm.com/cmsis) |
| Linker script | The linker script is device specific. The complete linker script for a project can be composed of several files, with one file to specify the memory layout of the device and other files to define the settings required for gcc itself. The installation of GNU Tools for ARM Embedded Processors already provided an example linker script to make it easier. |
| Library files | This included the runtime libraries provided by the toolchain (typically included in the installation). You can also add additional custom libraries if needed. |

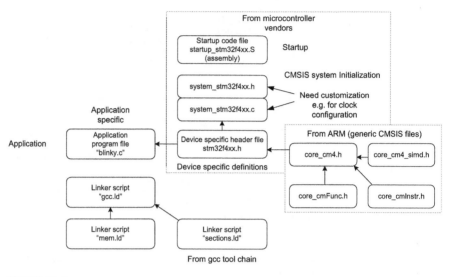

**FIGURE 17.2**

Example project with CMSIS-Core

The device-specific header file stm32f4xx.h defines all the peripheral registers so that you don't have to spend a long time creating peripheral definitions. The system_stm32f4xx.c provides the SystemInit() function that initializes the clocking system such as PLL and clock control registers.

Apart from the program files, you also need the linker script to define the memory layout of the executable image. The main linker script "gcc.ld" simply pulls in two other linker scripts:

```
/* Contents of gcc.ld */
INCLUDE "mem.ld"
INCLUDE "sections.ld"
```

- "mem.ld": This file defines the memory map (flash and SRAM) of the microcontroller you used.
- "sections.ld": This file defines the layout of information inside the executable image.

The "mem.ld" for STM32F4xx is defined as:

```
/* Specify the memory areas */
MEMORY
{
  FLASH (rx)   : ORIGIN = 0x08000000, LENGTH = 1024K
  RAM (xrw)    : ORIGIN = 0x20000000, LENGTH = 112K
}
```

The file "sections.ld" is already included in the GNU Tools for ARM Embedded Processors installation (e.g., <installation_directory>\share\gcc-arm-none-eabi\samples). You can use this file as is.

## 17.3 Creating a simple blinky project

The installation of GNU Tools for ARM® Embedded Processors only provides command line tools. You can invoke the compilation using the command line, make file (for Linux platform), batch file (for Windows platform), or using a third-party IDE. We will first demonstrate how to create a project using a batch file.

Assuming that we place the files listed in Figure 17.2 in a project directory, and the generic CMSIS include files in a subdirectory called CMSIS/Include, we can invoke the compilation and link process with a simple batch file:

```
rem Simple batch file for compiling the blinky project
rem Note the use of "^" symbol below is to allow multi-line commands
in Windows batch file.

set OPTIONS_ARCH=-mthumb -mcpu=cortex-m4
set OPTIONS_OPTS=-Os
set OPTIONS_COMP=-g -Wall
set OPTIONS_LINK=-Wl,--gc-sections,-Map=map.rpt,-lgcc,-lc,-lnosys -
ffunction-sections -fdata-sections
set SEARCH_PATH=CMSIS\Include
set LINKER_SCRIPT=gcc.ld
set LINKER_SEARCH="C:\Program Files (x86)\GNU Tools ARM Embedded\4.7
2012q4\share\gcc-arm-none-eabi\samples\ldscripts"

rem Compile the project
arm-none-eabi-gcc                          ^
   %OPTIONS_COMP% %OPTIONS_ARCH%           ^
   %OPTIONS_OPTS%                          ^
   -I %SEARCH_PATH% -T %LINKER_SCRIPT%     ^
   -L %LINKER_SEARCH%      ^
   %OPTIONS_LINK%          ^
   startup_stm32f4xx.S     ^
   blinky.c                ^
   system_stm32f4xx.c      ^
   -o blinky.axf
if %ERRORLEVEL% NEQ 0 goto end

rem Generate disassembled listing for debug/checking
arm-none-eabi-objdump -S blinky.axf > list.txt
if %ERRORLEVEL% NEQ 0 goto end
```

```
rem Generate binary image file
arm-none-eabi-objcopy -O binary blinky.axf blinky.bin
if %ERRORLEVEL% NEQ 0 goto end

rem Generate Hex file (Intel Hex format)
arm-none-eabi-objcopy -O ihex blinky.axf blinky.hex
if %ERRORLEVEL% NEQ 0 goto end

rem Generate Hex file (Verilog Hex format)
arm-none-eabi-objcopy -O verilog blinky.axf blinky.vhx
if %ERRORLEVEL% NEQ 0 goto end
```

Please note that apart from the assembly startup code files, all the other source files are identical to the blinky example in Chapter 15 and Chapter 16. The availability of CMSIS-Core enables much better software portability and reusability. Please refer to section 15.3 for detailed information about the source code.

The compilation and link process is carried out by arm-none-eabi-gcc. The rest of the compilation steps are optional. We added these steps to demonstrate how to create a binary file, hex file, and disassembled listing file.

## 17.4 Overview of the command line options

The GNU Tools for ARM® Embedded Processors can be used with a wide range of ARM processors, including Cortex®-M processors and Cortex-R processors. In the example in section 17.2 we used Cortex-M4 (without a floating point unit). You can specify which target processor is to be used and/or which architecture is to be used.

Table 17.3 lists target processor command line options.

Table 17.4 lists target architecture command line options.

Some of the other commonly used options are listed in Table 17.5.

By default the GNU C compiler uses a run-time library called Newlib. This library provides very good performance, but at the same time has larger code size. In version 4.7 of the GNU Tools for ARM Embedded Processors a new feature called Newlib-nano was introduced. It is optimized for size and can produce much smaller binary code. For example, with standard Newlib the blinky (binary image file) is 3700 bytes, and this reduced to just 1536 bytes when Newlib-nano is used.

There are a couple of areas that need attention when using Newlib-nano:

- Please note that --specs=nano.specs is a linker option. You must include this option in the linker option if the compiling and linking stages are separated.

**Table 17.3** Compilation Target Processor Command Line Options

| Processor | GCC Command Line Option |
|---|---|
| Cortex-M0+ | -mthumb -mcpu=cortex-m0plus |
| Cortex-M0 | -mthumb -mcpu=cortex-m0 |
| Cortex-M1 | -mthumb -mcpu=cortex-m1 |
| Cortex-M3 | -mthumb -mcpu=cortex-m3 |
| Cortex-M4 (no FPU) | -mthumb -mcpu=cortex-m4 |
| Cortex-M4 (soft FP) | -mthumb -mcpu=cortex-m4 -mfloat-abi=softfp -mfpu=fpv4-sp-d16 |
| Cortex-M4 (hard FP) | -mthumb -mcpu=cortex-m4 -mfloat-abi=hard -mfpu=fpv4-sp-d16 |

**Table 17.4** Compilation Target Architecture Command Line Options

| Architecture | Processor | GCC Command Line Option |
|---|---|---|
| ARMv6-M | Cortex-M0+, Cortex-M0, Cortex-M1 | -mthumb -march=armv6-m |
| ARMv7-M | Cortex-M3 | -mthumb -march=armv7-m |
| ARMv7E-M (no FPU) | Cortex-M4 | -mthumb -march=armv7e-m |
| ARMv7E-M (soft FP) | Cortex-M4 | -mthumb -march=armv7e-m -mfloat-abi=softfp -mfpu=fpv4-sp-d16 |
| ARMv7E-M (hard FP) | Cortex-M4 | -mthumb -march=armv7e-m -mfloat-abi=hard -mfpu=fpv4-sp-d16 |

- Formatted input/output of floating point numbers are implemented as weak symbols. When using %f in printf or scanf, you have to pull in the symbol by explicitly specifying the "-u" command option:

```
-u _scanf_float
-u _printf_float
```

For example, to output a float, the command line is:

```
$ arm-none-eabi-gcc --specs=nano.specs -u _printf_float
$(OTHER_OPTIONS)
```

# 17.5 Flash programming

After the program image has been generated, we need to test it by downloading the image into the flash memory of the microcontroller for testing. However, the GNU

**Table 17.5** Commonly Used Compilation Switches

| Options | Descriptions |
|---|---|
| "-mthumb" | Specifies Thumb instruction set |
| "-c" | Compile or assemble the source files, but do not link. Object file is generated for each source file. This is used when you have a project setup that separates compile and link stages. |
| "-S" | Stop after the stage of compilation proper; do not assemble. The output is in the form of an assembler code file for each non-assembler input file specified. |
| "-E" | Stop after the preprocessing stage. The output is in the form of preprocessed source code, which is sent to the standard output. |
| "-Os" | Optimization level - It can be from optimization level 0 ("-O0") to 3 ("-O3"), or can be "-Os" for size optimization. |
| "-g" | Include debug information |
| "-D*macro*" | User defined preprocessing macro |
| "-Wall" | Enable all warnings |
| "-I <directory>" | Include directory |
| "-o <output file>" | Specify output file |
| "-T <linker script>" | Specify linker script |
| "-L <ld script path>" | Specify search path for linker script |
| "-Wl,option1,option2" | "-Wl" passes options to linker. It can provide multiple options, separate by commas. |
| "--gc-sections" | Remove sections that are not used. Be careful with this option because it could also remove sections that are indirectly referenced. You can check linker map report to see what is removed and use KEEP() function in the linker script to ensure that certain data/code are not removed. |
| "-lgcc" | Link against libgcc.a |
| "-lc" | Instructs the linker to search in the system-supplied standard C library for functions not supplied by your own source files. This is the default choice, and is opposition of the "-nostdlib" option which force the linker NOT to search in the system-supplied libraries. |
| "-lnosys" | Specific no semihosting (use libnosys.a for linking). If semihosting is required, for example, using RDI monitor for semihosting support, you can use "--specs=rdimon.specs -lrdimon." |
| "-lm" | Link with math library |
| "-Map=map.rpt" | Generate map report file (map.rpt is the filename of the report) |
| "-ffunction-sections" | Put every function in its own section. Use with "--gc-sections" to reduce code size. |
| "-fdata-sections" | Put each data in its own section. Use with "--gc-sections" to reduce code size. |

**Table 17.5** Commonly Used Compilation Switches—*Cont'd*

| Options | Descriptions |
|---|---|
| "--specs=nano.specs" | Use Newlib-nano runtime library (introduced in version 4.7 of GNU Tools for ARM Embedded Processors). |
| "-fsingle-precision-constant" | Treat a floating point constant as single precision constant instead of implicitly converting it to double precision |

Tools for ARM® Embedded Processors do not include any flash programming support, so you need to use third-party tools to handle the flash programming. There are a number of options, discussed in the following sections.

### 17.5.1 Using Keil MDK-ARM

If you have access to Keil™ MDK-ARM and a supported debug adaptor (e.g., ULINK2, or if the development board has a supported debug adaptor), you can use the flash programming feature in Keil MDK-ARM to program the image created into the flash memory.

To use Keil MDK-ARM to program your program image, the file extension of the executable needs to be changed to .axf.

The next step is to create a μVision project in the same directory (typically the project name should be the same as the executable, e.g., "blinky"). In the project creation wizard, select the microcontroller device you use. There is no need to add any source file to the project. When the project wizard asks whether it should copy the default startup code, you should select "no" to prevent the original startup file for gcc from being overwritten.

Set up the debug options to use your debug adaptor (for debug and flash programming; see Chapter 15). By default the flash programming algorithm should be set up correctly by the project creation wizard.

Once the program image (e.g., blinky.axf) has been built, you can click the flash programming button 🔽 on the toolbar. The compiled image will then be programmed into the flash memory. After the image is programmed, you can optionally start a debug session using the μVision debugger to debug your program.

### 17.5.2 Using third-party flash programming utilities

There are many different flash programming utilities available. A common one is the CoFlash from coocox.org. This flash programming tool supports Cortex®-M microcontrollers from a number of major microcontroller vendors and a number of debug adaptors.

When CoFlash is started, it first displays the Config tab. Set up the microcontroller device and debug adaptor as required. Figure 17.3 shows the configurations used with the STM32F4Discovery board.

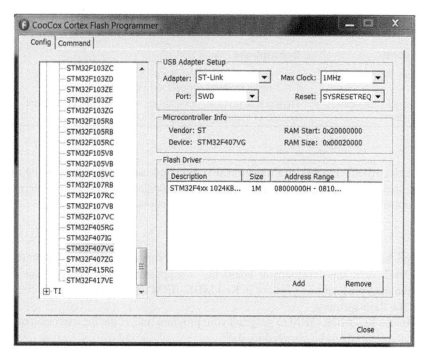

**FIGURE 17.3**

CoFlash configuration screen for STM32F4 Discovery board

Then switch to the command tab (Figure 17.4), where you can select the program image (can be binary or executable image ."elf"), and then you can click on the "Program" button to start the flash programming.

Use a third-party IDE together with GNU Tools for ARM® Embedded Processors. See section 17.6 on using Keil™ MDK, and section 17.7 on using CoIDE.

## 17.6 Using Keil™ MDK-ARM with GNU tools for ARM Embedded Processors

The μVision IDE in the Keil™ MDK-ARM can be used with gcc. When you click on the 🔩 (Components, Environment and Books) button on the toolbar and select the "Folders/Extensions" tab, you can select between using the ARM® C compiler and using GNU C compiler (Figure 17.5).

Once the toolchain path is set up, you can then add your program files to the projects by using the Keil MDK normally. Some of the project settings such as debug, trace, and flash programming are the same as the normal MDK environment. However, other project setting dialogs are different and are GNU toolchain specific.

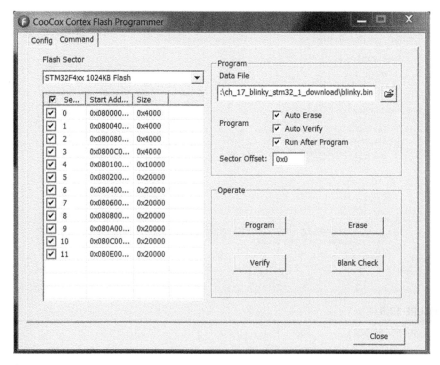

**FIGURE 17.4**

CoFlash command screen for STM32F4 Discovery board

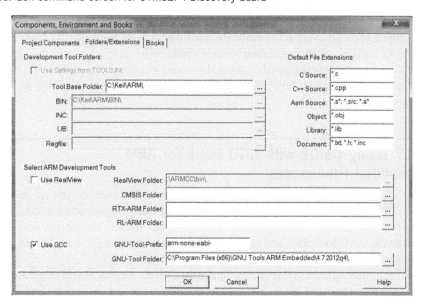

**FIGURE 17.5**

Keil MDK-ARM support for the use of GNU toolchain

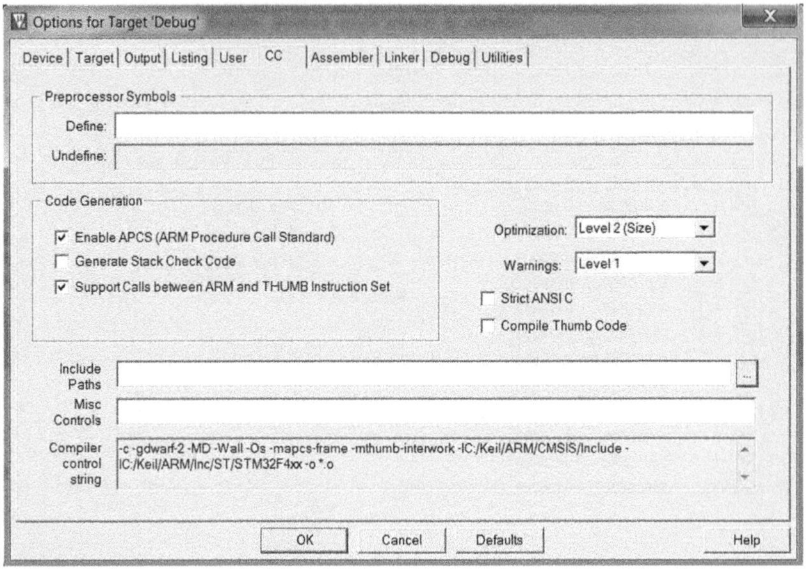

**FIGURE 17.6**

C compiler settings

For example, the C compile option settings (Figure 17.6) are different from the options available for the ARM C compiler (Figure 15.24). Assembler options are shown in Figure 17.7 and linker options are shown in Figure 17.8.

Please note that the generic CMSIS-Core include files do not need to be added to the project because μVision automatically adds the location of the CMSIS-Core files in the Keil MDK-ARM installation to the include path.

Once the project is compiled, you can download and debug the application. Note that some of the source-level debugging features might not be available in this environment.

## 17.7 Using CoIDE with GNU tools for ARM® Embedded Processors

The CoIDE is a popular choice for many users of the GNU toolchain. You can download it from the CooCox website (http://www.coocox.org). It supports a good number of the current Cortex®-M microcontrollers on the market. The CoIDE does not include the GNU toolchain, so the GNU toolchain still needs to be downloaded and installed separately.

After installing the GNU toolchain, and then CoIDE, the first step is to set up the GNU toolchain path in CoIDE. This can be done by accessing the "Select Toolchain Path" from the pull-down menu (Project → Select Toolchain Path) (see Figure 17.9).

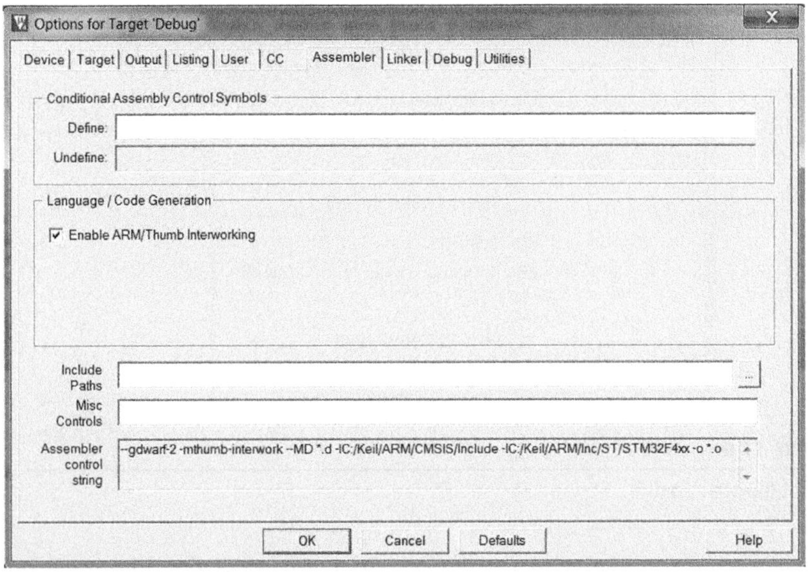

**FIGURE 17.7**

Assembler settings

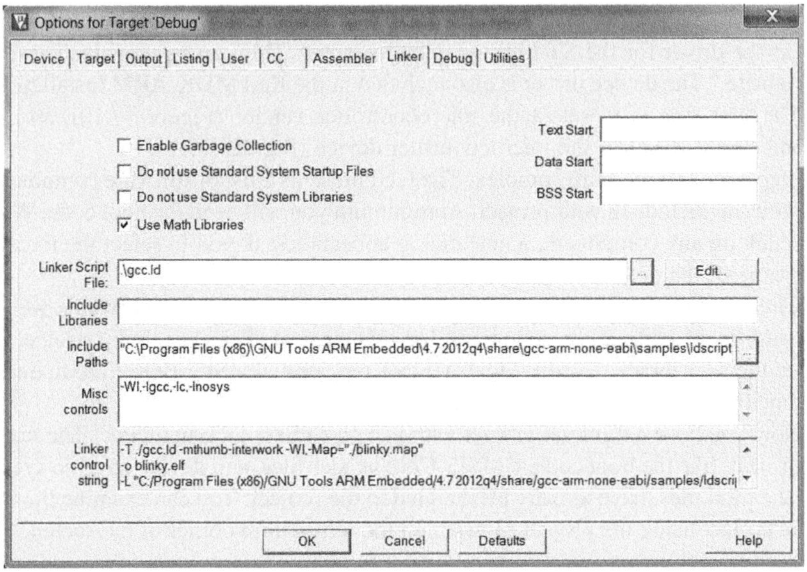

**FIGURE 17.8**

Linker settings

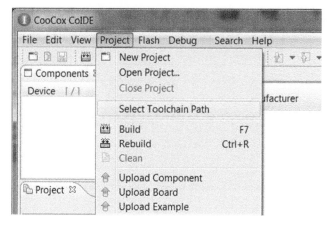

**FIGURE 17.9**

Select toolchain path

For example, in the system with GNU Tools for ARM Embedded Processors version 4.7, the selected path location is: C:\Program Files (x86)\GNU Tools ARM Embedded\4.7 2012q4\bin.

The process of setting up a new project is very easy. Many of the steps are shown as on-screen instructions. Here we will reuse the blinky project for STM32F4 Discovery board that we developed on the Keil™ MDK-ARM and IAR Embedded Workbench. To use this board with CoIDE, we also need to download and install the device driver for the ST-Link v2 debug adaptor. This can be downloaded from ST website.[4] The device driver is also included in the Keil MDK-ARM installation.[5]

The first step is to select the microcontroller vendor (Figure 17.10), and the second step is to select the microcontroller device (Figure 17.11).

Step 3 looks a bit more complex. The GUI presents a list of software components that you can include in your project. At minimum you will need the boot code. When you click on any component, a new dialog appears to ask you to select the location and name of the project (Figure 17.12).

Click on "yes" and create your project in a suitable folder. Afterwards we can add additional software components for this project. For the blinky project we select the C Library, Cortex-M4 CMSIS-Core, and CMSIS BOOT components (Figure 17.13).

Now you have a minimal project with just an endless loop in main.c. The rest of the project, like the boot code, CMSIS-Core header files, and device-specific system initialization files, have already been added to the project. You can examine the files in the project using the project browser on the left bottom corner of the screen. You

[4]http://www.st.com/internet/evalboard/product/251168.jsp (bottom of the Design Support tab)
[5]Typically located in C:\Keil\ARM\STLink.

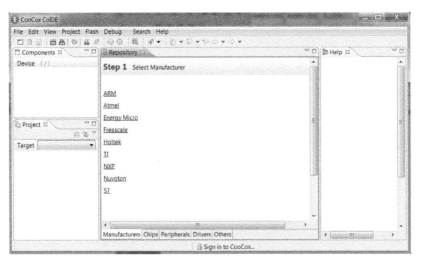

**FIGURE 17.10**

Step 1 — Select manufacturer

can also add additional files to the project by right-clicking on the project and selecting "Add files."

We can now copy the contents of the blinky.c we used in the previous example into main.c. A few additional edits of the source code are needed. First, the startup

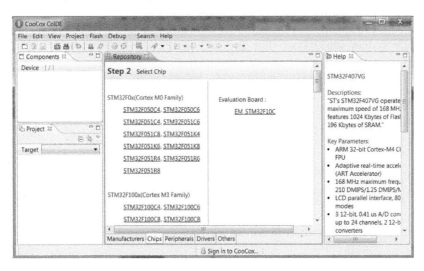

**FIGURE 17.11**

Step 2 — Select chip

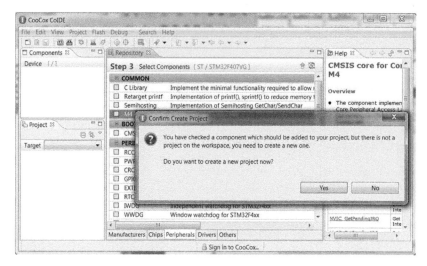

**FIGURE 17.12**

Step 3 — Select components

code (startup_stm32f4xx.c) does not include the call to SystemInit(), so we need to add this either in the startup code or in the beginning of the main.c.

Secondly, for the case of the STM32F4 Discovery board, we need to edit the system initialization code (system_stm32f4xx.c) to set the PLL_M parameter to 8 (as the board is using an 8MHz crystal). For other microcontroller boards you might also need to adjust the system clock settings accordingly.

In this example we add SystemInit() call in the beginning of main.c (line 11) and include system_stm32f4xx.h (line 2):

```
Modification of main.c (line and line 11)
 1:#include "stm32f4xx.h"
 2:#include "system_stm32f4xx.h"
 3:
 4:void Delay(uint32_t nCount);
 5:
 6:int main(void)
 7:{
 8:    SCB->CCR |= SCB_CCR_STKALIGN_Msk; // Enable double-word stack
       alignment
 9:    //(recommended in Cortex-M3 r1p1, default in Cortex-M3 r2px and
       Cortex-M4)
10:
11:    SystemInit();
       ...
```

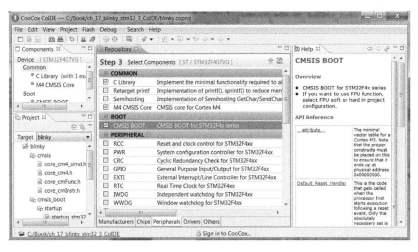

**FIGURE 17.13**

Step 3 — Components selected

Before we compile the project, we can review and modify some of the project settings such as optimization level and debug adaptor settings. These can be accessed by clicking on the "Configuration" icon on the toolbar (Figure 17.14), or by right-clicking on Blinky in the project window and selecting "Configuration."

Once the project settings are adjusted (e.g., optimizations, debug adaptor), we can compile the project using one of the following methods:

- Pull-down menu: "Project → Build,"
- Hot key F7, or
- Clicking on the "Build" button on the toolbar.

The compilation should complete with the display shown in Figure 17.15. When the compilation process is completed, we can start the debug session by clicking on the "Start Debug" icon on the toolbar, or by using Ctrl-F5 to start the debugger (Figure 17.16). The debugger screen has additional icons for debug operations (Figure 17.17).

## 17.8 Commercial gcc-based development suites

While you can use free versions of gcc to develop and debug your applications, there are a number of commercial development toolchains based on gcc that often offer a lot of additional features. In addition, using commercial toolchains gives you product support services that you do not get with free toolchains. For example, if a bug is found in a part of the toolchain, a commercial tool vendor can often develop a fix for you, but with free toolchains you do not have such an advantage. This is often critical for project development.

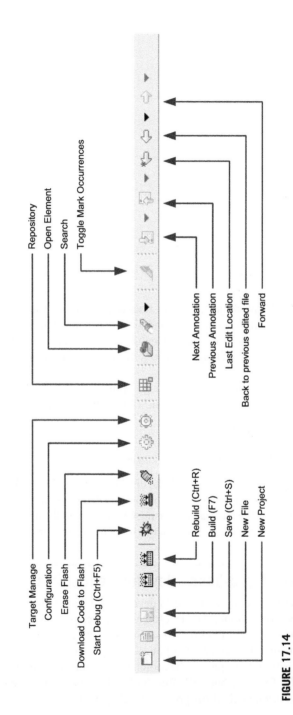

**FIGURE 17.14**

Icons on the ColDE toolbar

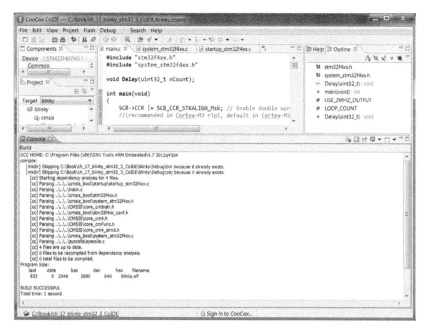

**FIGURE 17.15**

Compilation complete message

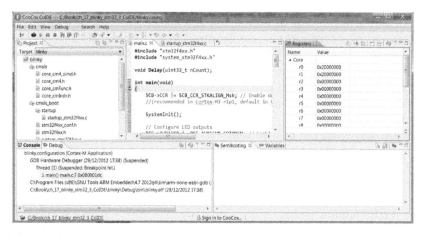

**FIGURE 17.16**

Debugger screen

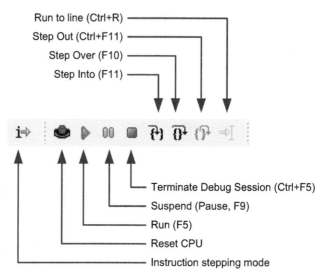

**FIGURE 17.17**

Icons on the debugger toolbar

## 17.8.1 Atollic TrueSTUDIO for ARM®

Atollic TrueSTUDIO for ARM is one of the commercial development suites that is based on the gcc and ECLIPSE IDE. The TrueSTUDIO product provides a complete solution for the majority of users:

- GNU toolchain including compiler, linker, etc.
- Eclipse-based IDE and project management
- Automatic linker script generation and easy-to-use project development flow
- Supports over 1100 ARM microcontroller devices, including flash programming support
- Example projects for over 80 development boards, with over 1000 example projects via Atollic TrueSTORE, a system that enables TrueSTUDIO users to download, install, and compile examples from a repository with just one mouse click
- Basic debugging, can be used with:
  - Segger J-Link
  - ST-LINK from STMicroelectronics (e.g., STM32F4 Discovery board)
  - Third-party gdbserver
  - OSJTAG
  - P&E Multilink probes
- Cortex®-M3/Cortex-M4 real-time trace with Serial Wire Viewer (SWV)
  - Data trace, with real-time data timeline oscilloscope and history log
  - Event trace (e.g., exception history)

- Instrumentation Trace (e.g., using SWV for printf via Instrumentation Trace Macrocell, ITM)
- Advanced debug features
  - OS-aware debugging for most popular OSs
  - Execution profiling with statistical PC sampling and SWV trace
  - Multi-core debug support

Utilizing the SWV feature in the Cortex-M3 and Cortex-M4 processors, True-STUDIO provides a wide range of analysis features such as real-time "animated" timeline charts (where the charts scrolls automatically in real-time with execution/tracing progress). For example, it can visualize variable changes or changes in memory locations over a period of time.

In addition, TrueSTUDIO also provides various features for project management such as:

- Integrated source code reviews
- Integrated bug database clients (support Bugzilla, Trac, etc.)
- Integrated version control system client (e.g., GIT, subversion, CVS)
- Integrated Fault Analysis

More demanding users can extend TrueSTUDIO by adding optional add-on products including:

- Atollic TrueINSPECTOR — A static source code analysis tool, which can detect potential coding problems. It also allows you to check your code for MISRA-C (2004) compliance, and provides code complexity analysis.
- Atollic TrueANALYZER — A dynamic test tool for measuring various coverage metrics. It can highlight which parts of the program have not been tested, including untested conditions in conditional code.
- Atollic TrueVERIFIER — A software test automation tool that can analyze the program code and generate a test suite for various functions inside the code. It also automatically compiles, downloads, and executes the test suite to the target board.

Similar to other commercial development suite vendors, Atollic also provides a cut-down version of TrueSTUDIO for free. The Atollic TrueSTUDIO for ARM Lite[6] is limited to 32KB code size for ARMv7-M, and limited to 8KB code size for ARMv6-M (i.e., Cortex-M0, Cortex-M0+, Cortex-M1). This free version provides almost all of the features available in the professional version including IDE, compiler, and debugger (which includes advance debug features such as SWV real-time trace).

If you are interested in trying out the Atollic TrueSTUDIO, on the Atollic website there is a whitepaper called "Embedded development using the GNU toolchain for ARM processors,"[7] which might be a good starting point.

---

[6]http://www.atollic.com/index.php/download/truestudio-for-arm
[7]http://www.atollic.com/index.php/whitepapers

### 17.8.2 Red Suite

Red Suite from Code Red Technologies (http://www.code-red-tech.com, recently acquired by NXP) is a fully featured development suite for ARM®-based microcontrollers, which includes all the tools necessary to develop high-quality software solutions in a timely and cost-effective fashion. It provides a comprehensive C/C++ programming environment, and the Red Suite IDE is based on the popular Eclipse IDE with many ease-of-use and microcontroller-specific enhancements, like syntax-coloring, source formatting, function folding, online and offline integrated help, extensive project management automation, and integrated source repository support (CVS integrated or subversion via download).

It contains the following features:

- Wizards that create projects for all supported microcontrollers
- Automatic linker script generation including support for microcontroller memory maps
- Direct download to flash when debugging
- Inbuilt Flash programmer
- Built-in datasheet browser
- Support for Cortex®-M, ARM7TDMI™, and ARM926-EJ based microcontrollers

With Cortex-M3 and Cortex-M4 based microcontrollers, Red Suite can take advantage of its advanced features, including full support for Serial Wire Viewing (SWV) through a feature called Red Trace. Red Trace enables a high level of visualization of what is happening in the target device.

The debugger includes a peripheral viewer that provides complete visibility of all registers and bit fields in all target peripherals in a simple tree-structured display. A powerful processor-register viewer is provided which gives access to all processor registers and provides smart formatting for complex registers such as flags and status registers.

In addition, Code Red Technology also provides a free version of the GNU toolchain called LPCXpresso, which works with NXP LPCXpresso development boards.

### 17.8.3 CrossWorks for ARM®

CrossWorks for ARM® is a C, C++, and assembly development suite from Rowley Associates (http://www.rowley.co.uk/arm/index.htm). It contains an IDE called CrossStudio which integrates the GNU toolchain. The source-level debugger in CrossStudio can work with a number of debug adaptors including CrossConnect for ARM (from Rowley Associates) and third-party in-circuit debugger hardware such as the SEGGER J-Link and Amontec JTAGkey.

CrossWorks for ARM is available in various editions, including non-commercial, low-cost packages (personal and educational licenses).

# Input and Output Software Examples

# 18

## 18.1 Producing outputs

In the last few chapters we have demonstrated simple programs that toggle LEDs on a development board on different toolchains. However, just having LEDs toggling can only represent a very limited amount of information and is often insufficient for communication. Having more ways of connecting your microcontroller to the outside world will certainly be more fun in learning embedded programming. In section 2.8 we briefly mentioned a number of ways to allow your microcontroller to communicate with external systems. In this chapter, we will see how to implement some of these basic communication methods.

A common task for beginners is to generate a simple output message of "Hello world!" In C language, this is commonly handled with a "printf" statement. Under the hood, the message output can be redirected to different forms of communication interfaces. Typically this is known as re-targeting. For example, it is very common to retarget printf to a UART during embedded software development.

In some cases, the printf statement can directly interact with a debugger connected to the microcontroller; this is called semi-hosting. Semi-hosting might also support accesses to file I/O and other system resources. Compared to re-targeting, semi-hosting can be easier to use with some tools, but is limited to debug environments and can also effect the application (e.g., speed, code size).

Some of the re-targeting and semi-hosting solutions support bidirectional communications (e.g., UART), and some of them are unidirectional (e.g., LCD display). However, in most cases you can create application code that interacts with input devices and does not need to rely on re-targeting/semi-hosting for data input.

## 18.2 Re-targeting to the Instrumentation Trace Macrocell (ITM)

### 18.2.1 Overview

One of the useful debug features in the Cortex®-M3 and Cortex-M4 processors is the Instrumentation Trace Macrocell (ITM). The ITM contains 32 stimulus port registers, and a write to these register can produce trace packets that output through the single pin Serial Wire Viewer (SWV) interface or the multi-bit Trace Port interface. The ITM feature allows you to export printf messages via SWV by re-targeting printf to the ITM. Please note that if the debug adaptor does not support trace, you cannot use the ITM to export printf messages.

The programmer's model of the ITM is available in Appendix G. However, CMSIS-Core contains three access functions to allow the ITM feature to be accessed easily:

- `uint32_t ITM_SendChar(uint32_t ch)` — output a character through ITM stimulus port #0.
- `int32_t ITM_ReceiveChar(void)` — Receive a character. Returns −1 if no data is available.
- `int32_t ITM_CheckChar(void)` — Returns 1 if a character is available, otherwise returns 0.

The ITM communication channels are output only. In order to provide bidirectional communication, CMSIS-Core added the "ITM_ReceiveChar" and "ITM_CheckChar" functions that allow the debugger and application to communicate using a global variable called "ITM_RxBuffer." When this variable is set to 0x5AA55AA5, it indicates that the buffer is empty.

When the debugger console interface receives a character from user input, it checks if the ITM_RxBuffer variable is empty. If yes, it puts the character into this buffer, otherwise it waits.

The program code running on the microcontroller can check the status of the buffer using the ITM_CheckChar() function call, which receives the character if data is available. The ITM_ReceiveChar() function also sets ITM_RxBuffer back to 0x5AA55AA5 to indicate that the buffer is ready for the next data.

To make use of these ITM functions in re-targeting, we need to define some additional functions that are toolchain specific.

### 18.2.2 Keil™ MDK-ARM

In Keil™ MDK-ARM (or other ARM® toolchains such as DS-5™ Professional), the function that needs to be implemented to support printf is "fputc."

```
/* Short version of retarget.c - Minimum code to support simple printf in Keil
MDK-ARM */

/****************************************************************/
/* Minimum retarget functions for
   ARM DS-5 Professional / Keil MDK */
/****************************************************************/

#pragma import(__use_no_semihosting_swi)
#include "stm32f4xx.h"
#include <stdio.h>

struct __FILE { int handle; /* Add whatever you need here */ };
FILE __stdout;
FILE __stdin;

int fputc(int ch, FILE *f) {
 return (ITM_SendChar(ch));
}

void _sys_exit(int return_code) {
label: goto label; /* endless loop */
}
```

A few extra lines of code are also needed (e.g., _sys_exit()) in the retarget code. We name this file as retarget.c, and add this to our "Hello World" project (Figure 18.1).

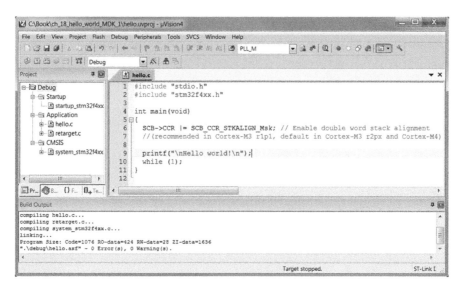

**FIGURE 18.1**

Simple Hello World project in Keil MDK

To get the message displayed properly, we need to ensure the debug and trace settings are correct. When clicking on the "Trace" tab of debug adaptor settings, you might see a screen similar to the one shown in Figure 18.2.

In the trace settings for SWV operations, it is important to make sure that the clock setting is correct. Here we show the speed as 168MHz, which is the the operation speed of the processor in this example project. (Note: Strictly speaking the Cortex®-M3/M4 TPIU (Trace Port Interface Unit) can operate at a clock speed different from the main CPU clock, so the label here really should be "TPIU Source Clock" rather than "Core Clock.")

We also need to enable the trace and double check that the ITM stimulus port #0 is enabled (by default Keil MDK enables all stimulus ports, but this can be different in other development tools).

If we are going to capture trace using SWV mode (single pin, using SWO signal), we need to make sure that Serial Wire debug protocol is selected in the debug settings. If the JTAG debug protocol is used, then the SWO pin is used for TDO for JTAG operation and cannot be used for trace capture.

Once the debug and trace settings are set up and the program has compiled correctly, we can then test the program by starting a debug session. Inside the debug session, we need to enable the Debug (printf) Viewer before we start the program. This can be accessed by the pull-down menu "View → Serial Windows → Debug (printf) Viewer," as shown in Figure 18.3.

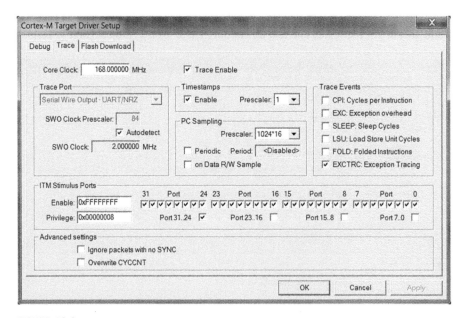

**FIGURE 18.2**

Trace settings to enable basic SWV operations

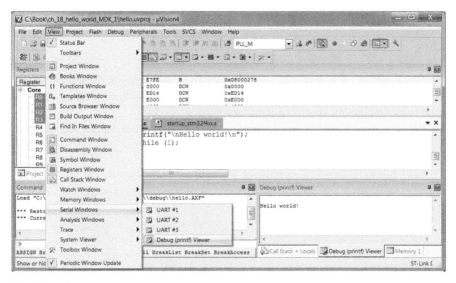

**FIGURE 18.3**

Accessing the Debug (printf) viewer and Hello world message display

After we start the program execution, we can see the "Hello world!" message displayed in the Debug (printf) viewer, as shown in the lower-right section of Figure 18.3.

The Debug (printf) viewer in Keil MDK-ARM supports bi-directional communication. To demonstrate this feature, we extend the Hello World program as follows:

```
/* Modified hello world to demonstrate bi-directional communication
in re-targeting */

#include "stm32f4xx.h"
#include "stdio.h"

int main(void)
{
char textbuffer[40]; // Text buffer

SCB->CCR |= SCB_CCR_STKALIGN_Msk; // Enable double word stack
alignment
//(recommended in Cortex-M3 r1p1, default in Cortex-M3 r2px and
Cortex-M4)
printf("\nHello world!\n");
while (1) {
  printf("Please enter text:");
  fgets(textbuffer, (sizeof(textbuffer)-1), stdin);
  printf("\nYou entered  :%s\n",textbuffer);
  }
}
```

The line "fgets" reads the user input (stdin) and stores it in the "textbuffer." Technically you could replace this line with "scanf":

```
scanf ("%s," &textbuffer[0]);
```

However, please note that in normal application development the use of "scanf" is not recommended because it does not check for buffer overflow. Alternatively you could also implement your own input function.

We need to update retarget.c to provide the "fgetc" function and the definition of ITM_RxBuffer, the receive data buffer:

```
/* Additional code in retarget.c to support user inputs */
volatile int32_t ITM_RxBuffer=0x5AA55AA5; // Initialize as EMPTY
int fgetc(FILE *f) {
  char tmp;
  while (ITM_CheckChar()==0); // Wait if buffer is empty
  tmp = ITM_ReceiveChar();
  if (tmp==13) tmp = 10;
  return (ITM_SendChar(tmp));
}
```

After adding this code to the project, the program can accept user inputs, as shown in Figure 18.4.

### 18.2.3 IAR Embedded Workbench

The same re-targeting operation can also be done in the IAR Embedded Workbench environment. With the IAR environment, the development suite can automatically insert the ITM code required by setting up semi-hosting configuration via ITM, as shown in Figure 18.5.

The application code needs to change slightly to use "gets" (or scanf) instead of "fgets." Please note that gets and scanf do not have buffer length checking, but this is fine for a debug environment.

**FIGURE 18.4**

Program execution result of re-targeting with user inputs

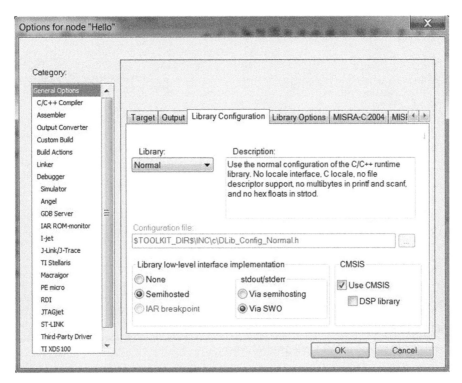

**FIGURE 18.5**

IAR semi-hosting via SWO

```
/* Hello world program for IAR Embedded Workbench */

#include "stm32f4xx.h"
#include "stdio.h"

int main(void)
{
  char textbuffer[40]; // Text buffer

  SCB->CCR |= SCB_CCR_STKALIGN_Msk; // Enable double word stack alignment
  //(recommended in Cortex-M3 r1p1, default in Cortex-M3 r2px and Cortex-M4)
  printf("\nHello world!\n");
  while (1) {
    printf("Please enter text:");
    gets(textbuffer);
    printf("\nYou entered  :%s\n",textbuffer);
    }
}
```

After the project has been created, you can start the debug session, and enable the Terminal I/O screen from the pull-down menu (View → Terminal I/O). When the program executes you can see the text displayed in the terminal I/O screen (at

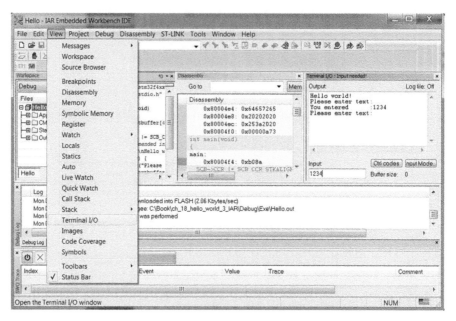

**FIGURE 18.6**

Terminal I/O screen in IAR Embedded Workbench

the top right-hand corner of Figure 18.6). The user input has to be entered into the input box at the bottom part of the Terminal I/O screen.

## 18.2.4 GCC

In gcc, theoretically you can implement a re-targeting function to re-direct printf to ITM. In normal gcc, the redirection of text message output is handled by implementing a "_write" function.

```
/**************************************************************/
/* Retarget functions for GNU Tools for ARM Embedded Processors*/
/**************************************************************/
#include <stdio.h>
#include <sys/stat.h>
#include "stm32f4xx.h"

__attribute__ ((used)) int _write (int fd, char *ptr, int len)
{
size_t i;
for (i=0; i<len;i++) {
ITM_SendChar(ptr[i]); // call character output function
}
return len;
}
/* Note: The "used" attribute is to work around a LTO (Link Time
Optimization) bug, but at the cost of increasing code size when not
used. Do not link this file when it is not used. */
```

However, the available support of the ITM and SWV features are dependent on the actual toolchain and the debugger and debug adaptor you use. In some cases, the toolchain you use might have specific semi-hosting features instead of ITM and SWV support. Please refer to the documentation of the toolchain you use for details.

## 18.3 Semi-hosting

In some development suites, printf messages can be transferred using the debug connection (Serial Wire/JTAG) instead of using trace (by ITM). This method is typically called semi-hosting. In some cases (e.g., ARM® DS-5™, IAR Embedded Workbench for ARM) it can perform additional functions like file I/O (e.g., a program running on the microcontroller may be able to open a file using "fopen" to access a file on the debug host). However, this does not replace some of the other functionalities of Cortex®-M3/M4 trace like data trace, exception trace, and profiling.

Semi-hosting requires the addition of an extra run-time code library to the executable image. Therefore if semi-hosting is supported, usually you should see additional project options for semi-hosting. For example, in the IAR Embedded Workbench, you can enable semi-hosting by setting the semi-hosting option in the "General" settings (Figure 18.7).

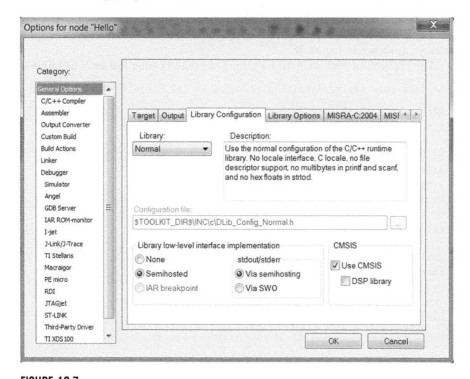

**FIGURE 18.7**

Semi-hosting option in IAR Embedded Workbench

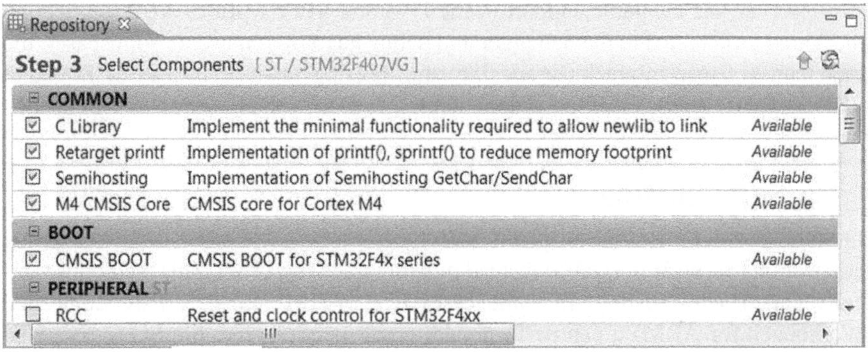

**FIGURE 18.8**

Including semi-hosting software components in CoIDE

During debugging, you can access the text message input output using the Terminal I/O window just like using ITM and SWV, as shown in Figure 18.6.

For CoIDE, the semi-hosting feature is also available. However, we have found that trying to implement the text input operation with scanf/gets in CoIDE 1.6.2 is a bit problematic, and ended up using a custom text string input function. To enable the semi-hosting feature, first we need to include semi-hosting software components in the project (as shown in Figure 18.8).

Then in the "Link" options in the project Configuration, select "Semi-hosting library" as shown in Figure 18.9.

The application code we used for testing the semi-hosting feature is as follows:

```
/* Test code for semi-hosting text I/O in CoIDE environment */

#include "stm32f4xx.h"
#include "stdio.h"

/* Semi-hosting functions */
extern char SH_GetChar();
extern void SH_SendChar(int ch);

/* A simple gets implementation with buffer length limit */
int my_gets(char dest[], int length)
{
unsigned int textlen=0; // Current text length
char ch; // current character
do {
 ch = SH_GetChar(); // Get a character from Semihosting
 switch (ch) {
   case 8: // Back space
    if (textlen>0) {
     textlen--;
     SH_SendChar(ch); // Back space
```

```
        SH_SendChar(' '); // Replace last character with space on console
        SH_SendChar(ch); // Back space again to adjust cursor position
        }
      break;
    case 13: // Enter is pressed
      dest[textlen] = 0; // null terminate
      SH_SendChar(ch); // echo typed character
    break;
    case 27: // ESC is pressed
      dest[textlen] = 0; // null terminate
      SH_SendChar('\n');
    break;
    default: // if input length is within limit and input is valid
     if ((textlen<length) &
       ((ch >= 0x20) & (ch < 0x7F))) // valid characters
        {
        dest[textlen] = ch; // append character to buffer
        textlen++;
        SH_SendChar(ch); // echo typed character
        }
     break;
    } // end switch
 } while ((ch!=13) && (ch!=27));
 if (ch==27) {
    return 1; // ESC key pressed
 } else {
    return 0; // Return key pressed
 }
}


int main(void)
{
 char TextBuffer[40];

 SCB->CCR |= SCB_CCR_STKALIGN_Msk; // Enable double word stack
alignment
 //(recommended in Cortex-M3 r1p1, default in Cortex-M3 r2px and
Cortex-M4)
 printf("Hello world!\n");
 while (1){
   printf("\nEnter text:\n");
   my_gets(&TextBuffer[0], sizeof(TextBuffer)-1);
   printf("\n%s\n", &TextBuffer[0]);
 }
}
```

When in the debug session, you can see the text message using the semi-hosting window, as shown in Figure 18.10.

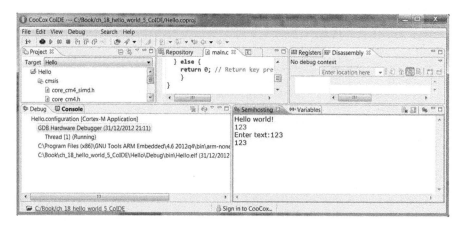

**FIGURE 18.9**

Semi-hosting option in Link configuration

**FIGURE 18.10**

Semi-hosting window display in CoIDE

Using semi-hosting for printf is usually much slower than using SWV or UART because in most tools the semi-hosting mechanism needs to halt the processor. This could result in unexpected side effects for some peripheral control operations (e.g., interrupt latency increases), but it has the advantage that it can cope with clock frequency changes in most implementations.

## 18.4 Re-targeting to peripherals

In some cases, you need to rely on peripherals for your information inputs and outputs. For example, some debug adaptors might not support Serial Wire Viewer (SWV) and therefore cannot collect information generated from ITM. As an alternative, peripherals like the UART can be a low-cost solution.

The Cortex®-M processors do not contain a UART interface. UART peripherals are added by chip designers, and different Cortex-M microcontrollers can have different UART designs and hence different programmer's models for the UART.

To demonstrate the use of UART for text input and output, we use the STM32VL (Value Line) Discovery Board (Figure 18.11). The STM32 microcontroller on this board is based on Cortex-M3. Since the debug adaptor on this board is based on an older version of ST-LINK, it does not support SWV.

There are several different ways to connect a UART interface of the microcontroller to a personal computer:

- Some microcontroller development boards include a debug adaptor that can also function as a USB to UART adaptor. In this case you need to install a USB virtual COM port device driver on your computer and then set up your UART terminal application to communicate with the virtual COM port. The setup process is development board-specific; please refer to the documentation of your development board for details.

**FIGURE 18.11**

STM32VL Discovery

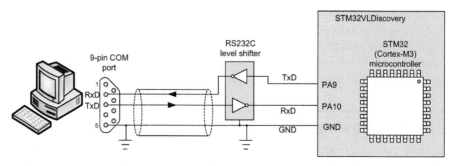

**FIGURE 18.12**

Connecting the UART interface to a PC via RS232 level shifter

- Some development boards already include a UART connector with the necessary level shifter circuit. In this case you can connect the UART interface of the personal computer to the UART connector of the board directly.
- Some development boards do not include a UART connector or level shifter. In order to test the UART's functionality, we need to add a simple level shifter circuit (e.g., MAX232) to the board to convert CMOS logic level to RS232 level (Figure 18.12). You can find ready-to-use RS232 converter modules in many online stores.

On the debug host side, many modern computers do not have a COM port anymore. In this case you need to get a USB to UART adaptor, and do not forget to install device drivers for these adaptors. You also need to have a terminal program to allow you to display and enter text. If you are using Windows XP you can find this functionality in the HyperTerminal application. From Windows 7 onwards, this program is not available, but you can find plenty of Terminal applications available online, including many free ones. For example, Putty and TeraTerm are two popular options.

In this example, we initialized USART 1 on the STM32 as 38400bps, 8-bit data, 1 stop bit, and no additional flow control. In order to use the USART, we also need to initialize some I/O pins (GPIO A pin 9 and pin 10). We also set the remaining unused I/O pin to analog inputs. The application code for the test is as follows:

```
/* uart_demo.c -- simple code to initialize UART and required GPIO
pins, then display "Hello World" message and echo user inputs */

#include "stm32f10x.h"
#include "stm32f10x_rcc.h"
#include "stm32f10x_gpio.h"
#include "stm32f10x_usart.h"
#include "stdio.h"

#define CFG_GPIO_PIN_MODE_INPUT              0
#define CFG_GPIO_PIN_MODE_OUTPUT_10MHZ       1
#define CFG_GPIO_PIN_MODE_OUTPUT_2MHZ        2
```

```
#define CFG_GPIO_PIN_MODE_OUTPUT_50MHZ              3
#define CFG_GPIO_PIN_CNF_INPUT_ANALOG               0x0
#define CFG_GPIO_PIN_CNF_INPUT_FLOATING_IN          0x4
#define CFG_GPIO_PIN_CNF_INPUT_PULL_UP_DOWN         0x8
#define CFG_GPIO_PIN_CNF_OUTPUT_GPO_PUSHPULL        0x0
#define CFG_GPIO_PIN_CNF_OUTPUT_GPO_OPENDRAIN       0x4
#define CFG_GPIO_PIN_CNF_OUTPUT_AFO_PUSHPULL        0x8
#define CFG_GPIO_PIN_CNF_OUTPUT_AFO_OPENDRAIN       0xC

//------------------------------
// Function declarations
void GPIO_start_setup(GPIO_TypeDef* GPIOx);
void GPIO_pin_config(GPIO_TypeDef* GPIOx, uint32_t pin_num, uint32_t
cfg, uint32_t val);

void UART_config(USART_TypeDef* USARTx, uint32_t BaudDiv);
void UART1_putc(char ch);
char UART1_getc(void);

int main(void)
{
  SCB->CCR |= SCB_CCR_STKALIGN_Msk; // Enable double word stack
  alignment
  //(recommended in r1p1, default in r2px)

  // Enable clocks to all GPIO ports for I/O port initialization
  // Also enable clock to USART1
  RCC->APB2ENR |=  RCC_APB2Periph_GPIOA | RCC_APB2Periph_GPIOB |
                  RCC_APB2Periph_GPIOC | RCC_APB2Periph_GPIOD |
                  RCC_APB2Periph_GPIOE | RCC_APB2Periph_USART1;

  // Set all GPIO port pins to analog inputs
  GPIO_start_setup(GPIOA);
  GPIO_start_setup(GPIOB);
  GPIO_start_setup(GPIOC);
  GPIO_start_setup(GPIOD);
  GPIO_start_setup(GPIOE);

  // Disable clocks to GPIO ports B, C, D and E, leaving A for
  operations
  RCC->APB2ENR &=  ~(RCC_APB2Periph_GPIOB | RCC_APB2Periph_GPIOC |
                    RCC_APB2Periph_GPIOD | RCC_APB2Periph_GPIOE);

  // Initialize USART1
  UART_config(USART1, 625); // 24MHz / 38400 = 625

  // Setup USART1 TXD (GPIOA.9) as output, use alternate function
```

```
    GPIO_pin_config(GPIOA, 9,
      (CFG_GPIO_PIN_MODE_OUTPUT_10MHZ|
       CFG_GPIO_PIN_CNF_OUTPUT_AFO_PUSHPULL),1);

    // Setup USART1 RXD (GPIOA.10) as input
    GPIO_pin_config(GPIOA, 10,
      (CFG_GPIO_PIN_MODE_INPUT| CFG_GPIO_PIN_CNF_INPUT_FLOATING_IN), 1);

    printf ("Hello World\n\r");

    while(1){ // echo user inputs
        UART1_putc(UART1_getc());
        }
}

// Initialize USART1 to simple polling mode (no interrupt)
void UART_config(USART_TypeDef* USARTx, uint32_t BaudDiv)
{
  USARTx->CR1 = 0;         // Disable UART during reprogramming
  USARTx->BRR = BaudDiv; // Set baud rate
  USARTx->CR2 = 0;         // 1 stop bit
  USARTx->CR3 = 0;         // interrupts and DMA disabled
  USARTx->CR1 = USART_WordLength_8b | USART_Parity_No |
                USART_Mode_Tx | USART_Mode_Rx | (1<<13); // Enable UART
  return;
}

// Output a character to USART1
void UART1_putc(char ch)
{ /* Wait until Transmit Empty flag is set */
  while ((USART1->SR & USART_FLAG_TXE) == 0);
  USART1->DR = ch; // send a character
  return;
}
// Read a character from USART. If no data received yet, wait
char UART1_getc(void)
{ /* Wait until Receive Not Empty flag is set */
  while ((USART1->SR & USART_FLAG_RXNE) == 0);
  return USART1->DR;
}

void GPIO_start_setup(GPIO_TypeDef* GPIOx)
{
  // Set all pins as analog input mode
  GPIOx->CRL = 0; // Bit 0 to 7, all set as analog input
  GPIOx->CRH = 0; // Bit 8 to 15, all set as analog input
  GPIOx->ODR = 0; // Default output value is 0
  return;
}
```

```
void GPIO_pin_config(GPIO_TypeDef* GPIOx, uint32_t pin_num, uint32_t
cfg, uint32_t val)
{
   uint32_t tmp_reg;       // temporary variable
   uint32_t shftval;       // number of bit to shift
   if (pin_num > 7) {      // Use CRH
    shftval = (pin_num - 8) << 2;
    /* multiplies by 4 because each pin take 4 bits of config info */
    tmp_reg = GPIOx->CRH;           // Read current configuration
    tmp_reg &= ~(0xF << shftval); // Clear old config
    tmp_reg |= cfg << shftval;     // Set new config
    if (val != 0){ // If val is non-zero, set output data bit to 1
      GPIOx->BSRR = 1<< pin_num; // set bit
      }
    else {
      GPIOx->BRR = 1<< pin_num;; // clear bit
      }
    GPIOx->CRH = tmp_reg;          // Write to config reg
    }
  else {      // Use CRL
    shftval = pin_num << 2;
    /* multiplies by 4 because each pin take 4 bits of config info */
    tmp_reg = GPIOx->CRL; // Read current configuration
    tmp_reg &= ~(0xF << shftval); // Clear old config
    tmp_reg |= ~(cfg << shftval); // Set new config
    if (val != 0){ // If val is non-zero, set output data bit to 1
      GPIOx->BSRR = 1<< pin_num; // set bit
      }
      else {
      GPIOx->BRR = 1<< pin_num;; // clear bit
      }
    GPIOx->CRL = tmp_reg;      // Write to config reg
    }
  return;
}
```

We also need to update "retarget.c" so that the text output is redirect to the UART:

```
/* retarget.c for UART text input/output in Keil MDK-ARM */

#include <stdio.h>
#include <time.h>
#include <rt_misc.h>
#pragma import(__use_no_semihosting_swi)


extern void UART1_putc(char ch);
extern char UART1_getc(void);
```

```
struct __FILE { int handle; /* Add whatever you need here */ };
FILE __stdout;
FILE __stdin;


int fputc(int ch, FILE *f) {
  UART1_putc(ch);
  return (ch);
}

int fgetc(FILE *f) {
  return ((int) (UART1_getc()));
}

int ferror(FILE *f) {
  /* Your implementation of ferror */
  return EOF;
}


void _ttywrch(int ch) {
  UART1_putc(ch);
}

void _sys_exit(int return_code) {
label: goto label; /* endless loop */
}
```

After the project is set up and compiled, the text "Hello World" is displayed on a terminal application running on the personal computer and the program running on the microcontroller then echoes the keys that typed in ("123"), as shown in Figure 18.13.

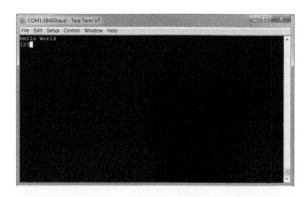

**FIGURE 18.13**

Text messages can be displayed on a debug host via a simple UART interface

In general the UART communication speed (e.g., 9600 to 115200 bps) is usually quite a bit slower than that of the SWV (over 1M bps), and this can mean that the application code might have to wait for the data transmission to be completed before sending the next one. However, you can utilize FIFO features in some of the UART peripherals to reduce the timing impact to the program code.

In real embedded applications a UART connection, is less likely and other display interfaces such as LCD modules are more common. Some microcontrollers have built-in LCD controllers and can be connected to the LCD directly. Many others need to connect to LCD modules. You can buy LCD modules with parallel or serial (e.g., SPI) interfaces. In this example, we will demonstrate the use of a common low-cost character LCD module as an output message display.

The 2 x 16 character LCD module has a built-in controller (Hitachi HD44780), and we will connect the module to the microcontroller with a simple parallel interface with 8-bit data, and three other control signals. In order to control the LCD module, we use the I/O pins on port C of the STM32VL Discovery board to connect to the LCD control signals. The I/O port arrangement is flexible. What is shown in Figure 18.14 is just an example; you can arrange the connection differently.

The module has two registers: The Instruction Register (IR) and the Data Register (DR). The LCD_RS signal is used to select which register to access, and the LCD_RW signal determines the direction of the transfer. Each transfer is enabled by the LCD_E signal (a strobe signal). Since the data connection is bidirectional, we need to enable and disable the tri-state buffers of the data pins according to the data transfer operations. For example, when writing to the IR register, the following steps are required:

1. Poll the IR register to make sure it is not busy
2. Set LCD_RW to 0 for write, set LCD_RS to 0 for IR
3. Set D7 to D0 output register value
4. Set D7 to D0 as output
5. Generate a high pulse on LCD_E
6. Set D7 to D0 as input

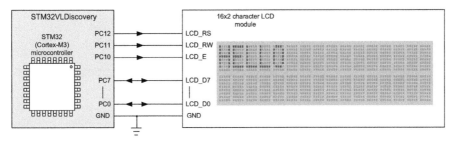

**FIGURE 18.14**

Connecting a character LCD module to STM32VL Discovery board

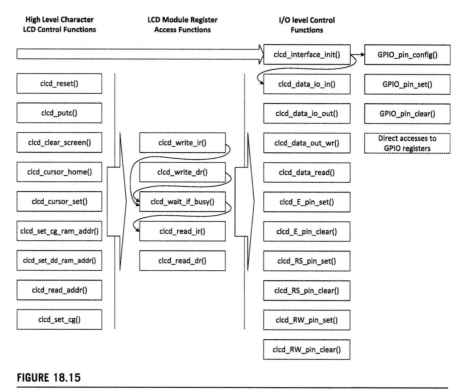

**FIGURE 18.15**

Hierarchy of LCD module control functions

In order to make the program code portable, the functions for controlling the LCD module are made hierarchical (Figure 18.15). In this way the character LCD control code can be ported to other microcontrollers easily by just changing the functions for the I/O signal controls.

After these functions have been created, you can use the high-level functions to control the character LCD module, for example, with the following code:

```
Program code to initialize the character LCD module and display
message on-screen

// Bit pattern for creating user character, in this case a smiley
const char smiley[8] = { 0x00, // 00000
                         0x11, // 10001
                         0x00, // 00000
                         0x00, // 00000
                         0x11, // 10001
                         0x0A, // 01010
                         0x04, // 00100
                         0x00}; // 00000
```

**FIGURE 18.16**

Printf message output to a low-cost character LCD module

```
// Initialize I/O port C for connection to Character LCD
clcd_interface_init();
// Reset Character LCD module
clcd_reset();
// Clear the screen
clcd_clear_screen();
// set first user character to smiley (max 8 user created characters
clcd_set_cg(0, &smiley[0]);
// at the end of this function, it automatically switch back to DD RAM
// with address = 0
printf ("Hello World ");
clcd_putc(0); // Display smiley character created eariler
// Set cursor location to 1st character of 2nd row
clcd_cursor_set(0,1);
printf ("0123456789012345");
```

When the program code is executed, the expected message is displayed on the character LCD module, as shown in Figure 18.16.

Re-targeting printf messages to hardware peripherals is also supported in other toolchains. In gcc, the required output function is _write (see section 18.2.4).

In the IAR Embedded Workbench for ARM, the low-level I/O routines for inputs and outputs are:

```
           /* Low-Level I/O Functions */
Output     size_t __write(int handle,const unsigned char *buf,size_t bufSize)
           {
              size_t i;
              for (i=0; i<bufSize;i++)
              {
              send_data(buf[i]);
              }
              return i;
           }
Input      size_t __read(int handle,unsigned char *buf,size_t bufSize)
           {
              size_t i;
              for (i=0; i<bufSize;i++)
              {
              // Wait for character available
              while(data_ready()==0);
              buf[i] = get_data(); // Get data
              }
              return i;
           }
```

For example, in order to allow the printf messages to be displayed on the character LCD module, the following "retarget.c" can be used:

```
/* Retarget.c for IAR Embedded Workbench for ARM to redirect printf to
character LCD */

#include <stdio.h>

extern void clcd_putc(char ch);

size_t __write(int handle,const unsigned char *buf,size_t bufSize)
{
    size_t i;
    for (i=0; i<bufSize;i++)
    {
        clcd_putc(buf[i]);
    }
    return i;
}
```

## CHAPTER OUTLINE

## 19.1 Introduction to embedded OSs

### 19.1.1 What are embedded OSs?

In Chapter 10 we covered various features of the Cortex®-M3 and Cortex-M4 processors that support the operations of embedded OSs. In general, an embedded OS can be anything from a simple task scheduler to a fully-featured OS like Linux.

Currently there are more than 30 embedded OSs available for Cortex-M processors. Most of these are simple OSs that enable multi-tasking, but some of them are application platforms that also provide additional software support such as a communication protocol stack (e.g., TCP/IP), a file system (e.g., FAT), or even a graphical user interface.

Many embedded OSs are RealTime Operating Systems (RTOS). This means that when a certain event occurs, it can trigger a corresponding task and that this must happen within a certain timeframe. An RTOS typically has a small memory footprint and provides very fast context switching (the time required to switch from one task to another).

Unlike OSs for personal computers or mobile computing devices (e.g., tablets), most embedded OS do not have any user interface, although user interface components (e.g., a GUI) can be added as application tasks running on the system. Also, Cortex-M processors cannot support fully-featured OSs like Linux or Windows, which require virtual memory system support.

---

**MEMORY MANAGEMENT UNIT (MMU) AND MEMORY PROTECTION UNIT (MPU)**

In application processors such as the Cortex-A processor family, the Memory Management Unit, "MMU," enables dynamic remapping of flat virtual address spaces seen by each process into physical address spaces on the system. Managing virtual memory can introduce large delays because address mapping information needs to be located and transferred from the memory (page table) to a hardware in the MMU (called Translation Lookaside Buffer, or TLB). As a result, operating systems that use virtual memory cannot guarantee real time responsiveness.

MPUs on Cortex-M processors only provide memory protection, and do not have the memory address translation requirement and therefore are suitable for real-time applications.

---

There is a special version of Linux called μCLinux that does not require an MMU and can work on those Cortex-M devices which have sufficient memory resources. Since a μCLinux system typically requires at least 2MB of SRAM, it is less popular in low-cost embedded systems because of the memory cost.

## 19.1.2 When to use an embedded OS

An embedded OS divides the available CPU processing time into a number of time slots and carries out different tasks in different time slots. Because the switching between time slots may happen hundreds of times per second, or more, it appears to the user that the processor executes several tasks in parallel.

Many applications do not require an embedded OS at all. The key benefit of using an embedded OS is to provide a scalable way of enabling several concurrent tasks to run in parallel. If the tasks are all fairly short and don't overlap each other most of the time, you can simply use an interrupt-driven arrangement to support multiple tasks.

There are a number of factors to consider when deciding whether to use an embedded OS or not:

- An embedded OS requires extra memory overhead. For example, it could take anything from 5KB of program memory space to over 100KB, depending on the features available in the OS.
- An embedded OS requires execution time overhead. For example, some processing time is required for context switching as well as task scheduling. Usually the execution time overhead is very small.
- Some of the embedded OS require license fees and/or royalty fees. Many others are free.
- Some embedded OS can only work with certain microcontroller devices, or can be toolchain-specific. If portability of the software code is important then you need to select an embedded OS which is supported on multiple platforms.

In general, as software code gets more complex, use of an embedded OS can make handling of multiple tasks much easier. Also, some embedded OS have additional safety features (e.g., stack space checking, MPU support), which can enhance the reliability of a system.

### 19.1.3 Role of CMSIS-RTOS

CMSIS-RTOS is an API specification. The CMSIS-RTOS itself is not a product but companies can build an RTOS based on CMSIS-RTOS. In Chapter 2 we gave an overview of the Cortex®-M Software Interface Standard (CMSIS). One of the projects within CMSIS is the CMSIS-RTOS. CMSIS-RTOS is an extension of existing RTOS designs to allow middleware to be designed in a way that can work with multiple RTOS products.

Some of the middleware products are quite complex and need to utilize task scheduling features in OSs to work. For example, a TCP/IP stack might run as a task inside a multi-tasking system and might need to spawn out additional child tasks when certain service requests are received. Traditionally these middleware (e.g., lightweight IP, lwIP) can include an OS emulation layer (Figure 19.1) that a software integrator needs to port when using a different OS.

The work of porting the OS emulation layer creates additional work for software developers, and sometimes the middleware vendors, and can increase project risks as the porting might not be straightforward.

CMSIS-RTOS was created to solve this issue. It can be implemented as an additional set of APIs or a wrapper for existing OS APIs. Since the API is standardized, middleware can be developed based on it and the product should, in theory, be able to work with any embedded OS that supports CMSIS-RTOS (Figure 19.2).

The RTOS products can still have their own native API interface, and application codes can still use those directly for additional features or for higher performance. This is good news for application developers because it saves a lot of time in porting

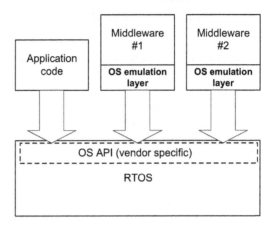

**FIGURE 19.1**

The need for OS emulation layer for middleware components

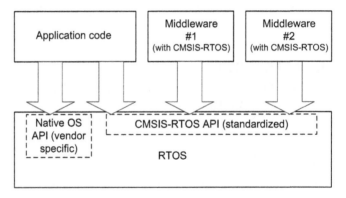

**FIGURE 19.2**

CMSIS-RTOS avoids the needs for OS emulation layer for each middleware component

middleware and reduces project risks. It is also good news for middleware vendors because it allows their products to work with more available OSs.

The CMSIS-RTOS also benefits RTOS vendors: As the amount of middleware that works with CMSIS-RTOS increases, having CMSIS-RTOS support in an embedded OS enables the OS product to work with more middleware. Also, as software in embedded systems increases in complexity and time-to-market becomes more important, porting of OS emulation layers for middleware will no longer be feasible for some projects because of the extra time needed and the associated project risk. CMSIS-RTOS enables RTOS products to reach these markets, which previous could only be covered by a few software platform solutions.

# 19.2 Keil™ RTX Real-Time Kernel

## 19.2.1 About RTX

The Keil™ RTX Real-Time Kernel is a royalty-free RTOS targeted at microcontroller applications. In older versions of Keil MDK-ARM, RTX is available as a precompiled library that is fully functional. In newer versions (mid-2012) of Keil MDK-ARM, the source code of the RTX kernel is also included in the installation.

The precompiled library is typically located in C:\Keil\ARM\RV31\LIB, and source code is typically located in C:\Keil\ARM\RL\RTX\SRC. In the C:\Keil\ARM\Boards directory, you can also find RTX examples for a number of Cortex®-M microcontroller boards.

Please also note that Keil RTX is now released under a simple, open source BSD license, so you can reuse and distribute RTX source code under the conditions described in the license document in the Keil MDK-ARM installation (C:\Keil\ARM\Hlp\license.rtf).

The Keil RTX kernel can be used as standalone RTOS or used with Keil Real-Time Library (RL-ARM, Figure 19.3), and can also work with third-party software products such as communication protocol stacks, data processing codecs, and other middleware.

## 19.2.2 Features overview

The RTX kernel is supported on all Cortex®-M processors in addition to traditional ARM® processors such as ARM7™ and ARM9™. It has the following features:

- Flexible scheduler: supporting pre-emptive, round-robin, and collaborative scheduling schemes
- Supports mailboxes, events (up to 16 per task), semaphores, mutex, and timers
- Unlimited number of defined tasks, with maximum of 250 active tasks at a time
- Up to 255 task priority levels
- Support for multi-threading and thread-safe operations
- Kernel-aware debug support in Keil™ MDK
- Fast context switching time

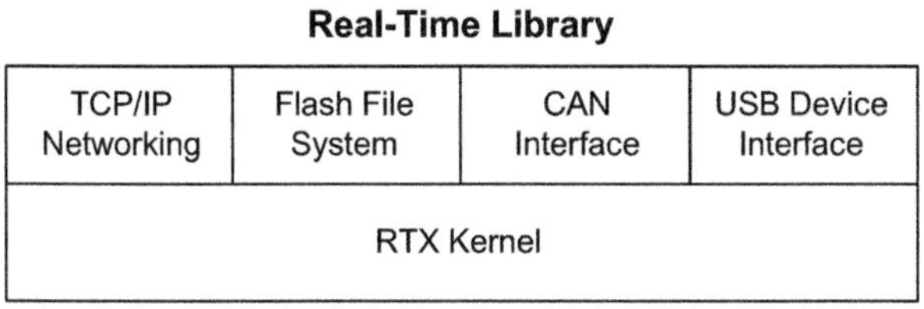

**Real-Time Library**

| TCP/IP Networking | Flash File System | CAN Interface | USB Device Interface |
|---|---|---|---|
| RTX Kernel | | | |

**FIGURE 19.3**

The RL-ARM product

- Small memory footprint (less than 4Kbytes for Cortex-M version, less than 5Kbytes for ARM7/ARM9)

In addition, the Cortex-M version of RTX kernel has the following features:

- SysTick timer support
- No interrupt lockout in Cortex-M versions (interrupt is not disabled by OS at any time)
- Since RTX has a very small memory footprint, it can be used even with Cortex-M microcontroller devices that have small memory capacity.

## 19.2.3 **RTX and CMSIS-RTOS**

In 2012, Keil™ released a trial version of the CMSIS-RTOS implementation for RTX. This design is to be finalized in 2013 so you can use RTX with CMSIS-RTOS API. At the time of writing (January, 2013), the CMSIS-RTOS RTX work has not been completed and the RTX included in the Keil MDK installation is still the old version, which does not have CMSIS-RTOS support. The RTX example code in the Keil MDK installation is also based on the previous proprietary APIs. At the moment, to use CMSIS-RTOS with RTX, you need to download the RTX implementation for CMSIS-RTOS from the Keil website separately at https://www.keil.com/demo/eval/rtx.htm.

Because the RTX source code in use at the time of writing is not the final version, please note that there is a small chance the CMSIS-RTOS RTX examples described in this chapter will need adjustment for the final version of CMSIS-RTOS RTX.

The CMSIS-RTOS package contains the source code, examples, and documentation. To make it easier for users, this CMSIS-RTOS package contains precompiled versions of the CMSIS-RTOS in the form of library. Table 19.1 listed the directories in RTX CMSIS-RTOS package from Keil® website.

**Table 19.1** Directory Structure in Current CMSIS-RTOS Package

| Directory | Content |
| --- | --- |
| Boards | CMSIS-RTOS RTX example projects for several evaluation boards. These examples are typically provided for several compilers. |
| Doc | Documentation for CMSIS-RTOS RTX. |
| Examples | Generic examples that show several features of CMSIS-RTOS RTX. These examples are typically provided for several compilers. |
| INC | The include files for CMSIS-RTOS RTX. cmsis_os.h is the central include file for user applications. |
| LIB | CMSIS-RTOS RTX Library files for ARMCC, GCC, and IAR Compiler. |
| SRC | Source code of the CMSIS-RTOS RTX Library along with project files for ARMCC, GCC, and IAR Compiler. |
| Templates | Templates for creating application projects with CMSIS-RTOS RTX. |

**Table 19.2** CMSIS-RTOS Precompiled Libraries

| Library File | Processor Configuration |
| --- | --- |
| LIB\ARM\RTX_CM0.lib | CMSIS-RTOS RTX Library for ARMCC Compiler, Cortex-M0 and M1, little-endian. |
| LIB\ARM\RTX_CM0_B.lib | CMSIS-RTOS RTX Library for ARMCC Compiler, Cortex-M0 and M1, big-endian. |
| LIB\ARM\RTX_CM3.lib | CMSIS-RTOS RTX Library for ARMCC Compiler, Cortex-M3 and M4 without FPU, little-endian. |
| LIB\ARM\RTX_CM3_B.lib | CMSIS-RTOS RTX Library for ARMCC Compiler, Cortex-M3 and M4 without FPU, big-endian. |
| LIB\ARM\RTX_CM4.lib | CMSIS-RTOS RTX Library for ARMCC Compiler, Cortex-M4 with FPU, little-endian. |
| LIB\ARM\RTX_CM4_B.lib | CMSIS-RTOS RTX Library for ARMCC Compiler, Cortex-M4 with FPU, big-endian. |

**Table 19.3** Additional CMSIS-RTOS Files Required for Projects

| File | Processor Configuration |
| --- | --- |
| INC\cmsis_os.h | CMSIS-RTOS header file for application code |
| Examples\*\RTX_Conf_CM.c | RTX Kernel System Configuration file – can be edited by user.<br>This file is coded with special tags in comments so that the RTX parameters can be edited easily with a Configuration Wizard. |
| INC\RTX_CM_LIB.h | RTX Kernel System Configuration code needed by RTX_Conf_CM.c |
| SRC\ARM\SVC_Tables.s | An assembly file to allow you to extend the SVC services available by adding a lookup Table of SVC functions. This is optional. |

The library files provided in the package include pre-compiled versions for various Cortex®-M processors, and are available in little endian and big endian versions for ARM®, gcc and IAR toolchains. For example, the library files for the ARM toolchain are shown in Table 19.2.

Alternatively, you can also use the source code directly from the SRC directory. In addition, you also need couple of additional files from the INC directory and SRC directory, as shown in Table 19.3.

## 19.2.4 Thread

In the CMSIS-RTOS, we use the term "thread" for each of the concurrent (parallel processing) programs. From an academic view, a task or a process could contain multiple threads. But here we will just look at a relatively simple case where each task has just one thread.

Each thread has a programmable priority level. In the RTX implementation the thread priority is an enumerated value. The CMSIS-RTOS has a number of pre-defined enumerations for thread priorities, and this is mapped into the signed numerical priority levels in the file cmsis_os.h:

```
/// Priority used for thread control.
/// \note MUST REMAIN UNCHANGED: \b osPriority shall be consistent in
every CMSIS-RTOS.
typedef enum {
  osPriorityIdle        = -3,   ///< priority: idle (lowest)
  osPriorityLow         = -2,   ///< priority: low
  osPriorityBelowNormal = -1,   ///< priority: below normal
  osPriorityNormal      =  0,    ///< priority: normal (default)
  osPriorityAboveNormal = +1,   ///< priority: above normal
  osPriorityHigh        = +2,   ///< priority: high
  osPriorityRealtime    = +3,   ///< priority: realtime (highest)
  osPriorityError       = 0x84  ///< system cannot determine priority
or thread has illegal priority
} osPriority;
```

Note that the thread priority level arrangement is completely separated from the interrupt priority.

In the RTX environment, each thread can be in one of the states shown in Table 19.4.

The thread state transition diagram is shown in Figure 19.4.

In a simple single core processor system, there can be only one thread in Running state at a time.

Unlike some other RTOSs, "main()" can be of the threads, dependent on the actual implementation of CMSIS-RTOS. If that is the case, we can create additional threads from "main()". If the "main()" thread is not needed at some stage any point,

**Table 19.4** Thread States in RTX Kernel

| State | Description |
|---|---|
| RUNNING | The thread is currently running. |
| READY | The thread is in the queue of threads which are ready to run (waiting for a time slot). When the current running thread is completed, RTX will select the next highest priority thread in the ready queue and start it. |
| WAITING | The thread has previously executed a function that indicate it is waiting for a delay request to complete or an event (signal/semaphore/mailbox/etc.) from another thread. It can switch from Waiting to Ready/Running (depending on task priority) when the specified event has occurred. |
| INACTIVE | The thread has not been started or the thread has been terminated. A terminated task can be re-created. |

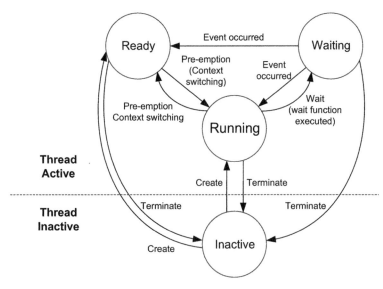

**FIGURE 19.4**

States of threads in CMSIS-RTOS

we can execute a wait function to put it in a waiting state, or even terminate it to prevent it from taking up execution time.

CMSIS-RTOS allows threads to execute in privileged state or unprivileged state. Please refer to the OS_RUNPRIV parameter in Table 19.6. Please note that with the current RTX implementation, if threads are configured to run in unprivileged state, "main()" will also start in unprivileged state. You can extend the SVC Handler service to support operations that require privileged state (e.g., access to NVIC or any registers in the System Control Space, SCS).

## 19.3 CMSIS-OS examples
### 19.3.1 Simple CMSIS-RTOS with two threads

The following examples are based on the Keil™ MDK-ARM development suite and CMSIS-RTOS RTX, using the STM32F4 Discovery board.

In the first example, we will look at a minimal setup with two threads: main() and a blinky thread. The threads each toggle an LED on the development board. To set up the first project, we use the precompiled version of CMSIS-RTOS RTX (library file RTX_CM4.lib) to simplify the compilation, as shown in Figure 19.5.

If you like, you can also use the source code version of the CMSIS-RTOS instead of using the precompiled library. In addition, we also need the files given in Table 19.5.

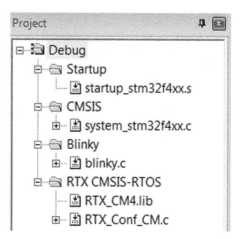

**FIGURE 19.5**

Project browser display with simple project

| Table 19.5 Additional Files Needed in the First CMSIS-RTOS Project Example | |
|---|---|
| **Files** | **Descriptions** |
| RTX_Conf_CM.c | RTX Kernel System Configuration file |
| cmsis_os.h | CMSIS-RTOS header file for application code |
| RTX_CM_lib.h | RTX Kernel System Configuration code needed by RTX_Conf_CM.c |

We also configure the RTX kernel option in the project option, as shown in Figure 19.6. This enables us to use the OS-aware debugging features later.

Previously (Table 19.5) we mentioned that the file RTX_Config_CM.c defines some of the configurations of the RTX kernel operations. This file is configurable by users. We can either edit this file directly in the program text editor, or we can use the Configuration Wizard. This file is coded in such a way that it can be recognized by the Configuration Wizard. By clicking on the "Configuration Wizard" tab at the bottom of the editor window, we can see the Configuration Wizard as shown in Figure 19.7.

Table 19.6 shows listed a number of options in RTX_Conf_CM.c.

The actual code for the first example is very simple.

```
/* Simple CMSIS-RTOS RTX example that use two threads (including
main()) to toggle two LEDs */
#include "stm32f4xx.h"
#include <cmsis_os.h>


/* Thread IDs */
osThreadId t_blinky;  // Declare a thread ID for blink
```

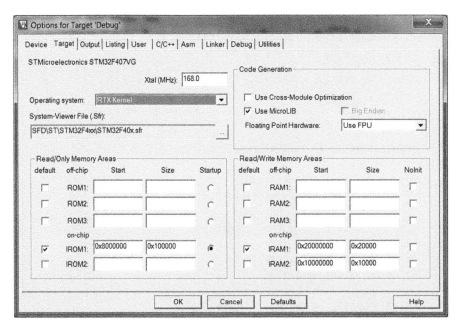

**FIGURE 19.6**

RTX Kernel project option

```c
/* Function declaration */
void blinky(void const *argument); // Thread
void LedOutputCfg(void);           // LED output configuration

// --------------------------------------------------------
// Blinky
// - toggle LED bit 12
// - Unprivileged Thread
void blinky(void const *argument) {
  while(1) {
    if (GPIOD->IDR & (1<<12)) {
      GPIOD->BSRRH = (1<<12); // Clear bit 12
    } else {
      GPIOD->BSRRL = (1<<12); // Set bit 12
    }
    osDelay(500);  // delay 500 msec
  }
}
```

```
// define blinky_1 as thread function
osThreadDef(blinky, osPriorityNormal, 1, 0);

// ----------------------------------------------------------
// - toggle LED bit 13
// - Unprivileged Thread
int main(void)
{
LedOutputCfg(); // Initialize LED output

// Create a task "blinky"
t_blinky = osThreadCreate(osThread(blinky), NULL);

// main() itself is another thread
while(1) {
    if (GPIOD->IDR & (1<<13)) {
    GPIOD->BSRRH = (1<<13); // Clear bit 13
    } else {
    GPIOD->BSRRL = (1<<13); // Set bit 13
    }
    osDelay(1000);   // delay 1000 msec
 }
} // end main
// ----------------------------------------------------------
void LedOutputCfg(void)
{
   // Configure LED outputs
   RCC->AHB1ENR |= RCC_AHB1ENR_GPIODEN; // Enable Port D clock
   // Set pin 12, 13, 14, 15 as general purpose output mode (pull-push)
   GPIOD->MODER |= (GPIO_MODER_MODER12_0 |
                    GPIO_MODER_MODER13_0 |
                    GPIO_MODER_MODER14_0 |
                    GPIO_MODER_MODER15_0 ) ;
   GPIOD->PUPDR = 0; // No pull up , no pull down
   return;
}
```

For each thread, there is an assoicated ID value with the data type osThreadId. This ID value is assigned when the thread is created and is needed for intertask communication, which will be demonstrated later. If no intertask communication is required, it is not necessary.

To create a new thread, we used the function osThreadCreate.

For each thread (apart from main), we also need to declare the function as a thread using osThreadDef. You can also define the priority of the thread using

**FIGURE 19.7**

Configuration Wizard

osThreadDef. During run-time the priority of a thread can also be changed dynamically using CMSIS-RTOS API.

After setting up the project, you can then compile and test the application. The two LEDs on the development should toggle at different speeds.

In other CMSIS-RTOS implementations, it is possible that the OS kernel does not start when the processor enters the "main()" program. In such cases you will need to start the OS kernel specifically. CMSIS-RTOS provides a predefined constant called osFeature_MainThread to indicate whether thread execution starts with the function "main()." If this is 1, then the OS kernel starts with "main()."

For example, you can use the following code to start the OS kernel conditionally:

```
int main(void)
{
  ...
#if (osFeature_MainThread==0)
  osKernelStart(osThread(blinky), NULL); // Start OS Kernel explicity
  // not required in RTX
```

**Table 19.6** CMSIS-RTOS RTX Options in RTX_Conf_CM.c

| Parameter | Descriptions | Default Value |
|---|---|---|
| OS_TASKCNT | Number of concurrent running threads: Defines max number of threads that will run at the same time. | 6 |
| OS_STKSIZE | Default Thread stack size [bytes] <64-4096> (needs to be a multiple of 8). It is used if the "osThreadDef" statement does not specify stack size (stacksz set to 0). | 200 |
| OS_MAINSTKSIZE | Main Thread stack size [bytes] <64-4096> (needs to be a multiple of 8). | 200 |
| OS_PRIVCNT | Number of threads with user-provided stack size <0-250> | 0 |
| OS_PRIVSTKSIZE | Total combined stack size [bytes] for threads with user-provided stack size <0-4096> (needs to be a multiple of 8). | 0 |
| OS_STKCHECK | Enable check for stack overflow for threads. Note that additional code reduces the Kernel performance. | 1 |
| OS_RUNPRIV | Processor mode for thread execution: 0 = Unprivileged mode, 1 = privileged mode. | 0 |
| OS_SYSTICK | Set to 1 to use Cortex®-M SysTick timer as RTX Kernel Timer. | 1 |
| OS_CLOCK | Defines the Timer clock frequency [Hz] <1-1000000000>. Typically this is the same as the processor clock frequency if SysTick is used. | 12000000 (12MHz) |
| OS_TICK | Defines the OS Timer tick interval [us] <1-1000000> | 1000 (1 ms) |
| OS_ROBIN | Set to 1 to enable Round-Robin Thread switching | 1 |
| OS_ROBINTOUT | Round-Robin Timeout [ticks] <1-1000> (valid if OS_ROBIN is 1) | 5 |
| OS_TIMERS | Enables user Timers | 0 |
| OS_TIMERPRIO | Timer Thread Priority (valid if OS_TIMERS is 1) 1. Low 2. Below Normal 3. Normal 4. Above Normal 5. High 6. Real-time (highest) | 5 |
| OS_TIMERSTKSZ | Timer Thread stack size [bytes] <64-4096> (needs to be a multiple of 8). | 200 |
| OS_TIMERCBQS | Timer Callback Queue size- Number of concurrent active timer callback functions. | 4 |
| OS_FIFOSZ | ISR FIFO Queue size (4 = 4 entries. Can be 4, 8, 12, 16, 24, 32, 48, 64, 96). ISR functions store requests to this buffer when they are called from the interrupt handler. | 16 |

```
#endif
 ...
```

or

```
int main(void)
{
  ...
 if (osFeature_MainThread==0) {
 osKernelStart(osThread(blinky), NULL); // Start OS Kernel explicity
 // not required in RTX
 }
  ...
```

The osThread(*name*) macro is used in the example for accessing a Thread definition. For example, when a function's input parameter needs to be a Thread (e.g., blinky), then we use osThread(*blinky*) to specify that the parameter is a Thread.

In this example, we also used a macro called osThreadDef(*name, priority, instances, stacksz*). This is used to create a Thread definition with the specified function, priority level, and stack size requirements of the thread. If the stack size requirement is set to 0, the default stack size is used, as defined by OS_STKSIZE in RTX_Config_CM.c.

Table 19.7 lists some of the commonly used functions for OS kernel management and Thread management.

Some of these functions use a enumeration type called osStatus. The definition of osStatus is listed in Table 19.8. Most of the functions will only be able to return a subset of these enumerations.

### 19.3.2 **Inter-thread communciation overview**

In most applications with an RTOS, there will be lots of interactions between threads. Instead of using shared data and polling loops to check the status of other tasks, or passing information, we should use the inter-thread communication features provided in the OS to make the operation more efficient. Otherwise, a thread waiting for input from another thread could stay in the READY queue for a long time and this can consume a lot of processing time.

Modern RTOSs typically provide a number of methods to support communications between threads. In CMSIS-RTOS, the supported methods include:

- Signal events
- Semaphores
- Mutex
- Mailbox/message

**Table 19.7** CMSIS-RTOS Functions for OS Kernel and Thread Management

|  | Function | Description |
|---|---|---|
| **osThreadID** | osThreadCreate(osThreadDef_t *thread_def, void *argument) | Create a thread and add it to Active Threads and set it to state READY. |
| **osThreadID** | osThreadGetId(void) | Return the thread ID of the current running thread. |
| **osStatus** | osThreadTerminate (osThreadId thread_id) | Terminate execution of a thread and remove it from Active Threads. |
| **osStatus** | osThreadSetPriority (osThreadId thread_id, osPriority priority) | Change priority of an active thread. |
| **osPriority** | osThreadGetPriority (osThreadId thread_id) | Get current priority of an active thread. |
| **osStatus** | osThreadYield (void) | Pass control to the next thread that is in state READY. |
| **osStatus** | osKernelStart (osThreadDef_t *thread_def, void *argument) | Start the RTOS Kernel and execute the specified thread. |
| **int32_t** | osKernelRunning(void) | Check if the RTOS kernel is already started. Returns 0 if the RTOS is not started. Returns 1 if started. |

**Table 19.8** osStatus Enumeration Definition

| osStatus Enumerator | Description |
|---|---|
| osOK | Function completed; no event occurred. |
| osEventSignal | Function completed; signal event occurred. |
| osEventMessage | Function completed; message event occurred. |
| osEventMail | Function completed; mail event occurred. |
| osEventTimeout | Function completed; timeout occurred. |
| osErrorParameter | Parameter error: a mandatory parameter was missing or specified an incorrect object. |
| osErrorResource | Resource not available: a specified resource was not available. |
| osErrorTimeoutResource | Resource not available within given time: a specified resource was not available within the timeout period. |
| osErrorISR | Not allowed in ISR context: the function cannot be called from interrupt service routines. |
| osErrorISRRecursive | Function called multiple times from ISR with same object. |
| osErrorPriority | System cannot determine priority or thread has illegal priority. |

| **Table 19.8** osStatus Enumeration Definition—*Cont'd* | |
|---|---|
| **osStatus Enumerator** | **Description** |
| osErrorNoMemory | System is out of memory: it was impossible to allocate or reserve memory for the operation. |
| osErrorValue | Value of a parameter is out of range. |
| osErrorOS | Unspecified RTOS error: run-time error but no other error message fits. |
| os_status_reserved | Reserved error value to prevent from enum down-size compiler optimization. |

In addition, there are additional features to support some of these communication methods such as memory pool management features, which are often used with mailboxes.

### 19.3.3 Signal event communication

In CMSIS-RTOS, each thread can have up to 31 signal events (depending on configuration via a macro called osFeature_Signals in RTX). A thread enters WAIT state when it executes the function osSignalWait. One of the input parameters, a 32-bit value called "signals," defines the signal events required to put the thread back to READY state. Each bit (apart from the MSB) of the "signals" parameter defines the signal events required, and if this parameter is set to 0 any signal event can put this thread back to READY state. Table 19.9 listed the CMSIS-RTOS functions for signal event communications.

The signal event functions osSignalSet, osSignalClear, and osSignalGet return 0x80000000 in case of incorrect parameters.

By default the "cmsis_os.h" in RTX specifies osFeature_Signals as 16. So it can work with 16 signal events (from 0x00000001 to 0x00008000).

Please note that signal flags used as events for waking up a thread from the WAITING state are cleared automatically. For example, in the following example, event flag 0x0001 is used to enable "main()" thread to send a signal to blinky event as shown in Figure 19.8.

```
/* Example code for simple signal event communication */

#include "stm32f4xx.h"
#include <cmsis_os.h>

/* Thread IDs */
osThreadId t_blinky; // Declare a thread ID for blink
/* Function declaration */
void blinky(void const *argument); // Thread
```

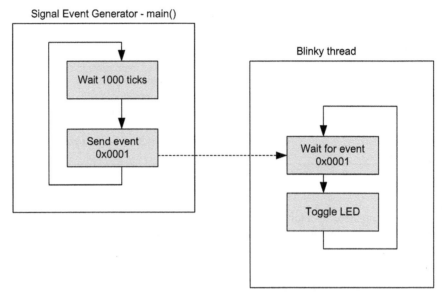

**FIGURE 19.8**

Simple signal event communication

| | **Function** | **Description** |
|---|---|---|
| **osEvent** | osSignalWait (int32_t signals, uint32_t millisec) | Wait for one or more Signal Flags to become signaled for the current RUNNING thread. If "signals" is non-zero, all specified signal flags need to be set to return to READY state. If "signals" is zero, any signal flag can put the thread back to READY. "millisec" is the timeout value. Set to `osWaitForever` in case of no time-out, or zero to return immediately |
| **int32_t** | osSignalSet (osThreadId thread_id, int32_t signal) | Set the specified Signal Flags of an active thread. |
| **int32_t** | osSignalClear (osThreadId thread_id, int32_t signal) | Clear the specified Signal Flags of an active thread. |
| **int32_t** | osSignalGet (osThreadId thread_id) | Get Signal Flags status of an active thread. |

**Table 19.9** Signal Event Functions

```
void LedOutputCfg(void);          // LED output configuration

// ------------------------------------------------------------
// Blinky
// - toggle LED bit 12
// - Unprivileged Thread
void blinky(void const *argument) {
 while(1) {
   osSignalWait(0x0001, osWaitForever);
   if (GPIOD->IDR & (1<<12)) {
   GPIOD->BSRRH = (1<<12); // Clear bit 12
   } else {
   GPIOD->BSRRL = (1<<12); // Set bit 12
   }
   }
 }

 // define blinky_1 as thread function
 osThreadDef(blinky, osPriorityNormal, 1, 0);

// ------------------------------------------------------------
// - toggle LED bit 13
// - Unprivileged Thread
int main(void)
{
 LedOutputCfg(); // Initialize LED output

 // Create a task "blinky"
 t_blinky = osThreadCreate(osThread(blinky), NULL);

 // main() itself is another thread
 while(1) {
   if (GPIOD->IDR & (1<<13)) {
   GPIOD->BSRRH = (1<<13); // Clear bit 13
   } else {
   GPIOD->BSRRL = (1<<13); // Set bit 13
   }
   osSignalSet(t_blinky, 0x0001); // Set Signal
   osDelay(1000);  // delay 1000 msec
   }
 } // end main
```

A thread can wait for multiple signal events and use osSignalGet() to determine what actions should be taken on return to READY state, as shown in Figure 19.9.

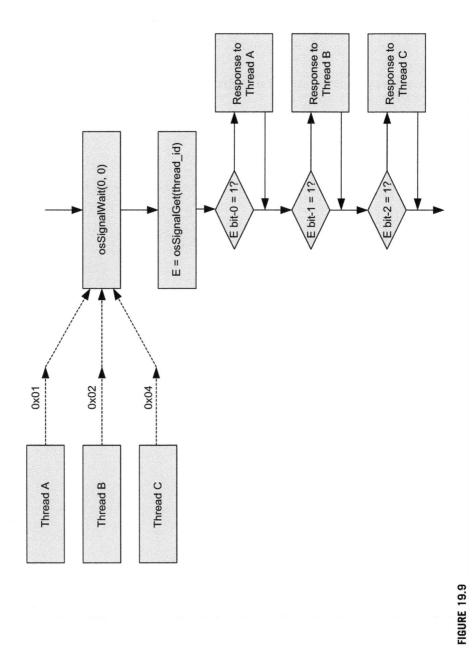

**FIGURE 19.9**

Using the osSignalGet function to detect which thread generated the signal

### 19.3.4 **Mutual Exclusive (Mutex)**

Mutual Exclusive, or commonly known as Mutex, is a common resource management feature in all types of OSs. Many resources in a processor system can only be used by one thread at a time. For example, a "printf" output communication channel (as shown in Figure 19.10) can only be used by one thread at a time.

Before using a Mutex, we first need to define a Mutex object using "osMutexDef(*name*)." When referencing a Mutex using the CMSIS-RTOS Mutex API, we need to use the "osMutex(*name*)" macro. Each Mutex also has an ID value that is needed by some of the Mutex functions. Table 19.10 listed the CMSIS-RTOS functions for Mutex operations.

In the following example, the program code contains two threads. Both of them use the ITM (Instrumentation Trace Macrocell) to output text messages.

```
/* Simple Mutex example — each printf statement is guarded by mutex
to make sure no two printf statements are executed concurrently */

#include "stm32f4xx.h"
#include <cmsis_os.h>
#include "stdio.h"

/* Thread IDs */
osThreadId t_blinky_id; // Declare a thread ID for blink
/* Declare Mutex */
osMutexDef(PrintLock); // Declare a Mutex for printf control
/* Mutex IDs */
osMutexId PrintLock_id; // Declare a Mutex ID for printf control

/* Function declaration */
void blinky(void const *argument); // Thread
void LedOutputCfg(void);           // LED output configuration

// ---------------------------------------------------------
// Blinky
// - toggle LED bit 12
// - Unprivileged Thread
void blinky(void const *argument) {
 while(1) {
   if (GPIOD->IDR & (1<<12)) {
   GPIOD->BSRRH = (1<<12); // Clear bit 12
   } else {
   GPIOD->BSRRL = (1<<12); // Set bit 12
   }
   osDelay(50);  // delay 50 msec
   osMutexWait(PrintLock_id, osWaitForever);
```

```
    printf ("blinky is running\n");
    osMutexRelease(PrintLock_id);
    }
  }
  // define blinky_1 as thread function
  osThreadDef(blinky, osPriorityNormal, 1, 0);

  // ---------------------------------------------------------
  // - toggle LED bit 13
  // - Unprivileged Thread
  int main(void)
  {
  LedOutputCfg(); // Initialize LED output
  // Create the printf control Mutex before starting blinky thread
  PrintLock_id = osMutexCreate(osMutex(PrintLock));
  osMutexWait(PrintLock_id, osWaitForever);
  printf ("\nMutex Demo\n");
  osMutexRelease(PrintLock_id);

  // Create a task "blinky"
  t_blinky_id = osThreadCreate(osThread(blinky), NULL);

  // main() itself is another thread
  while(1) {
    if (GPIOD->IDR & (1<<13)) {
    GPIOD->BSRRH = (1<<13); // Clear bit 13
    } else {
    GPIOD->BSRRL = (1<<13); // Set bit 13
    }
    osDelay(50);   // delay 50 msec
    osMutexWait(PrintLock_id, osWaitForever);
    printf ("main() is running\n");
    osMutexRelease(PrintLock_id);
    }
    } // end main
```

## 19.3.5 Semaphore

In some cases we would like to allow a limited number of threads to access certain resources. For example, a DMA controller might be able to support multiple DMA channels. Or a simple embedded server might be able to support a limited number of simultaneous requests due to memory size constraints. In these cases, we can use a semaphore instead of a Mutex.

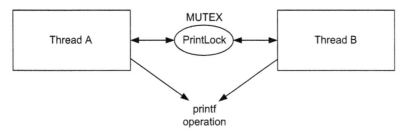

**FIGURE 19.10**

Using Mutex to control hardware resource sharing

**Table 19.10** Mutex Functions

|  | Function | Description |
|---|---|---|
| **osMutexId** | osMutexCreate(const osMutexDef_t *mutex_def) | Create and Initialize a Mutex object. |
| **osStatus** | osMutexWait (osMutexId mutex_id, uint32_t millisec) | Wait until a Mutex becomes available. |
| **osStatus** | osMutexRelease (osMutexId mutex_id) | Release a Mutex that was obtained by osMutexWait. |
| **osStatus** | osMutexDelete (osMutexId mutex_id) | Delete a Mutex that was created by osMutexCreate. |

The semaphore feature is very similar to Mutex. Whereas a Mutex permits just one thread to access to a shared resource at any one time, a semaphore can be used to permit a fixed number of threads to access a pool of shared resources. So a Mutex is a special case of a semaphore for which the maximum number of available tokens is 1.

A semaphore object needs to be initialized to the maximum number of available tokens, and each time a thread needs to use a shared resource, it uses the semaphore to check out a token and then checks it back in when it has finished using the resource. If the number of available tokens reaches zero, then all the available resources have been allocated and the next thread requesting the shared resource must wait for a token to become available.

In the following example, we create four threads that each toggle a LED on the development board, and use a semaphore to limit the number of active LEDs to 2.

Semaphore objects are defined using "osSemaphoreDef(name)." When referencing a semaphore object using the CMSIS-RTOS semaphore API, we need to use the "osSemaphore(name)" macro. Each semaphore also has an ID value that is needed by some of the semaphore functions, as shown in Table 19.11.

In the following example, the program code contains five threads, including main(). Four of them are used to toggle LEDs, and a semaphore is used to limit the number of LEDs that are turned on at any point in time to be two or fewer.

**Table 19.11** Semaphore Functions

|  | Function | Description |
|---|---|---|
| **osSemaphoreId** | osSemaphoreCreate(const osSemaphoreDef_t *semaphore_def, int32_t count) | Create and Initialize a semaphore object. |
| **int32_t** | osSemaphoreWait(osSemaphoreId semaphore_id, uint32_t millisec) | Wait until a semaphore becomes available. Returns number of available tokens or −1 in case of incorrect parameters |
| **osStatus** | osSemaphoreRelease(osSemaphoreId semaphore_id) | Release a semaphore that was obtained by osSemaphoreWait. |
| **osStatus** | osSemaphoreDelete(osSemaphoreId semaphore_id) | Delete a semaphore that was created by osSemaphoreCreate. |

```c
/* Semaphore example */

#include "stm32f4xx.h"
#include <cmsis_os.h>

/* Thread IDs */
osThreadId t_blinky_id1;
osThreadId t_blinky_id2;
osThreadId t_blinky_id3;
osThreadId t_blinky_id4;
/* Declare Semaphore */
osSemaphoreDef(two_LEDs);  // Declare a Semaphore for LED control
/* Semaphore IDs */
osSemaphoreId two_LEDs_id; // Declare a Semaphore ID for LED control

/* Function declaration */
void blinky(void const *argument); // Thread
void LedOutputCfg(void);    // LED output configuration

// ---------------------------------------------------------
// Blinky_1 - toggle LED bit 12
void blinky_1(void const *argument) {
 while(1) {
 // LED on
   osSemaphoreWait(two_LEDs_id, osWaitForever);
   GPIOD->BSRRL = (1<<12); // Set bit 12
   osDelay(500);           // delay 500 msec
```

```
    GPIOD->BSRRH = (1<<12); // Clear bit 12
    osSemaphoreRelease(two_LEDs_id);

    // LED off
    osDelay(500);              // delay 500 msec
    }
  }
// ------------------------------------------------------
// Blinky_2 - toggle LED bit 13
void blinky_2(void const *argument) {
 while(1) {
    // LED on
    osSemaphoreWait(two_LEDs_id, osWaitForever);
    GPIOD->BSRRL = (1<<13); // Set bit 13
    osDelay(600);              // delay 600 msec
    GPIOD->BSRRH = (1<<13); // Clear bit 13
    osSemaphoreRelease(two_LEDs_id);

    // LED off
    osDelay(600);    // delay 600 msec
    }
 }
// ------------------------------------------------------
// Blinky_3 - toggle LED bit 14
void blinky_3(void const *argument) {
 while(1) {
 // LED on
    osSemaphoreWait(two_LEDs_id, osWaitForever);
    GPIOD->BSRRL = (1<<14); // Set bit 14
    osDelay(700);              // delay 700 msec
    GPIOD->BSRRH = (1<<14); // Clear bit 14
    osSemaphoreRelease(two_LEDs_id);

    // LED off
    osDelay(700);    // delay 700 msec
    }
 }
// ------------------------------------------------------
// Blinky_4 - toggle LED bit 15
void blinky_4(void const *argument) {
 while(1) {
    // LED on
    osSemaphoreWait(two_LEDs_id, osWaitForever);
    GPIOD->BSRRL = (1<<15); // Set bit 15
    osDelay(800);              // delay 800 msec
```

```
    GPIOD->BSRRH = (1<<15); // Clear bit 15
    osSemaphoreRelease(two_LEDs_id);

    // LED off
    osDelay(800);    // delay 800 msec
    }
  }
}
// ----------------------------------------------------------
// define thread functions
osThreadDef(blinky_1, osPriorityNormal, 1, 0);
osThreadDef(blinky_2, osPriorityNormal, 1, 0);
osThreadDef(blinky_3, osPriorityNormal, 1, 0);
osThreadDef(blinky_4, osPriorityNormal, 1, 0);
// ----------------------------------------------------------
// - toggle LED bit 13
// - Unprivileged Thread
int main(void)
{
  LedOutputCfg(); // Initialize LED output
  // Create Semaphore with 2 tokens
  two_LEDs_id = osSemaphoreCreate(osSemaphore(two_LEDs), 2);

  // Create "blinky" threads
  t_blinky_id1 = osThreadCreate(osThread(blinky_1), NULL);
  t_blinky_id2 = osThreadCreate(osThread(blinky_2), NULL);
  t_blinky_id3 = osThreadCreate(osThread(blinky_3), NULL);
  t_blinky_id4 = osThreadCreate(osThread(blinky_4), NULL);

  // main() itself is another thread
  while(1) {
  osDelay(osWaitForever); // delay
  }
} // end main
```

### 19.3.6 Message queue

A message queue can be used to pass a sequence of data from one thread to another in a FIFO-like operation (Figure 19.11). The data can be of integer or pointer type.

Message queue objects are defined using "osMessageQDef(name, queue_size, type)." When referencing a message queue object using the CMSIS-RTOS API, we need to use the "osMessageQ(name)" macro. Each message queue also has an ID value that is needed by some of the message queue functions, as shown in Table 19.12.

In the following example, a number sequence 1, 2, 3, ... is sent from "main()" to another thread called "receiver."

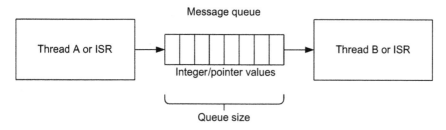

**FIGURE 19.11**

Message queue

| Table 19.12 | Message Queue Functions | |
| --- | --- | --- |
| | **Function** | **Description** |
| **osMessageQId** | osMessageCreate (const osMessageQDef_t *queue_def, osThreadId thread_id) | Create and Initialize a Message Queue. |
| **osStatus** | osMessagePut (osMessageQId queue_id, uint32_t info, uint32_t millisec) | Put a Message to a Queue. |
| **os_InRegs osEvent** | osMessageGet (osMessageQId queue_id, uint32_t millisec) | Get a Message or Wait for a Message from a Queue. |

```c
/* Simple message queue demo */

#include "stm32f4xx.h"
#include "stdio.h"
#include <cmsis_os.h>

/* Declare message queue */
osMessageQDef(numseq_q, 4, uint32_t); // Declare a Message queue
osMessageQId numseq_q_id;             // Declare a ID for message queue

/* Function declaration */
void receiver(void const *argument); // Thread
/* Thread IDs */
osThreadId t_receiver_id;

// --------------------------------------------------------
// Receiver thread
void receiver(void const *argument) {
 while(1) {
   osEvent evt = osMessageGet(numseq_q_id, osWaitForever);
```

```
    if (evt.status == osEventMessage) { // message received
    printf ("%d\n", evt.value.v); // ".v" indicate message as 32-bit value
    }
} // end while
}
// define thread function
osThreadDef(receiver, osPriorityNormal, 1, 0);
// --------------------------------------------------------
int main(void)
{
uint32_t i=0;
// Create Message queue
numseq_q_id = osMessageCreate(osMessageQ(numseq_q), NULL);

// Create "receiver" thread
t_receiver_id = osThreadCreate(osThread(receiver), NULL);

// main() itself is a thread that send out message
while(1) {
  i++;
  osMessagePut(numseq_q_id, i, osWaitForever);
  osDelay(1000);   // delay 1000 msec
}
} // end main
// --------------------------------------------------------
```

An additional example of using a message queue to pass pointers is given in section 19.3.8.

### 19.3.7 Mail queue

A mail queue (Figure 19.12) is very similar to a message queue, but the information being transferred consists of memory blocks that need to be allocated before putting data in, and freed after taking data out. Memory blocks can hold more information, for example, a data structure, whereas in a message queue the information transferred can only be a 32-bit value or a pointer.

The mail queue object is defined using "osMailQDef(*name, queue_size, type*)." When referencing a mail queue using CMSIS-RTOS API, we need to use the "osMailQ(*name*)" macro. Each mail queue also has an ID value that is needed by some of the mail queue functions, as shown in Table 19.13.

The following example showing how to use a mail queue to pass a block of memory containing a data structure with three elements.

```
/* Mail queue example */

#include "stm32f4xx.h"
```

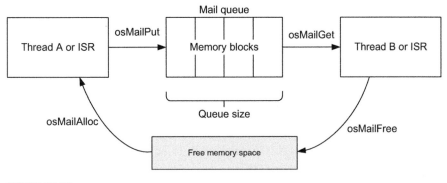

**FIGURE 19.12**

Mail queue

```
#include "stdio.h"
#include <cmsis_os.h>

typedef struct {
  uint32_t length;
  uint32_t width;
  uint32_t height;
} dimension_t;

/* Declare message queue */
osMailQDef(dimension_q, 4, dimension_t); // Declare a Mail queue
osMailQId dimension_q_id; // Declare a ID for Mail queue
```

**Table 19.13** Mail Queue Functions

|  | **Function** | **Description** |
|---|---|---|
| **osMailQId** | osMailCreate (const osMailQDef_t *queue_def, osThreadId thread_id) | Create and initialize a mail queue. |
| **void** * | osMailAlloc (osMailQId queue_id, uint32_t millisec) | Allocate a memory block from a mail. |
| **void** * | osMailCAlloc (osMailQId queue_id, uint32_t millisec) | Allocate a memory block from a mail and set memory block to zero. |
| **osStatus** | osMailPut (osMailQId queue_id, void *mail) | Put a mail to a queue. |
| **os_InRegs osEvent** | osMailGet (osMailQId queue_id, uint32_t millisec) | Get a mail from a queue. |
| **osStatus** | osMailFree (osMailQId queue_id, void *mail) | Free a memory block from a mail. |

```
/* Function declaration */
void receiver(void const *argument); // Thread
/* Thread IDs */
osThreadId t_receiver_id;

// ------------------------------------------------------------
// Receiver thread
void receiver(void const *argument) {
 while(1) {
   osEvent evt = osMailGet(dimension_q_id, osWaitForever);
   if (evt.status == osEventMail) { // mail received
   dimension_t *rx_data = (dimension_t *) evt.value.p;
   // ".p" indicate message as pointer
   printf ("Received data: (L) %d, (W), %d, (H) %d\n",
   rx_data->length,rx_data->width,rx_data->height);
   osMailFree(dimension_q_id, rx_data);
   }
 } // end while
}
 // define thread function
 osThreadDef(receiver, osPriorityNormal, 1, 0);
// ------------------------------------------------------------
int main(void)
{
 uint32_t i=0;
 dimension_t *tx_data;
 // Create Message queue
 dimension_q_id = osMailCreate(osMailQ(dimension_q), NULL);

 // Create "receiver" thread
 t_receiver_id = osThreadCreate(osThread(receiver), NULL);

 // main() itself is a thread that send out message
 while(1) {
   osDelay(1000);         // delay 1000 msec
   i++;
   tx_data=(dimension_t*)osMailAlloc(dimension_q_id,osWaitForever);
   tx_data->length = i; // fake data generation
   tx_data->width = i + 1;
   tx_data->height = i + 2;
   osMailPut(dimension_q_id, tx_data);
   }
 } // end main
// ------------------------------------------------------------
```

### 19.3.8 **Memory pool management feature**

CMSIS-RTOS has a feature called Memory Pool Management, which you can use to define a memory pool with a certain number of memory blocks and allocate these blocks during run-time.

The memory pool object is defined using "osPoolDef(*name, pool_size, type*)." When referencing a memory pool object using CMSIS-RTOS API, we need to use the "osPool(*name*)" define. Each memory pool also has an ID value that is needed by some of the memory pool functions, as shown in Table 19.14.

For example, we can repeat the data structure passing in the mail queue example using the message queue feature, and use the memory pool feature to manage the data block in the information transfer.

```
/* Example of message queue passing of data structures using memory
pool */

#include "stm32f4xx.h"
#include "stdio.h"
#include <cmsis_os.h>

typedef struct {
  uint32_t length;
  uint32_t width;
  uint32_t height;
} dimension_t;

/* Declare memory pool */
osPoolDef(mpool, 4, dimension_t);
osPoolId mpool_id;
```

**Table 19.14** Memory Pool Functions

|  | **Function** | **Description** |
|---|---|---|
| **osPoolQId** | osPoolCreate (const osPoolDef_t *pool_def) | Create and initialize a memory pool. |
| **void *** | osPoolAlloc (osPoolId pool_id) | Allocate a memory block from a memory pool. |
| **void *** | osPoolCAlloc (osPoolId pool_id) | Allocate a memory block from a memory pool and set memory block to zero. |
| **osStatus** | osPoolFree (osPoolId pool_id, void *block) | Return an allocated memory block back to a specific memory pool. |

```
  /* Declare message queue */
 osMessageQDef(dimension_q, 4, dimension_t); // Declare a message
queue
 osMessageQId dimension_q_id; // Declare a ID for message queue
 /* Note: Message queue has 4 entries, same as memory pool size */

 /* Function declaration */
 void receiver(void const *argument); // Thread
 osThreadId t_receiver_id;/* Thread IDs */

 // ----------------------------------------------------------
 // Receiver thread
 void receiver(void const *argument) {
  while(1) {
    osEvent evt = osMessageGet(dimension_q_id, osWaitForever);
    if (evt.status == osEventMessage) { // message received
    dimension_t *rx_data = (dimension_t *) evt.value.p;
      // ".p" indicate message as pointer
    printf ("Received data: (L) %d, (W), %d, (H) %d\n",
      rx_data->length,rx_data->width,rx_data->height);
    osPoolFree(mpool_id, rx_data);
    }
   } // end while
}
 // define thread function
 osThreadDef(receiver, osPriorityNormal, 1, 0);
 // ----------------------------------------------------------
 int main(void)
 {
 uint32_t i=0;
 dimension_t *tx_data;
 // Create Message queue
 dimension_q_id = osMessageCreate(osMessageQ(dimension_q), NULL);

 // Create Memory pool
 mpool_id = osPoolCreate(osPool(mpool));

 // Create "receiver" thread
 t_receiver_id = osThreadCreate(osThread(receiver), NULL);

 // main() itself is a thread that send out message
 while(1) {
   osDelay(1000);   // delay 1000 msec
   i++;
```

```
    tx_data = (dimension_t *) osPoolAlloc(mpool_id);
    tx_data->length = i; // fake data generation
    tx_data->width = i + 1;
    tx_data->height = i + 2;
    osMessagePut(dimension_q_id, (uint32_t)tx_data, osWaitForever);
    }
} // end main
// ------------------------------------------------------------
```

### 19.3.9 Generic wait function and time-out value

In all the previous examples we have used a generic function called osDelay (Table 19.15).

This is commonly used to put a thread in WAITING state. The input parameter is "millisec" (milli-second).

There is also an osWait function (Table 19.16). However, at the time of writing this function is not supported by the current version of CMSIS-RTOS RTX so cannot be demonstrated here.

In many CMSIS-API functions there is an input parameter called "millisec" to specify the waiting time; for example, osSemaphoreWait, osMessageGet, etc. In the normal value range it defines the time duration that will trigger a time-out, which causes the function to return. This parameter can be set to a constant definition called osWaitForever, which is defined as 0xFFFFFFFF in cmsis_os.h. When "millisec" is set to osWaitForever, the function will not time out.

When "millisec" is set to 0, the function returns immediately and does not wait. You can use the function return value to determine whether the required operation has succeeded or not.

It is undesirable and disallowed to enter WAITING state in any exception handler. As a result, when using CMSIS-RTOS APIs that have the millisec input parameter, the millisec parameter should be set to 0 so that they return immediately without stopping. Functions that are intended to create delay like osDelay should not be used in any interrupt handler.

### 19.3.10 Timer feature

In addition to the wait and delay functions, CMSIS-RTOS also supports Timer objects. A timer object can trigger the execution of a function. (Note: It is not a thread, although it is possible to send an event to a thread from that function.)

**Table 19.15** osDelay Function

|          | **Function**               | **Description**      |
|----------|----------------------------|----------------------|
| **osStatus** | osDelay (uint32_t millisec) | Wait for a time period |

**Table 19.16** osWait Function

|  | Function | Description |
|---|---|---|
| **os_InRegs osEvent** | osWait (uint32_t millisec) | Wait for Signal, Message, Mail, or Timeout. Return event that contains signal, message, mail information or error code. |

A Timer object can operate in periodic timer mode or one-shot mode. In periodic timer mode, the timer repeats its operation until it is deleted/terminated. In one-shot mode the timer triggers its function only once.

A Timer object is defined using "osTimerDef(name, type, *argument)." When referencing a timer object using CMSIS-RTOS API, we need to use "osTimer(name)" define. Each timer object also has an ID value that is needed by some of the timer functions, as shown in Table 19.17.

The following example shows simple use of a Timer object in both periodic mode and one-shot mode:

```
/* Example for timer objects. The 4 LEDs switch on 1 by 1 in sequence */

#include "stm32f4xx.h"
#include <cmsis_os.h>
/* Function declaration */
void toggle_led(void const *argument); // Toggle LED
void LedOutputCfg(void);               // LED output configuration

/* Declare Semaphore */
osTimerDef(LED_1, toggle_led); // Declare a Timer for LED control
osTimerDef(LED_2, toggle_led); // Declare a Timer for LED control
osTimerDef(LED_3, toggle_led); // Declare a Timer for LED control
osTimerDef(LED_4, toggle_led); // Declare a Timer for LED control
osTimerDef(LED_5, toggle_led); // Declare a Timer for LED control

/* Timer IDs */
osTimerId LED_1_id, LED_2_id, LED_3_id, LED_4_id,LED_5_id ;
// ----------------------------------------------------------
// For each round this function get executed 5 times,
// with argument = 1,2,3,4,5
void toggle_led(void const *argument)
{
  switch ((int)argument){
   case 1:
    GPIOD->BSRRL = (1<<12); // Set   bit 12
    osTimerStart(LED_2_id, 500);
```

```
   break;
  case 2:
   GPIOD->BSRRH = (1<<12); // Clear bit 12
   GPIOD->BSRRL = (1<<13); // Set   bit 13
   osTimerStart(LED_3_id, 500);
   break;
  case 3:
   GPIOD->BSRRH = (1<<13); // Clear bit 13
   GPIOD->BSRRL = (1<<14); // Set   bit 14
   osTimerStart(LED_4_id, 500);
   break;
  case 4:
   GPIOD->BSRRH = (1<<14); // Clear bit 14
   GPIOD->BSRRL = (1<<15); // Set   bit 15
   osTimerStart(LED_5_id, 500);
   break;
  default:
   GPIOD->BSRRH = (1<<15); // Clear bit 15
  }
}
// -------------------------------------------------------
int main(void)
{
 LedOutputCfg(); // Initialize LED output

 // Timers
 LED_1_id = osTimerCreate(osTimer(LED_1), osTimerPeriodic, (void *)1);
 LED_2_id = osTimerCreate(osTimer(LED_2), osTimerOnce,   (void *)2);
 LED_3_id = osTimerCreate(osTimer(LED_3), osTimerOnce,   (void *)3);
 LED_4_id = osTimerCreate(osTimer(LED_4), osTimerOnce,   (void *)4);
 LED_5_id = osTimerCreate(osTimer(LED_5), osTimerOnce,   (void *)5);

 osTimerStart(LED_1_id, 3000); // Start first timer

 // main() itself is another thread
 while(1) {
  osDelay(osWaitForever); // delay
  }
} // end main
```

If you are using CMSIS-RTOS RTX, when using timer objects, you should check that the configuration in RTX_Conf_CM.c has the OS_TIMERS parameter set to 1. You might also need to configure settings for the Timer thread.

**Table 19.17** Timer Functions

|  | Function | Description |
|---|---|---|
| **osTimerId** | osTimerCreate (const osTimerDef_t *timer_def, os_timer_type type, void *argument)) | Create and initialize a timer. |
| **osStatus** | osTimerStart (osTimerId timer_id, uint32_t millisec) | Start or restart a timer. |
| **osStatus** | osTimerStop (osTimerId timer_id) | Stop the timer. |
| **osStatus** | osTimerDelete (osTimerId timer_id) | Delete a timer that was created by osTimerCreate. |

## 19.3.11 Access privileged devices

Depending on the setting of CMSIS-RTOS RTX, "main()" can start in unprivileged state. In this case you cannot access any registers in the NVIC or the System Control Space (SCS), or some of the special registers in the processor core.

To enable "main()" and various threads to run in privileged state, you should set the OS_RUNPRIV parameter in RTX_Conf_CM.c to 1. However, there are many applications that require some threads to run in unprivileged state, for example, to enable the system to utilize memory protection features. In this case, it is very likely that you still want to execute some of the procedures in privileged state so that you can set up the NVIC or access other registers in SCS, or special registers in the processor.

In order to solve this problem, the CMSIS-RTOS RTX provides an extendable SVC mechanism. SVC #0 is used by the CMSIS-RTOS RTX, but other SVC services can be used by user-defined functions. The application code can use SVC calls to execute these user-defined functions inside the SVC handler, which executes in privileged state.

A SVC table code needs to be added to the project that carries out the SVC service look-up and defines the name of the user-defined SVC service.

```
SVC_table.s: Here we only added one user-defined SVC service, but you
can add more if needed. The name of the user-defined SVC service code
is called __SVC_1.
            AREA    SVC_TABLE, CODE, READONLY

            EXPORT SVC_Count

SVC_Cnt     EQU    (SVC_End-SVC_Table)/4
SVC_Count DCD     SVC_Cnt

; Import user SVC functions here.
            IMPORT __SVC_1
```

```
        EXPORT SVC_Table
SVC_Table
; Insert user SVC functions here. SVC 0 used by RTL Kernel.
        DCD __SVC_1                      ; user SVC function
SVC_End

    END
```

And inside the application code, we define user_defined_svc(void) as SVC #1, and implement __SVC_1, which is referenced in the SVC table.

```
/* Example of using SVC service to initialize a NVIC register */

#include "stm32f4xx.h"
#include <cmsis_os.h>

void __svc(0x01) user_defined_svc(void); // Define SVC #1 as
user_defined_svc

/* Thread IDs */
osThreadId t_blinky; // Declare a thread ID for blink
/* Function declaration */
void blinky(void const *argument); // Thread
void LedOutputCfg(void);           // LED output configuration

// ----------------------------------------------------------
// Blinky
// - toggle LED bit 12
// - Unprivileged Thread
void blinky(void const *argument) {
 while(1) {
   if (GPIOD->IDR & (1<<12)) {
     GPIOD->BSRRH = (1<<12); // Clear bit 12
   } else {
     GPIOD->BSRRL = (1<<12); // Set bit 12
   }
   osDelay(500);  // delay 500 msec
 }
 }

 // define blinky_1 as thread function
 osThreadDef(blinky, osPriorityNormal, 1, 0);
 // ----------------------------------------------------------
 // User defined SVC service (#1)
```

```
  // Note that the name must match the SVC service name defined in
 // SVC_Table.s
 void __SVC_1(void)
 {
  // add your NVIC/SCS initialization code here ...
  NVIC_EnableIRQ(EXTIO_IRQn);
  return;
 }
 // --------------------------------------------------------
 // - toggle LED bit 13
 // - Unprivileged Thread
 int main(void)
 {
  user_defined_svc(); // User defined SVC service (#1)
  LedOutputCfg();     // Initialize LED output

  // Create a task "blinky"
  t_blinky = osThreadCreate(osThread(blinky), NULL);

  // main() itself is another thread
  while(1) {
    if (GPIOD->IDR & (1<<13)) {
      GPIOD->BSRRH = (1<<13); // Clear bit 13
    } else {
      GPIOD->BSRRL = (1<<13); // Set bit 13
    }
    osDelay(1000);  // delay 1000 msec
  }
 } // end main
```

## 19.4 OS-aware debugging

In order to make debugging applications with an RTOS easier, the ITM stimulus port #31 (the last channel) is commonly reserved for OS events in debuggers. This allows the debugger to determine which task is being executed and which events have occurred.

For example, in Keil™ MDK-ARM the μVision debugger has a RTOS task and System view window and an Event Viewer window. They can be accessed using the pull-down menu in the debugger screen: "Debug → OS support → RTX Tasks and System, Debug → OS support → Event Viewer." To use these functions, the "RTX kernel" option needs to be set in the "Target" tab of the project settings (Figure 19.13).

You also need to have trace support in the debug adaptor (either Serial Wire Viewer or Trace Port interface).

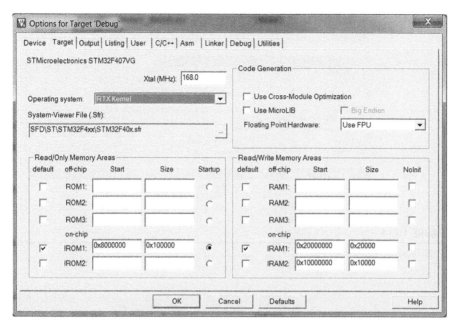

**FIGURE 19.13**

RTX Kernel option must be set to use OS-aware debug feature

During debugging, the RTX Tasks and System view provides a number of useful pieces of information about the current status of the OS kernel as well as some of the configuration details, as shown in Figure 19.14.

The Event viewer provides a time chart of which thread is currentlty executing, as shown in Figure 19.15.

## 19.5 Troubleshooting

Chapter 12 and Appendix I cover most of the common issues and troubleshooting techniques. Here are a few more areas that are more specific to embedded OS applications. If an application does not work properly, remember to check the items covered in the following sections.

### 19.5.1 Stack size and stack alignment

In a number of toolchains you can generate reports to see how much stack each thread requires. You should check this against the stack size setting of your project. This includes the stack size setting in startup code (e.g., Keil™ MDK) or linker configuration (e.g., IAR), the default stack size for main and thread, and stack size options in osThreadDef definitions.

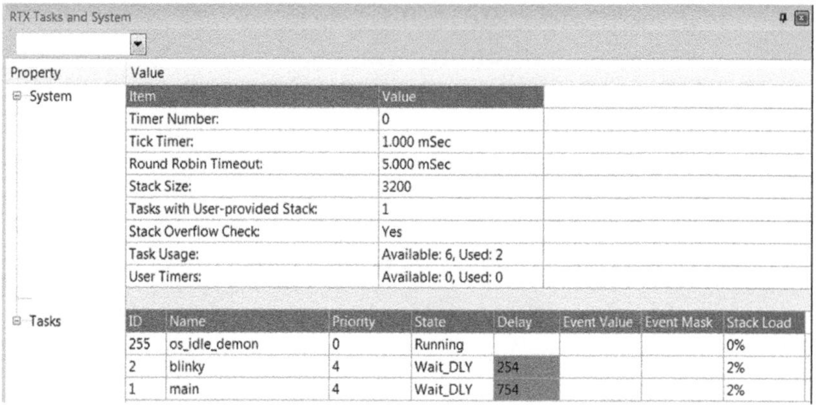

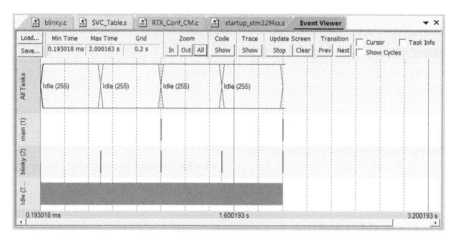

**FIGURE 19.14**

RTX Tasks and System view

**FIGURE 19.15**

Event viewer window

In addition, the stack size should be a multiple of 8. You might also need to check the linker report or memory map report to make sure that the stack areas are aligned to double-word boundaries.

## 19.5.2 Privileged level

If your embedded OS runs threads (or some of them) in unprivileged state, then these threads cannot access SCS areas such as NVIC registers. This can also affect access to the ITM because ITM stimulus ports can be configured to be privileged access only. Please refer to section 19.3.11 on how to extend SVC services in CMSIS-RTOS RTX.

### 19.5.3 Miscellaneous

When using CMSIS-RTOS features, remember to create the objects before using them (e.g., using the osXxxxCreate functions). The program code can compile without any issue when some of the create functions are accidentally omitted, but the results can be unpredictable.

When developing the examples ocasionally we have found that the RTX Tasks and System view and Event Viewer in μVision debugger stopped working. Simply unplugging the board and powering it again seems to fix the problem.

# Assembly and Mixed Language Projects

# 20

## 20.1 Use of assembly code in projects

For small projects, it is possible to develop the whole application in assembly language. However, this is rare in application development because:

- It is much harder as you need to have a good understanding of the instructions. It can be very complicated when the application requires complex data processing.
- Device drivers (e.g., header files) from microcontroller vendors are in C. If you need to access the peripherals in assembly, you need to create your own header files and driver libraries.
- Mistakes are not easy to spot and debug.
- Assembly program code is not portable. For example, different toolchains have different directives and syntax.

However, there are some situations where we may need to use assembly language for some parts of the project:

- To allow direct manipulation of stack memory (e.g., context-switching code in embedded OSs; see section 10.5)
- To optimize for maximum speed/performance or minimum program size for a specific task
- To reuse assembly code from old projects
- To learn about processor architecture

In most cases, even when a part of the application needs to be in assembly, other parts of the applications are still programmed in C or other high-level languages. This makes system initialization much easier and allows much better software portability.

There are various ways to add assembly code to a C project:

- The assembly code could be functions implemented in an assembly file and these functions can then be called from C code.
- The assembly code could be functions implemented within a C file using compiler-specific features (e.g., embedded assembler in ARM® toolchain, supported in Keil™ MDK-ARM, ARM DS-5™, and legacy toolchains).
- The assembly code can be instruction sequences inserted inside C code using inline assembler.
- Assembly instructions can be inserted inside C code using intrinsic, or CMSIS-Core functions.

In addition, some compilers also support idiom recognition features that recognize certain C constructs and translate them directly to sequences of one or more assembly instructions.

## 20.2 Interaction between C and assembly

Before we go into details of assembly language programming, let us first begin with some details on recommended practices that are needed to allow assembly code and C code to work together. For example, the way that input parameters and return results are passed between the calling function and the function being called (Table 20.1).

**Table 20.1** Simple Parameter Passing and Returning Value in a Function Call

| Register | Input Parameter | Return Value |
| --- | --- | --- |
| R0 | First input parameter | Function return value |
| R1 | Second input parameter | -, or return value (64-bit result) |
| R2 | Third input parameter | - |
| R3 | Fourth input parameter | - |

This is specified by an ARM® document called the Procedure Call Standard for the ARM Architecture (AAPCS, reference 8).

AAPCS covers the following areas:

- Register usage in function calls — A function or a subroutine should retain the values in R4–R11, R13, R14 (and S16–S31 for Cortex®-M4 with FPU). If these registers are changed during the function or the subroutine, the values should be saved on to the stack and be restored before return to the calling code.
- Parameters and return result passing — For simple cases, input parameters can be passed to a function using R0 (first parameter), R1 (second parameter), R2 (third parameter), and R3 (fourth parameter). In the case of Cortex-M4 with FPU, S0-S15 can also be used depending on the type of ABI selected (see section 13.4.4). Usually the return value of a function is stored in R0. If more than four parameters need to be passed to a function, the stack will be used (details can be found in AAPCS).
- Stack alignment — If an assembly function needs to call a C function, it should ensure that the currently selected stack pointer points to a double-word aligned address location (e.g., 0x20002000, 0x20002008, 0x20002010, etc.). This is a requirement of the EABI standard. Program code generated from an EABI-compliant C compiler can assume that the stack pointer is pointing to a double-word aligned location. If the assembly code does not call any C functions (either directly or indirectly), this is not strictly required.

When developing assembly functions to be called from C code, we need to make sure the contents of the "callee saved registers" are not changed. If these registers are used, their contents need to be pushed on to the stack and restored at the end of the function.

Similarly, assembly code calling a C function should ensure that the contents of the "caller saved registers" are saved before the function call because these registers could be changed by the function (Table 20.2).

We also need to be careful with the double-word stack alignment requirement. In ARM/Keil™ development tools, the assembler provides the REQUIRE8 directive to indicate if the function requires double-word stack alignment, and the PRESERVE8 directive to indicate that a function preserves the double-word alignment. This directive can help the assembler to analyze your code and generate warnings if a function that requires a double-word aligned stack frame is called by another function that does not guarantee double-word stack alignment. Depending on your application, these directives might not be required, especially for projects built entirely with assembly code.

Following the requirements outlined in AAPCS enables better software reusability and avoids potential problems when integrating your assembly code with program code or middleware from third parties.

**Table 20.2** Register Usages and Requirements in Function Calls

| Register | Function Call Behavior |
|---|---|
| R0-R3, R12, S0-S15 | Caller Saved Register – Contents in these registers can be changed by a function. Assembly code calling a function might need to save the values in these registers if they are required for operations in later stages. |
| R4-R11, S16-S31 | Callee Saved Register – Contents in these registers must be retained by a function. If a function needs to use these registers for processing, they need to be saved to the stack and restored before function return. |
| R14 (LR) | The value in the Link Register needs to be saved to the stack if the function contains a "BL" or "BLX" instruction (calling another function) because the value in LR will be overwritten when "BL" or "BLX" is executed. |
| R13 (SP), R15 (PC) | Should not be used for normal processing |

## 20.3 Structure of an assembly function

An assembly function can be very simple. For example, a function to add two input parameters can be as simple as:

```
My_Add ADDS R0, R0, R1 ; Add R0 and R1, result store in R0
       BX   LR          ; Return
```

To help improve clarity, we can add additional directives to indicate the start and end of a function. In ARM® toolchains (e.g., Keil™ MDK-ARM), the FUNCTION directive indicates the start of a function, and the ENDFUNC directive indicates the end of the function.

```
My_Add FUNCTION
       ADDS R0, R0, R1 ; Add R0 and R1, result store in R0
       BX   LR          ; Return
       ENDFUNC
```

A similar pair of directives is PROC and ENDP, which are synonyms for FUNCTION and ENDFUNC. Each FUNCTION directive must have a matching ENDFUNC directive and they must not be nested. FUNCTION and ENDFUNC, PROC, and ENDP are specific to ARM toolchains.

In the GNU toolchain, you should add .type to declare My_Add as a function:

```
       .type My_Add, %function
My_Add ADDS R0, R0, R1 ; Add R0 and R1, result store in R0
       BX   LR          ; Return
```

In most cases the program still works without the .type declaration, but when it is omitted, the LSB of the value when "My_Addr" label is referenced will be 0. For example, the following code will fail if My_Addr is not declared with type:

```
LDR R0,=My_Add /* This code will fail because LSB of R0 will be 0 */
BX  R0          /* which mean trying to switch to ARM state */
```

When used with ".long," for example, in the vector table, the LSB of the vector will also be zero, which can cause failure.

To put this function in an assembly code file (usually with the .s filename extension), you need to add additional directives to indicate the start of the program code and the type of memory where it is to be stored. For example, in the ARM toolchain you might write:

```
        PRESERVE8 ; Indicate the code here preserve
                  ; 8 byte stack alignment
        THUMB     ; Indicate THUMB code is used
        AREA      |.text|, CODE, READONLY ; Start of CODE area
My_Add  FUNCTION
        ADDS R0, R0, R1 ; Add R0 and R1, result store in R0
        BX   LR         ; Return
        ENDFUNC
        END             ; End of file
```

In the GNU toolchain:

```
        .text                /* text section */
        .syntax unified      /* Unified Assembly Syntax - UAL */
        .thumb               /* Thumb instruction set */
        .type My_Add, %function
        .global My_Add       /* Make My_Add visible from outside */
My_Add
        ADDS R0, R0, R1      /* Add R0 and R1, result store in R0 */
        BX   LR              /* Return */
        .end                 /* End of file */
```

In a more complex assembly function, more steps are required. In general, the structure of a function can be divided into the following stages:

- Prolog (saving register contents to the stack memory if necessary)
- Allocate stack space memory for local variables (decrement SP)
- Copy some of R0 to R3 (input parameters) to high registers (R8−R12) for later use (optional)
- Carry out processing/calculation
- Store result in R0 if a result is to be returned
- Stack adjustment to free space for local variables (increment SP)
- Epilog (restore register values from stack)
- Return

Most of these steps are optional; for example, prolog and epilog are not required if the function does not corrupt the contents in R4 to R11. The stack adjustments are also not required if there are sufficient registers for the processing. The following assembly function template illustrates some of these steps:

```
My_Func FUNCTION
        PUSH {R4-R6, LR}    ; 4 registers are pushed to stack
                            ; double word stack alignment is
                            ; preserved
        SUB  SP, SP,  #8    ; Reserve 8 bytes for local variables
        ; Now local variables can be accessed with SP related
        ; addressing mode
        ...                 ; Carry out processing
        MOVS R0, R5         ; Store result in R0 for return value
        ADD  SP, SP,  #8    ; Restore SP to free stack space
        POP  (R4-R6, PC}    ; epilog and return
        ENDFUNC
```

If the function calls another assembly or C function, the values in registers R0 to R3 and R12 could be changed by the called function. So unless you are certain that the function being called will not change these registers, you need to save the contents of these registers if they will be used later. Alternatively you might need to avoid using these registers for the data processing in your function.

## 20.4 Examples

### 20.4.1 Simple example with ARM® toolchains (Keil™ MDK-ARM, DS-5)

In this part we will create a very simple program to add the values $10+9+\cdots+1$. Here we reuse the assembly startup code (which is in assembly), and the SystemInit function (in C). Since we do not need the standard C startup code we can remove the stack and heap initialization code from the startup assembly code.

Inside the file simple_example.s, we have the following code:

```
; simple_example.s
  PRESERVE8 ; Indicate the code here preserve
            ; 8 byte stack alignment
  THUMB     ; Indicate THUMB code is used
  AREA     |.text|, CODE, READONLY   ; Start of CODE area
  EXPORT __main
  ENTRY
__main FUNCTION
     ; initialize registers
     MOV r0, #10  ; Starting loop counter value
     MOV r1, #0   ; starting result
     ; Calculated 10+9+8+...+1
```

```
loop
     ADD   r1, r0    ; R1 = R1 + R0
     SUBS  r0, #1    ; Decrement R0, update flag ("S" suffix)
     BNE   loop      ; If result not zero jump to loop
     ; Result is now in R1
deadloop
     B deadloop  ; Infinite loop
     ENDFUNC
     END             ; End of file
```

In the linker option, we use a modified linker script because there are no C libraries in the project.

```
; simple_example.sct
; *************************************************************
; *** Scatter-Loading Description File generated by uVision ***
; *************************************************************

LR_IROM1 0x08000000 0x00100000 {  ; load region size_region
ER_IROM1 0x08000000 0x00100000 { ; load address = execution address
   *.o (RESET, +First)
   ; *(InRoot$$Sections) ; This line is commented out because there
                            is no C lib
   .ANY (+RO)
 }
 RW_IRAM1 0x20000000 0x00020000 { ; RW data
   .ANY (+RW +ZI)
 }
}
```

After the project is built, we can test it in the debugger as normal.

The same project can be built using ARM® Development Studio 5 Professional (DS-5™). Please note that the command line options for DS-5 are slightly different from Keil™ MDK. In DS-5, this program can be assembled using:

```
$> armasm --cpu cortex-m4 -o simple_example.o simple_example.s
$> armasm --cpu cortex-m4 -o startup_stm32f4xx.o
   startup_stm32f4xx.s
```

The -o option specifies the output filename. For example, "simple_example.o" is an object file. The system initialization file (which is in C) is more complex. Since system_stm32f4xx.c needs CMSIS-Core header files and the STM32F4 header files, you need to specify the include file paths:

```
$> armcc -c -g -W --cpu cortex-m4 -I CMSIS/Include -I CMSIS/ST/
   STM32F4xx/Include -o system_stm32f4xx.o system_stm32f4xx.c
```

We then need to use the linker to create an executable image (ELF). This can be done by:

```
$> armlink --rw_base 0x20000000 --ro_base 0x08000000
simple_example.o startup_stm32f4xx.o system_stm32f4xx.o
"--keep=startup_stm32f4xx.o(RESET)" "--first=startup_stm32f4xx.
o(RESET)" --entry Reset_Handler --map -o simple_example.elf
```

Here, "–ro_base 0x08000000" specifies that the read-only region (program ROM) starts at address 0x08000000 (this is specific to the STM32F4 used in this example); "–rw_base 0x20000000" specifies that the read/write region (data memory) starts at address 0x20000000.

"--keep" ensures that the linker will not remove the vector table (RESET section in the startup code).

"--first" ensures that the linker places the vector table at the beginning of the image.

The "--map" option creates an image map, which is useful for understanding the memory layout of the compiled image.

Finally, we need to create the binary image:

```
$> fromelf --bin --output simple_example.bin simple_example.elf
```

To check that the image looks like what we wanted, we can also generate a disassembly listing file by:

```
$> fromelf -c --output simple_example.list simple_example.elf
```

If everything works fine, you can then load your ELF image or binary image into your hardware or instruction set simulator for testing.

## 20.4.2 Simple example with GNU tools for ARM-embedded processors

In gcc, the program code needs to be modified due to the following differences: labels in GNU assemblers are followed by a colon ( : ); comments are quoted with /* and */; directives are prefixed by a period ( . ).

```
/* simple_example.s for GCC */

    .text           /* text section */
    .syntax unified /* Unified Assembly Syntax - UAL */
    .thumb          /* Thumb instruction set */
    .type __main, %function
    .global __START /* Make __START visible from outside */
__START:
    /* initialize registers */
    MOV r0, #10  /* Starting loop counter value */
    MOV r1, #0   /* starting result */
    /* Calculated 10+9+8+...+1 */
```

```
loop:
    ADD  r1, r0   /* R1 = R1 + R0 */
    SUBS r0, #1   /* Decrement R0, update flag ("S" suffix) */
    BNE loop      /* If result not zero jump to loop */
    /* Result is now in R1 */
deadloop:
    B deadloop  /* Infinite loop */
    .end        /* End of file */
```

Then we modify the startup code slightly to remove pre-processing defines and compile and link the project together with the following batch file:

```
rem Batch file to compile and link the simple assembly example for gcc

set OPTIONS_ARCH=-mthumb -mcpu=cortex-m4
set OPTIONS_OPTS=-Os
set OPTIONS_COMP=-g -Wall
set OPTIONS_LINK=--gc-sections -Map=map.rpt
set SEARCH_PATH_1=CMSIS\Include
set SEARCH_PATH_2=CMSIS\ST\STM32F4xx\Include
set LINKER_SCRIPT=gcc.ld
set LINKER_SEARCH="C:\Program Files (x86)\GNU Tools ARM Embedded\4.7
2012q4\share\gcc-arm-none-eabi\samples\ldscripts"

rem Newlib-nano feature is available for v4.7 and after
rem set OPTIONS_LINK=%OPTIONS_LINK% --specs=nano.specs

arm-none-eabi-as %OPTIONS_ARCH% startup_stm32f4xx.s -o
startup_stm32f4xx.o
if %ERRORLEVEL% NEQ 0 goto end

arm-none-eabi-as %OPTIONS_ARCH% simple_example.s -o simple_example.o
if %ERRORLEVEL% NEQ 0 goto end

rem Compile the SystemInit
arm-none-eabi-gcc                                      ^
   %OPTIONS_COMP% %OPTIONS_ARCH%                       ^
   %OPTIONS_OPTS% -c                                   ^
   -I %SEARCH_PATH_1% -I %SEARCH_PATH_2% ^
   system_stm32f4xx.c                                  ^
   -o system_stm32f4xx.o
if %ERRORLEVEL% NEQ 0 goto end

rem Link
arm-none-eabi-ld -T %LINKER_SCRIPT% -L %LINKER_SEARCH% ^
```

```
      %OPTIONS_LINK% -o simple_example.elf
      system_stm32f4xx.o simple_example.o startup_stm32f4xx.o
 if %ERRORLEVEL% NEQ 0 goto end

 rem Generate disassembled listing for debug/checking
 arm-none-eabi-objdump -S simple_example.elf > list.txt
 if %ERRORLEVEL% NEQ 0 goto end

 rem Generate binary image file
 arm-none-eabi-objcopy -O binary simple_example.elf
simple_example.bin
 if %ERRORLEVEL% NEQ 0 goto end

 rem Generate Hex file (Intel Hex format)
 arm-none-eabi-objcopy -O ihex simple_example.elf simple_example.hex
 if %ERRORLEVEL% NEQ 0 goto end

 rem Generate Hex file (Verilog Hex format)
 arm-none-eabi-objcopy -O verilog simple_example.elf
simple_example.vhx
 if %ERRORLEVEL% NEQ 0 goto end

 :end
```

### 20.4.3 Accessing special registers

In the GNU toolchain, when accessing the special registers in assembly code, the name of the special registers must be in lowercase. For example:

```
 msr     control, r1
 mrs     r1, control
 msr     apsr, R1
 mrs     r0, psr
```

In the ARM® toolchain the special register names can either be lowercase or uppercase.

In C programming, you can access special registers using the CMSIS-Core API (see Appendix E, section E.4).

### 20.4.4 Data memory

For most applications, we will need a fair amount of SRAM for data storage. In most functions, we can use the stack for local variables. In the previous example in this chapter, we used the startup code to define the stack. For example, in the ARM® toolchain:

```
Stack_Size      EQU     0x00000400
                AREA    STACK, NOINIT, READWRITE, ALIGN=3
Stack_Mem       SPACE   Stack_Size
__initial_sp  ; Use for initial value of MSP in the vector table
```

And in gcc:

```
    .section .stack
    .align 3
#ifdef __STACK_SIZE
    .equ Stack_Size, __STACK_SIZE
#else
    .equ Stack_Size, 0xc00
#endif
    .globl __StackTop
    .globl __StackLimit
__StackLimit:
    .space Stack_Size
    .size __StackLimit, . - __StackLimit
__StackTop:
    .size __StackTop, . - __StackTop
```

Inside functions, we can modify the value of SP to reserve space for local variables. For example, in a function we might have three data variables "MyData1" (a word-size data variable), "MyData2" (a half word size data variable), and "MyData3" (a byte-size data variable). The total size required is 7 bytes, and since the stack pointer must be word aligned, we allocate two words.

In the following code, in order to make the byte and half-word transfers more efficient, we copy the adjusted SP value into a general register (R4) after the stack adjustment. Then we can access the local variable space in stack using R4-relative addressing:

```
MyFunction
    PUSH {R4, R5}
    SUB  SP, SP , #8 ; Allocate two words for space for local variables
    MOV  R4, SP ; Make a copy of SP to R4
    LDR  R5,=0x00001234
    STR  R5,[R4,#0] ; MyData1 = 0x00001234
    LDR  R5,=0x55CC
    STRH R5,[R4,#4] ; MyData2 = 0x55CC
    MOVS R5,#0xAA
    STRB R5,[R4,#6] ; MyData3 = 0xAA
    ...
    ADD  SP, SP, #8 ; Restore SP back to starting value to free space
    POP  {R4, R5}
    BX   LR
```

In addition, we might also need to allocate data space for global or static variables. In ARM toolchains, we can add a section called Data:

```
; -------------------------------------------------
; Allocate data variable space
          AREA  | Header Data|, DATA ; Start of Data definitions
          ALIGN 4
MyData4   DCD   0 ; Word size data
MyData5   DCW   0 ; half Word size data
MyData6   DCB   0 ; byte size data
; -------------------------------------------------
```

In gcc, this can be done by using .lcomm:

```
/* Data in LC, Local Common section */
.lcomm MyData4 4 /* A 4 byte data called MyData4 */
.lcomm MyData5 2 /* A 2 byte data called MyData5 */
.lcomm MyData6 1 /* A 1 byte data called MyData6 */
```

The *.lcomm* pseudo-op is used to create an uninitialized block of storage inside the "bss" region. The program code can then access this space using the defined labels *MyData4*, *MyData5* and *MyData6*.

### 20.4.5 Hello world

How can we talk about programming without trying out "Hello World"? In assembly language programming, it does take a bit more work, but getting a Hello World message output might not be as difficult as it seems. In this example we use the Instrumentation Trace Macrocell (ITM) to output the "Hello World" message.

In this program, we implement a few functions:

- Putc — Output a character in R0 to ITM (effective same as ITM_SendChar)
- Puts — Output a string pointed by R0

We also define the "Hello World" message as a constant null-terminated string, which is stored in the program memory.

```
; Hello World program
 PRESERVE8 ; Indicate the code here preserve
           ; 8 byte stack alignment
 THUMB     ; Indicate THUMB code is used
 AREA |.text|, CODE, READONLY ; Start of CODE area
 EXPORT __main
 ENTRY
__main  FUNCTION
```

```
      MOVS R0, #'\n'
      BL   Putc
MainLoop
      LDR R0,=HELLO_TXT ; Get address of the string
      BL Puts           ; Display string
      BL Delay          ; Delay
      B  MainLoop
      ENDFUNC


;----------·--------------------------------------------
Puts FUNCTION
      ; Subroutine to send string to display
      ; Input R0 = starting address of string.
      ; The string should be null terminated
      PUSH {R4, LR}  ; Save registers
      MOVS R4, R0
      ; Copy address to R1, because R0 will
      ; be used
PutsLoop      ; as input for Putc
      LDRB R0,[R4],#1
      ; Read one character and increment address (post index)
      CBZ R0, PutsLoopExit ; if character is null, goto end
      BL Putc   ; Output character to UART
      B  PutsLoop  ; Next character
PutsLoopExit
      POP {R4, PC} ; Return
      ENDFUNC
;--------------------------------------------------
; Send a character via ITM
ITM_BASE   EQU 0xE0000000
ITM_PORT0 EQU (ITM_BASE+0x000)
ITM_TER    EQU (ITM_BASE+0xE00)
ITM_TCR    EQU (ITM_BASE+0xE80)

Putc FUNCTION
      ; Function to display one character
      ; Input R0 - chacter to be displayed
      LDR   R1,=ITM_TCR ; 0xE0000E80
      LDR   R2,[R1]
      MOVS R3, #1 ; Check ITMENA bit. If 0, exit
      TST   R2, R3
      BEQ   PutcExit
      LDR   R1,=ITM_TER ; 0xE0000E00
```

```
        LDR   R2,[R1] ; Check Port 0 is enabled. If 0, exit
        TST   R2, R3
        BEQ   PutcExit
        LDR   R1,=ITM_PORT0 ; 0xE0000000
PutcWait
        LDR   R2,[R1] ; Read status
        CMP   R2, #0
        BEQ PutcWait
        STRB R0,[R1] ; Write a character
PutcExit
        BX LR
        ENDFUNC
;--------------------------------------
Delay FUNCTION
        LDR R0, =0x01000000
DelayLoop
        SUBS R0, R0, #1
        BNE DelayLoop
        BX LR
        ENDFUNC
;--------------------------------------
; Text to be displayed
HELLO_TXT
        DCB "Hello world\n", 0 ; Null terminated string
        ALIGN 4
        END   ; End of file
```

### 20.4.6 Displaying values in hexadecimal and decimal

By adding addition function we can also display values in decimal and hexadecimal form. Let us first look at hexadecimal, which is somehow simpler. The function first displays "0x," and then rotates the input value so that each nibble is placed in the lowest four bits, masked and converted to ASCII for display.

```
; Function to display Hexadecimal value
;----------------------------------------------------
PutHex FUNCTION
   ; Output register value in hexadecimal format
   ; Input R0 = value to be displayed
   PUSH  {R4 - R6, LR}
   MOV   R4, R0   ; Save register value to R4 because R0 is used
                  ; for passing input parameter
   MOV   R0,#'0'  ; Starting the display with "0x"
   BL    Putc
   MOV   R0,#'x'
```

```
  BL    Putc
  MOV   R5, #8           ; Set loop counter
  MOV   R6, #28          ; Rotate offset
PutHexLoop
  ROR   R4, R6           ; Rotate data value left by 4 bits
                         ; (right 28)
  AND   R0, R4,#0xF      ; Extract the lowest 4 bit
  CMP   R0, #0xA         ; Convert to ASCII
  ITE   GE
  ADDGE R0, #55          ; If larger or equal 10, then convert
                         ; to A-F
  ADDLT R0, #48          ; otherwise convert to 0-9
  BL    Putc             ; Output 1 hex character
  SUBS  R5, #1           ; decrement loop counter
  BNE   PutHexLoop       ; if all 8 hexadecimal character been
                         ; display then
  POP {R4-R6,PC}         ; return, otherwise process next 4-bit
  ENDFUNC
  ;-------------------------------------------------
```

The function PutDec is a bit more complex. The calculation starts from the smallest digit on the right first and each time the result is divided by 10 and the remainder is converted to ASCII code and placed in a text buffer.

```
; Function to display Decimal value
;-------------------------------------------------
PutDec FUNCTION
  ; Subroutine to display register value in decimal
  ; Input R0 = value to be displayed.
  ; Since it is 32 bit, the maximum number of character
  ; in decimal format, including null termination is 11
  ; According AAPCS, a function can change R0-R3, R12
  ; So we use these registers for processing.
  ; R0 - Input value
  ; R1 - Divide value (10)
  ; R2 - Divide result
  ; R3 - Remainder
  ; R12- Text buffer pointer
  PUSH {R4, LR}          ; Save register values
  MOV  R12, SP           ; Copy current Stack Pointer to R4
  SUB  SP, SP, #12       ; Reserved 12 bytes as text buffer
  MOVS R1, #0            ; Null character
  STRB R1,[R12, #-1]!    ; Put a null character at end of text buffer
                         ; buffer,pre-indexed
  MOVS R1, #10           ; Set divide value
```

```
PutDecLoop
  UDIV R2, R0, R1     ; R2 = R0 / 10 = divide result
  MUL  R4, R2, R1     ; R4 = R2 * 10 = Value - remainder (multiple of 10)
  SUB  R3, R0, R4     ; R2 = R0 - (R2 * 10) = remainder
  ADDS R3, #48        ; convert to ASCII (R2 can only be 0-9)
  STRB R3,[R12, #-1]!; Put ascii character in text buffer
                      ; pre-indexed
  MOVS R0, R2         ; Set R0 = Divide result and set Z flag if R4=0
  BNE  PutDecLoop     ; If R0(R4) is already 0, then there
                      ; is no more digit
  MOV  R0, R12        ; Put R0 to starting location of text
                      ; buffer
  BL   Puts           ; Display the result using Puts
  ADD  SP, SP, #12    ; Restore stack location
  POP  {R4, PC}       ; Return
  ENDFUNC
;-------------------------------------------------
```

When combining these functions, we can output the message display easily, even in assembly language.

## 20.4.7 NVIC interrupt control

It is also useful to have a few functions for interrupt control. For example, to enable an interrupt, we can use the following function EnableIRQ:

```
; A subroutine to enable an IRQ based on IRQ number
EnableIRQ FUNCTION
  ; Input R0 = IRQ number
  PUSH {R0-R2, LR}
  AND  R1, R0, #0x1F   ; Generate enable bit pattern for
                       ; the IRQ
  MOV  R2, #1
  LSL  R2, R2, R1      ; Bit pattern = (0x1 << (N & 0x1F))
  AND  R1, R0, #0xE0   ; Generate address offset if IRQ number
                       ; is above 31
  LSR  R1, R1, #3      ; Address offset = (N/32)*4 (Each word
                       ; has 32 IRQ enable)
  LDR  R0,=0xE000E100  ; SETEN register for external interrupt
                       ; #31-#0
  STR  R2, [R0, R1]    ; Write bit pattern to SETEN register
  POP  {R0-R2, PC}     ; Restore registers and Return
  ENDFUNC
```

Similarly, we can create a function DisableIRQ to disable an interrupt:

```
; A subroutine to disable an IRQ based on IRQ number
DisableIRQ FUNCTION
                    ; Input R0 = IRQ number
  PUSH {R0-R2, LR}
  AND   R1, R0, #0x1F  ; Generate enable bit pattern for
                       ; the IRQ
  MOV   R2, #1
  LSL   R2, R2, R1     ; Bit pattern = (0x1 << (N & 0x1F))
  AND   R1, R0, #0xE0  ; Generate address offset if IRQ number
                       ; is above 31
  LSR   R1, R1, #3     ; Address offset = (N/32)*4 (Each word
                       ; has 32 IRQ enable)
  LDR   R0,=0xE000E180 ; CLREN register for external interrupt
                       ; #31-#0
  STR   R2, [R0, R1]   ; Write bit pattern to SETEN register
  POP   {R0-R2, PC}    ; Restore registers and Return
ENDFUNC
```

To configure the priority of an interrupt, we can take advantage of the fact that the interrupt priority registers are byte addressable, making the coding much easier. For example, to set IRQ #4 priority level to 0xC0, we can use the following code:

```
; Setting IRQ #4 priority to 0xC0
LDR R0, =0xE000E400 ; External Interrupt Priority Register
                    ; starting address
MOVS R1, #0xC0      ; Priority level
STRB R1, [R0, #4]   ; Set IRQ #4 priority (Byte write)
```

In the Cortex®-M3 and Cortex-M4 processors, the width of the interrupt priority level fields is specified by chip manufacturers. The minimum width is 3 bits and the maximum is 8 bits. In CMSIS-Core, the width of a priority level field is specified by a parameter called __NVIC_PRIO_BITS in a device-specific header file. In some cases you might need to determine the implemented width at run-time. You can do this by writing 0xFF to one of the priority level registers and reading it back. In assembly programming you can do this with the following code:

```
; Determine the implemented priority width
LDR  R2,=0xE000E400 ; Priority Configuration register for
                    ; external interrupt #0
LDR  R1,=0xFF
STRB R1,[R2]        ; Write 0xFF (note : byte size write)
LDRB R1,[R2]        ; Read back (e.g. 0xE0 for 3-bits)
RBIT R1, R1         ; Bit reverse R1 (e.g. 0x07000000 for
                    ; 3-bits)
```

```
CLZ   R1, R1          ; Count leading zeros (e.g. 0x5 for 3-bits)
MOV   R0, #8
SUB   R0, R0, R1      ; Get implemented width of priority
                      ; (e.g. 8-5=3 for 3-bits)
MOVS  R1, #0x0
STRB  R1,[R2]         ; Restore priorty to reset value (0x0)
```

### 20.4.8 Unsigned integer square root

One mathematical calculation that is occasionally needed in embedded systems is square root. Since square root can only deal with positive numbers (unless complex numbers are used), the following example only handles unsigned integers. For the following implementation (Figure 20.1), the result is rounded to the next lower integer.

The corresponding program code is as follows:

```
simple_sqrt FUNCTION
 ; Input : R0
 ; Output : R0 (square root result)
 MOVW  R1, #0x8000 ; R1 = 0x00008000
 MOVS  R2, #0     ; Initialize result
```

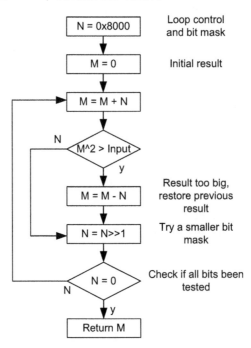

**FIGURE 20.1**

Simple square root implementation

```
simple_sqrt_loop
 ADDS  R2, R2, R1  ; M = (M + N)
 MULS  R3, R2, R2  ; R3 = M2
 CMP   R3, R0      ; If M2 > Input
 IT    HI          ; Greater Than
 SUBHI R2, R2, R1  ; M = (M - N)
 LSRS  R1, R1, #1  ; N = N >> 1
 BNE   simple_sqrt_loop
 MOV   R0, R2      ; Copy to R0 and return
 BX    LR          ; Return
 ENDFUNC
```

## 20.5 Mixed language projects

In some aspects, the majority of the projects we have covered in this book are mixed language projects. Most of the toolchains use assembly language for startup codes to give them higher flexibility in low level control such as stack manipulation.

### 20.5.1 Calling a C function from assembly

Inside assembly code, you can call an external C function. For example, the following C function has four input parameters and returns a 32-bit result:

```
int my_add_c(int x1, int x2, int x3, int x4)
{
 return (x1 + x2 + x3 + x4);
}
```

With the definition in AAPCS, x1=R0, x2=R1, x3=R2, and x4=R3. The result return is stored in R0. We also need to be aware that Caller Saved Registers (e.g., R0 to R3, and R12) can be changed by a C function. So if there is any data in these registers that are needed later, we need to save them first.

In the ARM® toolchain, you can call this function using the following code:

```
 MOVS   R0, #0x1 ; First parameter  (x1)
 MOVS   R1, #0x2 ; Second parameter (x2)
 MOVS   R2, #0x3 ; Third parameter  (x3)
 MOVS   R3, #0x4 ; Fourth parameter (x4)
 IMPORT my_add_c
 BL     my_add_c ; Call "my_add_c" function. Result store in R0
```

If this assembly code is written inside a C program file using embedded assembler or inline assembler in an ARM toolchain (Keil™ MDK-ARM, ARM DS-5™, or legacy ARM toolchain like RealView® Development Suite), you should use the __cpp keyword instead of IMPORT.

```
MOVS R0, #0x1 ; First parameter  (x1)
MOVS R1, #0x2 ; Second parameter (x2)
MOVS R2, #0x3 ; Third parameter  (x3)
MOVS R3, #0x4 ; Fourth parameter (x4)
BL  __cpp(my_add_c) ; Call "my_add_c" function. Result store in R0
```

The __cpp keyword is recommended for Keil MDK in accessing C or C++ compile-time constant expressions. For other toolchains the directive required can be different.

In GNU toolchain, you can use ".global" to enable a label in a different file to be visible.

## 20.5.2 Calling an assembly function from C

Let us reverse the process and implement the My_Add function in assembly, and call it from C code. Inside the assembly code, we need to make sure that if we need to modify callee saved registers (e.g., R4 to R11), we push these registers on to the stack first and then restore them before exiting the function.

We also need to save LR if this function is going to call another function because the value of LR will be changed when executing BL or BLX.

The My_Add function can be implemented as:

```
    EXPORT My_Add
My_Add FUNCTION
    ADDS R0, R0, R1
    ADDS R0, R0, R2
    ADDS R0, R0, R3
    BX  LR ; Return result in R0
    ENDFUNC
```

Inside the C program code, we need to declare My_Add function using extern:

```
extern int My_Add(int x1, int x2, int x3, int x4);
...
int y;
...
y = My_Add(1, 2, 3, 4); // call the My_Add function
```

If your assembly code needs to access some data variables in your C code, you can also use the "IMPORT" keyword (for ARM® toolchain) or ".global" (for GNU toolchain).

### 20.5.3 Embedded assembler (Keil™ MDK-ARM/ARM® DS-5™ professional)

In ARM® toolchains (including Keil™ MDK-ARM, DS-5™ Professional, etc.), a feature called embedded assembler allows you to implement assembly functions/subroutines inside a C file. To do this, you need to add the __asm keyword in front of the function declaration. For example, a function to add four integers can be written as:

```
__asm int My_Add(int x1, int x2, int x3, int x4)
{
   ADDS R0, R0, R1
   ADDS R0, R0, R2
   ADDS R0, R0, R3
   BX   LR ; Return result in R0
}
```

Inside embedded assembly code, you can import address values or data symbols using the __cpp keyword. For example:

```
__asm void function_A(void)
{
 PUSH {R0-R2, LR}
 BL __cpp(LCD_clr_screen)        ; Call a C function - method 1
 LDR  R0,=__cpp(&pos_x)          ; Get address of a C variable
 LDR  R0, [R0]
 LDR  R1,=__cpp(&pos_y)          ; Get address of a C variable
 LDR  R1, [R1]
 LDR  R2, =__cpp(LCD_pixel_set)  ; Import the address of a function
 BLX  R2                         ; Call the C function
 POP  {R0-R2, PC}
}
```

### 20.5.4 Inline assembler

The inline assembler feature is also available in ARM® C compilers. However, older versions of the ARM C compiler do not support inline assembler in Thumb state. Starting from ARM C Compiler 5.01, and Keil™ MDK-ARM 4.60, the inline assembler now supports Thumb-state code, with some limitations:

- It can be used only when targeting v6T2, v6-M, and v7/v7-M cores.
- The TBB, TBH, CBZ, and CBNZ instructions are not supported.
- As with previous versions, some system instructions such as SETEND are not permitted.

For example, you can use inline assembler in C code:

```
int qadd8(int i, int j)
{
 int res;
 __asm
 {
  QADD8 res, i, j
 }
 return res;
}
```

The GNU C compiler also supports inline assembler. The general syntax is as follows:

```
__asm (" inst1 op1, op2, ... \n"
       " inst2 op1, op2, ... \n"
       ...
       " instN op1, op2, ... \n"
       : output_operands  /* optional */
       : input_operands   /* optional */
       : clobbered_operands /* optional */
      );
```

In simple cases where the assembly instruction does not require parameters, it can be as simple as:

```
void Sleep(void)
{ // Enter sleep using WFI instruction
 __asm (" WFI\n");
 return;
}
```

If the assembly code requires input and output parameters, then you might need to define the input and output operands and the clobbered register lists if any other registers are modified by the inline assembly operation. For example, inline assembly code to multiply a value by 10 can be written as:

```
unsigned int DataIn, DataOut;
...
__asm(" movs r0, %0\n"
      " movs r3, #10\n"
      " muls r0, r0, r3\n"
      " movs %1, r0\n"
      :"=r (DataOut) : "r" (DataIn) : "cc", "r0", "r3");
```

In the code example, %0 is the first input parameter and %1 is the first output parameter. Since the operand order is: output_operands, input_operands, and clobbered_operands, "DataOut" is assigned to %0, and "DataIn" is assigned to %1. Since the code changes register R3, it needs to be added to the clobbered operand list.

More details of inline assembly in the GNU C compiler can be found online in the GNU toolchain documentation GCC-Inline-Assembly-HOWTO.

# 20.6 Intrinsic functions

In some cases, we need to use some instructions that cannot be generated using normal C code. Apart from using inline assembler, embedded assembler, or assembly language programming, one extra solution is to generate these special instructions with intrinsic functions.

There are two types of intrinsic functions:

*   CMSIS-Core intrinsic functions — See Appendix E, section E.5
*   Compiler-specific intrinsic functions — see Table 20.3 for some examples.

In general, CMSIS-Core intrinsic functions are preferred because they are portable. In some toolchains, the CMSIS-Core intrinsic is directly supported by

**Table 20.3** Examples of Intrinsic Functions provided in ARM C Compiler

| Assembly Instructions | ARM Compiler Intrinsic Functions |
| --- | --- |
| CLZ | unsigned char __clz(unsigned int val) |
| CLREX | void __clrex(void) |
| CPSID I | void __disable_irq(void) |
| CPSIE I | void __enable_irq(void) |
| CPSID F | void __disable_fiq(void) |
| CPSIE F | void __enable_fiq(void) |
| LDREX/LDREXB/LDREXH | unsigned int __ldrex(volatile void *ptr) |
| LDRT/LDRBT/LDRSBT/ LDRHT/ LDRSHT | unsigned int __ldrt(const volatile void *ptr) |
| NOP | void __nop(void) |
| RBIT | unsigned int __rbit(unsigned int val) |
| REV | unsigned int __rev(unsigned int val) |
| ROR | unsigned int __ror(unsigned int val, unsigned int shift) |
| SSAT | int __ssat(int val, unsigned int sat) |
| SEV | void __sev(void) |
| STREX / STREXB / STREXH | int __strex(unsigned int val, volatile void *ptr) |
| STRT/ STRBT / STRHT | void int __strt(unsigned int val, const volatile void *ptr) |
| USAT | int __usat(unsigned int val, unsigned int sat) |
| WFE | void __wfe(void) |
| WFI | void __wfi(void) |
| BKPT | void __breakpoint(int val) |

**Table 20.4** Example of Similarities between CMSIS-Core Intrinsic Functions and Compiler-specific Intrinsic Functions

| Instructions | CMSIS Intrinsic Function | ARM/Keil C Compiler Built-in Intrinsic Functions |
| --- | --- | --- |
| WFI (Wait For Interrupt) | __WFI(void) | __wfi(void) |
| WFE (Wait For Event) | __WFE(void) | __wfe(void) |
| SEV (Send Event) | __SEV(void) | __sev(void) |

the compiler. In other cases the Compiler intrinsic implementation is separate from CMSIS-Core.

Table 20.3 shows some of the intrinsic functions available in the ARM® C compiler (including Keil™ MDK-ARM and ARM DS-5™ Professional).

Please note that for some compilers, the name of some compiler-specific intrinsic functions can be very similar to CMSIS intrinsic functions. For example, some functions are uppercase in the CMSIS version and lowercase in the ARM/Keil C compiler-specific version, as seen in Table 20.4.

Cases where you need to use compiler-specific intrinsic functions instead of CMSIS intrinsic include the following:

- When you want to insert software breakpoints in the program code for debugging and you are using CMSIS-Core r3p1 or older versions. (Note: __BKPT(value) is available from r3p2). So if you need to insert a breakpoint in your application and you are using a C compiler from ARM (e.g., Keil MDK-ARM or ARM DS-5 Professional), you should use:
  - void __breakpoint(int val) // BKPT instruction for ARM/Keil tool chain
  - With other compilers the intrinsic can be different. Alternatively you can insert breakpoints using inline assembler.
- When you want to use unprivileged memory access instructions (LDRT, LDRHT, LDRBT, STRT, STRHT, STRBT). Again, there is no CMSIS-Core intrinsic for these instructions.

## 20.7 Idiom recognition

Some C compilers also provide a feature called idiom recognition. When the C code is constructed in a particular way, the C compiler automatically converts the operation into a special instruction or sequence of instructions. For example, the ARM® C compiler supports a number of idiom recognition patterns (Table 20.5).

**Table 20.5** Idiom Recognition in Keil MDK or ARM C Compiler for Cortex®-M

| Instruction | C Language Code that can be Recognized by Keil™ MDK or ARM® C Compiler |
|---|---|
| BFC | x.b=0; |
| BFI | x.b = n; |
| MLA | x += y*z; |
| MLS | c=c-a*b; |
| PKHBT | (a & 0xFFFF0000) ¦ (b&0x0000FFFF); |
| PKHTB | (a & 0xFFFF0000) ¦ ((b >> 1) & 0x0000FFFF); |
| REV16 | (((x&0xff)<<8)\|((x&0xff00)>>8)\|((x&0xff000000)>>8)\|((x&0x00ff0000)<<8)); |
| REVSH | ((i<<24)>>16)\|((i>>8)&0xFF); |
| SBFX | Use bit field feature in C/C++ (see section 23.7) |
| SMLABB | x16*y16 + z32; |
| SMLABT | x*(y>>16) + z32; |
| SMLATB | (x>>16)*y + z32; |
| SMLATT | (x>>16)*(y>>16) + z32; |
| SMLAWB | (((long long)x * y) >> 16)+ a); |
| SMLAWT | (((long long)x * (s16)(y>>16)) >> 16)+ a); |
| SMLAD | z + ((short)(x>>16)*(short)(y>>16)) + ((short)x*(short)y); |
| SMLADX | z + ((short)(x>>16)*(short)(y)) + ((short)x*(short)(y>>16)); |
| SMLAL | a += (s64)x * y; /* a is a s64 value */ |
| SMLALBB | a + x*(y>>16) ; /*a is long long, x and y are s16 */ |
| SMLALBT | a + x*(y>>16) ; /* a is long long, x is s16, y is s32 */ |
| SMLALTB | a + (x>>16)*y; /* a is long long, x is s32, y is s16 */ |
| SMLALTT | a + (x>>16)*(y>>16) ; /* a is long long, x and y are s32 */ |
| SMLAWB | (int)(((long long)x*y) >> 16) + z; /* x and z are int, y is short */ |
| SMMLA | (int)((((long long)i * j) + ((long long)a<<32)) >>32); |
| SMMLAR | (((((long long)a)<<32) + ((long long)i * j) + 0x80000000LL) >> 32; |
| SMMLS | (int)(((long long)a<<32 - ((long long)i * j))>>32); |
| SMMLSR | ((((long long)a)<<32) - (((long long)i * j) ) + 0x80000000LL) >> 32; |
| SMMUL | ((int)(((long long)i * j)>>32)); |
| SMMULR | (int)((((long long)i * j + 0x80000000LL) >> 32); |
| SMUAD | ((x>>16)*(y>>16)) + (((x<<16)>>16)*((y<<16)>>16)); /* x and y are s32 */ |
| SMULBB | x16 * y16; |
| SMULBT | (x16>>16) * y16; |
| SMULTB | x16 * (y16>>16); |
| SMULTT | (x16>>16) * (y16>>16); |
| SMULWB | (((long long)x * y) >> 16); |
| SMULWT | (((long long)x *(y>>16))>>16); |

*(Continued)*

**Table 20.5** Idiom Recognition in Keil MDK or ARM C Compiler for Cortex®-M—*Cont'd*

| Instruction | C Language Code that can be Recognized by Keil™ MDK or ARM® C Compiler |
|---|---|
| SSAT | (x < -8) ? -8: (x > 7 ? 7: x); |
| SXTAB | ((a<<24)>>24) + i; |
| SXTAH | ((a<<16)>>16) + i; |
| UBFX | Use bit field feature in C/C++ (see section 23.7) |
| UMAAL | (ull)u32 + u32 + ((ull)u32 *u32); |
| UMLAL | u64 a; a += ((u64)x * y); |
| USAT | a > 7 ? 7: (a < 0 ? 0: a); |
| UXTAB | ((unsigned)(a<<24)>>24) + i; |
| UXTAH | ((a<<16)>>16) + i; |

If the software is ported to a different C compiler without the same idiom recognition feature, the code will still compile because it uses standard C syntax, although the generated instruction sequence might be less efficient than using idiom recognitions.

# ARM® Cortex®-M4 and DSP Applications

The key new feature in the Cortex®-M4 processor compared to previous Cortex-M cores is the addition of DSP extensions. These instructions accelerate numerical algorithms and open the door to performing real-time signal processing operations directly on the Cortex-M4, without the need for an external digital signal processor. This chapter describes the DSP features of the Cortex-M4 and explains how to develop efficient code using these new extensions.

**The Definitive Guide to ARM® Cortex®-M3 and Cortex-M4 Processors.** http://dx.doi.org/10.1016/B978-0-12-408082-9.00021-X

The chapter starts with some motivation behind the DSP extensions, and provides a simple introduction to the feature using the dot product as an example. We then look more closely at the features and capabilities of a modern DSP — the Analog Devices SHARC processor. We continue with a detailed look at the Cortex-M4 instruction set, and provide tips and tricks for optimizing DSP code on this processor. The following chapter presents the CMSIS DSP library, an off-the-shelf optimized DSP library for the Cortex-M4 provided by ARM®.

## 21.1 DSP on a microcontroller?

Digital Signal Processing (DSP) includes a wide range of mathematically intensive algorithms. The umbrella term includes applications in audio, video, measurement, and industrial control, just to name a few. The first choice that comes to mind when looking for a processor to execute Digital Signal Processing applications on is the Digital Signal Processor, which shares the same acronym, and is also called a DSP. The architecture of a DSP is tuned to perform the mathematical operations found in these algorithms. However, in some sense, they are crippled savants that excel at certain focused operations while struggling with day-to-day tasks.

Microcontrollers, on the other hand, are general purpose and excel at control tasks: interfacing to peripherals, handling user interfaces, and general connectivity. As a result, microcontrollers have a wide range of peripherals which make it easy for them to interface with other ICs via common interfaces such as RS-232, USB, and Ethernet. Microcontrollers also have a long history of being embedded in portable products and a greater focus on minimizing power consumption. Microcontrollers, however, are not necessarily great at performing intense mathematical algorithms because they lack registers and a proper set of instructions that support these computations.

The recent boom in connected devices creates a need for products with both microcontroller and DSP features. This is evident in devices that handle multimedia content. They require both peripheral connectivity and some amount of DSP processing capability. Traditionally, these devices had two separate processors — a microcontroller and a DSP. However, with DSP extensions, many multimedia devices can be built around a single Cortex®-M4 processor. Therefore, the Cortex-M4 processor with DSP extensions solves the limitations of both traditional DSPs and microcontrollers. This yields lower power consumption, ease of integration, and most importantly, lower overall system cost.

## 21.2 Dot product example

This section discusses the salient features of DSPs with an eye toward how they improve overall performance. By way of example, we will look at the dot product operation that multiplies two vectors and accumulates the products, element by element:

$$z = \sum_{k=0}^{N-1} x[k]y[k]$$

Assume that the inputs $x[k]$ and $y[k]$ are arrays of 32-bit values and that we want to use a 64-bit representation for $z$. A C code implementation of the dot product is shown in Figure 21.1.

The dot product consists of a series of a multiplications and additions. This "multiply accumulate" or MAC operation is at the heart of many DSP functions.

Now consider the execution time of this algorithm on the Cortex®-M3. Fetching data from memory and incrementing the pointer takes 2 cycles:

```
xx = *x++; // 2 cycles
```

Similarly, the next fetch also takes 2 cycles:

```
yy = *y++; // 2 cycles
```

The multiplication and addition is of the form 32x32 + 64 and the Cortex-M4 can do this with a single instruction. However, on the Cortex-M3 , the same operation takes 3 to 7 cycles to execute depending the characteristics of the input values.

```
sum += xx * yy; // 3 to 7 cycles
```

The loop itself introduces an additional overhead. This usually involves decrementing the loop counter and then branching to the beginning of the loop. The standard loop overhead is 3 cycles. Thus, on the Cortex-M3, the inner loop of the dot product takes 10 to 18 cycles, with the execution time being data dependent.

```
int64 dot_product (int32 *x, int32 *y, int32 N) {
int32 xx, yy, int32 k;
int64 sum = 0;
      for(k = 0; k < N; k++) {
            xx = *x++;
            yy = *y++;
            sum += xx * yy;
      }
return sum;
}
```

**FIGURE 21.1**

Dot product implemented in C code

Let us take a look at the same code on the Cortex-M4. The code is identical to that shown in Figure 21.1 but with different cycle timing. On the Cortex-M4 we have:

```
xx = *x++;          // 2 cycles
yy = *y++;          // 1 cycle
sum += xx * yy;     // 1 cycle
(loop overhead)     // 3 cycles
```

Fetch operations that are preceded by a similar instruction take only one cycle. Therefore, in total we have 7 cycles for the code inside the loop. This is not only lower than the Cortex-M3 but also, as an added benefit, the run time is deterministic.

The loop overhead can be reduced by utilizing loop unrolling. For example, if you knew that the length of the vectors was a multiple of 4 samples, then you could unroll the loop by a factor of 4. Computing 4 samples would require $4 \times 4 + 3 = 19$ cycles. In other words, the dot product operation would require only 4.75 cycles per sample on the Cortex-M4, whereas on the Cortex-M3 it would take from 7.75 to 11.75 cycles per sample.

## 21.3 Architecture of a traditional DSP processor

Now let us consider implementing the dot product on a digital signal processor (DSP). We will use the SHARC processor from Analog Devices when illustrating the architecture and features of a modern DSP. The SHARC was selected because it is full-featured and has an easy to understand instruction set. On the SHARC, the dot product would be implemented in assembly as:

```
/* Zero out the accumulator */
MRF = 0;

/* Preload data values */
R4=DM(i2,1), R2=PM(i8,1);

/* Main loop */
LCNTR=R0, DO (PC, loop_end) UNTIL LCE;
loop_end:
    MRF = MRF + R2*R4 (SSF), R4=DM(i2,1), R2=PM(i8,1);

/* Final wrap up */
MRF = MRF + R2*R4 (SSF);
```

The inner loop contains a single instruction and completes in a single clock cycle:

```
MRF = MRF + R2*R4 (SSF), R4=DM(i2,1), R2=PM(i8,1);
```

This instruction performs 4 operations in parallel: 2 data fetches, multiplication, and addition. Note that the SHARC is able to fetch 2 data words simultaneously. This is possible because it has multiple blocks of internal memory and multiple memory buses allowing two memory accesses to occur in parallel. Furthermore, the SHARC memory fetches occur *in parallel* with computation and do not require separate instructions just for fetching data.

By combining memory accesses with computation, the SHARC can perform the fundamental dot product operation in a single cycle. The loop overhead is also mitigated using a feature called *zero overhead loops*. The loop instruction is:

```
LCNTR=R0, DO (PC, loop_end) UNTIL LCE;
```

This specifies that the inner loop should be performed R0 times, where R0 is a register. Once the loop is set up it executes R0 times without any additional overhead. A subtle point is that the data values that are fetched, R4 and R2, are not available until the *next* instruction. Thus the multiply accumulate MRF = MRF + R2*R4 uses the data values fetched from the *previous* instruction. The instruction prior to the loop:

```
R4=DM(i2,1), R2=PM(i8,1);
```

preloads the data that is used in the next instruction. The instruction after the loop:

```
MRF = MRF + R2*R4 (SSF);
```

completes the outstanding multiply accumulate. To compute a dot product of N elements the inner loop would be executed N−1 times. Using all of the features of the SHARC, an N point dot product can be executed in roughly N+2 instructions, a significant saving over the Cortex-M.

In our dot product example, we used a 64-bit accumulator to hold the result. Ideally, since $x[n]$ and $y[n]$ are 32-bit integers, the product will be 64-bits. Since we are adding N products, the bit width requirements expand further by $\log_2 N$ bits. Thus an even larger register — more than 64-bits — will be needed to hold the result without overflow. We glossed over this earlier in the dot product example, but if the processor cannot easily support variables greater than 64-bits (such as the Cortex-M), the headroom of intermediate computations will need to be carefully managed in order to prevent overflows.

To avoid overflow, the SHARC example uses a special accumulator register MRF, which provides guard bits. MRF is an 80-bit register and provides 16 guard bits when accumulating 64-bit results. The accumulators prevent overflow while maintaining the full precision of intermediate results.

In many cases, scaling intermediate signals to prevent overflow 100% of the time is impractical. Instead, some overflows should be expected and the processing designed to minimize their effects. To handle this, DSPs provide *saturating* arithmetic. Instead of wrapping around at overflow, which is standard behavior for twos complement arithmetic, DSPs can be configured to saturate at the maximum positive or minimum negative values. Consider the waveforms shown in Figure 21.2. In this example, the wave forms are represented as 16-bit integers and are limited to the range [-32768 to 32767]. The top plot shows the

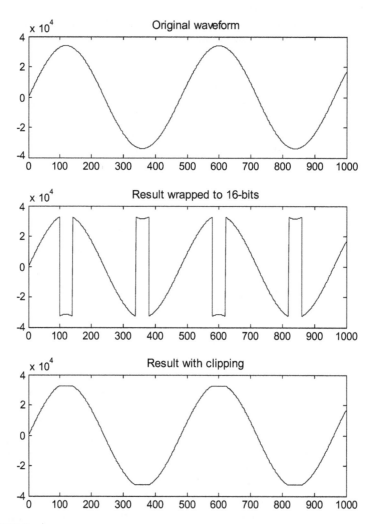

**FIGURE 21.2**

The effects of processing with and without saturation. The top plot shows the ideal result of processing, but it exceeds the allowable 16-bit range. The middle plot is the result wrapped to 16 bits and has severe distortion. The bottom plot is the result saturated to the allowable range and has mild distortion.

ideal result, which exceeds the allowable range. The middle plot shows how the result wraps when using standard twos complement addition. The bottom plot shows the result with saturation. The signals are slightly clipped but still recognizable as sine waves.

The Cortex®-M4 includes some basic saturating arithmetic. The processor provides saturating addition and subtraction instructions but no saturating MAC instructions. To perform a saturating MAC instruction, you have to separately multiply the two values and then perform a saturating addition. This takes one more cycle.

The SHARC also contains sophisticated address generators. Not only is moving data between registers and memory important, but updating the address pointers must also be done efficiently. In Figure 21.1, the address pointers are incremented after every memory fetch. Most modern processors can do the address increment in parallel with the memory fetch. DSPs take this a step further and allow any positive or negative increment value to be specified. Addresses are computed by dedicated address generators and each address generator has multiple registers associated with it.

Address generators also simplify addressing in circular buffers. Circular buffers are a type of FIFO (first in − first out buffer), in which the data in the buffer is never shifted (too wasteful). Instead, the input and output pointers move linearly through the buffer and when they reach the end, they wrap around. This wrapping effect is called circular addressing and the address generators on DSPs allow this to be done automatically without penalty. We will see circular addressing used later on in the FIR filter example. The address generators also have a special addressing mode (called bit-reversed addressing), which is particularly useful for FFT algorithms.

DSPs strive to have deterministic computation which is critical for real-time processing. In a real-time application, the key performance metric is not the average processing load but the peak processing load. For example, if you are processing audio at a sample rate of 44.1 kHz, you only have 1/44.1 kHz = 22.67 microseconds to complete the processing and be ready for the next sample. Variations in execution time are troublesome and require you to scale back computation in order to meet the real-time constraints.

Some DSPs also have the ability to perform the same operation on multiple pieces of data. This is referred to as SIMD, single instruction multiple data. SIMD can double or quadruple the overall numerical performance of a processor. For SIMD to work well, the processor needs a corresponding increase in memory throughout to match to the increase in numerical capabilities. The SHARC can load or store four 32-bit words to memory in a single cycle while simultaneously fetching an instruction on another bus. Very rarely is the SHARC slowed down by memory accesses.[1] 8- and 16-bit SIMD operations are part of the Cortex-M4 DSP

---

[1] This is true only if the data is stored in internal memory on the SHARC. There is a serious slowdown if data values have to be fetched from external memory. The SHARC does not have a cache to mitigate this.

extensions. Floating-point SIMD is only found on the higher performance ARM Cortex-A cores.

The SHARC also provides dedicated blocks of hardware for performing common sets of signal processing operations. These *hardware accelerators* effectively double or triple the overall throughput of the core by offloading tasks from the core processor to separate blocks of hardware. The SHARC has hardware accelerators for FIR filters, IIR filters, FFTs, and sample rate conversion.

The SHARC has a sophisticated DMA controller. This subsystem moves data around between memory blocks (internal and external) or between memory and peripherals, such as serial ports. The DMA controller allows the core to focus on computation rather than mundane tasks such as data movement.

As shown above, DSPs are optimized for numerically intensive tasks. Not only are the core and instruction set designed for parallelism, but they also have a large number of other features, which offload the main core. The Cortex-M4 adds some basic DSP features, such as a MAC instruction, but it is not a fully-fledged DSP. We will look more closely at the M4 instruction set next and provide guidelines and tricks for writing efficient code.

On the other hand, traditional DSPs are mostly optimized for maximum performance and can be significantly power hungry when compared to the Cortex-M4 based microcontrollers. For example, in order to support multiple operations in a single cycle, many traditional DSPs are based on VLIW (Very Large Instruction Word), which need to access 256-bit wide instruction memory (or wider) at the processor's clock speed. It might also require dual port memories or multiple blocks of fast on-chip SRAM in order to support the required data bandwidth, whereas the 16-bit/32-bit instruction size in Cortex-M processors allows much higher flexibility in program memory storage arrangement, and allow the processing tasks to be carried out with much lower power consumption.

In addition, many microcontroller vendors have included high performance DMA controllers and multi-layer AHBs (see example in Figure 6.5) in their microcontrollers to enhance the data throughput and bandwidth.

The interrupt overhead of the Cortex-M4 processor is also relatively low compared to a traditional DSP. For example, the ADI ADSP21160M DSP requires "Interrupt Dispatchers" code execution before and after the ISR and that can add hundreds of clock cycles. Even when an alternate processor register set (banked registers) is available, the overhead can still be over 30 clock cycles, as shown in Table 21.1.

## 21.4 Cortex®-M4 DSP instructions

This section provides an overview of the Cortex®-M4 DSP instructions. The goal is not to be exhaustive but to highlight which instructions are most often used in

**Table 21.1** ISR Overhead on ADI ADSP21160M[2]

|  | Interrupt Dispatcher Overhead before ISR | Interrupt Dispatcher Overhead after ISR |
|---|---|---|
| Normal | 183 | 109 |
| Fast | 40 | 26 |
| Super-fast (with alternate register set) | 34 | 10 |
| Final – use with user written assembly or when ISR included context saving | 24 | 15 |

practice. We want to show you the "forest" without spending too much time on individual "trees." The primary comprehensive references for the Cortex-M4 instruction set are the *Cortex-M4 Devices Generic User Guide* [Reference 3] and the *Cortex-M4 Technical Reference Manual* [Reference 5] — especially section 3.3.

We also advocate a C programming model for the Cortex-M4. Programming in C has many advantages for portability, ease of development, and maintainability. In many cases, the C compiler is able to choose the correct instructions without additional help. In some cases, we use *idioms*, short snippets of C code, which map directly to underlying Cortex-M4 instructions. The idioms have the advantage that they are fully portable. Finally, in many cases — particularly with SIMD — the compiler does not efficiently map C code to instructions and we have to provide the compiler with explicit guidance using *intrinsics*. This leads to non-ÁNSI C code and destroys portability between processors. Still, using intrinsics is preferred over assembly language programming.

### 21.4.1 Registers and data types

The Cortex®-M processor has a core register set containing 16 32-bit registers. The bottom 13 registers, R0–R12, are general purpose and can hold intermediate variables, pointers, function arguments, and return results. The upper three registers are reserved for the C compiler.

```
R0 to R12 — General purpose
R13 — Stack pointer    [reserved]
R14 — Link register    [reserved]
R15 — Program counter [reserved]
```

When optimizing code, you need to keep in mind that the register set can only hold 13 intermediate values. If you exceed this, the compiler will need to store

---

[2]Reference: Digital signal processor fundamentals and system design (www.coe.pku.edu.cn/tpic/2010913102418831.pdf)

intermediate values on the stack causing a slowdown in performance. Consider the dot product example from Figure 21.1. The registers required are:

```
x  — pointer
y  — pointer
xx — 32-bit integer
yy — 32-bit integer
z  — 64-bit integer [2 registers required]
k  — loop counter
```

In total, this function needs seven registers. The dot product is quite basic and already half of the registers are used.

The optional floating-point unit (FPU) on the Cortex-M4 has its own set of registers. This extension register file contains 32 single precision registers (32-bits each) labeled S0 to S31. One obvious advantage of floating-point code is that you have access to these additional 32 registers. R0 to R12 are still available to hold integer variables (pointers, counters, etc.) yielding a total of 45 registers that can be used by the C compiler. We will see later on that floating-point operations are typically slower than integer ones, but the extra registers can lead to more efficient code, especially for more complicated functions such as FFTs. Savvy C compilers make use of the floating-point registers when performing integer arithmetic. They store intermediate results to the floating-point register file, rather to the stack, since register-to-register transfers take 1 cycle as opposed to register-to-memory transfers, which may take 2 cycles.

To use the Cortex-M4 DSP instructions, start by including the main CMSIS file core_cm4.h. This defines a variety of integer, floating-point, and fractional data types as shown in the table below:

*Signed Integers*

```
int8_t  8-bit
int16_t 16-bit
int32_t 32-bit
int64_t 64-bit
```

*Unsigned Integers*

```
uint8_t  8-bit
uint16_t 16-bit
uint32_t 32-bit
uint64_t 64-bit
```

*Floating-Point*

```
float32_t single precision 32-bit
float64_t double precision 64-bit
```

*Fractional*

```
q7_t   8-bit
q15_t  16-bit
q31_t  32-bit
q63_t  64-bit
```

### 21.4.2 **Fractional arithmetic**

Fractional data types are commonly used in signal processing and are unfamiliar to many software programmers. We will spend a bit of time introducing them and their benefits in this section.

An N-bit signed integer using standard twos complement representation represents values in the range $[-2^{(N-1)}, 2^{(N-1)} - 1]$. A fractional N-bit integer has an implied division by $2^{(N-1)}$ and represents values in the range $[-1, 1 - 2^{-(N-1)}]$. If $I$ is the integer value then $F = I/2^{(N-1)}$ is the corresponding fractional value. 8-bit fractional values are in the range

$$\left[\frac{-2^7}{2^7}, \frac{2^7 - 1}{2^7}\right] \text{ or } \left[-1, 1 - 2^{-7}\right].$$

In binary, some common signed values in two's complement are represented as:

| | | | | |
|---|---|---|---|---|
| Maximum | = | 01111111 | = | $1 - 2^{-7}$ |
| Smallest positive | = | 00000001 | = | $2^{-7}$ |
| Zero | = | 00000000 | = | 0 |
| Smallest negative | = | 11111111 | = | $-2^{-7}$ |
| Minimum | = | 10000000 | = | $-1$ |

Similarly, 16-bit fractional values are in the range

$$\left[-1, 1 - 2^{-15}\right]$$

Note that the ranges of values are almost identical for the 8- and 16-bit fractional representations. This simplifies scaling in mathematical algorithms because you do not have to keep in mind a particular integer range but simply $-1$ to about $+1$.

You may be wondering why 8-bit fractional values are named "q7_t" rather than "q8_t". The reason is that there is an implied sign bit in the representation and there are actually only 7 fractional bits. The individual bit values are:

$$\left[S, \frac{1}{2}, \frac{1}{4}, \frac{1}{8}, \frac{1}{16}, \frac{1}{32}, \frac{1}{64}, \frac{1}{128}\right]$$

where S represents the sign bit. We can generalize fractional integers to include both integer and fractional bits. Qm.n refers to a fractional integer with 1 sign bit, m-1 integer bits, and n fractional bits. For example, Q1.7 is the familiar 8-bit data type (q7_t) mentioned above with 1 sign bit, no integer bits, and 7 fractional bits. Q9.7 is a 16 bit fractional integer with 1 sign bit, 8 integer bits, and 7 fractional bits. The bit values are:

$$\left[S, 128, 64, 32, 16, 8, 4, 2, 1, \frac{1}{2}, \frac{1}{4}, \frac{1}{8}, \frac{1}{16}, \frac{1}{32}, \frac{1}{64}, \frac{1}{128}\right]$$

The integer bits can be used as guard bits in signal processing algorithms.

Fractional values are represented as integers and stored in integer variables or registers. Addition of fractional values is identical to integer addition. Multiplication, however, is fundamentally different. Multiplying two N-bit integers yields a 2N-bit result. If you need to truncate the result back to N bits you typically take the low N bits.

Since fractional values are in the range $[-1 +1)$, multiplying two values will yield a result in the same range.[3] Let us examine what happens if you multiply two N-bit fractional values:

$$\frac{I_1}{2^{(N-1)}} \times \frac{I_2}{2^{(N-1)}}$$

You end up with a 2N-bit result

$$\frac{I_1 I_2}{2^{(2N-2)}}$$

where $I_1 I_2$ is standard integer multiplication. Since the result is 2N bits long and there is a factor of $2^{2N-2}$ in the denominator, this represents a Q2.(2N−2) number. To turn this into a Q1.(2N−1) fractional number it would be left shifted by 1 bit. Alternatively, this can be turned into a Q1.(N−1) bit number by either:

**1.** Left shifting by 1 bit and taking the high N bits, or
**2.** Right shifting by N−1 bits and taking the low N bits.

These operations describe a true fractional multiplication in which the result is back in the range $[-1 +1)$.

The Cortex®-M4 uses approach (1) for fractional multiplication but omits the left shift by 1 bit. Conceptually, the result is scaled down by 1 bit and lies in the range $[-1/2 +1/2)$. You have to keep this in mind while developing algorithms and at some point incorporate the missing bit shift.

### 21.4.3 SIMD data

The Cortex®-M4 also provides SIMD instructions that operate on packed 8- or 16-bit integers. A 32-bit register can hold either $1 \times$ 32-bit value, $2 \times$ 16-bit values, or $4 \times$ 8-bit values as illustrated below:

| 32-bit | | | |
|---|---|---|---|
| 16-bit | | 16-bit | |
| 8-bit | 8-bit | 8-bit | 8-bit |

---

[3]Well, almost. The only place where this falls apart is if you multiply $(-1) \times (-1)$ and obtain $+1$. This value is technically one LSB out of the allowable range and needs to be truncated to the largest allowable positive value.

Instructions which operate on 8- or 16-bit data types are useful for processing data, such as video or audio, which do not require full 32-bit precision.

To use the SIMD instructions from C code, you load values into int32_t variables and then invoke the corresponding SIMD intrinsic instructions.

### 21.4.4 Load and store instructions

Loading and storing 32-bit data values can be accomplished with standard C constructs. On the Cortex®-M3, each load or store instruction takes 2 cycles to execute. On the Cortex-M4 the first load or store instruction takes 2 cycles while subsequent loads or stores take 1 cycle. *Whenever possible, group loads and stores together to take advantage of this 1 cycle saving.*

To load or store packed SIMD data, define int32_t variables to hold the data. Then perform the loads and stores using the __SIMD32 macros supplied in the CMSIS library. For example, to load four 8-bit values in one instruction, use:

```
q7_t *pSrc *pDst;
int32_t x;
x = *__SIMD32(pSrc)++;
```

This also increments pSrc by a full 32-bit word (so that it points to the next group of four 8-bit values). To store the data back to memory, use:

```
*_SIMD32(pDst)++ = x;
```

The macro also applies to packed 16-bit data:

```
q15_t *pSrc *pDst;
int32_t x;
x = *__SIMD32(pSrc)++;
*__SIMD32(pDst)++ = x;
```

### 21.4.5 Arithmetic instructions

This section describes the Cortex®-M arithmetic operations, which are most frequently used in DSP algorithms. The goal is not to cover each and every Cortex-M arithmetic instruction but rather to highlight the ones that are most frequently used in practice. In fact, we cover only a fraction of the available commands. Our focus is on:

| | |
|---|---|
| Signed data | [Ignore unsigned] |
| Floating-point | |
| Fractional integers | [Ignore standard integer math] |
| Sufficient precision | [32-bit or 64-bit accumulators] |

We also ignore fractional operations with rounding, add/subtract variants, carry bits, and variants based on top or bottom placement of 16-bit words within a 32-bit register. We ignore these instructions because they are less frequently used and once the content of this section is understood, the other variants can be easily applied.

The instructions vary in how they are invoked by the C compiler. In some cases, the compiler determines the correct instruction to use based on standard C code. For example, fractional addition is performed using:

```
z = x + y;
```

In other cases, you have to use an *idiom*. This is a predefined snippet of C code that the compiler recognizes and maps to the appropriate single instruction. For example, to swap bytes 0 and 1, and 2 and 3 of a 32-bit word, use the idiom

```
((((x&0xff)<<8)¦((x&0xff00)>>8)¦((x&0xff000000)>>8)¦((x&0x00ff0000)
<<8));
```

The compiler recognizes this and maps it to a single REV16 instruction.

And finally, in some cases there is no C construct that maps to the underlying instruction and the instruction can only be invoked via an *intrinsic*. For example, to do a saturating 32-bit addition use:

```
z = __QADD(x, y);
```

In general, it is better to use idioms because they are standard C constructs and provide portability across processors and compilers. The idioms described in this chapter apply to Keil™ MDK, and the idioms map to precisely 1 Cortex-M instruction. Some compilers do not provide full support for idioms and may yield several instructions. Check your compiler documentation to see which idioms are supported and how they map to Cortex-M instructions.

### 32-bit integer instructions
ADD — 32-bit addition
Standard 32-bit addition without saturation, overflow may occur. Supports both int32_t and q31_t data types.

Processor support: M3 and M4. [1 cycle]
C code example:

```
q31_t x, y, z;
z = x + y;
```

SUB — 32-bit subtraction
Standard 32-bit subtraction without saturation, overflow may occur. Supports both int32_t and q31_t data types.

Processor support: M3 and M4. [1 cycle]

C code example:

```
q31_t x, y, z;
z = x - y;
```

## SMULL — long signed multiply

Multiplies two 32-bit integers and returns a 64-bit result. This is useful for computing products of fractional data while maintaining high precision.

Processor support: M3 [3 to 7 cycles] and M4 [1 cycle].

C code example:

```
int32_t x, y;
int64_t z;
z = (int64_t) x * y;
```

## SMLAL — long signed multiply accumulate

Multiplies two 32-bit integers and adds the 64-bit result to a 64-bit accumulator. This is useful for computing MACs of fractional data and maintaining high precision.

Processor support: M3 [3 to 7 cycles] and M4 [1 cycle].

C code example:

```
int32_t x, y;
int64_t acc;
acc += (int64_t) x * y;
```

## SSAT — signed saturation

Saturates a signed x integer to a specified bit position B. The result is saturated to the range:

$$-2^{B-1} \leq x \leq 2^{B-1} - 1$$

where $B = 1, 2, ..., 32$. The instruction is available in C code only via the intrinsic:

```
int32_t __SSAT(int32_t x, uint32_t B)
```

Processor support: M3 and M4 [1 cycle]

C code example:

```
int32_t x, y;
y = __SSAT(x, 16); // Saturate to 16-bit precision.
```

## SMMUL - 32-bit multiply returning 32-most-significant-bits

Fractional q31_t multiplication (if result is left shifted by 1 bit). Multiplies two 32-bit integers, generates a 64-bit result, and then returns the high 32 bits of the result.

Processor support: M4 only [1 cycle]

The instruction is available in C code via the following idiom:

```
(int32_t) (((int64_t) x * y) >> 32)
```

C code example:

```
// Performs a true fractional multiplication but loses the LSB of the
result
int32_t x, y, z;

z = (int32_t) (((int64_t) x * y) >> 32);
z <<= 1;
```

A related instruction is SMULLR, which rounds the 64-bit result of multiplication rather than simply truncating. The rounded instruction provides slightly higher precision. SMULLR is access via the idiom:

```
(int32_t) (((int64_t) x * y + 0x80000000LL) >> 32)
```

### SMMLA — 32-bit multiply with 32-most-significant-bit accumulate

Fractional q31_t multiply accumulate. Multiplies two 32-bit integers, generates a 64-bit result, and adds the high bits of the result to a 32-bit accumulator.

Processor support: M4 only [1 cycle]

The instruction is available in C code via the idiom:

```
(int32_t) (((int64_t) x * y + ((int64_t) acc << 32)) >> 32);
```

C code example:

```
// Performs a true fractional MAC
int32_t x, y, acc;

acc = (int32_t) (((int64_t) x * y + ((int64_t) acc << 32)) >> 32);
acc <<= 1;
```

Related instructions are SMMLAR, which includes rounding, and SMMLS, which performs subtraction rather than addition.

### QADD — 32-bit saturating addition

Adds two signed integers (or fractional integers) and saturates the result. Positive values are saturated to 0x7FFFFFFF and negative values are saturated to 0x80000000; wrap around does not occur.

The instruction is available in C code only via the intrinsic:

```
int32_t __QADD(int32_t x, uint32_t y)
```

Processor support: M4 only [1 cycle]
C code example:

```
int32_t x, y, z;
z = __QADD(x, y);
```

Related instruction:

QSUB — 32-bit saturating subtraction

## SDIV — 32-bit division

Divides two 32-bit values and returns a 32-bit result.

Processor support: M3 and M4 [2 to 12 cycles]
C code example:

```
int32_t x, y, z;
z = x / y;
```

### *16-bit integer instructions*

## SADD16 — Dual 16-bit addition

Adds two 16-bit values using SIMD. If overflow occurs then the result wraps around.

Processor support: M4 only [1 cycle]
C code example:

```
int32_t x, y, z;
z = __SADD16(x, y);
```

Related instructions:

SSUB16 — Dual 16-bit subtraction

## QADD16 — Dual 16-bit saturating addition

Adds two 16-bit values using SIMD. If overflow occurs then the result is saturated. Positive values are saturated to 0x7FFF and negative values are saturated to 0x8000.

Processor support: M4 only [1 cycle]
C code example

```
int32_t x, y, z;
z = __QADD16(x, y);
```

Related instructions:

QSUB16 — Dual 16-bit saturating subtraction

## SSAT16 — Dual 16-bit saturate

Saturates two signed 16-bit values to bit position B. The resulting values are saturated to the range:

$$-2^{B-1} \leq x \leq 2^{B-1} - 1$$

where B = 1, 2, …, 16. The instruction is available in C code only via the intrinsic:

```
int32_t __ssat16(int32_t x, uint32_t B)
```

Processor support: M4 [1 cycle]
C code example:

```
int32_t x, y;
y = __SSAT16(x, 12); // Saturate to bit 12.
```

## SMLABB — Q setting 16-bit signed multiply with 32-bit accumulate, bottom by bottom

Multiplies the low 16 bits of two registers and adds the result to a 32-bit accumulator. If an overflow occurs during the addition then the result will wrap.

Processor Support: M4 [1 cycle]

The instruction is available in C code using standard arithmetic operations:

```
int16_t x, y;
int32_t acc1, acc2;
acc2 = acc1 + (x * y);
```

## SMLAD — Q setting dual 16-bit signed multiply with single 32-bit accumulator

Multiplies two signed 16-bit values and adds both results to a 32-bit accumulator. (top * top) + (bottom * bottom). If an overflow occurs during the addition then the result will wrap. This is the SIMD version of SMLABB.

Processor Support: M4 [1 cycle]

The instruction is available in C code using the intrinsic:

```
sum = __SMLAD(x, y, z)
```

Conceptually, the intrinsic performs the operations:

```
sum = z + ((short)(x>>16) * (short)(y>>16)) + ((short)x * (short)y)
```

Related instructions are:

SMLADX — Dual 16-bit signed multiply add with 32-bit accumulate (top * bottom) + (bottom * top).

## SMLALBB — 16-bit signed multiply with 64-bit accumulate, bottom by bottom

Multiplies the low 16-bits of two registers and adds the result to a 64-bit accumulator.

Processor Support: M4 [1 cycle]

The instruction is available in C code using standard arithmetic operations:

```
int16_t x, y;
int64_t acc1, acc2;
acc2 = acc1 + (x * y);
```

See also SMLALBT, SMLALTB, SMLALTT

## SMLALD — dual 16-bit signed multiply with single 64-bit accumulator

Performs two 16-bit multiplications and adds both results to a 64-bit accumulator (top * top) + (bottom * bottom). If overflow occurs during the accumulation then the result wraps. The instruction is only available in C code via the intrinsic:

```
uint64_t __SMLALD(uint32_t val1, uint32_t val2, uint64_t val3)
```

C code example:

```
// Input arguments each containing 2 packed 16-bit values
// x[31:16] x[15:0], y[31:15] y[15:0]
uint32_t x, y;

// 64-bit accumulator
uint64_t acc;

// Computes acc += x[31:15]*y[31:15] + x[15:0]*y[15:0]
acc = __SMLALD(x, y, acc);
```

Related instructions are:

SMLSLD — Dual 16-bit signed multiply subtract with 64-bit accumulate

SMLALDX — Dual 16-bit signed multiply add with 64-bit accumulate. (top * bottom) + (bottom * top).

## 8-bit integer instructions
### SADD8 — Quad 8-bit addition
Adds four 8-bit values using SIMD. If overflow occurs then the result wraps around.

Processor support: M4 only [1 cycle]

C code example:

```
// Input arguments contain 4 8-bit values each:
// x[31:24] x[23:16] x[15:8] x[7:0]
// y[31:24] y[23:16] y[15:8] y[7:0]
int32_t x, y;

// Result also contains 4 8-bit values:
// z[31:24] z[23:16] z[15:8] z[7:0]
int32_t z;

// Computes without saturation:
// z[31:24] = x[31:24] + y[31:24]
// z[25:16] = x[25:16] + y[25:16]
//  z[15:8] = x[15:8] + y[15:8]
//   z[7:0] = x[7:0] + y[7:0]

z = __SADD8(x, y);
```

Related instructions:

SSUB8 — Quad 8-bit subtraction

### QADD8 — Quad 8-bit saturating addition
Adds four 8-bit values using SIMD. If overflow occurs then the result is saturated.
Positive values are saturated to 0x7F and negative values are saturated to 0x80.

Processor support: M4 only [1 cycle]

C code example:

```
// Input arguments contain 4 8-bit values each:
// x[31:24] x[23:16] x[15:8] x[7:0]
// y[31:24] y[23:16] y[15:8] y[7:0]
int32_t x, y;

// Result also contains 4 8-bit values:
// z[31:24] z[23:16] z[15:8] z[7:0]
int32_t z;

// Computes with saturation:
// z[31:24] = x[31:24] + y[31:24]
// z[25:16] = x[25:16] + y[25:16]
//  z[15:8] = x[15:8] + y[15:8]
//   z[7:0] = x[7:0] + y[7:0]

z = __QADD8(x, y);
```

Related instructions:

QSUB8 — Quad 8-bit saturating subtraction

### *Floating-point instructions*

The floating-point instructions on the Cortex-M4 are fairly straightforward and most of them are directly accessible via C code. The instructions execute natively if the Cortex-M4 processor includes the floating-point coprocessor. Most instructions execute in 1 cycle if the result is not used in the next instruction. If the result is needed by the next instruction, the instruction completes in 2 cycles. The floating-point instructions are IEEE 754 standard compliant.

If no coprocessor is present, then the instructions are emulated in software and execute much more slowly. We provide only a quick overview of the floating-point instructions here.

### VABS.F32 — floating-point absolute value

Computes the absolute value of a floating-point value.

Processor support: M4F only [1 or 2 cycles]

```
float x, y;
y = fabs(x);
```

### VADD.F32 — floating-point addition

Adds two floating-point values.

Processor support: M4F only [1 or 2 cycles]

```
float x, y, z;
z = x + y;
```

## VDIV.F32 — floating-point division
Divides two floating-point values.
  Processor support: M4F only [14 cycles]

```
float x, y, z;
z = x / y;
```

## VMUL.F32 — floating-point multiplication
Multiplies two floating-point values.
  Processor support: M4F only [1 or 2 cycles]

```
float x, y, z;
z = x * y;
```

## VMLA.F32 — floating-point multiply accumulate
Multiplies two floating-point values and adds the result to a floating-point accumulator.[4]
  Processor support: M4F only [3 or 4 cycles]

```
float x, y, z, acc;
acc = z + (x * y);
```

## VFMA.F32 — fused floating-point multiply accumulate
Multiplies two floating-point values and adds the result to a floating-point accumulator. The standard floating-point multiply accumulate (VMLA) performs two rounding operations; one after the multiplication and then a second one after the addition. The fused multiply accumulate maintains full precision of the multiplication result and performs a single rounding operation after the addition. This yields a slightly more precise result with roughly half the rounding error. The main use of the fused MAC is in iterative operations such as divisions or square roots.
  Processor support: M4F only [3 or 4 cycles]

```
float x, y, acc;
acc = 0;
__fmaf(x, y, acc);
```

## VNEG.F32 — floating-point negation
Multiplies a floating-point value by −1.
  Processor support: M4F only [1 or 2 cycles]

---

[4]We recommend not using the multiply accumulate instruction but rather separate multiplication and addition instructions because there is a possible 1 cycle savings. Refer to section 21.4.6 for details.

```
float x, y;
y = -x;
```

## VSQRT.F32 — floating-point square root
Computes the square root of a floating-point value.
   Processor support: M4F only [14 cycles]

```
float x, y;
y = __sqrtf(x);
```

## VSUB.F32 — floating-point subtraction
Subtracts two floating-point values.
   Processor support: M4F only [1 or 2 cycles]

```
float x, y, z;
z = x - y;
```

## 21.4.6 General Cortex®-M4 optimization strategies

This section builds upon the instruction set introduction of the previous section, and describes common optimization strategies that can be applied to DSP algorithms on the Cortex®-M4.

### Group load and store instructions
Load or store instructions on the Cortex-M4 take 1 or 2 cycles; 2 cycles if they are isolated and 1 cycle if they follow another load or store instruction. The idea is to group multiple load and store operations together into consecutive instructions. N consecutive load or store instructions take N+1 cycles and the benefits apply even when mixing load and store instructions. This optimization applies to both integer and floating-point operations.

### Examine the intermediate assembly code
The underlying DSP algorithm may appear straightforward and easy to optimize but compilers can get confused. Double-check the intermediate assembly output of the C compiler and make sure that that the proper assembly instructions are being used. Also check the intermediate code to see if registers are being properly used or if intermediate results are being stored on the stack. If something does not look right, double-check your compiler settings or refer to your compiler documentation.

   To generate intermediate assembly output files under MDK, go to the "Listings" tab of the target options window and check "Assembly Listing." Then rebuild the project.

### Enable optimization
This may seem obvious, but it is worth mentioning. The code generated by a compiler varies greatly between debug mode and optimization mode. The code

generated in debug mode is made for ease of debugging and not for execution speed.

Compilers also provide various levels of optimization. In Keil™ MDK, the options provided are:

Level 0 — Minimum optimization. Turns off most optimizations. It gives the best possible debug view and the lowest level of optimization.

Level 1 — Restricted optimization. Removes unused inline functions and unused static functions. Turns off optimizations that seriously degrade the ability to debug the code.

Level 2 — High optimization. Resulting code may be difficult to debug. This is the default optimization level.

Level 3 — Maximum optimization. Aggressive optimization including: high-level scalar optimizations, loop unrolling, more aggressive inlining of functions, reordering of instructions, and optimization across source files.

For best performance you typically use the highest optimization level, -O3. However, we have found that when code is carefully written to group load and store operations together (per section 20.4.6.1), using -O3 may reorder instructions leading to worse performance. You may have to experiment with the optimization level to determine the best performance for a particular algorithm.

### Performance considerations in floating-point MAC instructions

Conceptually, the MAC instruction computes a multiplication followed immediately by an addition. Since the addition requires the result of the multiplication, there is a 1 cycle stall and the minimum cycle count for the MAC instruction is 3 cycles. By splitting up the MAC into separate multiplication and addition — and by properly scheduling the instructions, for example, by interleaving two sets of MAC operation sequences — it is possible to perform a floating-point MAC in 2 cycles. *When optimizing for speed, it is good practice to avoid the floating-point MAC and only use separate arithmetic operations.* The penalty is an increase in code size, and being an inability to take the advantage of the slightly better accuracy in the fused MAC, but the code size increase is usually negligible, and in most cases has no impact on the results.

### Loop unrolling

The Cortex-M4 has a 3-cycle overhead per loop iteration. Unrolling a loop by a factor of N effectively reduces the loop overhead to 3/N cycles per iteration. This can be a considerable saving especially if the inner loop consists of only a few instructions. Loop unrolling also allows you to group load and store instructions together as well as to reorder floating-point instructions to avoid the 1-cycle stall penalty in such operations.

You can either manually unroll a loop by repeating a set of instructions or by having the compiler do it for you. The Keil MDK compiler supports *pragmas,* which guide the operation of the compiler. For example, to instruct the compiler to unroll a loop use:

```
#pragma unroll
for(i= 0; i < L; i++)
    {
    ...
    }
```

By default, the loop will be unrolled by a factor of 4. This pragma can be used with for, while, and do-while loops. Specifying #pragma unroll(N) causes the loop to be unrolled N times.

Be sure to examine the generated code to make sure that loop unrolling is effective. *When unrolling, the key thing to watch for is register usage.* Unrolling the loop too much may exceed the number of registers available, and intermediate results will be stored on the stack leading to a degradation of performance.

### Focus on the inner loop

Many DSP algorithms contain multiple nested loops. The processing in the inner loop is executed most frequently and should be the target of the optimization work. Savings in the inner loop are essentially multiplied by the outer loop counts. Only once the inner loop is in good shape should you consider optimizing the outer loops. Many engineers spend time optimizing non-time critical code with little performance benefits to show for their efforts.

### Inline functions

There is some overhead associated with each function call. If your function is small and executed frequently, consider inserting the code of the function in-line instead to eliminate the function call overhead.

### Count registers

The C compiler uses registers to hold intermediate results. If the compiler runs out of registers then results are placed on the stack. As a result, costly load and store instructions are needed to access the data on stack. When developing an algorithm it is good practice to start with pseudo-code in order to count the number of registers needed. In your count, be sure to include pointers, intermediate numerical values and loop counters. Often, the best implementation is the one that uses the fewest intermediate registers.

Be especially careful with register usage in fixed-point algorithms. The Cortex-M4 has only 13 general-purpose registers available for storing integer variables. The floating-point unit, on the other hand, adds 32 floating-point registers to the 13 general-purpose registers. The large number of floating-point registers makes it easy to apply loop unrolling and grouping of load and store instructions.

### Use the right amount of precision

The Cortex-M4 provides several multiply accumulate instructions, which operate on 32-bit numbers. Some provide 64-bit results (e.g., SMLAL) and others provide 32-bit results (e.g., SMMUL). Although they both execute in a single instruction, SMLAL requires two registers to hold the 64-bit result and may execute more slowly — especially if you run out of registers. In general, 64-bit intermediate results are preferred, but check the generated code to make sure that the implementation is efficient.

### 21.4.7 Instruction limitations

The set of DSP instructions on the Cortex®-M4 is quite comprehensive but there are a few gaps that make it different from some of the fully featured DSP processors.

- Saturating fixed-point arithmetic is only available for additions and subtractions but not for MAC instructions. For performance reasons you will often use the fixed-point MAC and need to scale down intermediate operations to avoid overflows.
- No SIMD MAC support for 8-bit values; use 16-bit instead.

## 21.5 Writing optimized DSP code for the Cortex®-M4

This section shows how to use the optimization guidelines and DSP instructions to develop optimized code. We look at the Biquad filter, FFT butterfly, and FIR filter. For each, we start with generic C code and then map it to the Cortex®-M4 DSP instructions while applying the optimization strategies.

### 21.5.1 Biquad filter

A Biquad filter is a second-order recursive or IIR filter. Biquad filters are used throughout audio processing for equalization, tone controls, loudness compensation, graphic equalizers, crossovers, etc. Higher-order filters can be constructed using cascades of second order Biquad sections. In many applications, Biquad filters are the most computationally intensive part of the processing chain. They are also similar to the PID controllers used in control systems, and many of the techniques developed in this section apply directly to PID controllers.

A Biquad filter is a linear time invariant system. When the input to the filter is a sine wave, the output is a sine wave at the same frequency but with a different magnitude and phase. The relationship between the input and output magnitudes and phases is called the *frequency response* of the filter. A Biquad filter has five coefficients and the frequency response is determined by these coefficients. By changing coefficients it is possible to realize lowpass, highpass, bandpass, shelf, and notch filters. For example, Figure 21.3 shows how the magnitude response of an audio "peaking filter" varies in response to coefficient changes. Filter coefficients are typically generated by design equations[5] or by using tools such as MATLAB.

---

[5]Useful design formulas for computing Biquad coefficients for a wide number of filter types have been provided by Robert Bristow-Johnson. See http://www.musicdsp.org/files/Audio-EQ-Cookbook.txt.

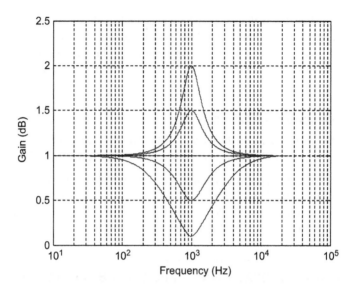

**FIGURE 21.3**

Typical magnitude response of a Biquad filter. This type of filter is called a "peaking filter" and boosts or cuts frequency around 1 kHz. The plots shows several variants with center gains of 0.1, 0.5, 1.0, 1.5, and 2.0

The structure of a Biquad filter implemented in Direct Form I is shown in Figure 21.4. The input $x[n]$ arrives and feeds a two sample delay line. The boxes labeled $z^{-1}$ represent a one sample delay. The left side of the figure is the feed-forward processing and the right hand side is the feedback processing. Because the Biquad filter includes feedback it is also referred to as a *recursive* filter. A Direct

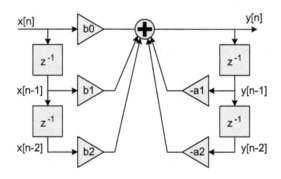

**FIGURE 21.4**

Direct Form I Biquad filter. This implements a second-order filter and is a building block used in higher order filters

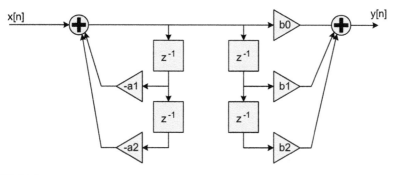

**FIGURE 21.5**

In this figure the feed-forward and feedback portions of the filter have been exchanged. The two delay chains receive the same input and can be combined as shown in the next figure

Form I Biquad filter has five coefficients, four state variables, and takes a total of five MACs per output sample.

Since the system shown in Figure 21.4 is linear and time invariant, we can switch the feed-forward and feedback sections as shown in Figure 21.5. With this change, the delay lines for the feedback and feed-forward sections both take the same input and can thus be combined. This leads to the structure shown in Figure 21.6, which is called Direct Form II. The Direct Form II filter has five coefficients, two state variables, and requires a total of five MACs per output sample. *Direct Form I and Direct Form II filters are mathematically equivalent.*

The obvious advantage of the Direct Form II filter over Direct Form I is that it requires half the number of state variables. Other benefits of each structure are more subtle. If you study the Direct Form I filter you will notice that the input state variables hold delayed versions of the input. Similarly, the output state variables contain delayed versions of the output. Thus, if the gain of the filter does not exceed 1.0 then the state variables in Direct Form I will never overflow.[6] The state variables in the Direct Form II filter, on the other hand, bear no direct relation to the input or output of the filter. In practice, the Direct Form II state variables have a much higher dynamic range than the inputs and outputs of the filter. Thus, even if the gain of the filter does not exceed 1.0 it is possible for the state variables in Direct Form II to exceed 1.0. Due to this property, Direct Form I implementations are preferred for fixed-point implementations (better numerical behavior) while Direct Form II is preferred for floating-point implementations (fewer state variables).

Standard C code for computing a single stage of a Biquad filter implemented using Direct Form II is shown below. The function processes a total of blockSize

---

[6]This is not 100% true; there are cases where you can still have overflows. Nevertheless, this rule of thumb is still useful.

samples through the filter. The input to the filter is taken from the buffer inPtr[] and the output is written to outPtr[]. Floating-point arithmetic is used:

```
// b0, b1, b2, a1, and a2 are the filter coefficients.
// a1 and a2 are negated.
// stateA, stateB, and stateC represent intermediate state
   variables.

for (sample = 0; sample < blockSize; sample++)
  {
    stateA = *inPtr++ + a1*stateB + a2*stateC;
    *outPtr++ = b0*stateA + b1*stateB + b2*stateC;
    stateC = stateB;
    stateB = stateA;
  }


// Persist state variables for the next call
state[0] = stateB;
state[1] = stateC;
```

The intermediate state variables stateA, stateB, and stateC are shown in Figure 21.6.

Next, we will examine the inner loop of the function and see how many cycles are required. We will break down the operations into individual Cortex®-M4 instructions:

```
stateA = *inPtr++;    // Data fetch [2 cycles]
stateA += a1*stateB;  // MAC with result used in next inst [4 cycles]
stateA += a2*stateC;  // MAC with result used in next inst [4 cycles]
out = b0*stateA;      // Mult with result used in next inst [2 cycles]
out += b1*stateB;     // MAC with result used in next inst [4 cycles]
```

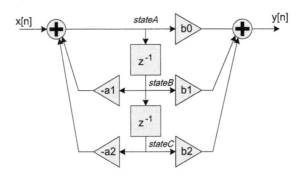

**FIGURE 21.6**

The Direct Form II Biquad structure. This requires 5 multiplications but only two delays. It is preferred when doing floating-point processing

```
out += b2*stateC;      // MAC with result used in next inst [4 cycles]
*outPtr++ = out;       // Data store [2 cycles]
stateC = stateB;       // Register move [1 cycle]
stateB = stateA;       // Register move [1 cycle]
                       // Loop overhead [3 cycles]
```

All together, the inner loop of the generic C code requires a total of 27 cycles per sample to execute.

The first step in optimizing the function is to split the MAC instructions into separate multiplications and additions. Then reorder the computation so that the result of a floating-point operation is not required in the next cycle. Some additional variables are used to hold intermediate results:

```
stateA = *inPtr++      // Data fetch [2 cycles]
prod1 = a1*stateB;     // Mult [1 cycle]
prod2 = a2*stateC;     // Mult [1 cycle]
stateA += prod1;       // Addition [1 cycle]
prod4 = b1*stateB;     // Mult [1 cycle]
stateA += prod2;       // Add [1 cycle]
out = b2*stateC;       // Mult [1 cycle]
prod3 = b0*stateA      // Mult [1 cycle]
out += prod4;          // Add [1 cycle]
out += prod3;          // Add [1 cycle]
stateC = stateB;       // Register move [1 cycle]
stateB = stateA;       // Register move [1 cycle]
*outPtr++ = out;       // Data store [2 cycles]
                       // Loop overhead [3 cycles]
```

These changes reduce the inner loop of the Biquad from 27 to 18 cycles. This is a step in the right direction, but there are three techniques that can be applied to the Biquad for further improvement:

1. Eliminate the register moves by careful use of intermediate variables. The state variables in the structure from top to bottom are initially:

```
stateA, stateB, stateC
```

After the first output is computed the state variables are shifted right. Instead of actually shifting the variables we will reuse them in-place using the ordering:

```
stateC, stateA, stateB
```

After the next iteration, the state variables are reordered as:

```
stateB, stateC, stateA
```

And finally after the forth iteration, the variables are:

```
stateA, stateB, stateC
```

The cycle then repeats and it has a natural period of 3 samples. That is, if we start with:

```
stateA, stateB, stateC
```

then after 3 output samples are computed we will be back to:

```
stateA, stateB, stateC
```

2. Unroll the loop by a factor of 3 to reduce loop overhead. The loop overhead of 3 cycles will be amortized over 3 samples.
3. Group load and store instructions. In the code above, the load and store instructions are isolated and each require 2 cycles. By loading and storing several results, the second and subsequent memory accesses will take only 1 cycle.

The resulting code is now quite a bit longer due to loop unrolling:

```
in1 = *inPtr++;        // Data fetch [2 cycles]
in2 = *inPtr++;        // Data fetch [1 cycles]
in3 = *inPtr++;        // Data fetch [1 cycles]

prod1 = a1*stateB;     // Mult [1 cycle]
prod2 = a2*stateC;     // Mult [1 cycle]
stateA = in1+prod1;    // Addition [1 cycle]
prod4 = b1*stateB;     // Mult [1 cycle]
stateA += prod2;       // Add [1 cycle]
out1 = b2*stateC;      // Mult [1 cycle]
prod3 = b0*stateA;     // Mult [1 cycle]
out1 += prod4;         // Add [1 cycle]
out1 += prod3;         // Add [1 cycle]

prod1 = a1*stateA;     // Mult [1 cycle]
prod2 = a2*stateB;     // Mult [1 cycle]
stateC = in2+prod1;    // Addition [1 cycle]
prod4 = b1*stateA;     // Mult [1 cycle]
stateC += prod2;       // Add [1 cycle]
out2 = b2*stateB;      // Mult [1 cycle]
prod3 = b0*stateC;     // Mult [1 cycle]
out2 += prod4;         // Add [1 cycle]
out2 += prod3;         // Add [1 cycle]

prod1 = a1*stateC;     // Mult [1 cycle]
prod2 = a2*stateA;     // Mult [1 cycle]
stateB = in3+prod1;    // Addition [1 cyte1e]
prod4 = b1*stateC;     // Mult [1 cycle]
stateB += prod2;       // Add [1 cycle]
out3 = b2*stateA;      // Mult [1 cycle]
```

```
prod3 = b0*stateB;      // Mult [1 cycle]
out3 += prod4;          // Add [1 cycle]
out3 += prod3;          // Add [1 cycle]

outPtr++ = out1;        // Data store [2 cycles]
outPtr++ = out2;        // Data store [1 cycles]
outPtr++ = out3;        // Data store [1 cycles]
                        // Loop overhead [3 cycles]
```

Counting cycles we end up with 38 cycles to compute 3 output samples or 12.67 cycles per sample. The code presented here operates on vectors that are a multiple of 3 samples in length. For generality, the code needs another stage, which handles the remaining 1 or 2 samples; this is not shown.

Is it possible to optimize this further? How far can we take loop unrolling? The core arithmetic operations for the Biquad filter consists of 5 multiplications and 4 additions. These operations when properly ordered to avoid stalls take 9 cycles on the Cortex-M4. The memory load and store each take 1 cycle at best. Putting this together, the absolute lowest number of cycles for a Biquad is 11 cycles per sample. This assumes that all data loads and stores are 1 cycle and there is no loop overhead. If we unroll the inner loop still further we would find:

| Unroll by | Total Cycles | Cycles/Sample |
|---|---|---|
| 3 | 38 | 12.67 |
| 6 | 71 | 11.833 |
| 9 | 104 | 11.55 |
| 12 | 137 | 11.41 |

At some point, the processor runs out of intermediate registers to hold the input and output variables and no further gains are possible. Unrolling by 3 or 6 samples is a reasonable choice. Beyond this the gains are marginal.

## 21.5.2 Fast Fourier transform

The Fast Fourier Transform (FFT) is a key signal processing algorithm that is used in frequency domain processing, compression, and fast filtering algorithms. The FFT is actually a fast algorithm to compute the discrete Fourier transform (DFT). The DFT transforms an N-point time domain signal $x[n]$ into N separate frequency components $X[k]$, where each component is a complex value containing both magnitude and phase information. The DFT of a finite length sequence of length N is defined as:

$$X[k] = \sum_{n=0}^{N-1} x[n] W_N^{kn}, \; k = 0, 1, 2, ..., N-1$$

where $W_N^k$ is a complex value representing the $k^{th}$ root of unity:

$$W_N^k = e^{-j2\pi k/N} = \cos(2\pi k/N) - j\sin(2\pi k/N).$$

The inverse transform, which converts from the frequency domain back to the time domain is nearly identical:

$$x[n] = \frac{1}{N} \sum_{k=0}^{N-1} X[k] W_N^{-kn}, \; n = 0, 1, 2, \ldots, N-1$$

Directly implementing the formulas above would require $O(N^2)$ operations to compute all N samples of the forward or inverse transform. With the FFT we will see that this reduces to $O(N\log_2 N)$ operations. The savings can be substantial for large values of N, and the FFT has made possible many new signal processing applications. The FFT was first described by Cooley and Tukey[7] in 1965 and a good general reference for FFT algorithms can be found here.[8]

FFTs generally work best when the length N is a composite number that can be represented as a product of small factors:

$$N = N_1 \times N_2 \times N_3 \cdots N_m$$

The easiest algorithms to follow are those for the case in which N is a power of 2, and they are called radix-2 transforms. The FFT follows a "divide and conquer" algorithm, and an N point FFT is computed using two separate N/2 point transforms together with a few additional operations. There are two main classes of FFTs: decimation-in-time and decimation-in-frequency. Decimation-in-time algorithms compute an N-point FFT by combining N/2-point FFTs of the even and odd time domain samples. Decimation-in-frequency algorithms are similar and compute the even and odd frequency domain samples using two N/2-point FFTs. Both algorithms have a similar number of mathematical operations. The CMSIS library uses decimation-in-frequency algorithms and we will focus on these.

The first stage of an 8-point radix-2 decimation-in-frequency FFT is shown in Figure 21.7. The 8-point transform is computed using two separate 4-point transforms.

The multiplicative factor $W_N^k$ defined earlier appears above and the values are referred to as *twiddle factors*. For speed, twiddle factors are precomputed and stored in an array rather than being computed in the FFT function itself.

The decomposition continues and the 4-point FFTs are each decomposed into two 2-point FFTs. Then at the end, four 2-point FFTs are computed. The final structure is shown in Figure 21.8. Note that there are $\log_2 8 = 3$ stages in the processing.

---

[7]Cooley, James W.; Tukey, John W. "An algorithm for the machine calculation of complex Fourier series". Math. Comput. 19 (90): 297–301, 1965.
[8]C. S. Burrus and T. W. Parks. "DFT/FFT and Convolution Algorithms". Wiley, 1984.

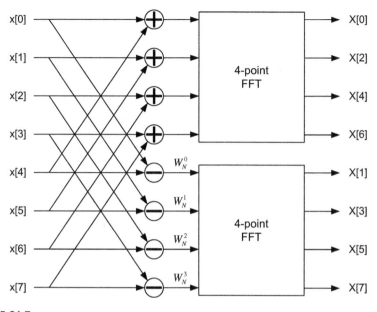

**FIGURE 21.7**

First stage of an 8-point radix-2 decimation-in-frequency FFT algorithm

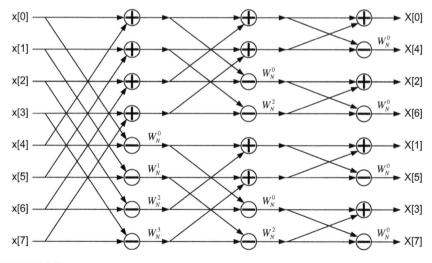

**FIGURE 21.8**

Overall structure of an 8-point FFT. There are 3 stages and each stage consists of 4 butterflies

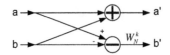

**FIGURE 21.9**

Single butterfly operation

Each stage consists of 4 *butterfly* operations and a single butterfly is illustrated in Figure 21.9.

Each butterfly includes a complex addition, subtraction, and multiplication. A property of the butterfly is that it can be done in place in memory. That is, fetch the complex values a and b, perform the operations, and then place the results back into memory in the same locations of the array. In fact, the entire FFT can be done in place with the output being generated in the same buffer that was used for the input.

The input in Figure 21.8 is in normal order and proceeds sequentially from x[0] to x[7]. The output after processing is scrambled, and the ordering is referred to as *bit-reversed order.* To understand the order, write the indexes 0 to 7 in binary, flip the bits, and convert back to decimal:

| 0 | → | 000 | → | 000 | → | 0 |
|---|---|-----|---|-----|---|---|
| 1 | → | 001 | → | 100 | → | 4 |
| 2 | → | 010 | → | 010 | → | 2 |
| 3 | → | 011 | → | 110 | → | 6 |
| 4 | → | 100 | → | 001 | → | 1 |
| 5 | → | 101 | → | 101 | → | 5 |
| 6 | → | 110 | → | 011 | → | 3 |
| 7 | → | 111 | → | 111 | → | 7 |

Bit-reversed ordering is a natural side effect of the in place processing. Most FFT algorithms (including the ones in the CMSIS DSP Library) provide the option of reordering the output values back into sequential order.

Butterflies are at the heart of FFT algorithms, and we will analyze and optimize the computation for a single butterfly in this section. The 8-point FFT requires $3 * 4 = 12$ butterflies. In general, a radix-2 FFT of length N has $\log_2 N$ stages each with $N/2$ butterflies, for a total of $(N/2)\log_2 N$ butterflies. The decomposition into butterflies yields the $O(N\log_2 N)$ operation count for FFTs. In addition to the butterflies themselves, the FFT requires indexing to keep track of which values should be used in each stage of the algorithm. In our analysis we will ignore this indexing overhead, but it must be accounted for in the final algorithm.

The C code for a floating-point butterfly is shown below. The variables index1 and index2 are the array offsets for the two inputs to the butterfly. The array x[] holds interleaved data (real, imag, real, imag, etc.):

The code also shows cycle counts for the various operations, and we see that a single butterfly takes 23 cycles on a Cortex®-M4. Looking more closely at the cycle count we see that 13 cycles are due to memory accesses and 10 cycles are due to arithmetic. By grouping the memory accesses together, the total number of cycles spent on memory accesses can be reduced to 11. Even so, on a Cortex-M4 a radix-2 butterfly is dominated by memory access and this will be true for the overall FFT algorithm. Little can be done to speed up the radix-2 FFT butterfly shown in Figure 21.10. Instead, in order to improve performance we need to consider higher radix algorithms.

In a radix-2 algorithm, we operate on 2 complex values at a time and there are a total of $\log_2 N$ stages of processing. At each stage we need to load N complex values, operate on them, and then store them back into memory. In a radix-4 algorithm, we operate on 4 complex values at a time and there are a total of $\log_4 N$ stages. This cuts memory accesses down by a factor of 2. Higher radixes can be considered as long as we do not run out of intermediate registers. We found that up to radix-4 butterflies can be efficiently implemented in fixed-point on a Cortex-M4 and radix-8 butterflies using floating-point.

A radix-4 algorithm is limited to FFT lengths, which are powers of 4: {4, 16, 64, 256, 1024, etc.} while a radix-8 algorithm is restricted to lengths {8, 64, 512, 4096, etc.} In order to efficiently implement any length, which is a power of 2, a *mixed-radix* algorithm is used. The trick is to use as many radix-8 stages as possible (they are the

```
// Fetch two complex samples from memory [5 cycles]
x1r = x[index1];
x1i = x[index1+1];

x2r = x[index2];
x2i = x[index2+1];

// Compute the sum and difference [4 cycles]
sum_r = (x1r + x2r);
sum_i = (x1i + x2i);

diff_r = (x1r - x2r);
diff_i = (x1i - x2i);

// Store sum result to memory [3 cycles]
x[index1] = sum_r;
x[index1+1] = sum_i;

// Fetch complex twiddle factor coefficients [2 cycles]
twiddle_r = *twiddle++;
twiddle_i = *twiddle++;

// Complex multiplication of the difference [6 cycles]
prod_r = diff_r * twiddle_r - diff_i * twiddle_i;
prod_i = diff_r * twiddle_i + diff_i * twiddle_r;

// Store back to memory [3 cycles]
x[index2] = prod_r;
x[index2+1] = prod_i;
```

**FIGURE 21.10**

C code implementation of a floating-point butterfly

most efficient) and then use a single radix-2 or radix-4 stage, as needed, to achieve the desired length. Here is how the various FFT lengths break down into butterfly stages:

| Length | Butterflies |
|--------|-------------|
| 16 | 2 x 8 |
| 32 | 4 x 8 |
| 64 | 8 x 8 |
| 128 | 2 x 8 x 8 |
| 256 | 4 x 8 x 8 |

The FFT function in the CMSIS DSP library uses this mixed radix approach for floating-point data types. For fixed-point, you must select either radix-2 or radix-4. In general, pick the radix-4 fixed-point algorithm if it supports the length you want.

Many applications also require that inverse FFT transforms be computed. Comparing the equations for the forward and inverse DFTs we see that the inverse transform has a scale factor of (1/N) and that the sign of the exponent of the twiddle factors is inverted. This leads to two different ways of implementing inverse FFTs:

1. Compute the forward FFT as before but use a new twiddle factor table. The new table is created using positive rather than negative exponents. This leads to twiddle factors simply being conjugated. Divide by N.
2. Keep the same twiddle factor table as before, but modify the FFT code to negate the imaginary portion of the twiddle factor table while performing twiddle factor multiplications. Then divide by N.

Both approaches above have some inefficiency. Approach (1) saves on code space but doubles the size of the twiddle factor table; (2) reuses the twiddle factor table but requires more code. Another approach is to use the mathematical relationship:

$$\text{IFFT}(X) = \frac{1}{N}\text{conj}(\text{FFT}(\text{conj}(X)))$$

This requires conjugating the data twice. The first (inner) conjugation is done at the start and the second conjugation (outer) can be combined with the division by N. The real overhead of this approach compared to (2) is roughly only the inner conjugate, which is acceptable.

When implementing an FFT in fixed-point it is crucial to understand the scaling and growth of values throughout the algorithm. A butterfly performs a sum and difference and it is possible for the values at the output of the butterfly to be double those at the input. In the worst case, values double every stage and the output is N times larger than the input. Intuitively, the worst case occurs if all input values equal 1.0. This represents a DC signal and the resulting FFT is all zeros except that bin $k = 0$ contains a value of N. In order to avoid overflow in fixed-point implementations, each butterfly stage must incorporate a scale of 0.5 as part of the addition and subtraction. This is in fact the scaling used by the fixed-point FFT functions in the CMSIS DSP library and the net result is that the output of the FFT is scaled down by $1/N$.

The standard FFT operates on complex data and there are variations for handling real data. Typically, a N-point real FFT is computed using a complex N/2 point FFT together with some additional steps. An excellent reference can be found here.[9]

### 21.5.3 FIR filter

The third standard DSP algorithm that we will consider is the FIR (Finite Impulse Response) filter. FIR filters occur in a variety of audio, video, control, and data analysis problems. FIR filters have several useful properties compared to IIR filters (like the Biquad):

1. Filters are inherently stable. This is true for all possible coefficients.
2. Linear phase can be achieved by making the coefficients symmetric.
3. Simple design formulas.
4. Well-behaved even when implemented using fixed-point.

Let $x[n]$ be the input to the filter at time $n$ and let $y[n]$ be the output. The output is computed using the difference equation:

$$y[n] = \sum_{k=0}^{N-1} x[n-k]h[k]$$

where $h[n]$ are the filter coefficients. In the difference equation above, the FIR filter has N coefficients:

$$\{h[0], \ h[1], \ \cdots, h[N-1]\}$$

and the output is computed using N previous input samples:

$$\{x[n], \ x[n-1], \ \cdots, x[n-(N-1)]\}$$

The previous input samples are called the *state variables*. Each output of the filter requires N multiplications and N−1 additions. A modern DSP can compute an N point FIR filter in roughly N cycles.

The most straightforward way of organizing the state data in memory is to use a FIFO as shown in Figure 21.11. When sample $x[n]$ arrives, the previous samples $x[n-1]$ through $x[n-N]$ are shifted down by one position and then $x[n]$ is written to the buffer. Shifting data like this is very wasteful and requires N−1 memory reads and N−1 memory writes per input sample.

---

[9]Matusiak, Robert, *Implementing Fast Fourier Transform Algorithms of Real-Valued Sequences with the TMS320 DSP Platform*, Texas Instruments Application Report SPRA291, August 2001.

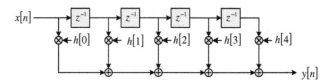

**FIGURE 21.11**

FIR filter implemented using a shift register. In practice this is rarely used since the state variables have to be shifted right whenever a new sample arrives

A better way to organize the data is to use a circular buffer as shown in Figure 21.12. The circular state index points to the oldest sample in the buffer. When sample $x[n]$ arrives, it overwrites the oldest sample in the buffer and then circularly increments. That is, it increments in normal fashion, and if the end of the buffer is reached it wraps around back to the beginning.

The standard C code for the FIR filter is shown in Figure 21.13. The function is designed to operate on a block of samples and incorporates circular addressing. The outer loop is over the samples in the block while the inner loop is over the filter taps as shown in the equation above.

In order to compute each output sample we have to fetch N state variables $\{x[n], x[n-1], \cdots, x[n-(N-1)]\}$ and N coefficients $\{h[0], h[1], \cdots, h[N-1]\}$ from memory. DSPs have been optimized to compute FIR filters. The state and coefficients can be fetched in parallel with the MACs and the corresponding memory pointers incremented. DSPs also have hardware support for circular addressing and

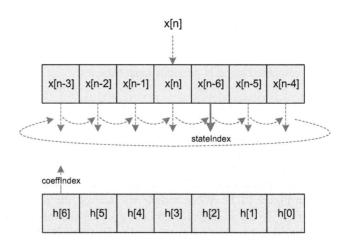

**FIGURE 21.12**

FIR filter implemented using a circular buffer for the state variables (top). The stateIndex pointer advances to the right and then circularly wraps when it reaches the end of the buffer. The coefficients are accessed in linear order

```
// Block-based FIR filter.
// N equals the length of the filter (number of taps)
// blockSize equals the number of samples to process
// state[] is the state variable buffer and contains the previous N
//   samples of the input
// stateIndex points to the oldest sample in the state buffer. It
//   will be overwritten with the most recent input sample.
// coeffs[] holds the N coefficients
// inPtr and outPtr point to the input and output buffers, respectively

for(sample=0;sample<blockSize;sample++)
{
 // Copy the new sample into the state buffer and then
    circularly wrap stateIndex
 state[stateIndex++] = inPtr[sample]
 if (stateIndex >= N)
  stateIndex = 0;

 sum = 0.0f;
 for(i=0;i<N;i++)
   {
    sum += state[stateIndex++] * coeffs[N-i];
    if (stateIndex >= N)
        stateIndex = 0;
   }
 outPtr[sample] = sum;
}
```

**FIGURE 21.13**

Standard C code for implementing an FIR filter by utilizing a circular buffer. The code processes a block of samples

can perform circular addressing without any overhead. Using these features together, a modern DSP can compute an N-point FIR filter in about N cycles.

The Cortex®-M4 will have difficulty implementing the code shown in Figure 21.13 efficiently. The Cortex-M4 does not have native support for circular addressing, and the bulk of time will be spent evaluating the if statement in the inner loop. A better approach is to use a FIFO for the state buffer and to shift in a block of input data. Instead of shifting the FIFO data every sample, just shift it once every block. This requires increasing the length of the state buffer by blockSize samples. This process is illustrated in Figure 21.14 for a block size of 4 samples. Input data is shifted in on the right side of the block. The oldest data then appears on the left-hand side. Coefficients continue to be time flipped as shown in Figure 21.12.

The coefficients in Figure 21.14 have $h[0]$ aligned with $x[n-3]$. This is the position needed to compute the first output $y[n-3]$ :

$$y[n-3] = \sum_{k=0}^{6} x[n-3-k]h[k]$$

To compute the next output, sample the coefficients conceptually shift over by one. This then repeats for all output samples. By using this block based approach with a FIFO state buffer we were able to eliminate the costly circular addressing from the inner loop.

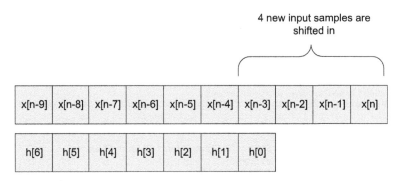

**FIGURE 21.14**

Work around for circular addressing. The size of the state buffer is increased by blockSize-1 samples, which in this example equals 3 samples

To optimize the FIR filter further we need to focus again on memory accesses. In the standard FIR implementation, N coefficients and N state variables are accessed for each output sample. The approach we take is to simultaneously compute multiple output samples and cache intermediate state variables in registers. In this example, we compute 4 output samples simultaneously.[10] A single coefficient is loaded and then multiplied by 4 state variables. This has the net effect of reducing memory accesses by a factor of 4. For the sake of simplicity and brevity, the code we show only supports block sizes that are a multiple of 4 samples. The CMSIS library, on the other hand, is general purpose and imposes no constraints on the length of the filter or block size. Even with these simplifications, the code is still quite involved.

---

[10]The number of outputs to simultaneously compute depends upon how many registers are available. In the CMSIS library Q31 FIR 3 samples are simultaneously computed. In the floating-point version, 8 samples are simultaneously computed.

```
/*
** Block based FIR filter. Arguments:
**   numTaps - Length of the filter. Must be a multiple of 4
**   pStateBase - Points to the start of the state variable array
**   pCoeffs - Points to the start of the coefficient array
**   pSrc - Points to the array of input data
**   pDst - Points to where the result should be written
**   blockSize - Number of samples to process. Must be a multiple of 4.
*/

void arm_fir_f32(
        unsigned int numTaps,
        float *pStateBase,
        float *pCoeffs,
        float *pSrc,
        float *pDst,
        unsigned int blockSize)
{
        float *pState;
        float *pStateEnd;
        float *px, *pb;
        float acc0, acc1, acc2, acc3;
        float x0, x1, x2, x3, coeff;
        unsigned int tapCnt, blkCnt;

        /* Shift the data in the FIFO down and store the new block of input data
        ** at the end of the buffer. */

        /* Points to the start of the state buffer */
        pState = pStateBase;

        /* Points ahead blockSize samples */
        pStateEnd = &pStateBase[blockSize];

        /* Unroll by 4 for speed */
        tapCnt = numTaps >> 2u;
        while(tapCnt > 0u)
        {
                *pState++ = *pStateEnd++;
                *pState++ = *pStateEnd++;
                *pState++ = *pStateEnd++;
                *pState++ = *pStateEnd++;

                /* Decrement the loop counter */
                tapCnt--;
        }

        /* pStateEnd points to where the new input data should be written */
        pStateEnd = &pStateBase[(numTaps - 1u)];
        pState = pStateBase;

        /* Apply loop unrolling and compute 4 output values simultaneously.
        * The variables acc0 ... acc3 hold output values that are being computed:
        *
        *   acc0 = b[numTaps-1]*x[n-numTaps-1]+b[numTaps-2]*x[n-numTaps-2] +
        *          b[numTaps-3]*x[n-numTaps-3]+ ... + b[0]*x[0]
        *   acc1 = b[numTaps-1]*x[n-numTaps]+b[numTaps-2]*x[n-numTaps-1] +
        *          b[numTaps-3]*x[n-numTaps-2] +... + b[0]*x[1]
        *   acc2 = b[numTaps-1]*x[n-numTaps+1]+b[numTaps-2]*x[n-numTaps] +
        *          b[numTaps-3]*x[n-numTaps-1] + ... + b[0]*x[2]
        *   acc3 = b[numTaps-1]*x[n-numTaps+2] + b[numTaps-2]*x[n-numTaps+1] +
        *          b[numTaps-3]*x[n-numTaps] + ... + b[0]*x[3]
        */

        blkCnt = blockSize >> 2;
```

**FIGURE 21.15**

Partially optimized floating-point FIR code. The example shows how the number of memory accesses can be reduced by simultaneously computing multiple results

```
/* Processing with loop unrolling. Compute 4 outputs at a time. */
while(blkCnt > 0u)
{
        /* Copy four new input samples into the state buffer */
        *pStateEnd++ = *pSrc++;
        *pStateEnd++ = *pSrc++;
        *pStateEnd++ = *pSrc++;
        *pStateEnd++ = *pSrc++;

        /* Set all accumulators to zero */
        acc0 = 0.0f;
        acc1 = 0.0f;
        acc2 = 0.0f;
        acc3 = 0.0f;

        /* Initialize state pointer */
        px = pState;

        /* Initialize coeff pointer */
        pb = pCoeffs;

        /* Read the first three samples from the state buffer:
    x[n-numTaps], x[n-numTaps-1], x[n-numTaps-2] */
        x0 = *px++;
        x1 = *px++;
        x2 = *px++;

        /* Loop unrolling. Process 4 taps at a time. */
        tapCnt = numTaps >> 2u;

        /* Loop over the number of taps. Unroll by a factor of 4.
    ** Repeat until we have computed numTaps-4 coefficients. */
        while(tapCnt > 0u)
        {
                /* Read the b[numTaps-1] coefficient */
                coeff = *(pb++);

                /* Read x[n-numTaps-3] sample */
                x3 = *(px++);

                /* p = b[numTaps-1] * x[n-numTaps] */
                p0 = x0 * coeff;

                /* p1 = b[numTaps-1] * x[n-numTaps-1] */
                p1 = x1 * coeff;

                /* p2 = b[numTaps-1] * x[n-numTaps-2] */
                p2 = x2 * coeff;

                /* p3 = b[numTaps-1] * x[n-numTaps-3] */
                p3 = x3 * coeff;

                /* Accumulate */
                acc0 += p0;
                acc1 += p1;
                acc2 += p2;
                acc3 += p3;

                /* Read the b[numTaps-2] coefficient */
                coeff = *(pb++);

                /* Read x[n-numTaps-4] sample */
                x0 = *(px++);
```

**FIGURE 21.15**

(*Continued*)

```
                      /* Perform the multiply-accumulate */
                      p0 = x1 * coeff;
                      p1 = x2 * coeff;
                      p2 = x3 * coeff;
                      p3 = x0 * coeff;
                      acc0 += p0;
                      acc1 += p1;
                      acc2 += p2;
                      acc3 += p3;

                      /* Read the b[numTaps-3] coefficient */
                      coeff = *(pb++);

                      /* Read x[n-numTaps-5] sample */
                      x1 = *(px++);

                      /* Perform the multiply-accumulates */
                      p0 = x2 * coeff;
                      p1 = x3 * coeff;
                      p2 = x0 * coeff;
                      p3 = x1 * coeff;
                      acc0 += p0;
                      acc1 += p1;
                      acc2 += p2;
                      acc3 += p3;

                      /* Read the b[numTaps-4] coefficient */
                      coeff = *(pb++);

                      /* Read x[n-numTaps-6] sample */
                      x2 = *(px++);

                      /* Perform the multiply-accumulates */
                      p0 = x3 * coeff;
                      p1 = x0 * coeff;
                      p2 = x1 * coeff;
                      p3 = x2 * coeff;
                      acc0 += p0;
                      acc1 += p1;
                      acc2 += p2;
                      acc3 += p3;

                      /* Read the b[numTaps-5] coefficient */
                      coeff = *(pb++);

                      /* Read x[n-numTaps-7] sample */
                      x3 = *(px++);

                      tapCnt--;
                  }

              /* Advance the state pointer to process the next
               * group of 4 samples */
              pState = pState + 4;

              /* Store the 4 results in the destination buffer. */
              *pDst++ = acc0;
              *pDst++ = acc1;
              *pDst++ = acc2;
              *pDst++ = acc3;

              blkCnt--;
          }
      }
```

**FIGURE 21.15**

(Continued)

The inner loop of the floating-point FIR filter takes 50 cycles and performs a total of 16 MACs. This is equivalent to 3.125 cycles/MAC. The CMSIS library takes this a step further and computes 8 intermediate sums leading to about 2.4 cycles/MAC in the inner loop.

The q15 FIR filter functions in the CMSIS library use similar memory optimizations as the floating-point function just presented. As a further optimization, the q15 functions apply the dual 16-bit SIMD capabilities of the M4. Two q15 functions are provided:

- arm_fir_q15() — uses 64-bit intermediate accumulators and the SMLALD and SMLALDX instructions.
- arm_fir_fast_q15() — uses 32-bit intermediate accumulators and the SMLAD and SMLADX instructions.

# Using the ARM® CMSIS-DSP Library

# 22

## CHAPTER OUTLINE

## 22.1 Overview of the library

The CMSIS-DSP library is a suite of common signal processing and mathematical functions that have been optimized for the Cortex®-M4 processor. The library is freely available as part of the CMSIS release from ARM® and includes all source code. The functions in the library are divided into several categories:

- Basic math functions
- Fast math functions
- Complex math functions
- Filters
- Matrix functions
- Transforms
- Motor control functions
- Statistical functions
- Support functions
- Interpolation functions

The library has separate functions for operating on 8-bit integers, 16-bit integers, 32-bit integers, and 32-bit floating point values.

The library has been optimized to take advantage of the DSP extensions found in the Cortex-M4 processor. Although the library is compatible with the Cortex-M0 and M3 processors, the functions have not been optimized for these cores; the functions operate properly but run more slowly.

The Definitive Guide to ARM® Cortex®-M3 and Cortex-M4 Processors. http://dx.doi.org/10.1016/B978-0-12-408082-9.00022-1

## 22.2 Pre-built binaries

The library includes pre-built binaries and projects files for a number of configurations and processor families. The projects files are for Keil™ μVision, and the supplied libraries were also built and tested under μVision. Select the library that matches your target processor. We suggest that you link against the Keil libraries even if you are using a different toolchain. The reason is that the libraries were optimized using Keil and provide the best performance. If you do need to rebuild the libraries for any reason, refer to the CMSIS-DSP library HTML documentation for details.

| Library Name | Project File | Processor | Endianness | Uses FPU |
|---|---|---|---|---|
| arm_cortexM4lf_math.lib | arm_cortexM4lf_math.uvproj | Cortex-M4 | Little | Yes |
| arm_cortexM4bf_math.lib | arm_cortexM4bf_math.uvproj | Cortex-M4 | Big | Yes |
| arm_cortexM4l_math.lib | arm_cortexM4l_math.uvproj | Cortex-M4 | Little | No |
| arm_cortexM4b_math.lib | arm_cortexM4b_math.uvproj | Cortex-M4 | Big | No |
| arm_cortexM3l_math.lib | arm_cortexM3l_math.uvproj | Cortex-M3 | Little | No |
| arm_cortexM3b_math.lib | arm_cortexM3b_math.uvproj | Cortex-M3 | Big | No |
| arm_cortexM0l_math.lib | arm_cortexM0l_math.uvproj | Cortex-M0 | Little | No |
| arm_cortexM0b_math.lib | arm_cortexM0b_math.uvproj | Cortex-M0 | Big | No |

## 22.3 Function naming convention

The functions in the library follow the naming convention:

arm_OP_DATATYPE

where OP is the operation performed and DATATYPE describes the operands:

- q7 — 16 bit fractional integers
- q15 — 16 bit fractional integers
- q31 — 32 bit fractional integers
- f32 — 32 bit floating point

For example:

```
arm_dot_prod_q7 — dot product of 8-bit fractional integers
arm_mat_add_q15 — matrix addition of 16-bit fractional integers
arm_fir_q31 — FIR filter with 32-bit fractional data and coefficients
arm_cfft_f32 — Complex FFT of 32-bit floating point values
```

## 22.4 **Getting help**

The library document is in HTML format and is located in the folder

```
CMSIS\Documentation\DSP\html
```

The file index.html is the main starting point.

## 22.5 **Example 1 — DTMF demodulation**

A standard touch tone dial pad with four rows and three columns is shown in Figure 22.1. Each row and column has a corresponding sine wave associated with it. When a button is pressed, the dial pad generates two sine waves; one based on the row index and another based on the column index. For example, if the number 4 is pressed then a 770 Hz sine wave is generated (corresponding to the row) plus a 1209 Hz sine wave (corresponding to the column). This signaling method is referred to as Dual-Tone Multi-Frequency (DTMF) signaling and is the standard signaling method used by analog phone lines.

In this example, we develop three different type ways of detecting the tones in the DTMF signal:

- FIR filter [q15]
- FFT [q31]
- Biquad Filter [float]

We'll focus on decoding one frequency, 697 Hz, and the example can be easily extended to decoding all seven tones. Other aspects of DTMF decoding, such as setting thresholds and decision making, are not covered. All of the code used in this example is shown in section 21.5.5.

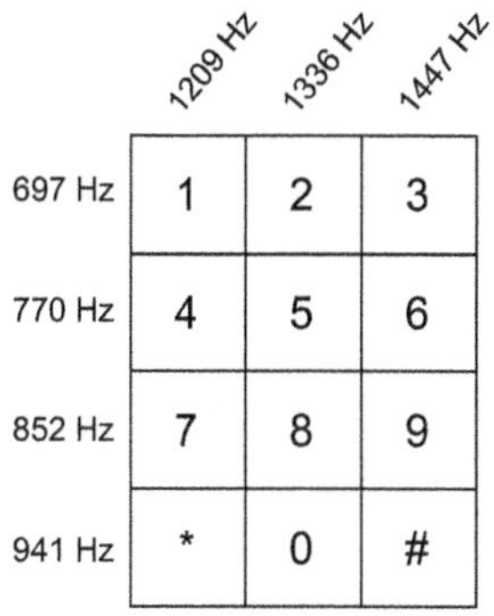

**FIGURE 22.1**

Keypad matrix in a DTMF signaling scheme. Each row and column has a corresponding sine wave that is generated when a button is pressed

The goal of the example is to show how to use various CMSIS-DSP functions and to handle different data types. As it turns out, the Biquad filter is computationally much more efficient than the FIR filter. The reason is that a single Biquad stage can be used to detect a sine wave rather than a 202 point FIR filter. The Biquad is roughly 40 times more efficient than the FIR. The FFT might seem like a good choice, especially since in a full DTMF implementation seven frequencies will need to be checked. Even taking all of this into account, the Biquad is still computationally more efficient than the FFT, and requires much less memory as well. In practice, most DTMF receivers use the Goetzel algorithm, which is computationally very similar to the Biquad filter used in this example.

## 22.5.1 Generating the sine wave

In a typical DTMF application, the input data will be taken from an A/D converter. In this example, we generate the input signal using math functions. The start of the example code generates a sine wave at 697 Hz or 770 Hz. The sine wave is first generated using floating point and then converted to Q15 and Q31 representations. All of the processing is performed at an 8 kHz sample rate that is the standard sample rate used in telephony applications; 512 samples of the sine wave are generated with an amplitude of 0.5.

## 22.5.2 Decoding using an FIR filter

The first approach we'll use for demodulating the DTMF signal is an FIR filter. The FIR filter will have a passband centered around 697 Hz and must be narrow enough to filter out the next closest frequency of 770 Hz. The following MATLAB code designs the filter:

```
SR = 8000; % Sample rate
FC = 697;  % Center frequency of the pass band, in Hz

NPTS = 202;

h = fir1(NPTS-1, [0.98*FC 1.02*FC] / (SR/2), 'DC-0' );
```

Some experimentation was needed in order to determine the correct length of the filter so that by 770 Hz there was sufficient attenuation. We found that a filter length of 201 points was sufficient. Since the CMSIS library requires an even length filter for the Q15 FIR filter function, the filter length was rounded up to 202 points. The impulse response of the designed filter is shown in Figure 22.2 and the magnitude response is shown in Figure 22.3.

The filter was designed to have a gain of 1.0 in the passband. The largest coefficient of the resulting filter has a value of about 0.019. If converted to 8-bit format (q7), the largest coefficient would only be about 2LSBs in size. The resulting filter would be severely quantized and unusable in 8-bits. At least 16-bits are required and

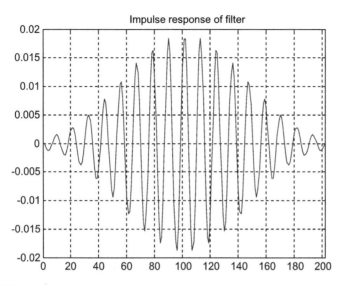

**FIGURE 22.2**

Impulse response of the FIR filter. The filter has a strong sine wave component at 697 Hz, which is the center of the passband

this example uses Q15 math. MATLAB was also used to convert the filter coefficients to Q15 format and write them to the console window. The coefficients were then copied into the Cortex®-M4 project:

```
hq = round(h * 32768);    // Quantize to Q15

fprintf(1, 'hfir_coeffs_q15 = {\n');
for i=1:length(hq);
 fprintf(1, '%5d', hq(i));
 if (i == length(hq))
  fprintf(1, '};\n');
 else
  fprintf(1, ', ');
  if (rem(i, 8) == 0)
   fprintf(1, '\n');
  end
 end
end
```

The CMSIS code to apply the FIR filter is straightforward. First, the function arm_fir_init_q15() is called to initialize the FIR filter structure. The function just checks to make sure that the filter length is even and greater than four samples and then sets a few structure elements. Next, the function processes the signal one

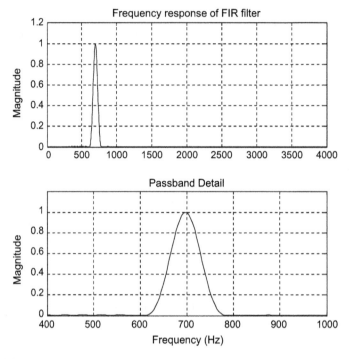

**FIGURE 22.3**

Frequency response of the filter. The top figure shows the magnitude response across the entire frequency band. The bottom figure shows the detail around the passband frequency of 697 Hz

block at a time. A macro **BLOCKSIZE** is defined as 32 and equals the number of samples that will be processed in each call to arm_fir_q15(). Each call generates 32 new output samples and these are written into the output buffer.

The output of the filter was computed for 697 Hz and 770 Hz inputs and the results are shown in Figure 22.4. The top plot shows the output when the input is 697 Hz. The sine wave falls into the middle of the band and has the expected output amplitude of 0.5. (Remember that the input sine wave has an amplitude of 0.5 and the filter has a gain of 1.0 in the center of the band.) The bottom plot shows the output when the input frequency is increased to 770 Hz. The output is greatly attenuated, as expected.

## 22.5.3 Decoding using an FFT

The next approach to decoding the DTMF tones that we'll explore uses an FFT. The advantage of using the FFT is that it provides a complete frequency representation of the signal and can be used to simultaneously decode all seven DTMF sine waves. We'll use a Q31 FFT for this example and a buffer of length 512 samples. Since the input data is real, we'll use a real transform.

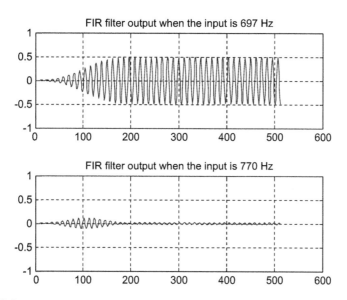

**FIGURE 22.4**

Output of the FIR filter for two different sine wave inputs. The top plot shows the output when the input is a 697 Hz sine wave, which is centered in the middle of the band. The bottom plot shows the output when the input is a 770 Hz sine wave and is properly attenuated

The FFT operates on an entire buffer of 512 samples. Several steps are involved. First, the data is windowed in order to reduce the transients at the edge of the buffer. There are several different types of windows; Hamming, Hanning, Blackman, etc., and the choice depends upon the desired frequency resolution and separation between neighboring frequencies. In our application, we used a Hanning, or raised cosine, window. The input signal is shown at the top of Figure 22.5 and the result after windowing is shown at the bottom. You can see that the windowed version decays smoothly to zero at the edges. All data is represented by Q31 values.

The ARM® CMSIS-DSP library function arm_rfft_init_q31 is applied to the data. The function produces complex frequency domain data. The function arm_cmplx_mag_q31 then computes the magnitude of each frequency bin. The resulting magnitude is shown in Figure 22.6. Since the FFT is 512 points long and the sample rate 8000 Hz, the frequency spacing per FFT bin is

$$\frac{8000}{512} = 15.625\,Hz$$

The largest magnitude is found at bin 45, which corresponds to a frequency of 703 Hz — the closest bin to 697 Hz.

If the input frequency is 770 Hz the FFT produces the magnitude plot shown in Figure 22.7. The peak occurs at bin 49, which corresponds to 766 Hz.

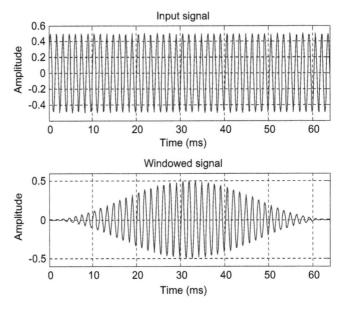

**FIGURE 22.5**

The top figure shows the input sine wave of 697 Hz. Some discontinuities at the signal edges can be seen and these could cause the peak frequency to be misidentified. The bottom figure shows the sine wave after it was windowed by a Hanning window

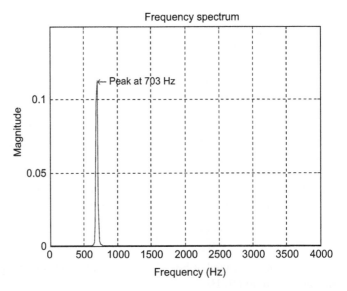

**FIGURE 22.6**

Magnitude of the FFT output. The peak frequency component occurs at 703 Hz, which is the bin closest to the actual frequency of 697 Hz

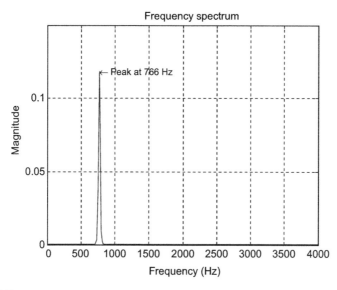

**FIGURE 22.7**

FFT output when the input frequency is 770

## 22.5.4 Decoding using a Biquad filter

The final approach taken was to use a second order IIR filter to do the tone detection. The approach is similar to the Goertzel algorithm that is used in most DSP based decoders. The Biquad filter is designed to have a pole near the unit circle at the desired frequency of 697 Hz and zeros at DC and Nyquist. This yields a narrow bandpass shape. The gain of the filter is adjusted so that it has a peak gain of 1.0 in the passband. By moving the pole closer to the unit circle the sharpness of the filter can be adjusted. We settled on placing the pole at a radius of 0.99 and an angle of:

$$\omega = 2\pi \left( \frac{697}{8000} \right)$$

The pole forms part of a complex conjugate pair and there is a matching pole at the negative frequency. The MATLAB code to generate the filter coefficients is shown below. The scaling by K creates a peak gain of 1.0 in the passband:

```
r = 0.99;

p1 = r * exp(sqrt(-1)*2*pi*FC/SR); % Pole location
p2 = conj(p1);                      % Conjugate pole
P = [p1; p2];                       % Make the array of poles

Z = [1; -1];                        % Zeros at DC and Nyquist

K = 1 - r;                          % Gain factor for unity gain
SOS = zp2sos(Z, P, K);              % Convert to biquad coeffs
```

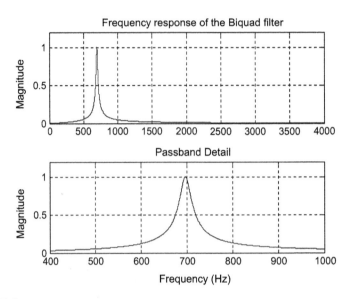

**FIGURE 22.8**

Frequency response of the IIR filter used in DTMF tone detection

The resulting frequency response of the filter is shown in Figure 22.8. The filter is sharp and passes frequencies within a very narrow band.

The CMSIS-DSP library has two versions of the floating point Biquad filter available: Direct Form I and Transposed Direct Form II. For floating point, the best version to use is always Transposed Direct Form II since it only requires two state variables per Biquad rather than four. With fixed point, you should always use Direct Form I, but in this case Transposed Direct Form II is the correct version to use.

The code for processing the Biquad filter is straightforward. This example uses a second order filter, which corresponds to a single Biquad filter stage. The filter has two associated arrays:

Coefficients — 5 values
State variables — 2 values

These arrays are defined at the top of the function and the coefficient array is set to the values computed by Matlab. The only change is that the feedback coefficients are negated compared to the standard MATLAB representation. Next, the function arm_biquad_cascade_df2T_init_f32() is called to initialize the Biquad instance structure and then we loop over the input data and process the data in blocks. Each call processes BLOCKSIZE = 32 samples through the filter and stores the result in the output array.

The output of the Biquad filter is shown in Figure 22.9. The top part of the figure shows the output when the input is at 697 Hz; the bottom shows the output when the input is 770 Hz. When the input is 697 Hz, the output builds up to the expected

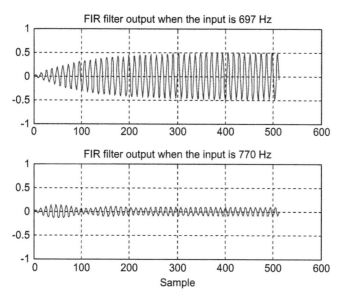

**FIGURE 22.9**

Output of the Biquad DTMF detection filter

amplitude of 0.5. When the input is a 770 Hz sine wave, there is still a bit of signal in
the output and the rejection is not as great as the FIR result shown in Figure 22.9.
Still, the filter does a reasonable job of discerning the various signal components.

## 22.5.5 Example DTMF code

```
#include "stm32f4xx.h"
#include <stdio.h>

#include "arm_math.h"

#define L 512
#define SR 8000
#define FREQ 697
// #define FREQ 770
#define BLOCKSIZE 8

q15_t inSignalQ15[L];
q31_t inSignalQ31[L];
float inSignalF32[L];

q15_t outSignalQ15[L];
float outSignalF32[L];
```

```
q31_t fftSignalQ31[2*L];
q31_t fftMagnitudeQ31[2];

#define NUM_FIR_TAPS 202
q15_t hfir_coeffs_q15[NUM_FIR_TAPS] = {
    -9,  -29,  -40,  -40,  -28,   -7,   17,   38,
    49,   47,   29,    1,  -30,  -55,  -66,  -58,
   -31,    9,   51,   82,   91,   72,   29,  -28,
   -82, -117, -119,  -84,  -20,   57,  124,  160,
   149,   91,    0,  -99, -176, -206, -175,  -88,
    33,  153,  235,  252,  193,   72,  -80, -217,
  -297, -293, -199,  -40,  141,  289,  358,  323,
   189,   -9, -213, -364, -412, -339, -161,   73,
   294,  436,  453,  336,  114, -149, -376, -499,
  -477, -312,  -51,  233,  456,  548,  480,  269,
   -27, -320, -525, -579, -462, -207,  113,  404,
   580,  587,  422,  131, -201, -477, -614, -572,
  -362,  -45,  287,  534,  626,  534,  287,  -45,
  -362, -572, -614, -477, -201,  131,  422,  587,
   580,  404,  113, -207, -462, -579, -525, -320,
   -27,  269,  480,  548,  456,  233,  -51, -312,
  -477, -499, -376, -149,  114,  336,  453,  436,
   294,   73, -161, -339, -412, -364, -213,   -9,
   189,  323,  358,  289,  141,  -40, -199, -293,
  -297, -217,  -80,   72,  193,  252,  235,  153,
    33,  -88, -175, -206, -176,  -99,    0,   91,
   149,  160,  124,   57,  -20,  -84, -119, -117,
   -82,  -28,   29,   72,   91,   82,   51,    9,
   -31,  -58,  -66,  -55,  -30,    1,   29,   47,
    49,   38,   17,   -7,  -28,  -40,  -40,  -29,
    -9,   0};

q31_t hanning_window_q31[L];

q15_t hfir_state_q15[NUM_FIR_TAPS + BLOCKSIZE] = {0};

float biquad_coeffs_f32[5] = {0.01f, 0.0f, -0.01f,
1.690660431255413f, -0.9801f};
float biquad_state_f32[2] = {0};

/* -------------------------------------------------------------
```

```
main program
*---------------------------------------------------------------*/


int main (void) {                        /* execution starts here      */
 int i, samp;
 arm_fir_instance_q15 DTMF_FIR;
 arm_rfft_instance_q31 DTMF_RFFT;
 arm_cfft_radix4_instance_q31 DTMF_CFFT;
 arm_biquad_cascade_df2T_instance_f32 DTMF_BIQUAD;

 // Generate the input sine wave
 // The signal will have an amplitude of 0.5 and a frequency of FREQ Hz
 // Create floating point, Q31, and Q7 versions.

 for(i=0; i<L; i++) {
  inSignalF32[i] = 0.5f * sinf(2.0f * PI * FREQ * i / SR);
  inSignalQ15[i] = (q15_t) (32768.0f * inSignalF32[i]);
  inSignalQ31[i] = (q31_t) ( 2147483647.0f * inSignalF32[i]);
 }

 /* ----------------------------------------------------------------
 ** Process with FIR filter
 ** ---------------------------------------------------------- */

 if (arm_fir_init_q15(&DTMF_FIR, NUM_FIR_TAPS, &hfir_coeffs_q15[0],
                      &hfir_state_q15[0], BLOCKSIZE) !=
ARM_MATH_SUCCESS) {
     // error condition
     // exit(1);
 }

 for(samp = 0; samp < L; samp += BLOCKSIZE) {
     arm_fir_q15(&DTMF_FIR, inSignalQ15 + samp, outSignalQ15 + samp,
     BLOCKSIZE);
 }

 /* ------------------------------------------------------------
 ** Process with a floating point Biquad filter
 ** ------------------------------------------------------------*/

 arm_biquad_cascade_df2T_init_f32(&DTMF_BIQUAD, 1, biquad_coeffs_f32,
                                  biquad_state_f32);

 for(samp = 0; samp < L; samp += BLOCKSIZE) {
```

```
    arm_biquad_cascade_df2T_f32(&DTMF_BIQUAD, inSignalF32 + samp,
                                outSignalF32 + samp, BLOCKSIZE);

}

/* ------------------------------------------------------------------
** Process with Q31 FFT
** ------------------------------------------------------------- */

// Create the Hanning window. This is usually done once at the
// start of the program.

for(i=0; i<L; i++) {
 hanning_window_q31[i] =
   (q31_t) (0.5f * 2147483647.0f * (1.0f - cosf(2.0f*PI*i / L)));
}

// Apply the window to the input buffer
arm_mult_q31(hanning_window_q31, inSignalQ31, inSignalQ31, L);

arm_rfft_init_q31(&DTMF_RFFT, &DTMF_CFFT, 512, 0, 1);

// Compute the FFT
arm_rfft_q31(&DTMF_RFFT, inSignalQ31, fftSignalQ31);

arm_cmplx_mag_q31(fftSignalQ31, fftMagnitudeQ31, L);
}
```

## 22.6 Example 2 — least squares motion tracking

Tracking the motion of an object is a common problem that occurs in many applications. Examples include navigation systems, exercise equipment, video game controllers, and factory automation. Noisy measurements of past positions of the object are available and they need to be combined to estimate future positions. One approach to solving this problem is to combine multiple noisy measurements in order to estimate the underlying trajectory (position, velocity, and acceleration) and then project this into the future.

Consider an object under constant acceleration. The motion of the object as a function of time $t$ will be:

$$x(t) = x_0 + v_0 t + a t^2$$

where:

$x_0$ is the initial position
$v_0$ is the initial position
$a$ is the acceleration

In this example we will assume that the acceleration is constant, but the approach used can be extended to time varying acceleration.

Assume that we have measurements of past location of the object at times $t_1$, $t_2$, ..., $t_N$. Assume that the position measurements are noisy and that we only know an approximate position. Place the measurements and times into column vectors:

$$x = \begin{bmatrix} x_1 \\ x_2 \\ \vdots \\ x_N \end{bmatrix} \quad t = \begin{bmatrix} t_1 \\ t_2 \\ \vdots \\ t_N \end{bmatrix}$$

The overall equation that relates the measurements to the unknowns $x_0$, $v_0$, and $a$ is then:

$$\begin{bmatrix} x_1 \\ x_2 \\ \vdots \\ x_N \end{bmatrix} = x_0 + v_0 \begin{bmatrix} t_1 \\ t_2 \\ \vdots \\ t_N \end{bmatrix} + a \begin{bmatrix} t_1^2 \\ t_2^2 \\ \vdots \\ t_N^2 \end{bmatrix}$$

This expression can be computed using matrix multiplication:

$$x = Ac$$

where:

$$A = \begin{bmatrix} 1 & t_1 & t_1^2 \\ 1 & t_2 & t_1^2 \\ \vdots & \vdots & \vdots \\ 1 & t_N & t_N^2 \end{bmatrix}$$

and:

$$c = \begin{bmatrix} x_0 \\ v_0 \\ a \end{bmatrix}.$$

Since there are three unknowns, at least three measurements are required to compute the result vector $C$. In most cases, there are many more measurements than unknowns and the problem is overdetermined. One standard solution to this problem is to do a least squares fit. The solution $\hat{c}$ is the one that minimizes the error between the N estimated positions and the actual N measurement positions. The least squares solution can be found by solving the matrix equation:

$$\hat{c} = \left(A^T A\right)^{-1} A^T x$$

We can solve equations of this type using the matrix functions within the CMSIS-DSP library. A matrix in the CMSIS library is represented using a data structure. For floating point data, the structure is:

```
typedef struct
{
uint16_t numRows;  /**< number of rows of the matrix.  */
```

```
  uint16_t numCols;   /**< number of columns of the matrix. */
  float32_t *pData;   /**< points to the data of the matrix. */
} arm_matrix_instance_f32;
```

Essentially the matrix structure keeps track of the size of the matrix (numRows, numCols) and contains a pointer to the data (pData). Element (R, C) of the matrix is stored at location:

```
pData[R*numRows + C]
```

in the array. That is, the array contains the first row of data, followed by the second row of data, and so on. You can either manually initialize the matrix instance structure yourself by setting internal fields, or you can use the function arm_mat_init_f32(). In practice it is easier to manually initialize the matrix.

The overall code that computes the least squares solution is shown in section 22.6.1. The top of the function allocates memory for all of the pData arrays used by the matrices. Matrices $t$ and $x$ are initialized with actual data while all other matrices are initially set to zero. After this, the individual matrix instance structures are initialized. Note, that multiple matrices are defined so that intermediate results can be stored. We define:

A — matrix $A$ from above
AT — transpose of $A$
ATA — the product $A^T A$
invATA — the inverse of $A^T A$
B — the product $(A^T A)^{-1} A^T$
c — the final result from above

The start of the main function initializes the values for matrix A. After this, several matrix math functions are called to finally arrive at the result vector c. The result contains three elements and they can be seen by inspecting cData within the debugger. We find that:

$$x_0 = c[0] = 8.7104$$
$$v_0 = c[1] = 38.8748$$
$$a = c[2] = -9.7923$$

The original input data and the resulting fit are shown in Figure 22.10. The thin line is the measured data showing the random noise, and the thick line is the resulting data fit. The figure shows that the underlying least squares fit is quite accurate and the fit could be used to extrapolate the measured data into the future.

## 22.6.1 Example least squares code

```
#include "arm_math.h"   /* Main include file for CMSIS DSP      */

#define NUMSAMPLES 51    /* Number of measurements               */
#define NUMUNKNOWNS 3    /* Number of unknowns in polynomial fit */
```

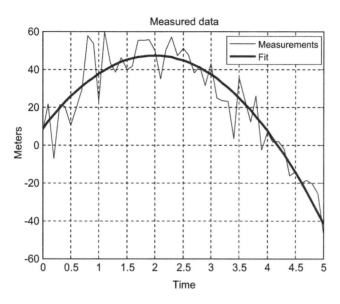

**FIGURE 22.10**

Raw measurements (thin line) and resulting least squares fit (thick line)

```
// Allocate memory for the matrix arrays. Only t and x have initial
data defined.

// Contains the times at which the data was sampled. In this examples,
the data
// is evenly spaced but this is not required for the least squares
fit.
float32_t tData[NUMSAMPLES] =
{
  0.0f, 0.1f, 0.2f, 0.3f, 0.4f, 0.5f, 0.6f, 0.7f,
  0.8f, 0.9f, 1.0f, 1.1f, 1.2f, 1.3f, 1.4f, 1.5f,
  1.6f, 1.7f, 1.8f, 1.9f, 2.0f, 2.1f, 2.2f, 2.3f,
  2.4f, 2.5f, 2.6f, 2.7f, 2.8f, 2.9f, 3.0f, 3.1f,
  3.2f, 3.3f, 3.4f, 3.5f, 3.6f, 3.7f, 3.8f, 3.9f,
  4.0f, 4.1f, 4.2f, 4.3f, 4.4f, 4.5f, 4.6f, 4.7f,
  4.8f, 4.9f, 5.0f
};

// Contains the noisy position measurements
float32_t xData[NUMSAMPLES] =
{
  7.4213f, 21.7231f, -7.2828f, 21.2254f, 20.2221f, 10.3585f, 20.3033f,
29.2690f,
```

```
   57.7152f, 53.6075f, 22.8209f, 59.8714f, 43.1712f, 38.4436f,
46.0499f, 39.8803f,
   41.5188f, 55.2256f, 55.1803f, 55.6495f, 49.8920f, 34.8721f,
50.0859f, 57.0099f,
   47.3032f, 50.8975f, 47.4671f, 38.0605f, 41.4790f, 31.2737f,
42.9272f, 24.6954f,
   23.1770f, 22.9120f, 3.2977f, 35.6270f, 23.7935f, 12.0286f, 25.7104f,
-2.4601f,
   6.7021f, 1.6804f, 2.0617f, -2.2891f, -16.2070f, -14.2204f,
-20.1870f, -18.9303f,
   -20.4859f, -25.8338f, -47.2892f
  };

  float32_t AData[NUMSAMPLES * NUMUNKNOWNS];
  float32_t ATData[NUMSAMPLES *NUMUNKNOWNS];
  float32_t ATAData[NUMUNKNOWNS * NUMUNKNOWNS];
  float32_t invATAData[NUMUNKNOWNS * NUMUNKNOWNS];
  float32_t BData[NUMUNKNOWNS * NUMSAMPLES];
  float32_t cData[NUMUNKNOWNS];

  // Array instance structure initialization. For each instance, the
form is:
  // MAT = {numRows, numCols, pData};

  // Column vector t
  arm_matrix_instance_f32 t = {NUMSAMPLES, 1, tData};

  // Column vector x
  arm_matrix_instance_f32 x = {NUMSAMPLES, 1, xData};

  // Matrix A
  arm_matrix_instance_f32 A = {NUMSAMPLES, NUMUNKNOWNS, AData};

  // Transpose of matrix A
  arm_matrix_instance_f32 AT = {NUMUNKNOWNS, NUMSAMPLES, ATData};

  // Matrix product AT * A
  arm_matrix_instance_f32 ATA = {NUMUNKNOWNS, NUMUNKNOWNS, ATAData};

  // Matrix inverse inv(AT*A)
  arm_matrix_instance_f32 invATA = {NUMUNKNOWNS, NUMUNKNOWNS,
invATAData};

  // Intermediate result invATA * AT
  arm_matrix_instance_f32 B = {NUMUNKNOWNS, NUMSAMPLES, BData};
```

```
// Solution
arm_matrix_instance_f32 c = {NUMUNKNOWNS, 1, cData};


/*------------------------------------------------------------------
** main program
**----------------------------------------------------------------*/

int main (void) {
 int i;
 float y;

 y = sqrtf(xData[0]);
 cData[0] = y;


 // Fill in the values for matrix A. Each row contains:
 // [1.0f t t*t]
 for(i=0; i<NUMSAMPLES; i++) {
  AData[i*NUMUNKNOWNS + 0] = 1.0f;
  AData[i*NUMUNKNOWNS + 1] = tData[i];
  AData[i*NUMUNKNOWNS + 2] = tData[i] * tData[i];
 }

 // Transpose
 arm_mat_trans_f32(&A, &AT);

 // Matrix multplication AT * A
 arm_mat_mult_f32(&AT, &A, &ATA);

 // Matrix inverse inv(ATA)
 arm_mat_inverse_f32(&ATA, &invATA);

 // Matrix multiplication invATA * x;
 arm_mat_mult_f32(&invATA, &AT, &B);

 // Final result.
 arm_mat_mult_f32(&B, &x, &c);

 // Examine cData in the debugger to see the final values
}
```

# Advanced Topics

# 23

## CHAPTER OUTLINE

## 23.1 Decisions and branches

### 23.1.1 Conditional branches

Very often in programming we need to handle conditional branches based on some complex decisions. For example, a conditional branch might depend on the value of an integer variable. If the range of the variable is small, for example, 0 to 31, there is a way to simplify the program code to make the decision-making stage more efficient.

Let us see an example. If we want to print out the value of an input integer if the value is a prime number in the range 0 to 31, the simplest code would be:

```
void is_a_prime_number(unsigned int i)
{
 if ((i==2) ||  (i==3) ||  (i==5) ||  (i==7) ||
    (i==11) || (i==13) || (i==17) || (i==19) ||
    (i==23) || (i==29) || (i==31)) {
    printf ("- %d\n", i);
    }
 return;
}
```

However, this can lead a very long-branch tree after compilation (see disassembled code below):

```
is_a_prime_number
  0x080002ca:  2802    .(  CMP  r0,#2
  0x080002cc:  d013    ..  BEQ  0x80002f6 ; branch_simple + 44
  0x080002ce:  2803    .(  CMP  r0,#3
  0x080002d0:  d011    ..  BEQ  0x80002f6 ; branch_simple + 44
  0x080002d2:  2805    .(  CMP  r0,#5
  0x080002d4:  d00f    ..  BEQ  0x80002f6 ; branch_simple + 44
  0x080002d6:  2807    .(  CMP  r0,#7
  0x080002d8:  d00d    ..  BEQ  0x80002f6 ; branch_simple + 44
  0x080002da:  280b    .(  CMP  r0,#0xb
  0x080002dc:  d00b    ..  BEQ  0x80002f6 ; branch_simple + 44
  0x080002de:  280d    .(  CMP  r0,#0xd
  0x080002e0:  d009    ..  BEQ  0x80002f6 ; branch_simple + 44
  0x080002e2:  2811    .(  CMP  r0,#0x11
  0x080002e4:  d007    ..  BEQ  0x80002f6 ; branch_simple + 44
  0x080002e6:  2813    .(  CMP  r0,#0x13
  0x080002e8:  d005    ..  BEQ  0x80002f6 ; branch_simple + 44
  0x080002ea:  2817    .(  CMP  r0,#0x17
  0x080002ec:  d003    ..  BEQ  0x80002f6 ; branch_simple + 44
  0x080002ee:  281d    .(  CMP  r0,#0x1d
  0x080002f0:  d001    ..  BEQ  0x80002f6 ; branch_simple + 44
  0x080002f2:  281f    .(  CMP  r0,#0x1f
  0x080002f4:  d103    ..  BNE  0x80002fe ; branch_simple + 52
  0x080002f6:  4601    .F  MOV  r1,r0
  0x080002f8:  a01e    ..  ADR  r0,{pc}+0x7c ; 0x8000374
  0x080002fa:  f000b93d ..=. B.W  __2printf ; 0x8000578
  0x080002fe:  4770  pG  BX  lr
```

We can reduce this by encoding the condition into a binary pattern and use this for branch decision:

```
void branch_method1(unsigned int i)
{
 /* Bit pattern is
    31:0 - 1010 0000 1000 1010 0010 1000 1010 1100 = 0xA08A28AC */
 if ((1<<i) & (0xA08A28AC)) {
    printf ("- %d\n", i);
 }
 return;
}
```

By doing this, the generated code is much shorter:

```
branch_method1
 0x080003c0: 2101     .!    MOVS r1,#1
 0x080003c2: 4a24     $J    LDR r2,[pc,#144] ; [0x8000454] = 0xa08a28ac
 0x080003c4: 4081     .@    LSLS r1,r1,r0
 0x080003c6: 4211     .B    TST r1,r2
 0x080003c8: bf08     ..    IT EQ
 0x080003ca: 4770     pG    BXEQ lr
 0x080003cc: 4601     .F    MOV r1,r0
 0x080003ce: a01f     ..    ADR r0,{pc}+0x7e ; 0x800044c
 0x080003d0: f000b950 ..P. B.W __2printf ; 0x8000674
```

This method can also be used when the conditional branch is dependent on a number of binary inputs. For example, in a software FSM design, you might need to determine the next state based on a number of binary inputs. The following code merges 4 binary inputs into an integer and uses it for a conditional branch:

```
void branch_method2(unsigned int i0, unsigned int i1, unsigned int i2,
unsigned int i3,unsigned int i4,unsigned int i)
{
 unsigned int tmp=0;
 if (i0) tmp=1;
 if (i1) tmp|=2;
 if (i2) tmp|=4;
 if (i3) tmp|=8;
 if (i4) tmp|=0x10;
 if ((1<<tmp) & (0xA08A28AC)) {
    printf ("- %d\n", i);
 }
 return;
}
```

If you are programming in assembler, you can also use the following method, which is one instruction shorter. Instead of using a bit mask that needs to be shifted left, the bit pattern is shifted right and the required bit is shifted into the carry flag. The status of the carry flag can then be used for the conditional branch (Figure 23.1).

```
branch_method3
 PUSH {R4,LR} ; Push two registers to ensure double word stack alignment
 LDR R1, =0xA08A28AC
 ADDS R2, R0, #1 ; Minimum shift is 1 bit
 LSRS R1, R1, R2
 BCC  branch_method3_exit
 BL __cpp(branch_method3_printf)
branch_method3_exit
 POP {R4, PC}
```

For a wider input data range, you can expand the bit pattern into an array. For example, to determine whether an input of 0 to 127 is a prime number, we could use the following code:

```
void branch_method4(unsigned int i)
{
 /* Bit pattern is
  31: 0 - 1010 0000 1000 1010 0010 1000 1010 1100 = 0xA08A28AC
  63:32 - 0010 1000 0010 0000 1000 1010 0010 0000 = 0x28208A20
  95:64 - 0000 0010 0000 1000 1000 0010 1000 1000 = 0x02088288
 127:96 - 1000 0000 0000 0010 0010 1000 1010 0010 = 0x800228A2
 */
 const uint32_t bit_pattern[4] = {0xA08A28AC,
  0x28208A20, 0x02088288, 0x800228A2};
 uint32_t i1, i2;
 i1 = i & 0x1F;        // Bit position
 i2 = (i & 0x60) >> 5;// Mask index
 if ((1<<i1) & (bit_pattern[i2])) {
    printf ("- %d\n", i);
 }
 return;
}
```

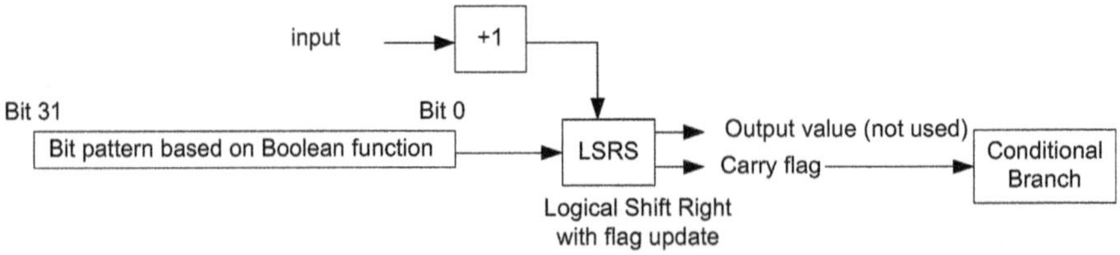

**FIGURE 23.1**

Conditional branch based on a 5-bit integer input

### 23.1.2 **Complex decision tree**

In many cases the decision tree can have many different destinations. Two of the important instructions in the Cortex®-M3 and Cortex-M4 processors used for this purpose are the table branch instructions (TBB and TBH).

We examined the Unsigned Bit Field eXtract (UBFX) and Table Branch (TBB/TBH) instructions in Chapter 5. These two instructions can work together to form a very powerful branching tree. This capability is very useful in data communication applications, where the data sequence can have different meanings with different headers. For example, let's say that the following decision tree based on Input A is to be coded in assembler (Figure 23.2).

```
DecodeA
 LDR  R0,=A          ; Get the value of A from memory
 LDR  R0,[R0]
 UBFX R1, R0, #6, #2  ; Extract bit[7:6] into R1
 TBB  [PC, R1]
BrTable1
 DCB ((P0       -BrTable1)/2) ; Branch to P0       if A[7:6] = 00
 DCB ((DecodeA1-BrTable1)/2) ; Branch to DecodeA1 if A[7:6] = 01
 DCB ((P1       -BrTable1)/2) ; Branch to P1       if A[7:6] = 10
 DCB ((DecodeA2-BrTable1)/2) ; Branch to DecodeA1 if A[7:6] = 11
DecodeA1
 UBFX R1, R0, #3, #2 ; Extract bit[4:3] into R1
 TBB  [PC, R1]
BrTable2
 DCB ((P2 -BrTable2)/2) ; Branch to P2 if A[4:3] = 00
 DCB ((P3 -BrTable2)/2) ; Branch to P3 if A[4:3] = 01
 DCB ((P4 -BrTable2)/2) ; Branch to P4 if A[4:3] = 10
 DCB ((P4 -BrTable2)/2) ; Branch to P4 if A[4:3] = 11
DecodeA2
 TST R0, #4 ; Only 1 bit is tested, so no need to use UBFX
 BEQ P5
 B   P6
P0 ... ; Process 0
P1 ... ; Process 1
P2 ... ; Process 2
P3 ... ; Process 3
P4 ... ; Process 4
P5 ... ; Process 5
P6 ... ; Process 6
```

This code completes the decision tree in a short assembler code sequence. If the branch target addresses are at a larger offset, some of the TBB instructions would have to be replaced by TBH instructions.

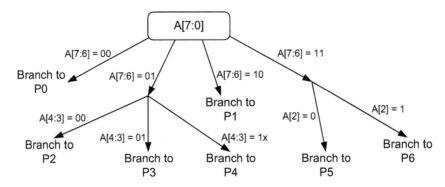

**FIGURE 23.2**

Bit Field Decoder: Example use of UBFX and TBB instruction

## 23.2 Performance considerations

To get the best out of the Cortex®-M3 and Cortex-M4 processors, a few aspects need to be considered. First, we need to avoid memory wait states. During the design stage of the microcontroller or SoC, the designer should optimize the memory system design to allow instruction and data accesses to be carried out at the same time, and use 32-bit memories, if possible.

For developers, the memory map should be arranged so that program code is executed from the code region and the majority of data accesses (apart from literal data) are carried out via the system bus. In this way, data accesses can be carried out at the same time as instruction fetches. Due to the design of the internal bus interconnect inside the Cortex-M3/M4 processor (the Bus Matrix), program execution from the system buses (address 0x20000000 or above) is slower than program execution from the CODE region (0x0 to 0x1FFFFFFF).

If possible, avoid using unaligned transfers. An unaligned transfer might take two or more Advanced High-Performance Bus (AHB) transfers to complete and therefore reduces performance, so plan your data structure carefully and avoid using packed structures (_pack) if possible. Normally C compilers do not generate unaligned data. In assembly language with ARM® tools, you can use the ALIGN directive to ensure that a data location is aligned.

If possible, limited your function calls to 4 input parameters or fewer. When there are more than 4 input parameters, the extra parameters need to be passed via stack memory and will take longer to set up and access. If there is a lot of information to be transferred, try to group the data into a structure and pass a pointer to the data structure to reduce the number of parameters.

If possible, the interrupt vector table should also be put into the code region, and the program stack placed in the SRAM or other RAM region at 0x20000000 or above. Thus, vector fetch and stacking can be carried out at the same time. If both vector table and stack are located in the SRAM region, extra clock cycles might result in increased interrupt latency because both vector fetch and stacking could share the same system bus (unless the stack is located in the code region, which uses a D-Code bus).

Most of you will be using C language for development, but for those who are using assembly language, you can use a few tricks to speed up parts of the program.

Use memory access instructions with offset addressing. When multiple memory locations in a small region are to be accessed, instead of writing:

```
LDR R0, =0xE000E400 ; Set interrupt priority #3,#2,#1,#0
LDR R1, =0xE0C02000 ; priority levels
STR R1,[R0]
LDR R0, =0xE000E404 ; Set interrupt priority #7,#6,#5,#4
LDR R1, =0xE0E0E0E0 ; priority levels
STR R1,[R0]
```

You can reduce the program code to the following:

```
LDR R0, =0xE000E400 ; Set interrupt priority #3,#2,#1,#0
LDR R1, =0xE0C02000 ; priority levels
STR R1,[R0]
LDR R1,=0xE0E0E0E0  ; priority levels
STR R1,[R0,#4]      ; Set interrupt priority #7,#6,#5,#4
```

The second store uses an offset from the first address and hence reduces the number of instructions.

Combine multiple memory accesses into Load/Store Multiple instructions (LDM/STM). The preceding example can be further reduced by using STM instruction as follows:

```
LDR   R0,=0xE000E400 ; Set interrupt priority base
LDR   R1,=0xE0C02000 ; priority levels #3,#2,#1,#0
LDR   R2,=0xE0E0E0E0 ; priority levels #7,#6,#5,#4
STMIA R0, {R1, R2}
```

You can also improve performance by making use of the addressing mode features available. For example, when reading a lookup table:

```
Read_Table
 ; Input R0 = index
 LDR R1,=Look_up_table ; Address of the lookup table
 LDR R1, [R1]          ; Get lookup table base address
 LSL R2, R0, #2        ; Times 4 (each item in the table is 4 bytes)
 ADD R2, R1            ; Get actual address (base + offset)
 LDR R0, [R2]          ; Read table
```

```
BX  LR                  ; Return
ALIGN 4
Look_up_table
DCD 0x12345678
DCD 0x23456789
...
```

You can reduce the code significantly:

```
Read_Table
  ; Input R0 = index
  LDR R1,=Look_up_table    ; Address of the lookup table
  LDR R1, [R1]             ; Get lookup table base address
  LDR R0, [R1, R0, LSL #2] ; Read table with base + (index << 2)
  BX  LR                   ; Return
  ALIGN 4
Look_up_table
DCD 0x12345678
DCD 0x23456789
...
```

In some cases, you should use IF-THEN (IT) instruction blocks to replace small conditional branches with an if-then-else structure. Since the Cortex-M3 and Cortex-M4 are pipelined processors, a branch penalty happens when a branch operation is undertaken. If the conditional branch operation is used to skip a few instructions, this can be replaced by an IT instruction block, which might save a few clock cycles. However, you need to check the clock cycle saving on a case-by-case basis. For example, in the following example code you will not be able to save any clock cycles by using IT instruction:

| Using Condition Branch | Using IT |
|---|---|
| CMP R0, R1 ; 1 cycle<br>BNE Label ; 2 cycles or 1 cycle<br>MOVS .... ; 1 cycle<br>MOVS .... ; 1 cycle<br>MOVS .... ; 1 cycle<br>MOVS .... ; 1 cycle<br>Label<br>; Branch taken (condition is EQ)<br>3 cycle, not taken is 6 cycles | CMP R0, R1  ; 1 cycle<br>ITTTT EQ   ; 1 cycle<br>MOVEQ .... ; 1 cycle<br>MOVEQ .... ; 1 cycle<br>MOVEQ .... ; 1 cycle<br>MOVEQ .... ; 1 cycle<br>; If no IT folding, both paths<br>take 6 cycles. Performance<br>can get worst |

If an operation can be carried out by either two Thumb® instructions or a single Thumb-2 instruction, the Thumb-2 instruction method should be used, because it gives a shorter execution time, despite the fact that the memory size is the same.

## 23.3 **Double-word stack alignment**

In applications that conform to AAPCS, it is necessary to ensure that the stack pointer value at function entry should be aligned to a double-word address. To achieve this requirement, the stacking of registers during exception handling must be adjusted accordingly. This is a configurable option on the Cortex®-M3 and Cortex-M4 processors. To enable this feature, the STKALIGN bit in the Configuration Control Register (CCR) in the System Control Block (SCB) needs to be set (see Table F.2.6 in Appendix F). For example, if a CMSIS-compliant device driver is used in a C language project:

```
SCB->CCR = SCB->CCR | SCB_CCR_STKALIGN_Msk;
// SCB_CCR_STKALIGN_Msk =0x200
```

If the project is in C but CMSIS is not used:

```
#define NVIC_CCR *((volatile unsigned long *) (0xE000ED14))
NVIC_CCR |= 0x200; /* Set STKALIGN in NVIC */
```

This can also be done in assembly language:

```
LDR R0,=0xE000ED14 ; Set R0 to be address of CCR
LDR R1, [R0]
ORR R1, R1, #0x200 ; Set STKALIGN bit
STR R1, [R0]       ; Write back to CCR
```

When the STKALIGN bit is set during exception stacking, bit 9 of the stacked xPSR (combined Program Status Register) is used to indicate whether a stack pointer adjustment has been made to align the stack pointer address. When unstacking, the stack pointer (SP) adjustment checks bit 9 of the stacked xPSR and adjusts the SP accordingly.

To prevent stack data corruption, the STKALIGN bit must not be changed within an exception handler. This can cause a mismatch of stack pointer location before and after the exception.

This feature is available from Cortex-M3 revision 1 onward. Early Cortex-M3 products based on revision 0 do not have this feature. In Cortex-M3 revision 2 and later, and Cortex-M4, this feature is enabled by default, whereas in Cortex-M3 revision 1 this needs to be turned on by software (Table 23.1).

This feature should be used if AAPCS-compliance is required.

## 23.4 **Various methods for semaphore implementation**

In general it is recommended that you use exclusive access instructions to implement semaphores. However, there are also other possibilities.

**Table 23.1** Availability of Double-word Stack Alignment for Exception Handling

| Processor | Double Word Stack Alignment |
|---|---|
| Cortex-M3 revision 0 | Not available – need specific C compiler option for AAPCS compliant |
| Cortex-M3 revision 1 (r1p0/r1p1) | Available – need to be enabled by software |
| Cortex-M3 revision 2 (r2p0/r2p1) | Available – enabled by default |
| Cortex-M4 all revisions | Available – enabled by default |

### 23.4.1 Using SVC services for semaphores

Due to the exception priority structure in the Cortex®-M processors, you can only have one instance of SVC exception happen at a time: You cannot execute an SVC instruction inside an SVC handler, or when the current priority is the same or higher than the SVC exception. Otherwise a fault exception will be triggered.

You can therefore implement semaphore control accesses using SVC services. In this way you can guarantee that only one application task in a system is accessing the semaphore. This method can be used if the code needs to be used on Cortex-M0/M0+/M1 processors, which do not support exclusive accesses.

### 23.4.2 Use bit-band for semaphores

It is possible to use the bit-band feature to carry out Mutex (mutual exclusive) operations, provided that the memory system supports locked transfers or only one bus master is present on the memory bus. With bit-banding, it is possible to implement the semaphore in normal C-code, but the operation is different from using exclusive access. To use bit-banding for resource allocation control, a memory location (such as word data) within a bit-band memory region is used, and each bit of this variable indicates that the resource is used by a certain task.

Since the bit-band alias writes are locked READ-MODIFY-WRITE transfers (the bus master cannot be switched to another one between the transfers), provided that all tasks only change the lock bit representing themselves, the lock bits of other tasks will not be lost, even if two tasks try to write to the same memory location at the same time. Unlike using exclusive accesses, it is possible for a resource to be "locked" simultaneously by two tasks for a short period of time until one of them detects the conflict and releases the lock (Figure 23.3).

Using bit-banding for semaphores works only if all the tasks in the system change only the lock bit they are assigned to using the bit-band alias. If any of the tasks change the lock variable using a normal write, the semaphore can fail if another task sets a lock bit just before the write to the lock variable. The previous lock bit set by the other task will be lost.

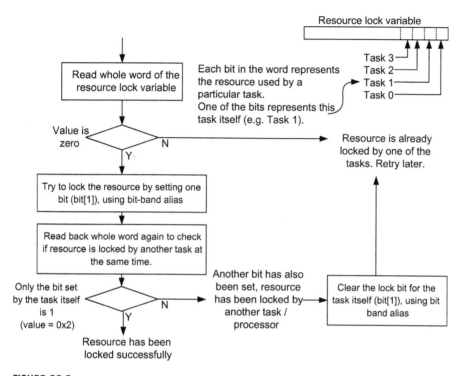

**FIGURE 23.3**

Mutex implemented using the bit-band feature (Task 1 trying to set a lock)

## 23.5 **Non-base Thread enable**

In the Cortex®-M3 and Cortex-M4 processors, it is possible to switch a running interrupt handler from privileged level to user access level. This is needed when the interrupt handler code is part of a user application and should not be allowed to have privileged access. This feature is enabled by the Non-base Thread Enable (NONBASETHRDENA) bit in the Configuration Control register (CCR).

---

**USE THIS FEATURE WITH CAUTION**

---

Because of the need to manually adjust the stack and modify the stacked data, this feature should be avoided in normal application programming. If it is necessary to use this feature, it must be done very carefully, and the system designer must ensure that the interrupt service routine is terminated correctly. Otherwise, it could cause some interrupts with the same or lower priority levels to be masked.

To use this feature, an exception handler redirection is involved. The vector in the vector table points to a handler that starts running in privileged mode but switches to unprivileged mode in the middle. As a result, this redirect handler needs to be placed in a memory region that is accessible in both privileged and unprivileged states. In this example, we create a SysTick Handler routine that needs to be run at unprivileged state called User_SysTick_Handler. Then the SysTick Handler (which the SysTick vector points to can be implemented as follows:

```
__asm void SysTick_Handler(void)
{ // Redirect handler - no Floating point instruction
 PUSH {R4, LR} ; Push 2 words for double word stack alignment
 SVC 0          ; A SVC function to change from privileged to
                ; unprivileged mode
 BL __cpp(User_SysTick_Handler)
 SVC 1          ; A SVC function to change back from user to
                ; privileged mode
 POP {R4, PC}  ; Return
}
```

The SVC handler is divided into three parts as follows:

Step 1: Determine the parameter when calling SVC.

Step 2: for SVC service #0, enable the NONBASETHRDENA, adjust the user stack and EXC_RETURN value, then return to the redirect handler in user mode, using the process stack.

Step 2: for SVC service #1 disable the NONBASETHRDENA, restore the user stack pointer position, then return to the redirect handler in privileged mode, using the main stack.

The SVC handler is shown in the code that follows. The SVC service #0 needs to create a stack frame in the Process Stack so that the unprivileged part of the SysTick Handler uses the Process Stack.

```
__asm void SVC_Handler(void)
{
 ; Extract SVC number
 TST  LR, #0x4 ; Test EXC_RETURN bit 2
 ITE  EQ  ; if zero then
 MRSEQ R0, MSP  ; Get correct stack pointer to R0
 MRSNE R0, PSP
 LDR  R1, [R0, #24] ; Get stacked PC
 LDRB.W R0, [R1, #-2] ; Get SVC parameter at stacked PC minus 2
 CBZ  R0, svc_service_0 ; if zero, branch to SVC service 0
 CMP  R0, #1
 BEQ  svc_service_1  ; if one, branch to SVC service 1
 B.W  Unknown_SVC_Request
```

```
; -----------------------------------------------------------
svc_service_0 ; Service to switch handler from
                 ; privileged mode to unprivileged mode
                 ; and use Process Stack. For this service,
                 ; stack framce must be in Main Stack because it is
                 ; called from an exception handler.
 MRS  R0, PSP      ; Adjust PSP to create space for new stack frame
                 ; and make a back up of PSP due to the
 TST  R0, #0x4     ; Check PSP to see if it is double word aligned
 ITE  EQ
 MOVSEQ R3, #0x20 ; No Padding
 MOVWNE R3, #0x24 ; Padding needed.
 LDR  R1, =__cpp(&svc_PSP_adjust)
 STRB R3, [R1]     ; Record PSP adjustment for use in SVC #1
 SUBS R0, R0, R3   ; PSP = PSP - 0x20 or 0x24
                 ; Make sure Process Stack is double word aligned
 MSR  PSP, R0      ; Copy back to PSP
 MOVS R1, #0x20    ; Copy stack frame from main stack to
                 ; process stack. This stack frame is 8 words
                 ; because SysTick_Handler has not execute
                 ; any FP instruction yet.
svc_service_0_copy_loop
 SUBS R1, R1, #4
 LDR  R2, [SP, R1] ; Read data
 STR  R2, [R0, R1] ;
 CMP  R1, #0
 BNE  svc_service_0_copy_loop

 LDR  R1,[R0, #0x1C]  ; Changed stacked xPSR so that IPSR=0
 MOVW R2, #0x3FF      ; Clear IPSR, stack alignment bit
 BIC  R1, R1, R2
 STR  R1,[R0, #0x1C]  ; Clear stacked IPSR of user stack to 0
 LDR  R0, =0xE000ED14 ; Set Non-base thread enable in CCR
 LDR  r1,[r0]
 ORR  r1, #1
 STR  r1,[r0]
 MRS  R0, CONTROL     ; Set CONTROL[0] so Thread run in unprivileged
                    ; state
 ORRS R0, R0, #1
 MSR CONTROL, R0
 ORR  LR, #0x1C       ; Change LR to return to thread,using PSP,
                    ; 8 words stack frame
 BX  LR
; -----------------------------------------------------------
```

```
        svc_service_1              ; Service to switch handler back from
                                   ; unprivileged mode to
                                   ; privileged mode
        MRS   R0, PSP              ; Update stacked PC in original privileged
                                   ; stack so that it
        LDR   R1,[R0, #0x18]       ; return to the instruction after 2nd
                                   ; SVC in the redirect handler
        STR   R1,[SP, #0x18]       ;
        MRS   R0, PSP              ; Adjust PSP back to what it was
                                   ; before 1st SVC
        LDR   R1, =__cpp(&svc_PSP_adjust)
        LDRB  R1, [R1]
        ADDS  R0, R0, R1
        MSR   PSP, R0
        LDR   R0, =0xE000ED14      ; Clear Non-base thread enable in CCR
        LDR   r1,[r0]
        BIC   r1, #1
        STR   r1,[r0]
        MRS   R0, CONTROL          ; Clear CONTROL[0]
        BICS  R0, R0, #1
        MSR   CONTROL, R0
        ORR   LR, #0x10            ; Return using 8 word stack frame
        BIC   LR, #0xC             ; Return to handler mode, using main stack
        BX    LR
Unknown_SVC_Request               ; Output error message
        BL    __cpp(Unknown_SVC_Request_Msg)
        B   .
        ALIGN
        }
```

The SVC services are used because the only way you can change the Interrupt Program Status register (IPSR) is via an exception entry or return. Other exceptions, such as software-triggered interrupts, could be used, but they are not recommended because they are imprecise and could be masked, which means that there is a possibility that the required stack copying and switch operation is not carried out immediately. The sequence of the code is illustrated in Figure 23.4, which shows the stack pointer changes and the current exception priority.

In Figure 23.4, the manual adjustment of the PSP inside the SVC service is highlighted by circles indicated by dotted lines.

## 23.6 Re-entrant Interrupt Handler

Over the years there have been questions from various forums about how to get an interrupt service to work in a re-entrant arrangement on the Cortex®-M

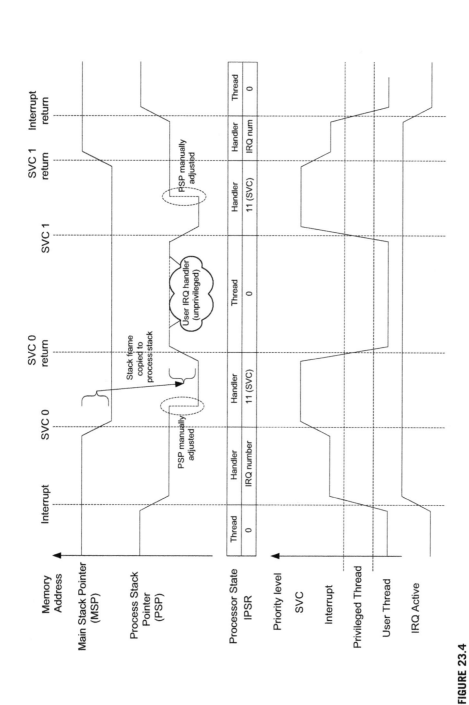

**FIGURE 23.4**

Operation of NONBASETHREADENA to enable a part of an exception handler to run in unprivileged state

microcontrollers. Due to the priority level mechanism of the Cortex-M processors, exceptions (including interrupts) are not designed to support re-entrant operations.

For example, when a timer interrupt is triggered and the ISR gets executed, the current priority level is set to the priority level of the timer interrupt. All exceptions with the same or lower priority will be blocked. Therefore if the timer interrupt is triggered again during the ISR execution, the timer interrupt pending status will be set and the ISR will be invoked again after the executing ISR is completed.

For a few users, this is a big problem as their existing software code relies on the ability to trigger the exception handler in a recursive manner, and they want to port their software from classic ARM® processors (e.g., ARM7TDMI™) to the Cortex-M. Re-entrant interrupts are possible on classic ARM processors like ARM7TDMI because the processor itself does not have any concept of interrupt priority.

There is a software workaround. You can create a wrapper for your Interrupt Handler so that it executes in Thread state, which can be interrupted by the same interrupt itself. The wrapper code contains two parts: the first part is the interrupt handler that switches itself back to Thread state and executes the ISR task, and the second part is an SVC exception handler that restores the state and resumes the original thread.

---

### USE THIS WORKAROUND WITH CAUTION

In general applications, use of re-entrant interrupts should be avoided. This allows a very high number of nested interrupt levels and can therefore cause stack overflow. The re-entrant interrupt mechanism demonstrated here also requires the priority to be the lowest exception in the system. Otherwise the processor can trigger a fault when the re-entrant interrupt is invoked during a lower-priority ISR.

---

The operation of the re-entrant interrupt code is shown in Figure 23.5.

```
Program code to allow SysTick Handler "Reentrant_SysTick_Handler()"
to be reentrant
/*-------------------------------------------*/
__asm void SysTick_Handler(void)
{
#if (__CORTEX_M >= 0x04)
#if (__FPU_USED == 1)
  ; The following 3 lines are for Cortex-M4 with FPU only
  TST   LR, #0x10     ; Test bit 4, if zero, need to trigger stacking
  IT    EQ
  VMOVEQ.F32  S0, S0  ; Trigger lazy stacking stacking
#endif
#endif
```

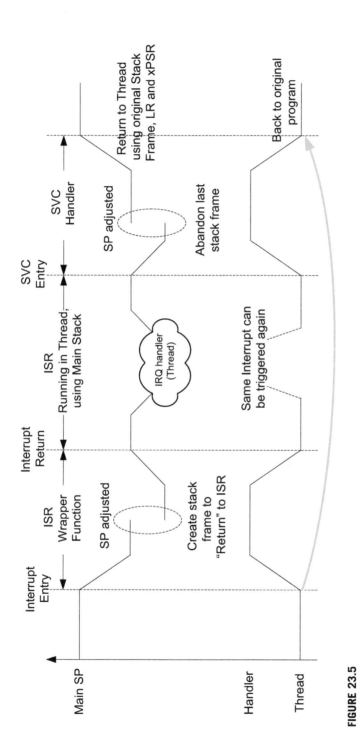

**FIGURE 23.5**

Using additional wrapper code to run an ISR in Thread to allow a re-entrant interrupt

```
    ; Now we are in Handler mode, using Main Stack, and
    ; SP should be Double word aligned
    MRS   R0, PSR
    PUSH {R0, LR}        ; Need to save PSR and LR in stack
    SUB   SP, SP , #0x20 ; Reserve 8 words for dummy stack frame for
return
    LDR   R0,=SysTick_Handler_thread_pt
    STR   R0,[SP, #24]
    LDR   R0,=0x01000000 ; xPSR
    STR   R0,[SP, #28]
    LDR   R0,=0xFFFFFFF9 ; Return to Thread with Main Stack, 8 word
stack frame
    MOV   LR, R0
    BX    LR
  SysTick_Handler_thread_pt
    BL    __cpp(Reentrant_SysTick_Handler)
    ; Block SysTick from being triggered just before SVC
    LDR   R0,=0xE000ED23 ; Address of SysTick priority level
    LDR   R0,[R0]
    MSR   BASEPRI, R0    ; Block SysTick from being triggered
    ISB                  ; Instruction Synchronisation Barrier
    SVC   0              ; Use SVC to return to original Thread
    B     .              ; Should not return here
  }
  __asm void SVC_Handler(void)
  {
    MOVS    R0, #0       ; Enable SysTick again
    MSR     BASEPRI, R0
    ISB     ; Instruction Synchronisation Barrier
#if (__CORTEX_M >= 0x04)
#if (__FPU_USED == 1)
    ; The following 3 lines are for Cortex-M4 with FPU only
    TST     LR, #0x10    ; Test bit 4, if zero, need to trigger stacking
    IT      EQ
    VMOVEQ.F32  S0, S0   ; Trigger lazy stacking stacking
#endif
#endif
    ; Extract SVC number
    TST     LR, #0x4  ; Test EXC_RETURN bit 2
    ITE     EQ        ; if zero then
    MRSEQ   R0, MSP      ; Get correct stack pointer to R0
    MRSNE   R0, PSP
    LDR     R1, [R0, #24]  ; Get stacked PC
    LDRB.W  R0, [R1, #-2]  ; Get SVC parameter at stacked PC minus 2
```

```
    CBZ    R0, svc_service_0 ; if zero, branch to SVC service 0
    B      Unknown_SVC_Request
svc_service_0
    ; Reentrant code finished, we can discard the current stack frame
    ; and restore the original stack frame. However, the current
    ; stack frame could be 8 words or 26 words.
    TST    LR,  #0x10       ; Test EXC_RETURN bit 4
    ITE    EQ
    ADDEQ  SP, SP, #104 ; LR Bit 4 was 0, 26 words in stack frame
    ADDNE  SP, SP, #32  ; LR Bit 4 was 1, 8 words in stack frame
    POP    {R0, R1}
    MSR    PSR, R0
    BX     R1
Unknown_SVC_Request
    BL   __cpp(Unknown_SVC_Request_Msg)
}
void Unknown_SVC_Request_Msg(unsigned int svc_num)
{  /* Display Error Message when SVC service is not known */
    printf("Error: Unknown SVC service request %d\n", svc_num);
    while(1);
}
/*----------------------------------------*/
void Reentrant_SysTick_Handler(void)
{
    printf ("[SysTick]\n");

    if (SysTick_Nest_Level < 3){
        SysTick_Nest_Level++;
        SCB->ICSR |= SCB_ICSR_PENDSTSET_Msk;  //Set pending status for
SysTick
        __DSB();
        __ISB();
        Delay(10);
        SysTick_Nest_Level--;
        } else {
        printf ("SysTick_Nest_Level = 3\n");
    }
    printf ("leaving [SysTick]\n");
    return;
}
```

Notes:

- The priority level of the reentrant interrupt handler need to be the lowest priority compared to the rest of the interrupts and exceptions.

- The use of BASEPRI prevents SVC handler to tail chain the reentrant interrupt handler because the interrupt handler changes SP, which stops SVC handler to access its normal stack frame.
- The handlers also force the deferred lazy stacking to take place to ensure FPU contexts are saved in nested ISR with FPU operations.

## 23.7 Bit Data Handling in C

In C or C++ you can define bit fields and proper use of this feature can help to generate more efficient code in bit data and bit field handling. For example, when dealing with I/O port control tasks, you can define a data structure of bits and a union in C to make coding easier:

```
Helper C structure and union definition in bit-data handling
typedef struct /* structure to define 32-bits */
{
uint32_t bit0:1;
uint32_t bit1:1;
uint32_t bit2:1;
uint32_t bit3:1;
uint32_t bit4:1;
uint32_t bit5:1;
uint32_t bit6:1;
uint32_t bit7:1;
uint32_t bit8:1;
uint32_t bit9:1;
uint32_t bit10:1;
uint32_t bit11:1;
uint32_t bit12:1;
uint32_t bit13:1;
uint32_t bit14:1;
uint32_t bit15:1;
uint32_t bit16:1;
uint32_t bit17:1;
uint32_t bit18:1;
uint32_t bit19:1;
uint32_t bit20:1;
uint32_t bit21:1;
uint32_t bit22:1;
uint32_t bit23:1;
uint32_t bit24:1;
uint32_t bit25:1;
```

```
  uint32_t bit26:1;
  uint32_t bit27:1;
  uint32_t bit28:1;
  uint32_t bit29:1;
  uint32_t bit30:1;
  uint32_t bit31:1;
} ubit32_t;          /*!< Structure used for bit access */

typedef union
{
   ubit32_t ub;      /*!< Type used for unsigned bit access */
   uint32_t uw;      /*!< Type used for unsigned word access */
} bit32_Type;
```

You can then declare variables using the newly created data type. For example:

```
bit32_Type foo;
foo.uw = GPIOD->IDR; // .uw access using word size
if (foo.ub.bit14) {  // .ub access using bit size
    GPIOD->BSRRH = (1<<14); // Clear bit 14
} else {
  GPIOD->BSRRL = (1<<14); // Set bit 14
}
```

In the above example the compiler generates a UBFX instruction to extract the bit value required. If the bit fields are defined as signed integers, the SBFX instruction will be used.

You can also declare a pointer to the register:

```
volatile bit32_Type * LED;

LED = (bit32_Type *) (&GPIOD->IDR);
if (LED->ub.bit12) {
    GPIOD->BSRRH = (1<<12); // Clear bit 12
} else {
  GPIOD->BSRRL = (1<<12); // Set bit 12
}
```

Please note that writing to a bit or bit field in this kind of code can result in a software read-modify-write sequence being generated by the C compiler. For I/O control this might be undesirable, because if another bit is changed by an interrupt handler that takes place between the read and write operations, the bit change made by the interrupt handler could be overwritten after the interrupt return, as described in Figure 6.12 in Chapter 6.

A bit field can have multiple bits. For example, the complex decision tree described in section 23.1.2 can be written in C like this:

```c
typedef struct
{
uint32_t bit1to0:2;
uint32_t bit2    :1;
uint32_t bit4to3:2;
uint32_t bit5    :1;
uint32_t bit7to6:2;
} A_bitfields_t;

typedef union
{
  A_bitfields_t ub; /*!< Type used for bit access */
    uint32_t    uw;     /*!< Type used for word access */
} A_Type;

void decision(uint32_t din)
{
   A_Type A;
   A.uw = din;
   switch (A.ub.bit7to6) {
   case 0:
     P0();
     break;
   case 1:
   switch (A.ub.bit4to3) {
       case 0:
         P2();
         break;
       case 1:
         P3();
         break;
       default:
         P4();
       break;
   };
   break;
   case 2:
    P1();
    break;
   default:
```

```
    if (A.ub.bit2) P6();
    else P5();
    break;
 }
 return;
}
```

## 23.8 Startup code

Most of the examples in this book use startup code (or boot code) that is coded in assembly language. It is possible to have the startup code written in C. However, this requires importing compiler-specific symbols and in some case compiler-specific directives. So the C startup code is still tool chain dependent.

For example, the projects for CoIDE (Chapter 17) use C startup code. In ARM® Application Note 179 Cortex®-M3 Embedded Software Development (reference 10), you can also find an example of a vector table written in C.

In a typical software development environment, the example software package provided by the microcontroller vendor would have the startup code and header files for various toolchains. This means that you do not have to worry about creating your own startup code and header files for the microcontroller devices.

Starting from CMSIS-Core v1.3, the System Initialization function `SystemInit()` is called from the startup code. This change allows the `SystemInit()` function to initialize external memory interface controller(s) before starting the C run-time startup code. In this way, you can place the stack and heap memories used by the C program at an external memory location.

The initial value for the Main Stack Pointer (MSP), however, still needs to point to a RAM region that does not require initialization, because some of the exceptions (e.g., NMI, HardFault) could happen at the beginning of the boot process.

## 23.9 Stack overflow detection

There are various methods for analyzing stack usage and detecting stack overflow.

### 23.9.1 Stack analysis by toolchain

First, many software development toolchains can generate reports on stack usage. For Keil™ MDK-ARM, the stack usage can be found in a generated HTML file, and for IAR Embedded Workbench, the stack usage can be generated by enabling stack analysis (see section 12.9.2).

Be aware that due to the space required by exception stack frames, you need to reserve more stack space than the application code requires. For systems without an OS, with everything running on the main stack, each level of nested interrupt can add an extra 8 words (or 26 words, if floating point unit is present and enabled), just for the stack frames, plus the stack space needed by the exception handlers and potentially extra padding space for double-word stack alignment.

For a system with an OS, it is likely that each thread stack (which uses the process stack pointer) only needs to support one level of exception stack frame space. Stack space for nested exceptions is located on the main stack.

### 23.9.2 Stack analysis by trial

Traditionally, in a debug environment, you can fill the stack space with a certain pattern (e.g., 0xDEADBEEF is a commonly used one), run the program for a period of time, and see how much of the stack memory has been changed and from there you can estimate the stack size needed. This method, however, might not be able to reach the worst-case scenarios, and therefore a significant memory size margin might need to be added to avoid stack overflow. For example, the conditions for reaching the maximum number of nested interrupt levels can be difficult to produce in a lab environment.

### 23.9.3 Stack overflow detection by stack placement

In many toolchains, the layout of information inside the SRAM can be viewed as Figure 23.6.

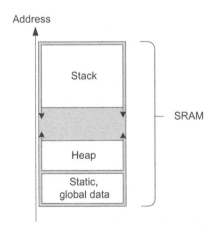

**FIGURE 23.6**

Typical data layout inside SRAM memory

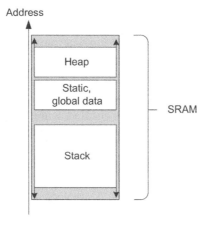

**FIGURE 23.7**

Alternate stack layout for stack overflow detection

Since stack usage grows downwards and heap usage grows upwards, this arrangement allows the free memory space to be used either by stack or by the heap if the memory size usage is a bit more than expected. Most of the time the stack and heap size used is smaller than the worst case, so this gives us a big sparse memory margin to avoid the problem of running out of stack space or heap space.

However, for stack overflow detection, we can rearrange the memory layout so that the stack is placed at the bottom, as shown in Figure 23.7. When the stack usage exceeds the allowed size, the stack access would cause a fault exception that can be detected easily. However, the fault handler will need to change the value of SP immediately so that it points to a valid space. Otherwise the fault handler might not be able to run. Also, as soon as the fault is triggered, the system is unrecoverable because the register contents could be lost when the exception handler starts.

### 23.9.4 Using MPU

For a system with an MPU, it is possible to set up the MPU configuration to define the allowed stack area and trigger a Memory Management fault if stack overflow is detected. Since it is possible to bypass the MPU in HardFault handler (or by using the FAULTMASK feature to escalate Memmange Fault to HardFault level), the fault handler can have additional stack space to deal with the error.

### 23.9.5 Using DWT and Debug Monitor Exception

Alternatively, a system that does not have a debugger connected can use the DWT to set a watchpoint at the end of the stack region and enable the Debug Monitor

Exception, which triggers when the selected address is accessed. However, if the system is connected to a debugger, the debug host might need to use the DWT comparators for various debug features, and this can cause a conflict.

### 23.9.6 Stack checking in OS context switching

Many embedded OSs, such as Keil™ RTX, support the stack-checking feature. At each context switch, the stack usage is checked against allowed stack size and an error is triggered if the stack size used by the thread exceeds the allowed value. However, if the thread only consumes a large stack space for a very short period of time, this stack overflow error might not be detected.

## 23.10 Flash patch feature

The Flash Patch and Breakpoint (FPB) unit of the Cortex®-M3 and Cortex-M4 processors provides a function called "flash patch." It allows the program code in a microcontroller design with mask ROM or OTP ROM to be patched via a small programmable memory (e.g., flash or EEPROM). We briefly covered this feature in section 14.6.2. Here we will see how to get it to work.

For the majority of microcontroller products which use flash for program memory, the flash patch feature is not needed because the program memory can be updated easily. For some low-cost devices, the program memory could be implemented as a mask ROM programmed by the chip manufacturer, or One-Time Programmable (OTP) ROM, which can only be programmed once. Since the ROM contents cannot be changed by the user, if a software bug is found after the devices are programmed, fixing the code can be costly.

In some cases, a microcontroller with mask ROM or OTP ROM will also include a small user-programmable ROM such as flash, EEPROM, etc. Since the size of this ROM is small, the impact to the device cost is relatively small when compared to replacing a whole batch of unusable devices programmed with firmware that has a serious defect.

If you believe that your project might require a flash patch, you should include a conditional branch in the startup code so that it can check to see if patch data is available (Figure 23.8). To do this, you need to know what the values are in the flash memory if nothing is programmed in it (typically it would be 0xFFFFFFFF). For example:

```
    AREA |.text|, CODE, READONLY


; Reset handler
Reset_Handler PROC
    EXPORT Reset_Handler    [WEAK]
    IMPORT SystemInit
    IMPORT __main
```

```
;-----------------------------------------
; Check if patch config present
LDR    R0, =0x00200000; Flash memory address (example)
LDR    R1,[R0]
LDR    R2,=0xFFFFFFFF ; Value if nothing there
CMP    R1, R2
ITT    NE               ; if not equal, setup patch
ORRNE  R0, R0, #1       ; Set LSB -> 0x01200001
BLXNE  R0               ; Call patch config code
;-----------------------------------------
; Normal startup
LDR    R0, =SystemInit
BLX    R0
LDR    R0, =__main
BX     R0
ENDP
```

The FPB has only a small number of comparators and therefore can only patch a limited number of program locations. So if a function has a bug, the most likely arrangement is to patch the branch instructions that called the function rather than trying to patch the function (Figure 23.9). Or if the required change is small enough

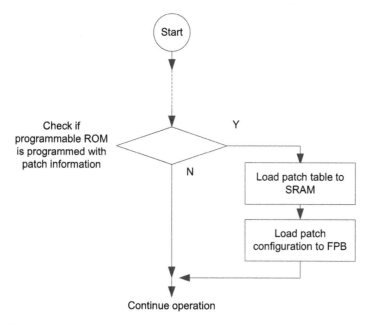

**FIGURE 23.8**

Simplified program flow for applying patch

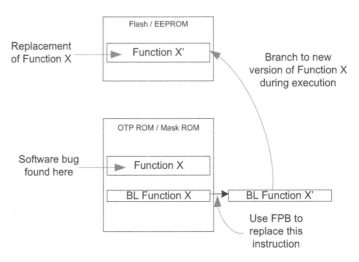

**FIGURE 23.9**

Patch a function call to replace a function with a new one

that only a couple of bytes in the program code need to be patched, you can also apply the patch directly.

When the processor reads an address in the program ROM and the address matches one of the FPB comparator setups and if the comparator is set to the REMAP function, then the access is remapped to an address location in SRAM, as specified by the REMAP register in the FPB (Figure 23.10).

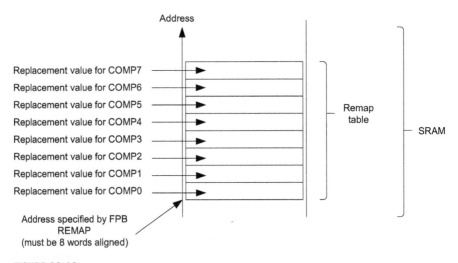

**FIGURE 23.10**

Values are replaced by contents in remap table in SRAM

After you have developed the replacement for the incorrect function, the next step is to create the setup code for FPB. You need to carry out the following tasks:

1. Set up the replacement values (patches) into a remap table in SRAM. The remap table contains a maximum of eight words and must be placed in a 32-byte aligned location in the SRAM region. The first six words are used for patching instructions, and the last two are used for patching literal data.
2. Set up the comparators in FPB. In maximum configurations of Cortex-M3 and Cortex-M4 processors there are eight comparators. The values programmed are the address values of the instructions to be patched. The first six are for patching instructions and the last two are used for patching literal data.
3. Set up the FPB REMAP register to point to the remap table in SRAM.
4. Enable the FPB.

For example, if there are six instructions and two literal data need to be patched, and the patch table is located at SRAM address 0x20010000:

```
#define HW_REG32(addr) (*((volatile uint32_t *)(addr)))
#define PATCH_TBL 0x20010000
typedef struct
{
  __IO uint32_t CTRL;    /*!< Flash Patch Control Register  */
  __IO uint32_t REMAP;   /*!< Flash Patch Remap Register   */
  __IO uint32_t COMP[8]; /*!< Flash Patch Comparator Register #0 - #7
*/
} FPB_TypeDef;

#define FPB_BASE    0xE0002000UL
#define FPB         ((FPB_TypeDef *) FPB_BASE)

void fpb_setup(void)
{
const unsigned int patch_addr={ .... }; // Addresses to be patched
const unsigned int patch_value={ .... }; // replacement values
  for (i=0;i<8;i++) {
     HW_REG32[PATCH_TBL + 4*i] = patch_value[i]; // Setup patch table
     // Setup FPB comparators — LSB is set to 1 to enable
     FPB->COMP[i] = patch_addr[i] | 0x1; // Instruction address
     }
 FPB->REMAP = PATCH_TBL; // Set remap table base
 FPB->CTRL = 3; // Enable
 return;
}
```

If the instruction being patched is 32 bits in size and is unaligned and if both half-words need to be changed, then we need to use two comparators in the FPB and two entries in the remap table to patch this instruction.

Please note that the flash patch feature uses the same hardware for breakpoint functionality. So if the microcontroller is connected to a debugger, the debugger can overwrite the patch configuration and therefore a device with the flash patch setup cannot be debugged.

The flash patch is a processor implementation-specific feature and is not available on the Cortex-M0 and Cortex-M0+ processors.

## 23.11 Revision versions of the Cortex®-M3 and Cortex-M4 processors

### 23.11.1 Overview

The Cortex®-M3 processor (r0p0) was first released to our silicon partners at the end of 2005 and silicon became available in 2006. Since then a number of update releases have been made. In ARM® processor products, very often you see the code rXpY, or even rXpY-nnrelm, where X is the major release revision, and Y is a minor version change. There can be additional changes (e.g., updates due to documentation updates, EDA tool support files updates) that can cause additional revision code changes in the nn and m fields. Typically the first release is r0p0-00rel0.

Currently, the Cortex-M3 processor is at revision r2p1, and Cortex-M4 processor is at r0p1. However, some of the Cortex-M3 microcontrollers on the market are still based on r1p1.

The revisions for the Cortex-M3 processor include: r0p0, r1p0, r1p1, r2p0, and r2p1.

The revisions for the Cortex-M4 processor include: r0p0, r0p1.

You can determine the revision of the Cortex-M3 or Cortex-M4 processor you are using by checking the CPUID. (See Table 9.11, section 9.7.)

In general, the changes for each release are documented in the Technical Reference Manual (TRM) released by ARM.

### 23.11.2 Changes from Cortex®-M3 r0p0 to r1p0/r1p1

Products based on Cortex®-M3 revision 1 have been available since the third quarter of 2006. For revision 1, the visible changes in the programmer's model and development features include the following:

- From revision 1, the stacking of registers when an exception occurs can be configured such that it is forced to begin from a double-word aligned memory address. This is done by setting the STKALIGN bit in the Nested Vectored Interrupt Controller (NVIC) Configuration Control register.

- For that reason, the NVIC Configuration Control register includes the STKALIGN bit.
- Release r1p1 includes the new AUXFAULT (Auxiliary Fault) status register (optional).
- Additional features include data-value matching being added to the DWT.
- ID register value changes to update the revision fields.

Changes invisible to end users include the following:

- The exported memory attribute for CODE memory space is hardwired to be cacheable, allocated, non-bufferable, and non-shareable. This affects the I-CODE Advanced High-Performance Bus (AHB) and the D-CODE AHB interface but not the System bus interface. The change only affects caching and buffering behavior outside the processor (e.g., level 2 cache or memory controllers with cache). The processor internal write buffer behavior does not change, and this modification has no effect on most microcontroller products.
- Support for a simpler bus multiplexing operation mode between I-CODE AHB and D-CODE AHB. Under this operation mode, the I-CODE and D-CODE buses can be merged using a simple bus multiplexer (r0p0 requires a larger bus matrix component). This can lower the total gate count.
- Added new output port for connection to the AHB Trace Macrocell (HTM, a CoreSight™ debug component from ARM®) for complex data trace operations.
- Debug components or debug control registers can be accessed even during system reset; only during power-on reset are those registers inaccessible.
- The Trace Port Interface Unit (TPIU) has support for Serial-Wire Viewer (SWV). This allows trace information to be captured with low-cost hardware.
- In revision 1, the VECTPENDING field in the NVIC Interrupt Control and Status register can be affected by the C_MASKINTS bit in the NVIC Debug Halting Control and Status register. If C_MASKINTS is set, the VECTPENDING value could be zero if the mask is masking a pending interrupt.
- The JTAG-DP debug interface module has been changed to the Serial Wire JTAG-Debug Port (SWJ-DP) module. This debug interface module supports both Serial Wire and JTAG debug protocol.

Since revision 0 of the Cortex-M3 processor does not have a double-word stack alignment feature in its exception sequence, some compiler tools, such as ARM DS-5™, RealView® Development Suite (RVDS), and the Keil™ MDK-ARM, have special options to allow software adjustment of stacking for exception handlers, which allow the developed application to be Embedded-Application Binary Interface (EABI) compliant. This can be important if the exception handler code requires EABI-compliant behavior.

## 23.11.3 Changes from Cortex®-M3 r1p1 to r2p0

In mid-2008, Revision 2 (r2p0) of the Cortex®-M3 was released to silicon vendors. Products using revision 2 arrived on the market in 2009. Revision 2 has a number of new features, most of them targeted at reducing power consumption and offering better debug flexibility. Changes that are visible in the programmer's model include the following:

- Sleep features: Wakeup Interrupt Controller (WIC) and State Retention Power Gating (SRPG) support added. In order to allow the processor system to reach an even lower power state in sleep mode, the WIC interface has been added to the NVIC and SRPGS support is included. (See section 9.2.8 for details.) At the system design level, the existing sleep features have also improved. In revision 2, it is possible for the wake up of the processor to be delayed. This allows more parts of the chip to be powered down (e.g., flash memory; Figure 23.11) and the power management system can resume program execution when the system is ready. This is needed in some microcontroller designs where some parts of the system are powered down during sleep, as it might take some time for the voltage supply to stabilize after power is restored.
- Default Configuration of Double-word Stack Alignment: The double-word stack alignment feature for exception stacking is now enabled by default (silicon vendors may select to retain revision 1 behavior). This reduces the startup overhead for most C applications (removing the need to set the STKALIGN bit in the NVIC Configuration Control register).
- Debug features: Watchpoint-triggered data trace in the DWT now supports tracing of read transfers only and write transfers only. This can reduce the trace data bandwidth required as you can specify the data are traced only when changed, or only when the data are read.
- This provides higher flexibility in implementing debug features. For example, the number of breakpoints and watchpoints can be reduced to reduce the size of the core in very low power designs. Better support for multi-processor debugging. A new interface has been introduced to allow simultaneous restart and single stepping of multiple processors (not visible in the programmer's view).
- Auxiliary Control Register: An Auxiliary Control register is added to NVIC to allow fine tuning of the processor's behavior. For example, for debugging purposes, it is possible to switch off the write buffers in Cortex-M3 so that bus faults will be synchronous to the memory access instruction (precise). The details of the Auxiliary Control register are covered in section 9.9 and Appendix F.3.2.

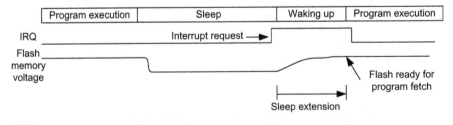

**FIGURE 23.11**

Sleep extension capability added in Cortex-M3 revision 2

- ID Register Values Updates: Various ID registers in NVIC and debug components have been updated.
- Additional design optimizations.

Revision 2 provides various improvements for embedded system designers. First, it means lower power consumption for the embedded products and better battery life. When WIC mode deep sleep is used, only a very small portion of the design needs to be active. Also, in designs targeted at ultra-low power, the silicon vendor can reduce the size of the design by reducing the number of breakpoints and watchpoints.

Second, it provides better flexibility in debugging and troubleshooting. Besides the improved data trace feature that can be used by debugger, we can also use the new Auxiliary Control register to force write transfers to be non-bufferable, in order to pinpoint the faulting instruction, or disable interrupts during multi-cycle instructions so that each multiple load/store instruction will be completed before the exception is taken. This can make the analysis of memory contents easier. For systems with multiple Cortex-M3 processors, revision 2 also brings the capability for simultaneous restarting and stepping of multiple cores.

In addition, revision 2 has a number of internal optimizations to allow higher performance and better interface features. This allows silicon vendors to develop faster Cortex-M3 products with more features. However, there are a few things that embedded programmers need to be aware of. They are as follows:

- Double-word stack alignment for exception stack frame: The exception stack frames will be aligned to double-word memory location by default. Some assembly applications written for revision 0 or revision 1 that use stacks to transfer data to exception handlers could be affected. The exception handler should determine whether stack alignment has been carried out by reading bit 9 of the stacked Program Status Register (PSR) in the stack frame, then it can determine the address of the stacked data before the exception. Alternatively, the application can program the STKALIGN bit to 0 to get the same stacking behavior as revision 0 and revision 1. EABI-compliant applications (e.g., C Code compiled using an EABI-compliant compiler) are not affected.
- The SysTick Timer might stop in deep sleep: If the Cortex-M3 microcontroller includes power-down features, or the core clocks are completely stopped in deep sleep mode, then the SysTick Timer might stop during deep sleep. Embedded applications that use an OS will need to use a timer external to the processor core to wake up the processor for event scheduling.
- Debug and power-down feature: Depending on the actual chip design, the power down feature could be disabled when the processor is connected to a debugger. Alternatively, the chip can be designed in a way that the debug connection could be lost when the microcontroller enters certain sleep modes. This is because the debugger needs to access the processor's debug registers during a debug session (e.g., to allow the core to be halted). For testing power down operations, ideally the device under test should be disconnected from the debugger.

## 23.11.4 Changes from Cortex®-M3 r2p0 to r2p1

After the Cortex®-M4 processor was released, many updates from the Cortex-M4 project were also fed back to the Cortex-M3 processor product:

- The Vector Table Offset Register (VTOR) has been increased by two bits to enable more flexible placement of the vector table.
- The Trace system includes global timestamp support for better trace stream correlation. As a part of this change the ETM protocol is changed from ETM v3.4 to v3.5 to include global timestamp support.
- The Cortex-M3 Trace Port Interface Unit (TPIU) has changed slightly so that when the trace interface is operated in Trace Port mode and there is no data, half sync packets may be inserted to ensure a partial frame is completed. This is different from previous versions that use NULL ID and bytes of 0x00 data.
- Watchpoints no longer occur if the transaction is aborted by the MPU.
- The revision information in various ID registers updated and ROM table ID changes.
- Additional changes have been implemented, which are only visible to silicon designers; such as better AHB Lite compliance, additional configuration options, etc.

## 23.11.5 Changes from Cortex®-M4 r0p0 to r0p1

There are only a few changes from r0p0 and r0p1. Apart from the revision ID value changes, the rest are only visible to silicon designers such as better AHB Lite compliance and configuration options.

# Software Porting

24

## CHAPTER OUTLINE

The Definitive Guide to ARM® Cortex®-M3 and Cortex-M4 Processors. http://dx.doi.org/10.1016/B978-0-12-408082-9.00024-5

## 24.1 Overview

Software porting is a common task for many software engineers. Even if the source code of a project is written in C, there can still be fair amount of work when porting the code due to:

- Different peripherals
- Different memory map
- Different ways of handling interrupts
- Tool chain specific C language extensions

CMSIS-Core and the architecture consistency between various Cortex®-M processors make software migration between different Cortex-M devices much easier. However, very often we also need to port software from other architectures to ARM® Cortex-M, or from classic ARM processors such as ARM7TDMI™ to Cortex-M. In this chapter, we will cover these areas.

## 24.2 Porting software from 8-bit/16-bit MCUs to Cortex®-M MCUs

### 24.2.1 Architectural differences

There are many architectural differences between common 8-bit/16-bit architectures and ARM® architectures. For example, the size of data types can be different, as shown in Table 24.1.

**Table 24.1** Data Size Comparison between ARM and 8-bit/16-bit Microcontrollers

| Data Type | 8-bit/16-bit Microcontrollers | ARM Architecture |
|---|---|---|
| char | 8-bits | 8-bits |
| short int | 16-bits | 16-bits |
| integer | 16-bits | 32-bits |
| pointers | 8/16/24-bits | 32-bits |
| float | 32-bits | 32-bits |
| double | 32-bits | 64-bits |

The differences can affect the program code in various ways, such as integer overflow behavior, program size, etc. For example, a program with an integer array might need to change in order to retain the same memory size for the array:

```
const int mydata = {0x1234, 0x2345 ....};
```

might change to:

```
const short int mydata = {0x1234, 0x2345 ....};
```

For floating point handling, if you want to retain 32-bit precision you might need to change the code to make sure that the floating point operations are all single precision, especially if you want to take advantage of the floating point unit in the Cortex®-M4 processor. For example, the code:

```
X=T*atan(T2*sin(X)*cos(X)/(cos(X+Y)+cos(X-Y)-1.0));
```

Should be changed to:

```
X=T*atanf(T2*sinf(X)*cosf(X)/(cosf(X+Y)+cosf(X-Y)-1.0F));
```

Alternatively you can choose to use double-precision calculation if a higher accuracy benefits your application. However, this can increase code size and execution time.

Another area of difference when comparing against 8-bit and 16-bit architectures is how data are stored in memory. The first one is data alignment:

In an 8-bit processor, the memory system is 8-bits wide and there is no data alignment concern. However, in ARM Cortex-M microcontroller systems the memory is 32-bit, so a piece of data can be aligned or unaligned (see section 6.6 and Figure 6.6). By default, a C compiler does not generate unaligned data. If a data structure is defined with elements of various sizes, it might need to insert padding space to keep data elements aligned (Figure 24.1).

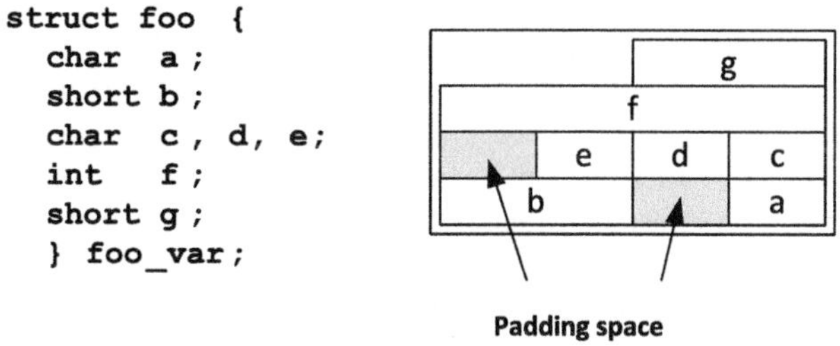

**FIGURE 24.1**

Padding space can be present in structure

The padding space in a structure can have several different impacts. For example:

The total data memory size could increase if you have an array of structures.

- Data structure code with hardcoded address offsets for data elements might fail.
- Memory copy code with hardcoded structure sizes might fail.

In general, program code that is written in a portable way (e.g., using "sizeof()" instead of hard coding the size) can avoid most of the issues. You might also want to rearrange the elements inside the structure to avoid the extra padding space.

The second area related to data storage is about the way that local variables are stored. Some 8-bit architectures place local variables in static memory locations in the SRAM if all the registers are used. In the ARM architecture, local variables are typically placed in the stack memory if all the registers are used. Since each time a function is called the stack could be at a different address, the local variables do not have a static memory location.

The advantage of using the stack for local variable is that if the function is not active, its local variables do not take up memory space. However, some of the debugging techniques that rely on local variables having a static location will not work. For those cases, you might need to add the "static" keyword when declaring the local variable, or change it to a global variable.

## 24.2.2 Common modifications

When porting applications from these microcontrollers to the Cortex®-M, modifications to the software typically involve:

- Startup code and vector table — Different processor architectures have different startup code and interrupt vector tables. Usually the startup code and the vector table will have to be replaced.
- Stack allocation adjustment — With the Cortex-M processors, the stack size requirement can be very different from an 8-bit or 16-bit architecture. In addition, the methods to define stack location and stack size are also different from 8-bit and 16-bit development tools.
- Architecture specific/toolchain-specific C language extensions — Many of the C compilers for 8-bit and 16-bit microcontrollers support a number of C language extensions. These include special data types like Special Function Registers (SFR) and bit data in 8051, or various "#pragma" statements in various C compilers.
- Interrupt control — In 8-bit and 16-bit microcontroller programming, the interrupt configuration is usually done by writing directly to interrupt control registers. When porting the applications to the ARM® Cortex-M processor family, such code should be converted to use the CMSIS-Core interrupt control functions. For example, enable and disable of interrupts can be converted to "__enable_irq()" and "__disable_irq()." Configuration of individual interrupts can be handled by various NVIC functions in CMSIS-Core.

- Peripheral programming — In 8-bit and 16-bit microcontroller programming, peripheral control is usually handled by programming the registers directly. When using ARM microcontrollers, many microcontroller vendors provide device-driver libraries to make using the microcontroller easier. You can use these library functions to reduce software development time, or write to the hardware registers directly if preferred. If you prefer to program the peripherals by accessing the registers directly, it is still beneficial to use the header files in the device driver library, as these have all the peripheral registers defined and can save you time preparing and validating the code.

- Assembly code and inline assembly — Obviously all the assembly and inline assembly code will need to be rewritten. In most cases, you can rewrite the required function in C when the application is ported to Cortex-M microcontrollers.

- Unaligned data — Some 8-bit or 16-bit microcontrollers might support unaligned data. In normal situations, C compilers do not generate unaligned data unless we use the __packed attribute when declaring data. Unaligned data handling is less efficient than aligned data in Cortex-M3 and Cortex-M4 and is not supported in Cortex-M0/M0+. As a result, some data structure definitions or pointer manipulation code might need to be changed for better portability and efficiency. If necessary, we can still apply the __packed attribute in data structures to support unaligned data elements inside.

- Adjustment of code due to data size differences — As described in section 24.2.1, integers in most 8-bit and 16-bit processors are 16-bit, while in ARM architectures integers are 32-bit. For example, when porting a program file from these processors to the ARM architecture, we might want to change "int" in the code to use "short int" or "int16_t" (in "stdint.h," introduced in C99) so that the size remains unchanged.

- Floating point — As described in section 24.2.1, a program which uses floating point calculations might need to be modified when porting from 8-bit/16-bit architecture to the ARM architecture.

- Adding fault handlers — In many 8-bit and 16-bit microcontrollers, there are no fault exceptions. While embedded applications can operate without any fault handlers, adding fault handlers can help an embedded system to recover from error conditions (e.g., data corruption caused by voltage drop or electromagnetic interference).

### 24.2.3 Memory size requirements

One of the areas mentioned in section 24.2.2 is the stack memory. After porting to the ARM® architecture, the required stack size could increase or decrease, depending on the application. The stack size might increase because:

- Each register push takes 4 bytes of memory in ARM, while in 16-bit or 8-bit each register push take 2 bytes or 1 byte.

- In ARM programming, local variables are often stored in the stack, while in some architectures local variables might be defined in a separate data memory area.

On the other hand, the stack size could decrease because:

- With 8-bit or 16-bit architecture, multiple registers are required to hold large data items and often these architectures have fewer registers compared to ARM, so more stacking would be required.
- More powerful addressing modes in ARM means address calculations can be carried out on the fly without taking up register space. The reduction of register use for an operation can reduce the stacking requirement.

Overall, the total RAM size required could decrease significantly after porting because most local variables do not take up SRAM space when the function is not active. Also, with more registers available in the ARM processor's register bank compared to some other architectures, some of the local variables might only need to be stored in the register bank instead of taking up memory space.

Depending on the application types, the program memory requirement in ARM Cortex®-M is often lower than 8-bit microcontrollers and most 16-bit microcontrollers. So when you port your applications from these microcontrollers to an ARM Cortex-M microcontroller, you might be able to use a microcontroller device with smaller flash memory size. The reduction of the program memory size is often caused by:

- More efficient handling of 16-bit and 32-bit data (including integers, pointers)
- More powerful addressing modes
- Some memory access instructions can handle multiple data, including PUSH and POP

There can be exceptions — for applications that contain only a small amount of code, the code size in ARM Cortex-M microcontrollers could be larger compared to 8-bit or 16-bit microcontrollers because:

- A Cortex-M microcontroller might have a much larger vector table due to more interrupts.
- The C startup code for a Cortex-M processor might be larger. If you are using ARM development tools like Keil™ MDK or Development Suite 5 (DS-5™), switching to the MicroLIB run-time library might help to reduce the code size.

### 24.2.4 Non-applicable optimizations for 8-bit or 16-bit microcontrollers

Some optimization techniques used in 8-bit/16-bit microcontroller programming are not required on ARM® processors. In some cases, these optimizations might result in extra overhead due to architecture differences. For example, many 8-bit microcontroller programmers use byte variables as loop counters for array accesses:

```
unsigned char i; /* use 8-bit data to avoid 16-bit processing */
char a[10], b[10];
for (i=0;i<10;i++) a[i] = b[i];
```

When compiling the same program on ARM processors, the compiler will have to insert a UXTB instruction to replicate the overflow behavior of the array index ("i"). To avoid this extra overhead we should declare "i" as integer "int," "int32_t" or "uint32_t" for best performance.

Another example is the unnecessary use of casting. For example, the following code uses casting to avoid the generation of a 16x16 multiply operation in an 8-bit processor:

```
unsigned int x, y, z;
z = ((char) x) * ((char) y); /* assumed both x and y must
                                be less than 256 */
```

Again, such a casting operation will result in extra instructions in the ARM architecture. Since Cortex®-M processors can handle a 32x32 multiply with 32-bit result in a single instruction, the program code can be simplified into:

```
unsigned int x, y, z;
z = x * y;
```

## 24.2.5 Example — migrate from 8051 to ARM® Cortex®-M

In general, since most applications can be programmed entirely in C on the Cortex-M microcontrollers, the porting of applications from 8-bit/16-bit microcontrollers is usually straightforward and easy. Here we will see some simple examples of the modifications required.

### Vector table

In the 8051, the vector table contains a number of JMP instructions that branch to the start of the interrupt service routines (as shown in left hand side of table 24.2). In some development environments, the compiler might create the vector table for you automatically. In ARM®, the vector table contains the initial value of the main stack pointer and starting addresses of the exception handlers (right hand side of table 24.2). The vector table is part of the startup code, which is often provided by the development environment. For example, when creating a new project in the Keil™ MDK project wizard, it will offer to copy and add the default startup code, which contains the vector.

### Data type

In some cases, we need to modify the data type so as to maintain the same program behavior, as shown in Table 24.3.

**Table 24.2** Vector Table Porting

| 8051 | Cortex-M |
|---|---|
| ```
org 00h
      jmp     start
org 03h ; Ext Int0 vector
ljmp handle_interrupt0
org 0Bh ; Timer 0 vector
ljmp handle_timer0
org 13h ; Ext Int1 vector
ljmp handle_interrupt1
org 1Bh ; Timer 1 vector
ljmp handle_timer1
org 23h ; Serial
interrupt
ljmp handle_serial0
org 2bh ; Timer 2 vector
ljmp handle_timer2
``` | ```
__Vectors DCD __initial_sp ; Top of Stack
          DCD Reset_Handler ; Reset Handler
          DCD NMI_Handler ; NMI Handler
          DCD HardFault_Handler ; Hard Fault
          DCD MemManage_Handler ; MPU Fault
          DCD BusFault_Handler ; Bus Fault
          DCD UsageFault_Handler; Usage Fault
          DCD 0,0,0,0 ; Reserved
          DCD SVC_Handler ; SVCall Handler
          DCD 0,0 ; Reserved
          DCD PendSV_Handler ; PendSV Handler
          DCD SysTick_Handler ; SysTick Handler
          ; External Interrupts
          DCD WWDG_IRQHandler ; Window WatchDog
          ...
``` |

**Table 24.3** Data Type Change during Software Porting

| 8051 | Cortex-M |
|---|---|
| int my_data[20]; // array of 16-bit values | short int my_data[20]; // array of 16-bit values |

Some function calls might also need to be changed if we want to ensure only single precision floating point is used, as shown in Table 24.4.

Some special data types in 8051 are not available on the Cortex-M: bit, sbit, sfr, sfr16, idata, xdata, bdata. They are compiler-specific and are not supported on the ARM architecture.

### Interrupt

Interrupt control code in 8051 is normally written using direct accesses to SFRs. They need to be changed to CMSIS-Core functions when porting to ARM Cortex-M microcontrollers, as shown in Table 24.5.

The interrupt service routine will also require minor modifications. Some of the special directives used by an interrupt service routine will need to be removed when the application code is ported to a Cortex-M microcontrollers, as shown in Table 24.6.

### Sleep mode

Entering sleep mode is different too. In 8051, sleep mode can be entered by setting the IDL (idle) bit in PCON. In Cortex-M, you can use vendor-specific functions provided in the device-driver library, or use the WFI instruction directly as shown in Table 24.7 (but this will not give you the best low power optimization).

## 24.3 Porting software from ARM7TDMI™ to Cortex®-M3/M4

### 24.3.1 Overview of the hardware differences

The ARM7TDMI™ is a very successful and popular processor for microcontrollers. Currently it is still shipping in large volumes, and is used by many designers. In some cases some of these designers decided to migrate from ARM7TDMI to Cortex®-M microcontrollers.

There are a number of characteristic differences between ARM7-based systems and Cortex-M3/M4-based systems (e.g., memory map, interrupts, Memory Protection Unit [MPU], system control, and operation modes).

### Memory map

The most obvious target of modification in porting programs between different microcontrollers is their memory map differences. In the ARM7™, memory and peripherals can be located at almost any address, whereas the Cortex-M3 and Cortex-M4 processors have a predefined memory map. Memory address differences are usually resolved at the compile and link stages. Peripheral code porting could be more time consuming because the programmer's model for the peripheral could be completely different. In that case, device-driver code might need to be completely rewritten, or alternatively, the code changed to use new device driver library code from the microcontroller vendors.

**Table 24.4** Floating Point C Code Change during Software Porting

| 8051 | Cortex-M |
|---|---|
| `Y=T*atan(T2*sin(Y)*cos(Y)/(cos(X+Y)+cos(X-Y)-1.0));` | `Y=T*atanf(T2*sinf(Y)*cosf(Y)/(cosf(X+Y)+cosf(X-Y)-1.0F));` |

**Table 24.5** Interrupt Control Change during Software Porting

| 8051 | Cortex-M |
|---|---|
| `EA=0; /* Disable all interrupts */`<br>`EA = 1; /* Enable all interrupts */`<br>`EX0=1; /* Enable Interrupt 0 */`<br>`EX0 = 0; /* Disable Interrupt 0 */`<br>`PX0 = 1; /* Set interrupt 0 to high priority*/` | `__disable_irq(); /* Disable all interrupts */`<br>`__enable_irq(); /* Enable all interrupts */`<br>`NVIC_EnableIRQ(Interrupt0_IRQn);`<br>`NVIC_DisableIRQ(Interrupt0_IRQn);`<br>`NVIC_SetPriority(Interrupt0_IRQn, 0);` |

**Table 24.6** Interrupt Handler Change during Software Porting

| 8051 | Cortex-M |
|---|---|
| ```void timer1_isr(void) interrupt 1    using 2 { /* Use register bank 2 */    ...;    return; }``` | ```__irq void timer1_isr(void) {    ...;    return; }``` |

**Table 24.7** Sleep Mode Control Change During Software Porting

| 8051 | Cortex-M |
|---|---|
| `PCON = PCON | 1; /* Enter Idle mode */` | `__WFI(); /* Enter sleep mode */` |

**Table 24.8** Mapping of ARM7TDMI Exceptions and Modes to the Cortex-M3 or Cortex-M4 Processor

| Modes and Exceptions in the ARM7 | Corresponding Modes and Exceptions in the Cortex-M3 |
|---|---|
| Supervisor (default) | Privileged, Thread |
| Supervisor (software interrupt) | Privileged, SVC |
| FIQ | Privileged, interrupt |
| IRQ | Privileged, interrupt |
| Abort (prefetch) | Privileged, bus fault exception |
| Abort (data) | Privileged, bus fault exception |
| Undefined | Privileged, usage fault exception |
| System | Privileged, Thread |
| User | User access (non-privileged), Thread |

Many ARM7 products provide a memory remap feature so that the vector table can be remapped to SRAM after boot-up. In the Cortex-M3 or Cortex-M4 microcontrollers, the vector table can be relocated using the VTOR register so that memory remapping is no longer needed. Therefore, the memory remap feature might be unavailable in many Cortex-M3 and Cortex-M4 microcontroller products.

Big endian support in the ARM7 is different from such support in the Cortex-M3 and Cortex-M4. Program files can be recompiled to the new big endian system, but hardcoded lookup tables might need to be converted during the porting process.

In ARM720T, and some later ARM® processors like ARM9™, a feature called "high vectors" (or "Hivecs") is available, which allows the vector table to be relocated to 0xFFFF0000. Although it is often used for other purposes, this feature was introduced to support Windows CE and is not available in any of the current Cortex-M processors.

### Interrupts

The second target is the difference in the interrupt controller being used. Program code for control of the interrupt controller, such as enabling or disabling interrupts, will need to be changed because the NVIC has a different programmer's model. In addition, new code is required for setting up interrupt priority levels and vector addresses for various interrupts. In most cases you can utilize the NVIC control functions included in CMSIS-Core. This makes your software much more portable.

Interrupt wrapper code for nested interrupt handling can be removed. In the Cortex-M processors, the NVIC has built-in nested interrupt handling.

The interrupt return method is also changed. This requires modification of interrupt return in assembler code. With the Cortex-M processors, C Handlers can be normal C functions and do not require special compile directives.

Enable and disable of interrupts, previously done by modifying Current Program Status Register (CPSR), must be replaced by setting up the Interrupt Mask register. In addition, in the ARM7TDMI, it is possible to re-enable interrupts at the same time as returning from an interrupt handler due to the restoration of CPSR from SPSR (Saved Program Status Register). In the Cortex-M processors, if interrupts are disabled during an interrupt handler by setting PRIMASK, FAULTMASK, or BASEPRI, the mask registers should be cleared manually before interrupt return. Otherwise, the mask registers are still set and interrupts will not be re-enabled.

In the Cortex-M3 and Cortex-M4 processors, some registers are automatically saved by the stacking and unstacking mechanisms. Therefore, some of the software stacking operations could be reduced or removed. However, in the case of the Fast Interrupt request (FIQ) handler, traditional ARM cores have separate registers for FIQ (R8–R11). Those registers can be used by the FIQ without the need to push them on to the stack. In the Cortex-M processors, these registers are not stacked automatically, so when an FIQ handler is ported to the Cortex-M processor, either the registers being used by the handler must be changed or a stacking step will be needed.

There are also differences in error handling. The Cortex-M3 and Cortex-M4 processors provide various fault status registers so that the cause of faults can be located. In addition, new fault exception types are defined in the Cortex-M processors (e.g., stacking and unstacking faults, memory management faults, and hard faults). Therefore, the fault handlers will need to be rewritten.

### MPU

The Memory Protection Unit (MPU) will also need code to configure and control it. Microcontroller products based on the ARM7TDMI/ARM7TDMI-S do not have MPUs, so moving the application code to the Cortex-M3 or Cortex-M4 microcontrollers should not be a problem. However, products based on the ARM720T have a Memory Management Unit (MMU), which has different functionality to the MPU in Cortex-M3 and Cortex-M4 processors. If an application uses the MMU to support a virtual memory system, it cannot be ported to the Cortex-M3 as the MPU does not support address translation.

### System control

System control is another key area to look into when you're porting applications. The Cortex processors have built-in instructions for entering sleep mode. In addition, the device-specific system controller inside Cortex-M3 and Cortex-M4 microcontroller products is likely to be completely different from that of the ARM7 products, so the code that handles system management features will need to be rewritten.

### Operation modes

In the ARM7, there are seven operation modes; in the Cortex-M3 and Cortex-M4 processors, these have been changed to a different scheme (see Table 24.8).

A normal Interrupt Request (IRQ) can be used to replace the FIQ in the ARM7 because in Cortex-M processors, we can configure the priority for any particular interrupt to be the highest; thus it will be able to preempt other exceptions, just like the FIQ in the ARM7. However, due to the difference between banked FIQ registers in the ARM7 and the stacked registers in the Cortex-M processors, the registers being used in the FIQ handler must be changed, or the registers used by the handler must be saved to the stack manually.

### Differences between FIQ and non-maskable interrupt

Many engineers might expect the FIQ in the ARM7 to be directly mapped to the Non-Maskable Interrupt (NMI) in the Cortex-M processors. In some applications this is possible, but a number of differences between the FIQ and the NMI need special attention when you're porting applications using the NMI as an FIQ.

First, the NMI cannot be disabled, whereas on the ARM7, the FIQ can be disabled by setting the F-bit in the CPSR. So in a Cortex-M system it is possible for an NMI handler to start right at boot-up time, whereas in the ARM7 the FIQ is disabled at reset.

Second, in the Cortex-M processors you cannot use SVC in an NMI handler, whereas you can use a software interrupt (SWI) in an FIQ handler on the ARM7. During execution of an FIQ handler on the ARM7, it is possible for other exceptions to take place (except FIQ and IRQ, because the I and F bits are set automatically when the FIQ is served). However, on the Cortex-M processors, a fault exception inside the NMI handler can cause the processor to lock up.

### 24.3.2 Assembly language files

Porting assembly files depends on whether the code is written for ARM® state or Thumb state.

### Thumb state

If the code is written for Thumb state, porting is much easier. In most cases, the file can be reused without a problem. However, a few Thumb instructions in the ARM7™ are not supported in the Cortex®-M3 and Cortex-M4 as follows:

- Any code that tries to switch to ARM state.
- The SWI instruction is replaced by SVC (note that the code for parameter passing and result return will need to be updated).
- Finally, make sure that the program accesses the stack only in full descending stack operations. It is possible, though uncommon, to implement a different stacking model (e.g., full ascending) on an ARM7.

### ARM state

The situation for ARM code is more complicated. There are several scenarios as follows:

- *Vector table*: In the ARM7, the vector table starts from address 0x0 and consists of branch instructions. In the Cortex-M processors, the vector table contains the initial value for the stack pointer followed by the reset vector address and then by the addresses of all the other exception handlers. Due to these differences, the vector table will need to be completely rewritten. Normally, the startup code you get from the microcontroller vendors should include the vector table so you do not have to create the vector table yourself.
- *Register initialization*: In the ARM7, it is often necessary to initialize the banked registers for different modes. For example, there are banked stack pointers (R13), link registers (R14), and SPSRs for each of the exception modes in the ARM7. Since the Cortex-M processor has a different programmer's model, the register initialization code will have to be changed. In fact, the register initialization code on the Cortex-M processors will be much simpler because there is no need to switch the processor into a different mode. In most simple applications without an OS, you can just use the Main Stack Pointer for the whole project, so you do not have to initialize multiple stack pointers as in ARM7.
- *Mode switching and state switching codes*: Since the operating mode scheme in the Cortex-M processors is different from that of the ARM7, the code for mode switching needs to be removed or changed. The same applies to ARM/Thumb state switching code.
- *Interrupt enabling and disabling*: In the ARM7, IRQ interrupts can be enabled or disabled by clearing or setting the I-bit in the CPSR. In the Cortex-M processors, this is done by clearing or setting an Interrupt Mask register, such as PRIMASK or FAULTMASK. Furthermore, there is no F-bit in the Cortex-M processors because there is no FIQ input.
- *Coprocessor accesses*: There is no coprocessor support on the current range of Cortex-M processors, so this kind of operation cannot be ported.

- *Interrupt handler and interrupt return*: In the ARM7, the first instruction of the interrupt handler is in the vector table, which normally contains a branch instruction to the actual interrupt handler. In the Cortex-M processors, this step is no longer needed. For interrupt returns, the ARM7 relies on manual adjustment of the return program counter. In the Cortex-M processors, the correctly adjusted program counter is saved to the stack and the interrupt return is triggered by loading the special value EXC_RETURN into the program counter. Instructions such as MOVS and SUBS should not be used as interrupt returns on the Cortex-M processors. Because of these differences, interrupt handlers and interrupt return codes need modification during porting. Because you can use normal C functions for interrupt handling, it might be easier to recode the interrupt handlers in C.
- *Nested interrupt support code*: In the ARM7, when a nested interrupt is needed, usually the IRQ handler will need to switch the processor to system mode or SVC mode before re-enabling interrupts. This is not required in the Cortex-M processors.
- *FIQ handler*: If an ARM7 FIQ handler is to be ported to a Cortex-M interrupt, you might need to add an extra step to save the contents of R8–R11 to stack memory. In the ARM7, R8–R12 are banked, so the FIQ handler can skip the stack push for these registers. However, on the Cortex-M processors, R0–R3 and R12 are saved on to the stack automatically, but R8–R11 are not.
- *SWI handler*: The SWI instruction is replaced by SVC. However, when porting a SWI handler to SVC, the code to extract the parameters passed with the SWI instruction needs to be updated. The calling SVC instruction address can be found in the stacked PC, which is different from the SWI in the ARM7, where the program counter address has to be determined from the link register.
- *SWP instruction (swap)*: There is no Swap Instruction (SWP) in the Cortex-M processors. If SWP was used for semaphores, they will need to be recoded using the exclusive access instructions. This requires rewriting the semaphore code. If the instruction was used purely for data transfers, this can be replaced by multiple memory access instructions.
- *Access to CPSR and SPSR*: The CPSR in the ARM7 is replaced with combined Program Status registers (xPSR) in the Cortex-M processors and the SPSR has been removed. If the application needs to access the current values of processor flags, the program code can be replaced with a read access to the APSR. If an exception handler would like to access the Program Status register (PSR) before the exception takes place, it can find the value on the stack, because the value of xPSR is automatically saved to the stack when an interrupt is accepted. So there is no need for an SPSR in the Cortex-M processors.
- *Conditional execution*: In the ARM7, conditional execution is supported for many ARM instructions, whereas most Thumb-2 instructions do not have the condition field inside the instruction coding. When porting these instructions to the Cortex-M3 and Cortex-M4 processors, the assembly tool might automatically convert these conditional instructions to use an IF-THEN (IT) instruction

block; alternatively, we can manually insert the IT instructions or insert branches to produce conditionally executed sequences. One potential issue with replacing conditional sequences with IT instruction blocks is that this could increase the code size and, as a result, could cause minor problems. For instance, load/store operations in another part of the program might then exceed the access range of the instruction.

- *Use of the program counter value:* When running ARM code on the ARM7, the read value of the PC during an instruction is the address of the current instruction plus 8. This is because the ARM7 has three pipeline stages and, when reading the PC during the execution stage, the program counter has already been incremented twice, 4 bytes at a time. When porting code that processes the PC value to the Cortex-M processors, since the code will be in Thumb, the offset of the program counter will only be 4.
- *Use of the value of R13:* In the ARM7, the stack pointer R13 is a full 32-bit value; in the Cortex-M processors, the lowest 2 bits of the stack pointer are always forced to zero. Therefore, in the unlikely case that R13 is used as a data register, the code has to be modified because the lowest 2 bits would be lost.

For the rest of the ARM program code, we can try to compile it as Thumb/Thumb-2 and see if further modifications are needed. For example, some of the pre-index and post-index addressing modes support by the ARM7 are not supported by the Cortex-M processors and have to be recoded into multiple instructions. Some of the code might involve long branch ranges or large immediate data values that cannot be compiled as Thumb code and so must be modified to Thumb-2 code manually.

### 24.3.3 C language files

Porting C program files is much easier than porting assembly files. In most cases, application code in C can be recompiled for the Cortex®-M processors without any problem. However, there are still a few areas that potentially need modification, as follows:

- *Inline assemblers:* Some C code might contain inline assembly that needs modification. This code can be easily located via the __asm keyword. In some older versions of ARM® C compilers, inline assembler is not supported and this might have to be changed to Embedded Assembler. (See section 20.5 for details.)
- *Interrupt handler:* In the C program you can use __irq to create interrupt handlers that work with the ARM7™. Due to the difference between the ARM7 and the Cortex-M exception models, such as saved registers and interrupt returns, depending on development tools being used, keywords indicating that a function is an interrupt handler (such as the __irq keyword) might need to be removed. In ARM development tools including Keil™ MDK-ARM or DS-5™, uses of __irq directive on the Cortex-M processor are allowed, and in general are recommended for reasons of clarity. In some other toolchains, however, you might need to remove some of these compiler-specific keywords.

- *Pragma directives:* ARM C compiler pragma directives like "#pragma arm" and "#pragma thumb" should be removed.

### 24.3.4 **Pre-compiled object files and libraries**

Most C compilers will provide pre-compiled object files for various function libraries and startup code. Many of those (such as startup code for traditional ARM® processor cores) cannot be used on the Cortex®-M processors due to the difference in operating modes and states. Some of them will have source code available and can be recompiled for Thumb-2. Refer to your tool vendor documentation for details.

### 24.3.5 **Optimization**

After getting the program to work with a Cortex®-M microcontroller, you might be able to further improve it to obtain better performance and lower memory use. A number of areas should be explored:

- *Use of Thumb-2 instructions:* For example, if a 16-bit Thumb instruction transfers data from one register to another and then carries out a data processing operation on it, it might be possible to replace the sequence with a single Thumb-2 instruction. This can reduce the number of clock cycles required for the operation.
- *Bit band:* If peripherals are located in bit-band regions, access to control register bits can be greatly simplified by accessing the bit via a bit-band alias.
- *Multiply and divide:* Routines that require divide operations, such as converting values into decimal for display, can be modified to use the divide instructions in the Cortex-M3 and Cortex-M4 processors. For multiplication of larger data, the multiply instructions in the Cortex-M3/M4, such as unsigned multiply long (UMULL), signed multiply long (SMULL), multiply accumulate (MLA), multiply and subtract (MLS), unsigned multiply accumulate long (UMLAL), and signed multiply accumulate long (SMLAL) can be used to reduce the complexity of the code.
- *Immediate data:* Some of the immediate data that cannot be coded in 16-bit Thumb instructions can be produced using 32-bit Thumb instructions. This means that you might be able to reduce to complexity in some code fragments by reducing the number of steps to set up an immediate data.
- *Branches:* Some long distance branches that cannot be coded in 16-bit Thumb code (usually ending up with multiple branch steps) can be coded with 32-bit Thumb instructions, reducing code size and branch overhead.
- *Boolean data:* Multiple boolean data items (either 0 or 1) can be packed into a single byte/half-word/word in bit-band regions to save memory space. They can then be accessed via the bit-band alias.
- *Bit-field processing:* The Cortex-M3 and Cortex-M4 processors provide a number of instructions for bit-field processing, including Unsigned Bit Field eXtract (UBFX), Signed Bit Field eXtract (SBFX), Bit Field Insert (BFI), Bit

Field Clear (BFC), and Reverse Bits (RBIT). They can simplify many code sequences for peripheral programming, data packet formation or extraction, and serial data communications.

- *IT instruction block:* Some short branches might be replaceable by an IT instruction block. This may avoid wasting clock cycles when the pipeline is flushed during branching.
- *ARM/Thumb state switching:* In some situations, ARM developers frequently divide code amongst source files so that some of them can be compiled to ARM code and others compiled to Thumb code. This is usually needed to get the right balance between code density and execution speed. With Thumb-2 features in the Cortex-M processors, this step is no longer needed, so some of the state switching overhead can be removed, producing short code, less overhead, and possibly fewer program files.

## 24.4 Porting software between different Cortex®-M processors
### 24.4.1 Differences between different Cortex®-M processors

In most cases, porting software between different Cortex®-M processors is relatively straightforward due to the consistency of the architecture. However, there are some differences between the various Cortex-M processors.

### Instruction set

One of the major differences between Cortex-M processors is the instruction set support. The Cortex-M processors are designed to be upward-compatible, so instructions available on Cortex-M0/M0+/M1 (ARMv6-M architecture) can also be used on Cortex-M3 and Cortex-M4 (ARMv7-M architecture) (Figure 24.2). In theory a binary program image compiled for ARMv6-M can run directly on a device with ARMv7-M. However, in practice the memory map and peripherals could be different and, in any case, it is best to recompile the code to take advantage of the additional instructions available.

When moving downwards, however, the program code will need to be recompiled. Also, assembly code (including inline assembly and Embedded Assembly) may also need to be modified. For example, when porting an application from a Cortex-M3 microcontroller to a Cortex-M0 microcontroller, the following instructions are not available.

### IT instruction block

- Compare and branch (compare and branch if zero [CBZ] and compare and branch if non-zero [CBNZ])
- Multiple accumulate instructions (multiply accumulate [MLA], multiply and subtract [MLS], signed multiply accumulate long [SMLAL], and unsigned

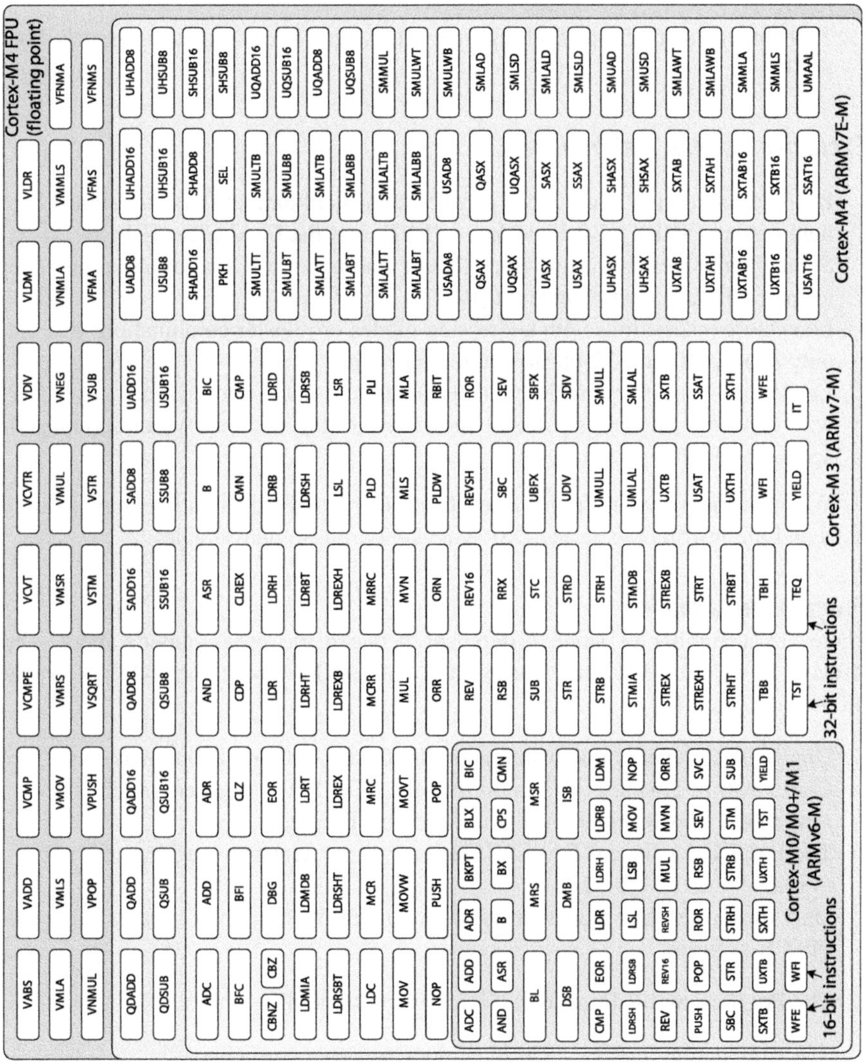

**FIGURE 24.2**

Instruction set of the Cortex-M processors

multiply accumulate long [UMLAL]) and multiply instructions with 64-bit results (unsigned multiply long [UMULL] and signed multiply long [SMULL])
- Hardware divide instructions (unsigned divide [UDIV] and signed divide [SDIV]) and saturation (signed saturate [SSAT] and unsigned saturate [USAT])
- Table branch instruction (Table Branch Half-word [TBH] and Table Branch Byte [TBB])

### Exclusive access instructions

- Bit field processing instructions (unsigned bit field extract [UBFX], signed bit field extract [SBFX], Bit Field Insert [BFI], and Bit Field Clear [BFC])
- Some data processing instructions (count leading zero [CLZ], rotate right extended [RRX], and reverse bit [RBIT])
- Load/store instructions with addressing modes or register combinations that are only supported with 32-bit instruction encoding
- Load/store instructions with translate (load word data from memory to register with unprivileged access [LDRT] and store word to memory with unprivileged access [STRT])

When porting software from Cortex-M4 to Cortex-M3, the floating point instructions and the instructions in the DSP extension are also unavailable.

### Programmer's model

There are a number of small differences between the programmer's model for ARMv7-M (Cortex-M3 and Cortex-M4) and ARMv6-M (Cortex-M0, Cortex-M0+ and Cortex-M1):

- Unprivileged level is not available on the Cortex-M0 and is optional on the Cortex-M0+. This also affects bit 0 of the CONTROL register, which is not available if unprivileged level is not present (Figure 24.3).
- The FAULTMASK and BASEPRI registers (for exception masking) are not available on ARMv6-M.
- Only the Cortex-M4 processor has the optional floating point register bank and FPSCR register.

The Program Status Register (PSR) also has some differences:

- The Application PSR in ARMv6-M does not have the Q bit.
- The GE bits are only available on the Cortex-M4 processor.
- The ICI/IT bits (Interrupt Continuable Instruction/IF-THEN) are not available on ARMv6-M.
- The width of the IPSR is only 6 bits in Cortex-M0/M0+/M1 processor because they only support up to 32 interrupts.

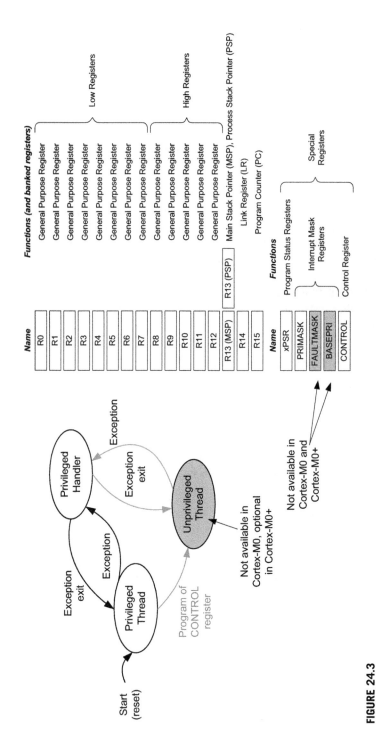

**FIGURE 24.3**

Programmer's model differences

### NVIC

The NVIC feature on the Cortex-M processor is configurable. This means that the number of interrupts supported and the number of programmable interrupt priorities can be decided by the chip manufacturers. Table 24.9 lists the differences in the NVIC in different Cortex-M processors.

### System-level features

There are some differences in the system-level features, as can be seen from Table 24.10.

### Low power features

The low power support at the processor level is identical, as shown in Table 24.11.

However, at chip-design level, different microcontrollers have different low power features and therefore the code for low power optimization usually needs to be changed when porting applications from one microcontroller to another.

### Debug and trace features

There are further differences in the debug and trace features, as can be seen in Table 24.12.

## 24.4.2 Required software changes

For microcontroller applications using CMSIS-compliant device libraries, in most cases you will need to:

- Replace the device driver header files
- Replace device specific startup code
- Adjust Interrupt Priority level if needed
- Adjust compilation options such as processor type, floating point options

To improve software portability, you should use the interrupt control functions provided in CMSIS-Core to set up interrupt configurations in the NVIC. If your application accesses the NVIC registers directly, you may need to adjust the source code during porting, because the NVIC in ARMv6-M does not allow byte or half-word accesses. For example, the definition of the priority level registers in the CMSIS-Core header files is different between ARMv7-M and ARMv6-M.

ARMv6-M does not have the Software Trigger Interrupt Register (NVIC->STIR). Therefore, in Cortex®-M0/M0+ processors, software needs to use the Interrupt Set Pending Register (NVIC->ISPR) to trigger an interrupt in software.

The programmer's model of the SysTick timer is basically the same. However, the SysTick timer value in the Cortex-M3 and Cortex-M4 is reset to 0 and in the Cortex-M0 and Cortex-M0+ the timer initial value can be undefined. As a result, when porting SysTick setup code, you must make sure that the program code initializes the SysTick timer value.

**Table 24.9** NVIC Feature Comparison

| Features | Cortex-M0 | Cortex-M0+ | Cortex-M3 | Cortex-M4 |
|---|---|---|---|---|
| Number of IRQ | 1 to 32 | 1 to 32 | 1 to 240 | 1 to 240 |
| System Exceptions | 5 (NMI, HardFault, SVC, PendSV, SysTick) | 5 | 9 (ARMv6-M system exceptions + 3 configurable fault handlers + debug monitor) | 9 |
| Programmable Priority levels | 4 | 4 | 8 to 256 | 8 to 256 |
| Priority Grouping | No | No | Yes | Yes |
| Masking registers | PRIMASK | PRIMASK | PRIMASK, FAULTMASK, BASEPRI | PRIMASK, FAULTMASK, BASEPRI |
| Vector Table Offset Register | No | Optional | Yes | Yes |
| Software Trigger Interrupt Register | No | No | Yes | Yes |
| Interrupt Active Status Registers | No | No | Yes | Yes |
| Dynamic priority change support | No | No | Yes | Yes |
| Register accesses | 32-bit | 32-bit | 8/16/32-bit | 8/16/32-bit |
| Double word stack alignment | Always enable | Always enable | Programmable | Programmable |

**Table 24.10** System Feature Comparison

| Features | Cortex-M0 | Cortex-M0+ | Cortex-M3 | Cortex-M4 |
|---|---|---|---|---|
| Privileged/unprivileged | No | Optional | Yes | Yes |
| SysTick Timer | Optional | Optional | Yes | Yes |
| MPU | No | Optional | Optional | Optional |
| Bit band | Not included in the processor, can be added at the system level | Not included in the processor, can be added at the system level | Optional | Optional |
| Single cycle I/O interface | No | Yes | No | No |
| Bus architecture | von Neumann | von Neumann | Harvard | Harvard |
| Fault Handling | HardFault | HardFault | HardFault + 3 other fault handlers | HardFault + 3 other fault handlers |
| Fault Status Registers | No (Debug FSR for debug only) | No (Debug FSR for debug only) | CFSR, HFSR, DFSR, AFSR | CFSR, HFSR, DFSR, AFSR |
| Self-reset | System Reset Request | System Reset Request | System Reset Request + VECTRESET | System Reset Request + VECTRESET |
| Unaligned accesses support | No | No | Yes | Yes |
| Exclusive access | No | No | Yes | Yes |

**Table 24.11** Low Power Feature Comparison

| Features | Cortex-M0 | Cortex-M0+ | Cortex-M3 | Cortex-M4 |
|---|---|---|---|---|
| Sleep modes | Sleep and deep sleep | Sleep and deep sleep | Sleep and deep sleep | Sleep and deep sleep |
| Sleep-on-exit | Yes | Yes | Yes | Yes |
| WIC support | Yes | Yes | Yes | Yes |
| SRPG support | Yes | Yes | Yes | Yes |
| Event support (e.g., SEV) | Yes | Yes | Yes | Yes |
| SEVONPEND | Yes | Yes | Yes | Yes |

Typically you might also need to adjust the clock frequency of the microcontroller and modify the code that utilizes the low power features of the microcontrollers. When moving from one Cortex-M microcontroller to another, there might be differences in terms of program execution speed. For example, when porting an application from a Cortex-M0 microcontroller to a Cortex-M3 microcontroller, you might be able to reduce the clock frequency of the microcontroller to achieve the same performance but with lower power consumption.

### 24.4.3 Embedded OS

In an application with an embedded OS, you might need to switch to a different version of the OS in order to allow it to work properly. For example, an embedded OS written for the Cortex®-M3 processor might work on the Cortex-M4 processor as long as the application does not use the floating point unit. But as soon as the floating point unit is used, it will have to provide context saving and restore for floating point register banks, as well as deal with extra information in the CONTROL register, EXC_RETURN value, and different stack frame sizes.

A Cortex-M0 application project running with an embedded OS might need some extra adjustment. In the Cortex-M0 processor, there is no unprivileged access level, as all application threads can access the NVIC and the registers in the System Control Space (SCS). When the application is ported to other Cortex-M processors, the OS might run the threads in unprivileged state by default and all the accesses to NVIC and SCS registers would be blocked. So you might need some adjustments in the project to avoid NVIC and SCS accesses in threads, or you could adjust the OS configuration to enable the threads to run in privileged state.

Some embedded OSs might utilize the MPU feature. The programmer's models of the MPUs in the Cortex-M3/M4 and Cortex-M0+ processors are mostly the same, but there are some minor differences. These are listed in section 11.8, Table 11.12. For example, the address bit field in the MPU Base Address Register for the Cortex-M3/M4 MPU allows you to define a MPU region as small as 32 bytes. In the

**Table 24.12** Debug and Trace Feature Comparison

| Features | Cortex-M0 | Cortex-M0+ | Cortex-M3 | Cortex-M4 |
|---|---|---|---|---|
| Debug interface | Typically either Serial wire or JTAG | Typically either Serial wire or JTAG | Typically both Serial Wire and JTAG | Typically both Serial Wire and JTAG |
| Program run control (Halting, resume, single step) | Yes | Yes | Yes | Yes |
| On-the-fly memory accesses | Yes | Yes | Yes | Yes |
| Debug monitor | No | No | Yes | Yes |
| Software breakpoint | Yes | Yes | Yes | Yes |
| Hardware breakpoint comparators | Up to 4 | Up to 4 | Up to 8 (6 instructions and 2 literal data) | Up to 8 (6 instructions and 2 literal data) |
| Hardware watchpoint comparators | Up to 2 | Up to 2 | Up to 4 | Up to 4 |
| Instrumentation trace | No | No | Yes | Yes |
| Data, event and profiling trace | No | No | Yes | Yes |
| Profiling counter | No | No | Yes | Yes |
| PC sampling register via trace connection | No | No | Yes | Yes |
| PC sampling register via debug connection | Yes | Yes | Yes | Yes |
| Instruction Trace | No | Optional Micro Trace Buffer (MTB) | Optional ETM | Optional ETM |
| Trace interface | No | MTB instruction trace via debug connection | Serial Wire Viewer (SWV) or Trace Port interface | Serial Wire Viewer (SWV) or Trace Port interface |
| Debug and Trace registers accesses from software (needed by debug monitor) | No | No | Yes | Yes |

Cortex-M0+ processor, the smallest supported region size is 256 bytes. However, by using the sub-region disable feature, you can create a 32-byte region with only minor modifications to the MPU setup code.

### 24.4.4 Creating portable program code for Cortex®-M processors

In some projects, we need to create program code that can be reused on various Cortex®-M processors including Cortex-M0, Cortex-M0+, Cortex-M3, and Cortex-M4. In order to maximize software reusability, there are a few areas that should be considered when developing embedded application code:

- Use CMSIS-Core functions to access processor features instead of directly accessing system registers.
- Avoid using any features that are limited to the Cortex-M3 and Cortex-M4 processors. For example, in CMSIS-Core, the Interrupt Active Status Register is not available in the Cortex-M0 and Cortex-M0+. Other features that are limited to Cortex-M3 and Cortex-M4 include: bit band, unaligned transfers, and dynamic interrupt priority change.
- When developing portable program code, we can enable the UNALIGN_TRP bit in the Configuration Control Register (SCB->CCR) to detect any unaligned data transfers and modify the code to prevent this when unaligned data accesses are found.
- When creating assembly code (e.g., inline assembly, embedded assembly) we also need to make sure that the instructions used are available on ARMv6-M.

# References

## Document

1. **ARMv7-M Architecture Reference Manual**
   ARM DDI 0403D, http://infocenter.arm.com/help/topic/com.arm.doc.ddi0403c/index.html
2. **Cortex-M3 Devices Generic User Guide**
   ARM DUI 0522A, http://infocenter.arm.com/help/topic/com.arm.doc.dui0552a/index.html
3. **Cortex-M4 Devices Generic User Guide**
   ARM DUI 0553A, http://infocenter.arm.com/help/topic/com.arm.doc.dui0553a/index.html
4. **Cortex-M3 Technical Reference Manual**
   ARM DDI 0337I, http://infocenter.arm.com/help/topic/com.arm.doc.ddi0337i/index.html
5. **Cortex-M4 Technical Reference Manual**
   ARM DDI 0439D, http://infocenter.arm.com/help/topic/com.arm.doc.ddi0439d/index.html
6. **ARM Compile toolchain Assembler Reference, version 4.1**
   ARM DUI 0489C, http://infocenter.arm.com/help/index.jsp?topic=/com.arm.doc.dui0489c/index.html
7. **ARM Compiler toolchain Compiler Reference, version 5.01**
   ARM DUI 0491G, http://infocenter.arm.com/help/topic/com.arm.doc.dui0491g/index.html
8. **AAPCS Procedure Call Standard for the ARM Architecture**
   ARM IHI 0042D, http://infocenter.arm.com/help/topic/com.arm.doc.ihi0042d/IHI0042D_aapcs.pdf
9. **A Programmer Guide to the Memory Barrier instruction for ARM Cortex-M Family Processor**
   ARM DAI0321A, http://infocenter.arm.com/help/topic/com.arm.doc.dai0321a/index.html
10. **Cortex-M3 Embedded Software Development**
    ARM DAI0179B, http://infocenter.arm.com/help/topic/com.arm.doc.dai0179b/index.html
11. **Cortex-M4(F)Lazy Stacking and Context Switching**
    ARM DAI0298A, http://infocenter.arm.com/help/topic/com.arm.doc.dai0298a/index.html
12. **ARM Compiler Toolchain version 5.01, Using the Compiler**
    ARM DUI0472G, http://infocenter.arm.com/help/topic/com.arm.doc.dui0472g/index.html
    (Bit-band command line option: http://infocenter.arm.com/help/topic/com.arm.doc.dui0472g/BEIFCGDI.html
    Bit-band attribute: http://infocenter.arm.com/help/topic/com.arm.doc.dui0472g/BEIJIHIJ.html)

13. **Procedure Call Standard for ARM Architecture**
    ARM    IHI    0042D,    http://infocenter.arm.com/help/topic/com.arm.doc.ihi0042d/
    IHI0042D_aapcs.pdf
14. **AMBA 3 AHB-Lite Protocol Specification**
    http://infocenter.arm.com/help/topic/com.arm.doc.ihi0033a/index.html
15. **AMBA 3 APB Protocol Specification**
    http://infocenter.arm.com/help/topic/com.arm.doc.ihi0024b/index.html
16. **CoreSight Technology System Design Guide**
    http://infocenter.arm.com/help/topic/com.arm.doc.dgi0012d/index.html
17. **AN210 − Running FreeRTOS on the KeilMCBSTM32 Board with RVMDK Evaluation Tools**
    http://infocenter.arm.com/help/topic/com.arm.doc.dai0210A/index.html
18. **AN234 − Migrating from PIC Microcontrollers to Cortex-M3**
    http://infocenter.arm.com/help/topic/com.arm.doc.dai0234a/index.html
19. **AN237 − Migrating from 8051 to Cortex Microcontrollers**
    http://infocenter.arm.com/help/topic/com.arm.doc.dai0237a/index.html
20. **Keil Application Note 202 − MDK-ARM Compiler Optimizations**
    http://www.keil.com/appnotes/docs/apnt_202.asp
21. **Keil Application Note 209 − Using Cortex-M3 and Cortex-M4 Fault Exceptions**
    http://www.keil.com/appnotes/docs/apnt_209.asp
22. **Keil Application Note 221 − Using CMSIS-DSP Algorithms with RTX**
    http://www.keil.com/appnotes/docs/apnt_221.asp
23. **AN33-Fixed Point Arithmetic on the ARM**
    http://infocenter.arm.com/help/topic/com.arm.doc.dai0033a/index.html
24. **Mastering stack and heap for system reliability**
    http://www.iar.com/Global/Resources/Developers_Toolbox/Building_and_debugging/
    Mastering_stack_and_heap_for_system_reliability.pdf
25. **ARM and Thumb-2 Instruction Set Quick Reference Card**
    http://infocenter.arm.com/help/topic/com.arm.doc.qrc0001m/index.html
    (This Quick Reference Card is not specific to the ISA used by the Cortex-M processor)

# Index

Note: Page numbers followed by "f" denote figures; "t" tables.